(3.4) $(a + b)(a - b) = a^2 - b^2$

(3.5) $a^2 - b^2 = (a + b)(a - b)$

(3.6) $x^3 + y^3 = (x + y)(x^2 - xy + y^2)$

(3.7) $x^3 - y^3 = (x - y)(x^2 + xy + y^2)$

(3.8) $acx^2 + (ad + bc)xy + bdy^2 =$
$$(ax + by)(cx + dy)$$

(4.1) If $ad = bc$, then $a/b = c/d$ and
conversely; $b \neq 0$, $d \neq 0$

(4.2) $\dfrac{a}{b} = \dfrac{an}{bn} = \dfrac{a \div m}{b \div m}$

(4.3) $\dfrac{n}{d} \div \dfrac{r}{s} = \dfrac{n}{d} \times \dfrac{s}{r}$

(4.4) $\dfrac{n}{d} = \dfrac{-n}{-d} = -\dfrac{-n}{d} = -\dfrac{n}{-d}$

(5.1) If $f(x)$, $g(x)$, and $h(x)$ are polynomials, then $f(x) = g(x)$ and
$f(x) + h(x) = g(x) + h(x)$ are
equivalent equations

(5.2) If the constant k is different from
zero, then $f(x) = g(x)$ is equivalent
to $k \cdot f(x) = k \cdot g(x)$
If $f(x)$, $g(x)$, and $h(x)$ are expressions, then for all values of x
for which the three expressions
are real $f(x) > g(x)$ is equivalent
to the statement in each of (5.3)
to (5.5).

(5.3) $f(x) + h(x) > g(x) + h(x)$

(5.4) $f(x) \cdot h(x) > g(x) \cdot h(x)$ for
$x \in \{x \mid h(x) > 0\}$

(5.5) $f(x) \cdot h(x) < g(x) \cdot h(x)$ for
$x \in \{x \mid h(x) < 0\}$
Similar statements can be made if
$f(x) < g(x)$.

(5.4a) If k is a positive constant, then
$f(x) > g(x)$ and $k \cdot f(x) > k \cdot g(x)$
are equivalent, as are $f(x) < g(x)$
and $k \cdot f(x) < k \cdot g(x)$

(5.5a) If c is a negative constant, then
$f(x) > g(x)$ and $c \cdot f(x) < c \cdot g(x)$
are equivalent, as are $f(x) < g(x)$
and $c \cdot f(x) > c \cdot g(x)$

(5.6) If $k(x)$ is a polynomial, then each
root of $f(x) = g(x)$ is also a root of
$k(x) \cdot f(x) = k(x) \cdot g(x)$

(7.1) $(a/b)^n = a^n/b^n$

(7.2) $(a^n)^m = a^{mn}$

(7.3) $a^{-t} = 1/a^t$; $a \neq 0$

(7.4) $a^{1/k} = \sqrt[k]{a}$

(7.5) $b^{j/k} = (b^j)^{1/k} = (b^{1/k})^j = (\sqrt[k]{b})^j$
provided $\sqrt[k]{b}$ is real

(7.6) $\sqrt[k]{ab} = \sqrt[k]{a}\,\sqrt[k]{b}$

(7.7) $\sqrt[k]{a/b} = \sqrt[k]{a}/\sqrt[k]{b}$

(7.8) $\sqrt[r]{\sqrt[u]{a}} = \sqrt[ru]{a}$

(8.1) The solution set of $(x + d)^2 = k$ is
$\{-d \pm \sqrt{k}\}$

(8.3) If $ax^2 + bx + c = 0$, then
$x = (-b \pm \sqrt{b^2 - 4ac})/2a$

(8.4) $af^2(x) + bf(x) + c = 0$ is a quadratic
equation

(8.5) $D = b^2 - 4ac$

(8.7) $r + s = -b/a$

(8.8) $rs = c/a$

(8.9) $ax^2 + bx + c = a(x - r)(x - s)$

(10.2) $D_n = \begin{vmatrix} a_{11} & a_{12} & \cdots & a_{1j} & \cdots & a_{1n} \\ a_{21} & a_{22} & \cdots & a_{2j} & \cdots & a_{2n} \\ \cdots & \cdots & \cdots & \cdots & \cdots & \cdots \\ a_{i1} & a_{i2} & \cdots & a_{ij} & \cdots & a_{in} \\ \cdots & \cdots & \cdots & \cdots & \cdots & \cdots \\ a_{n1} & a_{n2} & \cdots & a_{nj} & \cdots & a_{nn} \end{vmatrix}$

(10.3) $A_{ij} = (-1)^{i+j} m(a_{ij})$

(continued on back cover)

College Algebra

College Algebra

SEVENTH EDITION

PAUL K. REES
Louisiana State University

FRED W. SPARKS
Texas Tech University

CHARLES SPARKS REES
University of New Orleans

McGraw-Hill Book Company
New York   St. Louis   San Francisco   Auckland   Bogota   Düsseldorf
Johannesburg   London   Madrid   Mexico   Montreal   New Delhi   Panama
Paris   São Paulo   Singapore   Sydney   Tokyo   Toronto

College Algebra

234567890 VHVH 783210987

This book was set in Baskerville by Ruttle, Shaw & Wetherill, Inc.
The editors were A. Anthony Arthur and Shelly Levine Langman;
the designer was Merrill Haber;
the production supervisor was Joe Campanella.
Von Hoffmann Press, Inc.,
was printer and binder.

Library of Congress Cataloging in Publication Data

Rees, Paul Klein, date
    College algebra.

    Includes bibliographical references and index.
    1. Algebra. I. Sparks, Fred Winchell,
date joint author. II. Rees, Charles Sparks,
joint author. III. Title.
QA154.2.R44   1977   512.9   76-9829
ISBN 0-07-051716-9

Contents

v

3 SPECIAL PRODUCTS AND FACTORING   **58**

4 FRACTIONS   **71**

5 LINEAR AND FRACTIONAL EQUATIONS AND INEQUALITIES   **91**

9 SYSTEMS OF EQUATIONS AND OF INEQUALITIES 208

10 MATRICES AND DETERMINANTS 255

Preface

The topics presented in this, the seventh, edition of *College Algebra* are essentially the ones treated in the sixth and several earlier editions. They are the topics from algebra that are needed to pursue successfully courses in trigonometry and calculus with analytic geometry. Every chapter of the book has been carefully reviewed, and varying amounts of each have been rewritten for the sake of clarity and for the proper degree of rigor. Essentially all the problems are new.

The seventh edition contains only 18 chapters, as compared to 21 in the sixth edition. This is the result of combining the two chapters on matrices and determinants; placing the work on variation in the chapter on functions, relations, and graphs; and placing the work on partial fractions in the chapter on polynomial equations.

The material that has been rewritten or is new includes a considerable portion of the last half of Chapter 2, The Four Fundamental Operations; Sec. 4.3, The Lowest Common Denominator; Sec. 4.4, Addition of Fractions; Sec. 4.5, Multiplication of Fractions; Sec. 5.6, Nonlinear Inequalities; Sec. 5.7, Fractional Equations; Sec. 5.8, Equalities and Inequalites that Involve Absolute Values; Sec. 6.5, The Rectangular Coordinate System; Sec. 6.9, Inverse Functions; Sec. 6.10, Variation; Sec. 7.1, Nonnegative Integral Exponents; Sec. 7.3, Rational Exponents and Radicals; Sec. 7.4,

Laws of Radicals; Sec. 8.6, Equations in Quadratic Form; Sec. 8.7, Radical Equations; Sec. 9.2, Graph of an Equation in Two Variables; Sec. 9.3, Simultaneous Solution by Graphical Methods; Sec. 9.4, Independent, Inconsistent, and Dependent Equations; Sec. 9.8, Elimination by a Combination of Methods; Sec. 9.9, Symmetric Equations; Sec. 9.11, Linear Programming; most of Chapter 10, Matrices and Determinants; Sec. 11.13, A Word about Calculators; Sec. 12.1, Complex Numbers as Ordered Pairs; Sec. 12.7, Roots of Complex Numbers; Sec. 13.6, Identical Polynomials; Sec. 13.7, Conjugate Roots; Sec. 13.9, Descartes' Rule of Signs; Sec. 14.3, Sum of an Arithmetic Progression; Sec. 16.6, Ordered Partitions; Sec. 17.2, The rth Term of the Binomial Formula; Sec. 18.4, Dependent and Independent Events. Much of the rewritten material is very similar to that in our *Algebra and Trigonometry*.

We have, as usual, put the exercises a normal lesson apart, and the problems in groups of four similar ones. Most classes need be assigned only every fourth problem, but others may need more drill. Each exercise has a variety of graded problems which vary from the standard drill type to problems of a more challenging nature. Answers to three-fourths of the problems are given in the book, and the others are available in the Instructor's Manual. There is a review exercise at the end of each chapter that contains three or more exercises. There are about 80 regular and 16 review exercises, with a total of about 4,100 problems.

A list of axioms, definitions, and theorems is printed on the endpapers for convenient reference.

Some instructors may prefer to begin the course with Chapter 3 or 4, and to refer back to earlier chapters as the occasion demands.

The original authors wish to express their pleasure at having Charles Sparks Rees join us as a third author. He is the son of one of us and the namesake of the other.

Paul K. Rees
Fred W. Sparks
Charles Sparks Rees

College Algebra

CHAPTER ONE *The Number System of Algebra*

Mathematics is concerned with building logical structures. These structures are based on undefined terms and concepts. The definitions of other terms and concepts are based on these undefined elements, and then assumptions are made about the terms and concepts. These assumptions are called *axioms*. They are statements that we agree to accept without considering their truth or falsity. The structure is further developed by proving statements that are called *theorems*. Each step in every proof is based on the axioms and previously proven theorems.

Axioms

Theorems

We shall deal with the development, properties, and use of the real number system, including the application of the processes of arithmetic to letters and other symbols that represent numbers.

1.1 / SETS

One of the basic and useful concepts of mathematics is denoted by the word "set." This word is used every day in such phrases as "set of dishes," "set of chessmen," and "set of drawing instruments" and in other phrases referring to a *collection* of related objects. We shall assume that the reader knows the meaning of the word "collection," and we shall define a set as follows:

A *set* is a collection of well-defined objects called *elements*. By "well-defined" we mean that there is some criterion that enables us to say that an element belongs to the given set or does not belong to the given set.

The following examples illustrate the above definition. In each example, the set is designated by S and the criterion is stated.

S is the *choir* of All Saints Church. The criterion that determines the membership of the choir is the list of names selected by the choir director.

S is the set of all numbers less than 20 that are divisible by 3 and that are in the set of counting numbers 1, 2, 3, 4, and so on. In this case, the elements of S are 3, 6, 9, 12, 15, and 18, since no other counting number less than 20 is divisible by 3.

As implied above, we frequently use a capital letter to designate a set and employ both the listing method and the rule method to describe a set. In the listing method, we tabulate the elements of the set and enclose the tabulation in braces. In the rule method, we enclose a descriptive phrase in braces. The first two examples below illustrate the listing method, and the last two illustrate the rule method.

EXAMPLE 1   If S consists of the first three counting numbers, then $S = \{1, 2, 3\}$.

EXAMPLE 2   If Jones, Brown, Long, and Small are the only members of a football squad who play end position, and if E is the set of ends on the squad, then $E = \{$Jones, Brown, Long, Small$\}$.

EXAMPLE 3   The counting numbers that are less than 5 are 1, 2, 3, and 4. Now if we designate this set by S and the phrase "is less than" by $<$, and if we require that x stand for a counting number, we can represent the set in this way: $S = \{x$, such that $x < 5\} = \{1, 2, 3, 4\}$ or by the notation $S = \{x \,|\, x < 5\}$. The vertical bar in this notation is read "such that."

EXAMPLE 4   Suppose that x stands for a counting number such that $x + 3 = 5$ and S is the set of such counting numbers; then

$$S = \{x \,|\, x + 3 = 5\} = \{2\}$$

In this case, S contains only one element.

The notation in Examples 3 and 4 above is called a *set builder* and is the one most often used.

We shall now consider the sets $S = \{a, b, c, d, e\}$ and $T = \{a, c, e\}$. Here each element of T is an element of S. This situation illustrates the following definition:

Subset

Proper subset

If each element of a set T is an element of a set S, then T is a *subset* of S. Furthermore, if there are elements in S that are not in T, then T is a *proper* subset of S.

We use the notations $T \subseteq S$ to denote the fact that T is a subset of S, and $T \subset S$ to indicate that T is a proper subset of S.

Belongs to

If a is an element of the set S, we say that a *belongs to* S, and express this statement by the notation $a \in S$. The notation $b \notin S$ means that b does not belong to S.

Equality of sets

The set S is *equal* to the set T if each element of S belongs to T and each element of T belongs to S. In other words, $S = T$ if $S \subseteq T$ and $T \subseteq S$. For example, if $S = \{a, c, e, g\}$ and $T = \{g, c, a, e\}$, then $S = T$. Note that the order of the elements in the two sets does not affect the relation of equality.

Frequently, the same set of elements belongs to each of two sets. For example, if $S = \{1, 2, 3, 4, 5, 6\}$, $T = \{2, 4, 6, 8, 10\}$, and $R = \{2, 4, 6\}$, then $R \subset S$ and $R \subset T$. We call R the *intersection* of S and T. This illustrates the following definition:

Intersection of two sets

The *intersection* of S and T is designated by $S \cap T$† and is the set of all elements that belong to S and that also belong to T.

The following are examples of the intersection of two sets:

EXAMPLE 5　$\{x \mid x$ is a counting number divisible by $2\}$
$\cap \ \{x \mid x$ is a counting number less then $10\} = \{2, 4, 6, 8\}$

EXAMPLE 6　If $A = \{x \mid x$ is a councilman of Townsville$\}$ and $B = \{x \mid x$ is a member of the Kiwanis Club of Townsville$\}$, then $A \cap B = \{x \mid x$ is a councilman who is a Kiwanian$\}$.

If, in Example 6, no councilman is a Kiwanian, then $A \cap B$ contains no elements and is an example of the *empty set,* defined as follows:

Empty set

The *empty set,* or the *null set,* is any set that contains no elements, and it is designated by $\varnothing$ or $\{\quad\}$.

Other examples of the null set are:

EXAMPLE 7　$\{x \mid x$ is a woman who has been President of the United States$\}$

---

† The notation $S \cap T$ is read "the intersection of S and T."

EXAMPLE 8 $\{x \mid x$ is a two-digit counting number less than 10$\}$

EXAMPLE 9 $\{x \mid x$ is a former Governor of California$\} \cap \{x \mid x$ is a former Governor of Texas$\}$

Disjoint sets

If $S \cap T = \varnothing$, the sets S and T are called *disjoint sets.*

Another concept associated with the theory of sets is the complement of one set with respect to another. As an example, we shall consider the sets $A = \{x \mid x$ is a student in a given college$\}$ and $B = \{x \mid x$ is a member of the football squad of that college$\}$. The complement of B with respect to A is the set $C = \{x \mid x$ is a student of the college who is not on the football squad$\}$. This illustrates the following definition:

Complement of a set

The *complement* of the set B with respect to A is designated by $A - B$, and $A - B = \{x \mid x \in A$ and $x \notin B\}$.

In the example above the definition, $B \subset A$. This condition is not necessary for $A - B$ to have a meaning. For example, if $T = \{x \mid x$ is a redhead in college $C\}$ and $S = \{x \mid x$ is a member of the senior class of college $C\}$, then $T - S = \{x \mid x$ is a redhead not classified as a senior$\}$.

We shall next discuss the union of two sets. As an example, consider $S = \{1, 2, 3, 4, 5, 6\}$ and $T = \{2, 4, 6, 8, 10\}$. The elements 1, 3, and 5 belong to S but not to T; the elements 8 and 10 belong to T but not to S; and the elements 2, 4, and 6 belong to both S and T. Hence the elements of $V = \{1, 2, 3, 4, 5, 6, 8, 10\}$ are in S or are in T or are in both S and T. The set V is called the *union* of the sets S and T. This illustrates the following definition:

Union of two sets

The *union* of the sets S and T is the set whose elements are in S or in T or in both S and T, and it is designated by $S \cup T$.

The following examples illustrate the concepts of union and intersection when three sets are involved. In the examples we shall use the sets $A = \{a, b, c, d\}$, $B = \{a, c, e, g\}$, and $C = \{a, e, r, t\}$. The parentheses indicate the operation that is to be performed first.

EXAMPLE 10   Show that $(A \cup B) \cup C = A \cup (B \cup C)$.

Solution
$$A \cup B = \{a, b, c, d\} \cup \{a, c, e, g\} = \{a, b, c, d, e, g\}$$

Hence,
$$(A \cup B) \cup C = \{a, b, c, d, e, g\} \cup \{a, e, r, t\} = \{a, b, c, d, e, g, r, t\}$$

Furthermore,
$$B \cup C = \{a, c, e, g\} \cup \{a, e, r, t\} = \{a, c, e, g, r, t\}$$

Consequently,

$$A \cup (B \cup C) = \{a, b, c, d\} \cup \{a, c, e, g, r, t\}$$
$$= \{a, b, c, d, e, g, r, t\}$$

Therefore, $(A \cup B) \cup C = A \cup (B \cup C)$

Thus the associative property holds for the union of sets.

EXAMPLE 11   Show that $A \cup (B \cap C) = (A \cup B) \cap (A \cup C)$.

Solution $B \cap C = \{a, e\}$

Hence, $A \cup (B \cap C) = \{a, b, c, d\} \cup \{a, e\} = \{a, b, c, d, e\}$

Also, $A \cup B = \{a, b, c, d, e, g\}$

and $A \cup C = \{a, b, c, d, e, r, t\}$

Therefore,

$(A \cup B) \cap (A \cup C) = \{a, b, c, d, e\}$

and we have

$A \cup (B \cap C) = (A \cup B) \cap (A \cup C)$

Thus the distributive property holds for union over intersection of sets.

EXAMPLE 12   Show that $A \cap (B \cup C) = (A \cap B) \cup (A \cap C)$.

Solution $B \cup C = \{a, c, e, g, r, t\}$
$$A \cap (B \cup C) = \{a, b, c, d\} \cap \{a, c, e, g, r, t\}$$
$$= \{a, c\}$$
$$A \cap B = \{a, b, c, d\} \cap \{a, c, e, g\} = \{a, c\}$$
$$A \cap C = \{a, b, c, d\} \cap \{a, e, r, t\} = \{a\}$$
$$(A \cap B) \cup (A \cap C) = \{a, c\}$$

Hence, $A \cap (B \cup C) = (A \cap B) \cup (A \cap C)$

Thus the distributive property holds for intersection over union of sets.

Universal set   The totality of elements that are involved in any specific discussion or situation is called the *universal set* and is designated by the capital letter U. For example, the states in the United States are frequently classified into sets, such as the New England states, the Midwestern states, the Southern states, and in several other ways. Each of these sets is a subset of the universal set, which, in this example, is composed of all the states of the United

States.   Each of the various clubs, athletic teams, academic classes, and other groups whose members are students of a given college is a subset of the universal set composed of the entire student body of the college.

A method for picturing sets and certain relations between them was devised by an Englishman, John Venn (1834–1923).   The fundamental idea is to represent a set by a simple plane figure.   In order to illustrate the method, we shall use circles.   We shall represent the universal set U by a circle C of radius r, and we shall define U as the set of all points within and on the circumference of C.   We shall represent various subsets of U by circles wholly within the circle C.   Figure 1.1 illustrates the device.

Cartesian product

We shall define and discuss the *cartesian product* of two sets but must first define the concept of an ordered pair of elements.   The pair of elements (x, y) is an *ordered pair* if the position of each element in the pair associates a specific property to it.   For example, if D and R are given sets, and if we require that the first element x of (x, y) be an element of D, and the second element y belong to R, then (x, y) is an ordered pair.

Ordered pair

The *cartesian product* of two sets D and R is the set of all ordered pairs (x, y) that can be formed such that $x \in D$ and $y \in R$.   It is indicated by $D \times R$.   As an example, if $D = \{1, 2, 5\}$ and $R = \{3, 4\}$, then $D \times R$ is $\{(1, 3), (1, 4), (2, 3), (2, 4), (5, 3), (5, 4)\}$, and $R \times D$ is $\{(3, 1), (3, 2), (3, 5), (4, 1), (4, 2), (4, 5)\}$.   Thus we see that $D \times R$ and $R \times D$ are not the same.

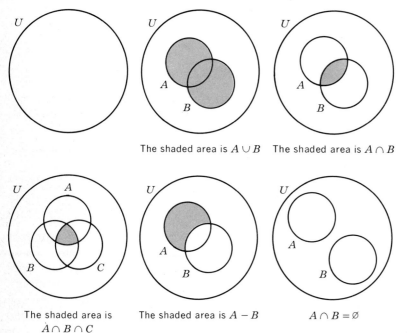

The shaded area is $A \cup B$ The shaded area is $A \cap B$

The shaded area is $A \cap B \cap C$ The shaded area is $A - B$ $A \cap B = \emptyset$

Figure 1.1

EXERCISE 1.1 Operations on Sets

Use the listing method to describe the sets in Probs. 1 to 4.

1 $\{x\,|\,x$ is a counting number less than 16 and divisible by 3$\}$

2 $\{x\,|\,x$ is a consonant in the word *vowel*$\}$

3 $\{x\,|\,x$ is a month of the year that has fewer than 31 days$\}$

4 $\{x\,|\,x$ is a former vice-president of the United States between 1955 and 1975 who became president$\}$

Use set-builder notation to designate the set in each of Probs. 5 to 8.

5 $\{2, 4, 6, 8\}$ 6 $\{2, 4, 8, 16\}$

7 $\{$McKinley, Harding, F. D. Roosevelt, Kennedy$\}$

8 $\{$violet, indigo, blue, green, yellow, orange, red$\}$

Find $A \cup B$, $A \cap B$, and $A - B$ for the sets in each of Probs. 9 to 16.

9 $A = \{1, 3, 5, 7, 10\}$, $B = \{2, 5, 10\}$ 10 $A = \varnothing$, $B = \{2, 4, 6, 8\}$

11 $A = \{1, 3, 5, 7, 9\}$, $B = \varnothing$

12 $A = \{2, 3, 5, 7, 11\}$, $B = \{2, 5, 11, 14\}$

13 $A = \{x\,|\,x$ is a letter in "orange"$\}$, $B = \{x\,|\,x$ is a letter in "red"$\}$

14 $A = \{x\,|\,x$ is a counting number less than 10$\}$,
$B = \{x\,|\,x$ is a one-digit counting number divisible by 3$\}$

15 $A = \{x\,|\,x$ is a vowel in "measure"$\}$, $B = \{x\,|\,x$ is a vowel in "cream"$\}$

16 $A = \{x\,|\,x$ is a student under 5 feet tall at K College$\}$,
$B = \{x\,|\,x$ is a female student under 5 feet tall at K College$\}$

Perform the operations called for in Probs. 17 to 24 if $A = \{x\,|\,x$ is a counting number less than 12$\}$, $B = \{x\,|\,x \in A$ and is divisible by 2$\}$, $C = \{x\,|\,x \in A$ and is divisible by 3$\}$, and $D = \{x\,|\,x \in A, x \notin B, x \notin C\}$.

17 $A \cup B, A \cap C, A - D$ 18 $B \cup C, B \cap C, B - A$

19 $A - B, A \cap B, A \cup C$ 20 $D - C, D \cap C, D \cup C$

21 $A \cap B \cap C \cap D, A \cup B \cup C, (A \cup C) - B$

22 $(A - B) \cup (C - D), (A \cup B) - (C \cup D), [A - (B \cup C)] - D$

23 $[B \cup (C \cap D)] - [(B \cup C) \cap (B \cup D)]$

24 $B \cap (C \cup D) - [(B \cap C) \cup (B \cap D)]$

25 If $A = \{4, 7\}$ and $B = \{2, 5, 8\}$, find $A \times B$.

26 If $A = \{1, 4, 8\}$ and $B = \{1, 4, 8\}$, find $B \times A$.

27 If $A = \{3, 4, 6\}$, $B = \{3, 6, 8\}$, and $C = \{2, 4, 6\}$, find $(A \times B) \cap (A \times C)$.

28 If A, B, and C are as in Prob. 27, find $(A \cap B) \times (A \cap C)$.

If A is the set of points inside and on a circle, and B is the set of points inside and on a triangle, construct Venn diagrams for Probs. 29 to 32 and shade the area that represents the requested set.

29   $A \cap B$, if A and B have points in common

30   $A \cap B$, if A and B have no points in common

31   $A \cup B$, if A and B have points in common

32   $A - B$, if A and B have points in common

1.2 / THE NATURAL NUMBERS

In Sec. 1.1 we used the term "counting number." This brings up two questions: What is a number? and, What is meant by counting? We shall not attempt to give a rigorous answer to these questions. We shall, however, discuss a situation that illustrates the concept. First, consider the sets $A = \{a, b, c, d, e\}$ and $B = \{$Tom, Dick, Harry, Joe, Jim$\}$. Here we have two different sets since the elements of A are letters and the elements of B are names. There is, however, one property that is common to the two sets: each element in B can be matched with one and only one element in A, and each element in A can be matched with one and only one in B. We call this matching

One-to-one correspondence   a *one-to-one correspondence,* and the correspondence can be set up in many ways. As an illustration, we shall match the elements of the sets in the order in which they appear, proceeding from left to right:

$$
\begin{array}{ccccc}
a & b & c & d & e \\
| & | & | & | & | \\
\text{Tom} & \text{Dick} & \text{Harry} & \text{Joe} & \text{Jim}
\end{array}
$$

We call this common property of the two sets the *number of elements* in each set. The Hindu-Arabic symbol for this number is 5, and the English name for it is "five." We say that the sets A and B are equivalent, and they illustrate the following definition:

Equivalent sets   Two sets are *equivalent* if there exists a one-to-one correspondence between the elements of the two sets.

We can use this concept of equivalent sets to illustrate the meaning of the term "counting number."

For this purpose, we shall assume that the Hindu-Arabic number symbols and their order of succession are known, and we shall use these symbols as elements of sets. Now we say that the number *one* is the number associated with the totality of sets that are equivalent to the set $\{1\}$, the number *two* is the number associated with all sets equivalent to $\{1, 2\}$, and so on. Similarly, if n is a counting number expressed in the Hindu-Arabic notation, then the number n is the number associated with all sets equivalent to the set $\{1, 2, 3, \ldots, n\}$, where the dots indicate that the succession of number symbols is continued from 3 to n.

The process of counting

The *process of counting* the elements of a set consists of establishing a one-to-one correspondence between the elements of the given set and the elements of $\{1, 2, 3, \ldots, n\}$, where the dots mean that the sequence of number symbols is continued until the correspondence is completed.

Natural numbers

The numbers used in counting are called *natural numbers*.

The natural numbers are used for defining the numbers in the real number system, which will be discussed in the next section.

1.3 / THE REAL NUMBER SYSTEM

Although the natural numbers may have sufficed for a primitive culture, the advance of civilization not only demanded but depended upon a progressive extension of the number system by the invention of other numbers. The extended system which we shall use in the first eight chapters of this book is called the *real number system*. In this section we shall define the various subsets of the set of real numbers and shall give a geometrical interpretation of each.

We shall use the straight line L and the unit length u in Fig. 1.2 for this purpose. We shall accept the terms "straight line," "point," "length," and "distance" as undefined terms. A *segment* of a straight line, or an *interval* on the line, is the portion of the line between two points on it. We shall say that two segments are equivalent or congruent if their lengths are the same. Furthermore, we shall use the fact that a geometrical method exists for dividing a line segment into any given integral number of equal parts, and we shall assume that any given length can be laid off on a straight line any desired natural number of times.

We choose the reference point P on the line L, a portion of which is shown in Fig. 1.2, and starting at P, we lay off successive intervals of length u to the right of P. Next, starting with the right end of the first interval to the right of P, we label the right ends of the intervals, proceeding progressively to the right with the natural numbers $1, 2, 3, \ldots$, where the dots mean that the sequence of numbers is continued indefinitely. In this way,

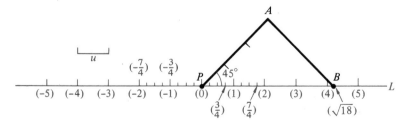

Figure 1.2

one and only one point on L is associated with each of the natural numbers. We shall use the notation (n) to refer to the point associated with the number n. For example, the point associated with 4 is denoted by (4).

Geometric addition
        One of the fundamental operations involving natural numbers is *addition*. We shall interpret this operation by use of the line L. The symbol indicating addition is $+$ and is read "plus." Thus, $a + b$ means that we are to add b to a. In order to obtain the point $(a+b)$, we start at the point (a), move (b) unit intervals to the right, and thus reach the point $(a+b)$. For example, to get the point $(3 + 2)$, we start at (3) and move two units to the right and reach the point (5). Hence we say that $3 + 2 = 5$, and we

Sum   call 5 the *sum* of 3 and 2.
        We next call attention to the fact that if a is a natural number and if we start at the point P on L in Fig. 1.2 and move a intervals to the right, we arrive at the point (a). So far, we have not associated a number with the point P, and since the natural numbers are assigned to other points on L, we cannot use a natural number for this purpose. Therefore we shall introduce a new number, *zero*, denoted by 0, and assign 0 to P. Now if we start at P, or (0), and move a units to the right, we arrive at the point (a). Consequently, we define the number zero as follows:

Zero   The number *zero*, denoted by 0, is the number such that $0 + a = a$, where a stands for any given number. Hence it is called the *additive identity*.
        We now define the operation $a + (-b)$ to mean that we start at (a) and move b units to the *left* of (a), as illustrated in Fig. 1.3. If (a) is to the right of the point (b) on L, then the point $a + (-b)$ is to the right of the point 0. If, however, (b) is to the right of (a), then the operation $a + (-b)$ brings us to a point that is to the left of (0), and so far we have associated no numbers with this portion of L. This will be our next task.
        We first note that the operations $1 + (-1)$, $2 + (-2)$, and in general, $a + (-a)$ bring us to the point zero. Hence, $1 + (-1) = 2 + (-2) = a + (-a) = 0$. We now define the negative of the number a as follows:

Negative of a number   The *negative of the number a* is the number $-a$ such that $a + (-a) = 0$.
        In order to associate a point on L with $-a$, we start at zero, lay off a units on L to the left of (0), and assign $-a$ to the left extremity of the ath

Reflection of a point   interval. We call the point $(-a)$ the *reflection* of the point (a) on L with respect to (0). Thus (-1) is the reflection of (1) with respect to zero, (-2) is the reflection of (2), and so on. In this way, we obtain the points in Fig. 1.2 that are labeled (-1), (-2), (-3), and so on.

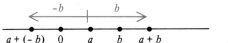

                                                                    Figure 1.3

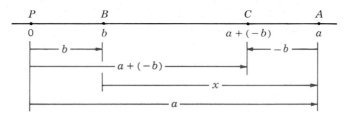

Figure 1.4

Integers

Positive integers
Negative integers

We shall now define the set of *integers* to be the set of numbers composed of the natural numbers, zero, and the negatives of the natural numbers. Hereafter we shall call the natural numbers the *positive integers* and the negatives of the natural numbers the *negative integers*. The number zero is an integer that is neither positive nor negative. In the terminology of sets, we have {integers} = {positive integers} ∪ {0} ∪ {negative integers}. Therefore, the set of positive integers is a proper subset of the set of integers, or {positive integers} ⊂ {integers}. Similarly, {0} ⊂ {integers} and {negative integers} ⊂ {integers}.

We shall agree that if p and q are integers, the statement $p > q$ means that (p) is to the right of (q) on L. Hence $5 > 3$, $2 > -6$, and $-3 > -7$. Similarly, $r < s$ means that (r) is to the left of (s) on L. Thus if a is a positive integer, then (a) is to the right of (0), so $a > 0$. Similarly, if a is a negative integer, $a < 0$.

Difference

Subtraction

We shall define the *difference* of a and b, designated by $a - b$, as the number x such that $b + x = a$. The procedure for finding x is called *subtraction,* and we shall show graphically by use of Fig. 1.4 that $x = a + (-b)$. In Fig. 1.4, the lengths of PB, PA, PC, BA, and AC are as indicated. Since $AC = -b$, $CA = b$, and it follows that $PB = CA = b$. Furthermore,

$$
\begin{aligned}
x &= BA \\
&= BC + CA \\
&= BC + PB \qquad \text{since } CA = PB \\
&= PC \\
&= a + (-b)
\end{aligned}
$$

Hence, $a - b = a + (-b)$.

EXAMPLE 1   Since $5 - 3 = 5 + (-3)$, then to get the difference $5 - 3$, we start at (5) and count three intervals to the left and arrive at (2). Hence, $5 - 3 = 2$.

EXAMPLE 2   To obtain the difference of 5 and 7, we have $5 - 7 = 5 + (-7)$; so we start at (5), count seven intervals to the left through (0), and arrive at (-2). Hence, $5 - 7 = -2$.

EXAMPLE 3   Similarly, the difference of -2 and 3 is $-2 - 3 = -5$.

The procedure for finding the difference of a and b when b is negative will be discussed in the next chapter.

The product of two integers a and b is indicated by $a \cdot b$, $a \times b$, $a(b)$, $(a)(b)$, or ab. The operation of obtaining the product is called *multiplication*. We shall first consider the case in which a is positive, and we shall illustrate the process with $a = 3$ and $b = 2$. We define the product $3 \cdot 2$ to be the sum of *three 2s*. Thus $3 \cdot 2 = 2 + 2 + 2 = 6$. Similarly, $2 \cdot 3 = 3 + 3 = 6$. Hence $3 \cdot 2 = 2 \cdot 3$. We readily can verify that $4 \cdot 5 = 5 \cdot 4$, $6 \cdot 9 = 9 \cdot 6$, and in general, $a \cdot b = b \cdot a$ if a and b are replaced by any two designated positive integers. However, at this point we cannot verify that $3(-2) = -2(3)$, since the latter product has not been defined. Nevertheless, we shall assume that if a and b are integers, then $a \cdot b = b \cdot a$. This is called the *com-*

Commutative   *mutative* property of multiplication and will be discussed more fully in the next chapter. By the commutative property, $-3 \cdot 2 = 2 \cdot -3$, and we define the latter product to be $-3 + (-3) = -6$. Hence $-3 \cdot 2 = -6$. This interpretation of multiplication does not suffice for $-2(-3)$. In the next chapter, however, we shall prove that if a and b are positive, then $-a(-b) = ab$. Hence $-2(-3) = 6$.

Since $3 \cdot a = a + a + a$ and $2 \cdot a = a + a$, it is logical to define $1 \cdot a$ as the number a; that is, $1 \cdot a = a$.

We shall next consider the question: If a and b are integers and $b \neq 0$, what number must be multiplied by b in order to obtain a? In other words, we seek the number x such that $bx = a$. The operation for determining the number x, if it exists, is called *division*, and this operation is the *inverse* of multiplication. The number x is called the *quotient* of a and b, and it is usually expressed in the form $\dfrac{a}{b}$ or a/b. A number expressed in either of these forms is called a *fraction*. Unless a is a multiple† of b, a/b is not an integer and hence is not associated with any point on L that represents an integer. We shall now show how to associate a point on L with a number of the type a/b, and we shall illustrate the method by use of $\frac{3}{4}$. We shall consider Fig. 1.5, which shows a portion of L on an enlarged scale, with the segment from 0 to 1 divided into four equal parts. Since the length of the segment from 0 to 1 is the unit u, the length of each of the four subdivisions of this segment is one-fourth of u. We now assign $\frac{1}{4}$, $\frac{2}{4}$, $\frac{3}{4}$, and $\frac{4}{4}$ to the points indicated in the figure. Hence, $(\frac{3}{4})$ is the right extremity of the third subdivision to the right of (0). Similarly, $(\frac{7}{4})$ is the right extremity of the seventh interval of length $\frac{1}{4}$ to the right of (0). We note that $\frac{4}{4}$ and 1 are associated with the same point, and so $\frac{4}{4} = 1$. We define the product

---

† If a and b are nonzero integers, then a is a multiple of b if an integer n exists such that $a = nb$.

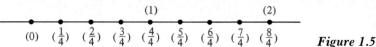

Figure 1.5

$4 \times \frac{1}{4}$ as $\frac{1}{4} + \frac{1}{4} + \frac{1}{4} + \frac{1}{4}$, and by our graphical interpretation of addition, this sum is $\frac{4}{4}$, or 1. Hence we have $4 \times \frac{1}{4} = 1$.

In general, to get the point associated with a/b, where a and b are positive integers, we first subdivide the interval from 0 to 1 on L into b equal parts. Each of these subintervals will be of length $1/b$. Then we lay off a of these intervals to the right of 0 and assign a/b to the right extremity of the ath interval. If $a < b$, then (a/b) is to the left of 1; if $a > b$, (a/b) is to the right of 1; and if $a = b$, (a/b) is the point 1.

Since (a/b) is the point that is the natural number a of the intervals of length $1/b$ to the right of (0), we call a/b a number, and since it is a ratio, we use the word "rational" to describe it. Since it is associated with a point to the right of (0), it is positive. The negative of a/b is $-(a/b)$, and the point on L associated with $-(a/b)$ is the reflection of (a/b) on L with respect to (0). The points $(\frac{3}{4})$, $(\frac{7}{4})$, $(-\frac{3}{4})$, and $(-\frac{7}{4})$ are shown in Fig. 1.2.

In the foregoing discussion, we assumed that a and b represented positive integers. We shall show in a later section that if a and b are both positive or both negative, then a/b is positive, and that if either of a or b is positive and the other is negative, then a/b is negative. We shall assume for the present that these statements are true, and we define a rational number as follows:

Rational number   A *rational*† number is a number that can be expressed as the quotient of two integers.

Since $1 \cdot a = a$, we have, by the definition of a quotient, $a = a/1$. Hence any integer a can be expressed as the quotient $a/1$. Consequently, the integers are included in the rational numbers, or in the terminology of sets, {integers} $\subset$ {rational numbers}.

We shall now show that there is a point on L to the right of 0 that cannot be associated with a rational number. For this purpose, we shall first define the *square* of the number a, designated by a^2, as $a \cdot a$. Hence $3^2 = 3 \cdot 3 = 9$. We shall also define the positive *square root* of the number n, designated by $\sqrt{n}$, as the positive number whose square is n. For example, $\sqrt{16} = 4$ since $4 \cdot 4 = 16$. The pythagorean theorem states that the sum of the squares of the two legs of a right triangle is equal to the square of the hypotenuse. It is also proved in plane geometry that if one acute angle of a right triangle is 45°, the other acute angle is 45° and the sides which form the right angle are equal. We shall now construct a straight-line segment originating at point P in Fig. 1.2 that makes an angle of 45°

---

† The word "rational" is derived from the word "ratio." The quotient a/b is also called the ratio of a to b. Hence a rational number is a number that expresses a ratio.

with the positive direction of L. On this line we lay off three unit intervals starting at P, and we label the upper end of the third interval A. Then at A we construct a perpendicular to this line that intersects L at B. Since the acute angle at P is $45°$, the acute angle at B is $45°$, and therefore $PA = AB = 3$. Then, by the pythagorean theorem, the length of the line segment from P to B is $\sqrt{3^2 + 3^2} = \sqrt{9 + 9} = \sqrt{18}$. Consequently, $\sqrt{18}$ is associated with the point B.

We shall now prove that $\sqrt{18}$ cannot be expressed as the quotient of two integers. For this purpose, we shall use the following definitions and assumptions, most of which will be discussed later.

Even, odd

1 An integer is *even* or *odd* according as it is or is not a multiple of 2.

2 If an even integer is multiplied by another integer, the product is an even integer. Conversely, if the product of two integers is even, then at least one of the integers is even.

3 The halves of equal numbers are equal.

4 The squares of two equal numbers are equal.

5 $(a \cdot b \cdot c)^2 = a^2 \cdot b^2 \cdot c^2$, and $\sqrt{a \cdot b \cdot c} = \sqrt{a} \cdot \sqrt{b} \cdot \sqrt{c}$

6 If d is the greatest integer that is a divisor of a and b and if $a/d = q$ and $b/d = p$, then $a/b = q/p$.

7 $\dfrac{ab}{2} = \dfrac{a}{2} \cdot b = a \cdot \dfrac{b}{2}$

We shall assume that $\sqrt{18}$ can be expressed as the quotient of two integers. Hence, by assumption 6, there exist two integers q and p such that $\sqrt{18} = q/p$, where q and p have no common integral divisor greater than 1. Then, by the definition of a quotient, we have $p \cdot \sqrt{18} = q$, and we complete the proof by the following steps:

$$(p \cdot \sqrt{18})^2 = q^2 \qquad \text{by assumption 4}$$
$$p^2 \cdot (\sqrt{18})^2 = q^2 \qquad \text{by assumption 5}$$
$$p^2 \cdot (18) = q^2 \qquad \text{by the definition of } \sqrt{18}$$

$p^2 \cdot (18)$ is an even integer by assumption 2, since 18 is an even integer. Hence, q^2 is an even integer, and since $q^2 = q \cdot q$, q is an even integer by assumption 2. Consequently, $q = 2n$, where n is an integer. Now, replacing q by $2n$ in $p^2 \cdot (18) = q^2$, we have

$$p^2 \cdot 18 = (2n)^2$$
$$= 4n^2 \qquad \text{by assumption 4}$$
$$p^2 \cdot 9 = 2n^2 \qquad \text{by assumption 3}$$

Since $2n^2$ is an even integer, $p^2 \cdot 9$ is an even integer. Therefore, since 9 is odd, p^2 is even, by assumption 2. Furthermore, since $p^2 = p \cdot p$, it follows from assumption 2 that p is even. Hence, if $\sqrt{18} = q/p$, then p and q are both even, and this contradicts the assumption that q and p have no common divisor greater than 1. Consequently, $\sqrt{18}$ cannot be expressed as the quotient of two integers.

We shall call the number $\sqrt{18}$ an *irrational number* and shall define this type of number more precisely after the next paragraph.

It is proved in arithmetic that some rational numbers that are not integers can be expressed as terminating decimals and that others can be expressed as nonterminating periodic decimals.† For example, $\frac{1}{2} = 0.5$, $\frac{3}{4} = 0.75$, and $\frac{42}{37} = 1.135135135\cdots$. By the statement $\frac{42}{37} = 1.135135135\cdots$, we mean that by annexing the cycle 135 a sufficient number of times, we obtain a decimal that differs from $\frac{42}{37}$ by a number that is less than any number that is chosen in advance. It is also proved in arithmetic that any terminating decimal can be expressed as the quotient of two integers, and in Chap. 14 we shall prove that any nonterminating periodic decimal also can be expressed as the quotient of two integers.

We now return to the discussion of $\sqrt{18}$. There is a process in arithmetic that enables us to express $\sqrt{18}$ approximately as a decimal with as many decimal places as is desired. The first six steps in this process yield 4, 4.2, 4.24, 4.242, 4.2426, and 4.24264. The process never terminates, and the decimal never becomes periodic, but each step yields a number whose square is nearer 18 than the square of the preceding number. Consequently, we say that $\sqrt{18} = 4.24264\cdots$, where the decimal can be continued indefinitely by a repeated application of the square root process.

Since every terminating decimal and every nonterminating periodic decimal can be expressed as the quotient of two integers and, furthermore, since the quotient of any two integers can be expressed as a terminating or a nonterminating periodic decimal, it seems reasonable to assume that a nonterminating nonperiodic decimal cannot be expressed as the quotient of two integers. It is proved in more advanced mathematics that this assumption is true. Therefore, such a number is not a rational number, and we call it an irrational number. This illustrates the following definition:

Irrational number

An *irrational number* is a number whose decimal representation is nonterminating and nonperiodic.

Real number system

We shall now define the *real number system* as the set of numbers composed of the set of rational numbers and the set of irrational numbers.

In the terminology of sets, the above definition can be stated thus: {real numbers} = {rational numbers} ∪ {irrational numbers}. Furthermore, since no rational number is equal to an irrational number, {irrational numbers} ∩ {rational numbers} = ∅ (or the empty set).

---

† By "nonterminating periodic decimal" we mean that following the decimal point or following a certain digit at the right of the decimal point, the decimal consists of an indefinite number of repetitions of the same cycle of integers. For example, $0.325325325\cdots$, $0.125343434\cdots$, and $42.137245245245\cdots$.

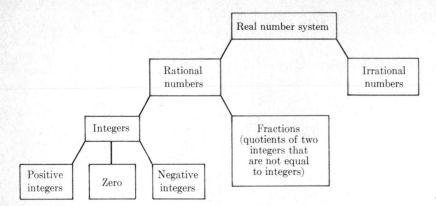

Figure 1.6

It is proved in more advanced mathematics that each point on the line L is associated with one and only one real number. Furthermore, as we have stated previously, the numbers associated with the points to the right of (0) are positive, those associated with points to the left of (0) are negative, and the number 0 is neither positive nor negative. Since a real number other than zero is either positive or negative, it is a *directed*, or *signed*, number. By the sign of the number a, we mean the direction of the point (a) from (0). For example, if the point (a) is to the right of (0), then a is positive; and if (a) is to the left of (0), a is negative. We shall agree that placing the negative sign, $-$, before a number changes the direction of the number. Thus the points (a) and $(-a)$ are on opposite sides of (0). If a is positive, $-a$ is negative; and if b is negative, $-b$ is positive.

Signed

In contrast to this interpretation of the negative sign, we shall agree that placing a positive sign, $+$, before a number does not affect the direction of the number. Consequently, $+n = n$ and $+(-n) = -n$.

Frequently we have occasion to refer to the absolute value of a number, and we shall define this concept as follows:

Absolute value

If a number n is positive, the *absolute value* of n, designated by $|n|$, is equal to n. If n is a negative number, then $|n| = -n$. If $n = 0$, then $|n| = 0$. For example, $|5| = 5$, $|-3| = 3$, and $|0| = 0$.

Figure 1.6 shows the composition of the real number system in diagrammatic form.

EXERCISE 1.2 **The Real Number System**

The set in each of Probs. 1 to 16 fits into one rectangle in Fig. 1.6 better than into any other. Name the proper rectangle.

1 $A = \{3, 7, 2, 5\}$ **2** $A = \{-2, -3, -4, -9\}$

3 $A = \{-1, 0, 3, 9\}$ **4** $A = \{13, 6, 0, -5\}$

5   $A = \{2/3, 5/8, 7/9\}$ 6 $A = \{-3/7, -5/9, -2/3\}$

7   $A = \{-8/3, 0, 2.7, 3\}$ 8 $A = \{-2, -3.8, -3/7, 0\}$

9   $A = \{1, 2/3, \sqrt{18}\}$ 10 $A = \{-2, 0, \sqrt{5}\}$

11   $A = \{-5, 4/3, \sqrt{17}\}$ 12 $A = \{0, 6, \sqrt{3}\}$

13   $A = \{x \mid x$ is the quotient of two integers$\}$

14   $A = \{x \mid x$ is a nonrepeating terminating decimal$\}$

15   $A = \{x \mid x$ is a repeating nonterminating decimal$\}$

16   $A = \{x \mid x$ is a nonrepeating nonterminating decimal$\}$

If $A = \{3, 7, 3/7, 5/9\}$, $B = \{3, 5, 6, 9\}$, and $C = \{0, 2/3, \sqrt{18}\}$, select the name from the real-number diagram that best describes the set in Probs. 17 to 20.

17   $A \cup B$ 18 $A \cap B$ 19 $A \cap C$ 20 $B \cup C$

Insert either $<$ or $>$ between each pair of numbers in Probs. 21 to 24 so as to make a true statement.

21   13   9   22   2 -5 23 $\frac{3}{7}$ 0.41   24   1.4 $\sqrt{2}$

     2   5   5 -2 $\frac{3}{5}$ 0.62   1.41 $\sqrt{2}$

     23   $\frac{47}{2}$ -5 -2 $\frac{7}{9}$ $-\frac{1}{2}$ 1.42 $\sqrt{2}$

Find the result of performing the operations indicated in each of Probs. 25 to 32.

25   $(-7) + 6$ 26 $(-3) + (-2)$ 27 $8 + (-3)$

28   $9 + (-7)$ 29 $|-7| + 6$ 30 $|(-3)| + |(-2)|$

31   $|8| + |(-3)|$ 32 $|9 + (-7)|$

EXERCISE 1.3 **REVIEW**

1   Designate {Truman, Barkley, Nixon, Johnson, Humphrey, Agnew, Ford, Rockefeller} by use of set-builder notation.

2   Designate {2, 4, 8, 16, 32} by use of set-builder notation.

3   Write out the elements of $\{x \mid x$ is a counting number less than 19 and divisible by 4$\}$.

4   Write out the elements of $\{x \mid x < 13$ and not divisible by 2 or 3$\}$.

5   If $A = \{2, 3, 7, 11, 13\}$, and $B = \{2, 9, 13\}$, find $A \cup B$, $A \cap B$, and $A - B$.

6   Find $A \cup B$, $A \cap B$, and $A - B$ if $A = \{x \mid x$ is a letter in "camel"$\}$ and $B = \{x \mid x$ is a letter in "meal"$\}$.

7   If $A = \{3, 4, 6, 7, 9, 11\}$, $B = \{4, 7, 11\}$, and $C = \{3, 7\}$, find $A \cap B \cap C$, $A \cap B \cup C$, and $A - (B \cup C)$.

8 If A, B, and C are as in Prob. 7, find $(A \cap B) \cup (B \cap C)$, $A - (B \cup C)$, and $(A - B) \cup C$.

9 If A, B, and C are as in Prob. 7, show that $[A \cap B] \cup C \neq A \cap [B \cup C]$.

10 Find $A \times B$ if $A = \{2, 7\}$ and $B = \{3, 4, 5\}$.

11 Find $B \times A$ if $B = \{x \mid x$ is a counting number between 4 and 8$\}$ and $A = \{x \mid x$ is a two-digit number less than 12$\}$.

12 If $A = \{4, 3, 1.5, \sqrt{2}\}$ and $B = \{\sqrt{5}, 5, 3, 1.7\}$, what type of number is $A \cap B$?

13 If $A = \{5, 9, 4/3, \sqrt{3}\}$ and $B = \{2, 3, 6, \sqrt{3}\}$, what type of number is $A - B$?

14 Find the value of each of $3 + (-2)$, $3 + |(-2)|$, and $|(-4) + (-3)|$.

15 Insert the proper one of $>$ and $<$ between $\sqrt{3}$ and 1.7, $\frac{5}{9}$ and $\frac{3}{5}$, $\sqrt{2}$ and 1.4.

CHAPTER TWO

The Four Fundamental Operations

We defined the real number system and gave a short discussion of addition, subtraction, multiplication, and division in Chap. 1, but left many questions unanswered. We must still answer such questions as the meaning of 0.5 and of $(-2)(-7)$. We cannot answer these questions or many more that may occur to the thoughtful reader by use of the number line L. However, in this chapter we shall develop the foundations for a logical structure that will enable us to answer the above questions and that will serve as the basis for the fundamental operations and for solving equations.

2.1 / A LOGICAL STRUCTURE

If we examine a book on plane geometry, we see that the text starts with some definitions and that these defini-

Axioms tions are followed by statements called *axioms*. The axioms are statements that are accepted without proof, and they deal with both the defined and the undefined terms and with the relations that exist between them. Every assumption about the defined or undefined terms is set forth in either a definition or an axiom. Then with the definitions and axioms as a basis,

Theorems statements called *theorems* are proved. The first theorem, of course, depends solely on the axioms and definitions. After a theorem is proved, however, it becomes a part of the material available for the proof of other theorems. We shall use this procedure in establishing the logical foundation for the algebra of numbers, and in the next section we shall define some of the terms that we shall employ.

2.2 / DEFINITIONS

We shall use the letters of the alphabet to represent real numbers. A letter used in this way is called a variable, which we define as follows:

Variable A *variable* is a symbol, usually a letter, that stands for, or that may be replaced by, a number from a specified set of numbers. The set of numbers is called the *replacement set*.

Replacement set

Constant A *constant* is a symbol whose replacement set contains only one number.

EXAMPLE 1 An upright cylindrical water tank is 10 ft high and has a radius of 6 ft. It is supplied by a pipe with an automatic valve that closes when the tank is full and opens when the tank is drained to a depth of 1 ft. The volume of the water in the tank is given by the formula

$$V = \pi r^2 d$$

This formula contains the constants and variables tabulated below.

Symbol	Replacement Set	Classification
π	{3.1416 approximately}	Constant
r (radius)	{6}	Constant
d (depth)	$\{d \mid 1 \le d \le 10\}$	Variable
V (volume)	$\{V \mid 36\,\pi \le V \le 360\pi\}$	Variable

Fundamental operations The four *fundamental operations* of algebra are addition, subtraction, multiplication, and division.

The sum of two numbers a and b is written $a + b$, the difference of a and b is expressed as $a - b$, the product of a and b is denoted by ab, and the quotient of a and b is expressed as a/b. We shall discuss these operations in more detail later.

Factor In the product ab, the numbers a and b are called *factors* of ab.

The product $a \cdot a$ is expressed a^2. Similarly, $a \cdot a \cdot a = a^3$, $a \cdot a \cdot a \cdot a = a^4$, and in general

$$\underbrace{a \cdot a \cdot a \cdots a}_{n \text{ factors}} = a^n$$

The numbers a^2, a^3, and a^n are called "a square," "a cube," and "the nth power of a," respectively.

Base and exponent   In the number a^n, a is called the *base,* and n is the *exponent* of the base.

The result obtained by combining two or more numbers by means of one or more of the four fundamental operations of algebra is called an

Expression   *expression.*

EXAMPLE 2   The following are expressions:

$$a + b \qquad 3a + bc \qquad \frac{a}{b} \qquad \frac{3a + 2b}{a + b} \qquad \left(\frac{x^2 + 2y^2}{3 + 4y}\right)\left(\frac{x^3 - 3y^4}{2 - 5y}\right)$$

An expression that does not involve addition or subtraction is called a

Monomial   *monomial.*

EXAMPLE 3   The expressions a, $2ab$, and $2ab/3bc$ are monomials.

Multinomial   The sum of two or more monomials is a *multinomial.*

EXAMPLE 4   The expression

$$2a + \frac{3b^2}{2bc} - \frac{4ab}{2bc} + \frac{1}{2}a^3$$

is a multinomial.

Each monomial in a multinomial, together with the sign that precedes

Term   it, is called a *term* of the multinomial.

A multinomial consisting of exactly two terms is a *binomial,* and a multinomial consisting of exactly three terms is a *trinomial.* If each term of a multinomial is an integral power of a number symbol or is the product of the integral powers of two or more number symbols, the multinomial is

Polynomial   called a *polynomial.*

EXAMPLE 5   Examples of polynomials are $2a^4 + a^3 - 4a^2 + 5a + 3$ and $2x^4y + 2x^3y - 4x^2y^2 + 2xy^3 - 4y^2$. However, $x^2/y + 3y^4$ is not a polynomial.

If a monomial is expressed as the product of two or more

Coefficient   symbols, each of the symbols is called the *coefficient* of the product of the others.

EXAMPLE 6   In the monomial $3ab$, 3 is the coefficient of ab, a is the coefficient of $3b$, b is the coefficient of $3a$, and $3a$ is the coefficient of b.   We call the 3 in $3ab$ the *numerical* coefficient.   Usually when we refer to the coefficient in a monomial, we mean the numerical coefficient.

Similar terms   Two monomials, or two terms, are called *similar* if they differ only in their numerical coefficients.

EXAMPLE 7   The monomials $3a^2b$ and $-2a^2b$ are similar, and the terms in $4(3a/5b) + 2(3a/5b)$ are similar.

2.3 / THE RELATION OF EQUALITY

In Chap. 1 we defined the relation of equality as applied to two sets, and we stated that two numbers are equal if they represent the same point on the line L.   We shall not attempt a general definition of this relation but shall state below several agreements, or axioms, dealing with it.   These axioms define the properties of the relation of equality, although they do not actually define the relation itself.   In the statements of the axioms, the letters used stand for real numbers

$a = a$	reflexive axiom	**(2.1)**
If $a = b$, then $b = a$	symmetric axiom	**(2.2)**
If $a = b$ and $b = c$, then $a = c$	transitive axiom	**(2.3)**

By use of Eqs. (2.2) and (2.3), we can prove our first theorem, which is stated below.

THEOREM   If $a = b$ and $c = b$, then $a = c$ **(2.4)**

Proof   If $c = b$, then $b = c$ by Eq. (2.2).   Hence, since $a = b$ and $b = c$, it follows that $a = c$ by Eq. (2.3).

If $a = b$, then $a + c = b + c$	addition axiom	**(2.5)**
If $a = b$, then $ac = bc$	multiplication axiom	**(2.6)**
If $a = b$, then a can be replaced by b in any statement involving algebraic expressions without affecting the truth or falsity of the statement	replacement axiom	**(2.7)**

We shall now state and prove a theorem that will be employed frequently in the remainder of this book.

THEOREM	**If $a = b$ and $c = d$, then $a + c = b + d$** (2.8)

Proof

$$a = b \qquad \text{given}$$
$$a + c = b + c \qquad \text{by Eq. (2.5)}$$
$$= b + d \qquad \text{by Eq. (2.7), since } c = d$$

The following example illustrates the use of the above axioms and theorems.

EXAMPLE   If $a = b$ and $c = d$, then $ar + cs = br + ds$.

Solution   Since $a = b$ and $c = d$, then

$$ar = br \qquad \text{by Eq. (2.6)}$$

and
$$cs = ds \qquad \text{by Eq. (2.6)}$$

Hence,
$$ar + cs = br + ds \qquad \text{by Eq. (2.8)}$$

2.4 / ADDITION

Closure   In Chap. 1 we demonstrated graphically that the sum of two integers is an integer, and we implied that the sum of two real numbers is a real number. This property of a set of numbers is called *closure*. More precisely, we say: If the sum of any two numbers in a set of numbers is an element of the set, then the set is said to be *closed* under the operation of addition. We shall assume that this property holds for the operation of addition in the set of real numbers and state the assumption below.

**If a and b are real numbers, there exists
a real number c such that $a + b = c$**   closure axiom for addition   (2.9)

The usual procedure for expressing $3a + 2b + 4a + 3b + a + 4b$ in the simplest form is to rearrange the terms as $3a + 4a + a + 2b + 3b + 4b$. Then we add the coefficients of the similar terms and get $8a + 9b$. This procedure brings up two questions: Why is the first expression equal to the expression with the terms rearranged? Why does $3a + 4a + a = 8a$? The answers to these questions depend upon the agreements or axioms dealing with addition, which we shall now state.

It is readily verified that $3 + 5 = 5 + 3$, that $4 + 7 = 7 + 4$, and that for any two numbers that we try, we obtain the same result regardless of the order in which the numbers are added. We shall assume that this is true for any two real numbers a and b and state the assumption below.

Commutative axiom
of addition

$$a + b = b + a \qquad \text{commutative axiom of addition} \qquad (2.10)$$

As indicated by the phrase following the statement of the axiom, this property of addition is called the *commutative* axiom of addition.

Our next axiom deals with the sum of three numbers and is known as the *associative* axiom. We may easily verify that $(7 + 4) + 9 = 11 + 9 = 20$, and that $7 + (4 + 9) = 7 + 13 = 20$. We assume that this property is true for any three numbers, and thus we have the following axiom:

Associative axiom
of addition

$$(a + b) + c = a + (b + c) \qquad \text{associative axiom of addition} \qquad (2.11)$$

This axiom, together with Eq. (2.10), enables us to say that when any two of three numbers $a + b + c$ are added and the third added to this sum, the result is the same regardless of the way in which the first two numbers are chosen. We can prove this statement by writing all possible ways in which this operation can be performed and then showing that each of them is equal to some one combination, such as $(a + b) + c$. As an example, we shall prove that $(b + c) + a = (a + b) + c$ and that $c + (b + a) = (a + b) + c$.

First Proof

$$(b + c) + a = a + (b + c) \qquad \text{by Eq. (2.10)}$$
$$= (a + b) + c \qquad \text{by Eqs. (2.2)† and (2.11)}$$

Second Proof

$$c + (b + a) = c + (a + b) \qquad \text{by Eq. (2.10)}$$
$$= (a + b) + c \qquad \text{by Eq. (2.10)}$$

The next axiom involves a combination of addition and multiplication and is illustrated by the following example:

$$(4 + 8)3 = 12 \times 3 \qquad \text{since } 4 + 8 = 12$$
$$= 36$$

Also

$$(4 \times 3) + (8 \times 3) = 12 + 24 = 36$$

Hence we can obtain the result of the operation $(4 + 8)3$ either by adding the numbers in the parentheses and multiplying the sum by 3 or by multiplying 4 and 8 by 3 and adding the products. This property is known as the *distributive property* of multiplication with respect to addition. We shall assume that this property holds for any three numbers, and we shall state the assumption in this way:

Distributive property
of multiplication with
respect to addition

$$(a + b)c = ac + bc \qquad \text{right-hand‡ distributive axiom} \qquad (2.12)$$

---

† The axiom shown by Eq. (2.2) enables us to interchange the members in the associative axiom shown by Eq. (2.11) and have $a + (b + c) = (a + b) + c$.

‡ This statement is called the "right-hand" distributive axiom because the factor c is at the right of the binomial $a + b$. Later we shall prove that the left-hand distributive theorem, $c(a + b) = ca + cb$, is true.

The axiom shown by Eq. (2.12) can be extended to cover situations in which the polynomial in the parentheses consists of more than two terms. As an example, we shall prove that

$$(a + b + d)c = ac + bc + dc$$

Proof By the statement following Eq. (2.11), $a + b + d = (a + b) + d$. Hence,

$$
\begin{aligned}
(a + b + d)c &= [(a + b) + d]c \\
&= (a + b)c + dc &&\text{by Eq. (2.12)} \\
&= ac + bc + dc &&\text{by Eq. (2.12)}
\end{aligned}
$$

The above axiom enables us to express the sum of two or more similar monomials as a monomial. For example,

$$
\begin{aligned}
3ab + 2ab + 4ab &= (3 + 2 + 4)ab &&\text{by Eq. (2.12)} \\
&= 9ab
\end{aligned}
$$

In Chap. 1 we defined the number zero, or 0, and the negative of a number, and we interpreted the definitions graphically. We shall repeat these definitions below.

Zero, denoted by the symbol 0, is the number such that $a + 0 = a$ for every number a. Consequently, since by Eq. (2.10) $a + 0 = 0 + a$, we have

$$a + 0 = 0 + a = a \qquad \text{additive identity} \qquad (2.13)$$

The negative of the number a is the number $-a$ such that $a + (-a) = 0$, and since by Eq. (2.10) $a + (-a) = -a + a$, we have

$$a + (-a) = -a + a = 0 \qquad (2.14)$$

The negative of a is also called the *additive inverse* of a.

We shall next prove two very useful theorems involving addition. The first theorem is called the *cancellation theorem* for addition and is stated as follows:

THEOREM If $a + b = a + c$, then $b = c$ cancellation theorem for addition   (2.15)

Proof

$$
\begin{aligned}
a + b &= a + c &&\text{given} \\
a + b + (-a) &= a + c + (-a) &&\text{by Eq. (2.5)}
\end{aligned}
$$

By Eq. (2.10), we have $a + b + (-a) = a + (-a) + b$ and $a + c + (-a) = a + (-a) + c$. Hence, by Eq. (2.4),

$$a + (-a) + b = a + (-a) + c$$

Hence, $0 + b = 0 + c$ since $a + (-a) = 0$

Therefore, $b = c$ since by (2.13) $0 + b = b$ and $0 + c = c$

The second theorem is:

THEOREM

$$\text{If } a + b = d \text{ and } a + c = d, \text{ then } b = c \tag{2.16}$$

Proof

$$a + b = d \quad \text{and} \quad a + c = d \qquad \text{given}$$
$$a + b = a + c \qquad\qquad\qquad \text{by Eq. (2.4)}$$
$$b = c \qquad\qquad\qquad\qquad \text{by Eq. (2.15)}$$

Difference of two numbers

The difference of the numbers a and b is expressed as $a - b$ and is defined to be the number x such that $a = b + x$. In other words,

$$\text{If } a - b = x, \text{ then } a = b + x \tag{2.17}$$

We shall now prove that if $a = b + x$,† then

THEOREM

$$a + (-b) = a - b \tag{2.18}$$

Proof

$$a = b + x \qquad\qquad\qquad \text{given}$$
$$a + (-b) = b + x + (-b) \qquad \text{by Eq. (2.5)}$$
$$= b + (-b) + x \qquad \text{by Eq. (2.10), the commutative axiom}$$
$$= [b + (-b)] + x \qquad \text{by Eq. (2.11), the associative axiom}$$
$$= 0 + x \qquad \text{since by Eq. (2.14) } b + (-b) = 0$$
$$= x \qquad \text{since by Eq. (2.13) } 0 + x = x$$

Hence, since $a - b = x$, we have

$$a + (-b) = a - b \qquad \text{by Eq. (2.4)}$$

Removal of parentheses

A theorem that is used in the *removal of parentheses* from an expression and in the insertion of parentheses in an expression is the following:

THEOREM

$$-(a + b) = -a - b \tag{2.19}$$

Proof

$$(a + b) + [-(a + b)] = 0 \qquad \text{by Eq. (2.14)}$$

Also

$$(a + b) + (-a) + (-b) = (-a) + (a + b) + (-b)$$
$$\text{by the commutative axiom shown by Eq. (2.10)}$$
$$= [(-a) + a] + b + (-b)$$
$$\text{by the associative axiom shown by Eq. (2.11)}$$
$$= 0 + b + (-b)$$
$$\text{since by Eq. (2.14) } -a + a = 0$$
$$= 0 \quad \text{since } b + (-b) = 0$$

Hence,
$$-(a + b) = (-a) + (-b) \qquad \text{by Eq. (2.16)}$$
$$= -a - b \qquad \text{by Eq. (2.18)}$$

† This theorem was proved graphically in Sec. 1.3. Here we shall prove it by means of the axioms.

This theorem can be extended to cover cases in which there are three or more terms in the parentheses.

As a final theorem in this section, we shall prove that

THEOREM
$$-(-a) = a \qquad (2.20)$$

Proof
$$-a + [-(-a)] = 0 \qquad \text{by Eq. (2.14)}$$

Also
$$(-a) + a \quad = 0 \qquad \text{by Eq. (2.14)}$$

Hence,
$$-(-a) \quad = a \qquad \text{by Eq. (2.15)}$$

2.5 / THE ORDER RELATIONS

We shall employ the properties of the order relations in the discussion of addition and multiplication. In Chap. 1 we stated that "greater than" is denoted by $>$ and "less than" by $<$. We also stated that $a > b$ if the point (a) is to the right of (b) on the number line. This interpretation is not a definition of the relation, and we shall not attempt to formulate one. We shall, however, state four axioms that completely determine the properties of the relation. If we accept the relation "greater than" as undefined, we can define the relation "less than" as follows:

If a and b are real numbers, then $b < a$ if and only if $a > b$ (2.21)

We shall now state the four basic axioms dealing with the order relations. In the statement of the axioms, the letters represent real numbers.

Exactly one of the statements
$a = b$, $a > b$, $a < b$ is true	trichotomy axiom	**(2.22)**
If $a > b$ and $b > c$, then $a > c$	transitivity axiom	**(2.23)**
If $a > b$, then $a + c > b + c$	additivity axiom	**(2.24)**
If $a > b$ and $c > 0$, then $ac > bc$;		
if $d < 0$, then $ad < bd$	multiplicativity axiom	**(2.25)**

We shall now prove the following theorem:

THEOREM
If $a > 0$, then $-a < 0$ (2.26)

Proof
$$
\begin{aligned}
a &> 0 & &\text{given} \\
a + (-a) &> 0 + (-a) & &\text{by Eq. (2.24)} \\
0 &> -a & &\text{since } a + (-a) = 0 \text{ and } 0 + (-a) = -a \\
-a &< 0 & &\text{by Eq. (2.21)}
\end{aligned}
$$

Consequently, if a is a positive number, then $-a$ is a negative number.

Positive number We now define a *positive number* as a number greater than zero and a *nega-*
Negative number *tive number* as a number less than zero. Therefore:

$$\text{If } a \text{ is a positive number, then } a > 0;$$
$$\text{if } b \text{ is a negative number, then } b < 0 \qquad (2.27)$$

We shall now prove a theorem that enables us to tell whether or not a is greater than b.

THEOREM $a > b$ **if and only if** $a - b > 0$ (2.28)

Proof

$a > b$	given
$a + (-b) > b + (-b)$	by Eq. (2.24)
$a - b > 0$	since $a + (-b) = a - b$ and $b + (-b) = 0$

Conversely, we shall assume that $a - b > 0$ and shall prove that $a > b$.

$a - b > 0$	assumed
$a + (-b) > 0$	since by Eq. (2.18) $a + (-b) = a - b$
$a + (-b) + b > 0 + b$	by Eq. (2.24)
$a + [(-b) + b] > 0 + b$	by the associative axiom shown by Eq. (2.11)
$a + 0 > 0 + b$	since $(-b) + b = 0$ by Eq. (2.14)
$a > b$	since by Eq. (2.13) $a + 0 = a$ and $b + 0 = b$

2.6 / LAW OF SIGNS FOR ADDITION

The extension of the number system to include negative numbers necessitates an extension of the notion of addition. We must define the meaning of the sum of two signed numbers and derive a law that enables us to decide whether the sum is positive or negative. The proof of this law (to be stated presently) involves the following theorem:

THEOREM   **If** $r = s$, **then** $-r = -s$ (2.29)

Proof

$-r + r = 0$	by Eq. (2.14)
$-r + s = 0$	replacing r by s
$-r + s + (-s) = 0 + (-s) = -s$	by the addition axiom shown by Eq. (2.5)
$-r + [s + (-s)] = -s$	by the associative axiom shown by Eq. (2.11)
$-r = -s$	by Eq. (2.13), since $s + (-s) = 0$

The statement of the law involves the concept of absolute value, and the student is advised to review the definition near the end of Sec. 1.3. We shall now state and prove the following theorem:

THEOREMS   If a and b are positive real numbers, then:

1   $a + b > 0$ and $|a + b| = |a| + |b|$
2   $-a - b < 0$ and $|-a - b| = |a| + |b|$
3   If $a > b$, then $a + (--b) > 0$ and $|a + (-b)| = |a| - |-b|$
4   If $a < b$, then $a + (-b) < 0$ and $|a + (-b)| = -|a| + |b|$

Proof of 1

$$a > 0 \qquad \text{since } a \text{ is positive}$$
$$a + b > 0 + b \qquad \text{by Eq. (2.24)}$$
$$a + b > b \qquad \text{by Eq. (2.13), since } 0 + b = b$$
$$a + b > 0 \qquad \text{by Eq. (2.23), since } b > 0$$

Also, since $a + b > 0$, then

$$|a + b| = a + b \qquad \text{by the definition of absolute value}$$
$$= |a| + |b| \qquad \text{since } |a| = a \text{ and } |b| = b$$

Proof of 2

$$-a - b = -(a + b) \qquad \text{by Eq. (2.19)}$$

Hence, since $a + b > 0$, $-(a + b) < 0$ by Eq. (2.26). Consequently, $-a - b < 0$. Furthermore, $|-a - b| = |-(a + b)| = a + b$ by the definition of absolute value. Hence, $|-a - b| = |a| + |b|$ since $a = |a|$ and $b = |b|$.

Proof of 3

$$a > b \qquad \text{given}$$
$$a + (-b) > b + (-b) \qquad \text{by the additivity axiom shown by Eq. (2.24), with } c = -b$$
$$a + (-b) > 0 \qquad \text{by Eq. (2.14), since } b + (-b) = 0$$

Moreover, by the definition of absolute value,
$$|a + (-b)| = a + (-b)$$
$$\qquad \text{since } a + (-b) \text{ is positive}$$
$$= a - b \qquad \text{by Eq. (2.18)}$$
$$= |a| - |-b| \qquad \text{since } a = |a| \text{ and } b = |-b|$$

Proof of 4   If $a < b$, then $b > a$ by Eq. (2.21). Hence,

$$b + (-b) > a + (-b) \qquad \text{by Eq. (2.24), with } c = -b$$

and
$$0 > a + (-b) \qquad \text{since } b + (-b) = 0 \text{ by Eq. (2.14)}$$

Consequently, $a + (-b) < 0 \qquad \text{by Eq. (2.21)}$

Furthermore,

$$|a + (-b)| = -[a + (-b)] \quad \text{by the definition of absolute value}$$
$$= -a - (-b) \quad \text{by Eq. (2.19)}$$
$$= -a + b \quad \text{since } -(-b) = +b \text{ by Eq. (2.20)}$$
$$= -|a| + |b| \quad \text{by the definition of absolute value}$$

If we state the above theorem in words, we have the following:

Law of signs for addition
The absolute value of the sum of two positive or two negative numbers is equal to the sum of their absolute values. The sum of the two numbers is positive if the two addends are positive, and the sum is negative if the two addends are negative. The absolute value of the sum of a positive number and a negative number is equal to the difference of their absolute values. The sum is positive if the number with the greater absolute value is positive and is negative if the number with the greater absolute value is negative.

The sign of a signed numeral is the sign that precedes it or is understood to precede it. For example, the sign of 3 is positive since a plus sign is understood to appear before it, and the sign of -3 is negative. Hence for two signed numerals, we may state the law of signs for addition as follows:

The sum of two signed numerals with the same sign is the sum of their absolute values preceded by the common sign of the addends. The sum of two signed numerals with different signs is the difference of their absolute values preceded by the sign of the addend with the greater absolute value.

EXAMPLE 1 $6 + 3 = 9$, since 6 and 3 are positive.

EXAMPLE 2 $(-2) + (-4) = -6$, since -2 and -4 are negative.

EXAMPLE 3 $-4 + 9 = 5$, since -4 and 9 have different signs and 9 has the greater absolute value.

EXAMPLE 4 $3 + (-8) = -5$, since 3 and -8 have different signs and -8 has the greater absolute value.

Binary operation
A *binary operation* in a set is a rule that assigns to each pair of elements of the set, taken in a definite order, a unique element of the set. The closure axiom shown by Eq. (2.9) states that if a and b are real numbers,

there exists a real number c such that $a + b = c$. Hence addition is a binary operation in the set of real numbers. Furthermore, since the sum of two positive integers is a positive integer, addition is a binary operation in the set of positive integers.

EXERCISE 2.1 **Axioms and Theorems**

By use of the axioms and theorems dealing with addition, equality, and the order relations, prove that the statements in Probs. 1 to 20 are true.

1   $a + (b + c) = b + (a + c)$
2   If $(b + c)a = ab + d$, then $ac = d$.
3   If $a + b = a$, then $b = 0$.
4   If $a + b = 0$, then $b = -a$.
5   If $(a + b) + c = a + (b + d)$, then $c = d$.
6   If $(a + b) + c = b + c$, then $a = 0$.
7   If $a + (b + c) = a$, then $b = -c$.
8   If $(a + b) + (c + d) = (a + e) + (c + f)$, then $b + d = e + f$.
9   If $a + (b + c) = d$ and $a - c = d$, then $b = -2c$.
10 If $(a + b) + (c + d) = 0$ and $(b + c) + (a - e) = 0$, then $d = -e$.
11 If $a + (b + c) = d$ and $a + b = d - e$, then $c = e$.
12 If $a + (b + c) = a$ and $d + (e + f) = d$, then $b - e = f - c$.
13 If $a > b$, $b > c$, and $d > 0$, then $ad > cd$.
14 If $a + b > a + c$, then $b > c$.
15 If $a + (b + c) > a + d$, then $b > d - c$.
16 If $a + b > c + d$, then $a - c > d - b$.
17 If $b > c$ and $ab > ac$, then $a > 0$.   HINT: Assume that $a = 0$ and then that $a < 0$, and show that each assumption leads to a conclusion that contradicts the hypothesis that $ab > ac$.
18 If $a > 0$ and $ab > ac$, then $b > c$.
19 If $a > 0$, $ab > bc$, and $bc > ac$, then $b > c$.
20 If $a > 0$, $b > 0$, $d > 0$ and $(a - c)b > db$, then $a > c$.   HINT: First prove that $a - c > d$.
21 Show that addition is a binary operation in the set of positive real numbers.
22 Give an example which shows that addition is not a binary operation in the set $\{x \mid 0 \le x \le 10\}$.
23 Give an example which shows that addition is not a binary operation in the set $\{x \mid x$ is an integer and $-6 \le x \le 1\}$.
24 Give an example which shows that the procedure for finding $a - b$ is not a binary operation in the set of positive integers.

2.7 / ADDITION OF MONOMIALS AND POLYNOMIALS

The sum of two or more similar monomials is equal to a monomial that is obtained by use of the right-hand distributive axiom shown by Eq. (2.12) or its extension. For example,

$$12ab + 8ab + 5ab = (12 + 8 + 5)ab \qquad \text{by Eq. (2.12)}$$
$$= 25ab$$

In this section we shall consider monomials that involve multiplication only, as illustrated by $3a$, $5x^2y$, and ab^3. We shall frequently encounter polynomials with all terms similar but in which some of the terms are preceded by the minus sign and others involve no numerical coefficient. In order to apply the distributive axiom to such polynomials, we shall assume† that the following statements are true:

$$a = 1 \cdot a$$
$$-(na) = -n \cdot a$$
$$0 \cdot a = 0$$

The procedure for adding two or more similar monomials and for combining by addition the similar terms in a polynomial is illustrated by the following examples.

EXAMPLE 1 $-5b + b + 6b = -5 \cdot b + 1 \cdot b + 6 \cdot b$
$$= (-5 + 1 + 6)b \qquad \text{by Eq. (2.12)}$$
$$= (-5 + 7)b = 2b \qquad \text{by the associative axiom shown by Eq.}$$
$$\text{(2.11) and the law of signs for addition}$$

EXAMPLE 2 $7x^2y - 3x^2y - 2xy^2 + 8xy^2 = 7 \cdot x^2y + (-3 \cdot x^2y) + (-2 \cdot xy^2) + 8 \cdot xy^2$
$$= [7 + (-3)]x^2y + (-2 + 8)xy^2$$
$$= (7 - 3)x^2y + (-2 + 8)xy^2$$
$$= 4x^2y + 6xy^2$$

After some practice, the first and second steps in problems similar to Example 2 can be performed mentally, and thus we can proceed directly to the result in the third step. This is illustrated in Example 3.

EXAMPLE 3 $2a - 3a + 6b + 4b - 7c + 9c = (2 - 3)a + (6 + 4)b + (-7 + 9)c$
$$= -a + 10b + 2c$$

In the polynomial $4x^3 + 5x^2y - 6xy^2 + 2x^3 - 2x^2y + 2xy^2 - 3x^3 + 4xy^2$, we

† We shall discuss these assumptions more fully in Sec. 2.9.

have three sets of monomials in which the elements of each set are similar. These sets are $\{4x^3, 2x^3, -3x^3\}$; $\{5x^2y, -2x^2y\}$; and $\{-6xy^2, 2xy^2, 4xy^2\}$. By repeated applications of the commutative axiom shown by Eq. (2.10), we can rearrange the terms in the polynomial so that the similar terms are together, and then we can apply the distributive axiom. We must remember that each monomial, together with the sign that precedes it, is a term of the polynomial and that when a term is shifted from one position to another, the sign must be carried along with the remainder of the term. We shall now employ the above procedure to express the polynomial as the sum of two monomials.

Addition of polynomials

$$4x^3 + 5x^2y - 6xy^2 + 2x^3 - 2x^2y + 2xy^2 - 3x^3 + 4xy^2$$
$$= 4x^3 + 2x^3 - 3x^3 + 5x^2y - 2x^2y - 6xy^2 + 2xy^2 + 4xy^2$$

by Eq. (2.10), the commutative axiom

$$= (4 + 2 - 3)x^3 + (5 - 2)x^2y + (-6 + 2 + 4)xy^2$$

by Eq. (2.12), the right-hand distributive axiom

$$= 3x^3 + 3x^2y + 0xy^2 = 3x^3 + 3x^2y$$

The above procedure is called *combining the similar terms* in a polynomial.

The sum of two or more polynomials is obtained by repeated applications of the commutative axiom and the distributive axiom. For example, to get the sum of $-4a + 2b - 5c$, $2a - 3c + 4b$, and $6c + 7a - 8b$, we begin by writing the given polynomials with each one below the preceding one, and with the terms rearranged so that similar terms are in the same column. Then we add the numerical coefficients in each column. In this way, we get

$$
\begin{array}{l}
-4a + 2b - 5c \\
\;\;2a + 4b - 3c \\
\;\;7a - 8b + 6c \\
\hline
\;\;5a - 2b - 2c \quad \text{sum}
\end{array}
$$

Frequently, we are required to add two or more polynomials when at least one of them contains one or more terms that are not similar to any term in at least one of the others. The method for dealing with such situations is illustrated by Example 4.

EXAMPLE 4   Add the polynomials $3x^2 + 4y^2 - 3xy + 7z^2$, $2x^2 + 4z^3$, and $4y^2 - 2z^2 - 2xy$.

Solution   We write the polynomials as shown below and perform the addition as indicated.

$$
\begin{array}{l}
3x^2 + 4y^2 - 3xy + 7z^2 \\
2x^2 \qquad\qquad\qquad\quad + 4z^3 \\
\qquad\;\; 4y^2 - 2xy - 2z^2 \\
\hline
5x^2 + 8y^2 - 5xy + 5z^2 + 4z^3
\end{array}
$$

2.8 / SUBTRACTION

The operation of subtracting the number b from the number a is the process of determining x such that $b + x = a$. The number a is called the *minuend*, b the *subtrahend*, and x the *difference* of a and b. We determine x by the method below:

Minuend
Subtrahend
Difference

$$b + x = a \qquad \text{given}$$
$$b + x + (-b) = a + (-b) \qquad \text{by Eq. (2.5)}$$
$$b + (-b) + x = a + (-b) \qquad \text{by Eq. (2.10), the commutative axiom}$$
$$0 + x = a + (-b) \qquad \text{since } b + (-b) = 0 \text{ by Eq. (2.14)}$$
$$x = a + (-b) \qquad \text{since } 0 + x = x$$

Thus, to subtract one number from another, we add the negative of the subtrahend to the minuend. The statement "a minus b," or $a - b$, means that b is to be subtracted from a. The procedure is illustrated by the following examples.

EXAMPLE 1   Subtract 4 from 6.

Solution
$$6 \text{ minus } 4 = 6 + (-4) = 2$$

EXAMPLE 2   Subtract $-16ab$ from $-20ab$.

Solution
$$-20ab \text{ minus } -16ab = -20ab + 16ab = -4ab$$

EXAMPLE 3   Subtract $-18x^2yz$ from $-12x^2yz$.

Solution
$$-12x^2yz \text{ minus } -18x^2yz = -12x^2yz + 18x^2yz = 6x^2yz$$

EXAMPLE 4   Subtract $3x - 2y - 9z$ from $5x + 3y - 6z$.

Solution   The procedure in solving Example 4 is to write the subtrahend below the minuend as indicated below, then mentally change the sign preceding each term of the subtrahend and proceed as in addition.

$$
\begin{array}{ll}
5x + 3y - 6z & \text{minuend} \\
\underline{3x - 2y - 9z} & \text{subtrahend} \\
2x + 5y + 3z & \text{difference}
\end{array}
$$

EXERCISE 2.2 Addition and Subtraction

Perform the indicated operations in Probs. 1 to 8.

1 $6 + 3 - 4$ **2** $9 - 2 + 5$ **3** $8 - 1 - 3$

4 $-7 - 2 + 6$ **5** $|9| + |-3|$ **6** $|8| - |4|$

7 $|7 - 11| + |6 - 4|$ **8** $|9 - 14| - |14 - 9|$

Add the three expressions in each of Probs. 9 to 24.

9 $2a - 3b - 5c$ **10** $-3p + 4q - 6r$ **11** $7x + 3y - 6z$

$\quad\quad 3a + 4b + 7c$ $\quad\quad\ 5p - 3q - 3r$ $\quad\quad\ -8x - 2y + 9z$

$\quad\quad -4a - 5b + 2c$ $\quad\quad\ 4p - q + 8r$ $\quad\quad\ 3x - y + 2z$

12 $2a - 3p + 7x$ **13** $5ab - 3bc + 6ac$ **14** $9xy - 7xz - 4yz$

$\quad\quad 3a + 5p - 8x$ $\quad\quad\ 3ab + 7bc - 4ac$ $\quad\quad\ 3xy + 4xz - 3yz$

$\quad\quad -4a + 4p + 3x$ $\quad\quad\ -7ab - 4bc - 3ac$ $\quad\quad\ -8xy + 5xz + 6yz$

15 $8pq + 5pr - 6qr$ **16** $7bq - 8bz - 4qz$

$\quad\quad\ pq - 8pr + 7qr$ $\quad\quad\ -9bq + 7bz - 3qz$

$\quad\quad -7pq + 3pr - qr$ $\quad\quad\ 2bq + bz + 7qz$

17 $2a + 3b - 4c, -3a - 6b + 3c, 5a - 4b + 10c$

18 $6x - 2y + 3z, -7x - 9y - 6z, 4x + 12y + 6z$

19 $5p + 7d - 3q, 3p - 8d - 6q, -7p + 3d + 9q$

20 $-7r + 5s - 4t, 8r - 6s + 7t, 2r + 3s - 2q$

21 $2ab^2 + 3ab + 5a^2b, 3ab^2 - 4ab - 3a^2b, 4ab^2 + 7ab - 8a^2b$

22 $4xy - 3xy^2 + 2xy^3, -7xy + 8xy^2 + 3xy^3, 2xy - 3xy^2 - 2xy^3$

23 $3p^2q + 4pq + 5pq^2, -7p^2q - 6pq + 3pq^2, 5p^2q + 4pq - 7pq^2$

24 $7r^2s + 2rs + 3s^2, -9r^2 - 7rs - 2s^2, 3r^2s + 5rs + s^2$

In each of Probs. 25 to 36, subtract the second number or expression from the first.

25 $18, 11$ **26** $81, 37$ **27** $28, -12$ **28** $-31, -42$

29 $2a + b - 3c, 3a + 2b + 4c$

30 $7a - 5b + 8c, 6a - 6b + 9c$

31 $5x + 2y - 6z, 7x - 2y + 3z$

32 $7a - 3k + 4p, 8a - 4k - 2p$

33 $3a^2 + 4b^2 - 5c^2, 2a^2 + 3b^2 - 6c^2$

34 $5x^2 - 7y^2 + 4z^2, 2x^2 - 8y^2 + 5z^2$

35 $9ax - 4ax^2 - 8a^2x, 2ax - 5ax^2 - 9a^2x$

36 $6p^2q^2 - 5pq^2 - 2pq, 6p^2q^2 - 7pq^2 - 3pq$

37 Subtract $2a + 3b - 2c$ from the sum of $a - b + c$ and $4a - 2b - 3c$.

38 Subtract $4x + 2y - 3z$ from the sum of $2x + 3y - 5z$ and $3x - 4y + 7z$.

39 Subtract the sum of $2a + 3p + 5x$ and $3a + 2p - 6x$ from $6a + 4p + x$.

40 Subtract the sum of $5x + 2y - 3a$ and $-2x + 3y - 4a$ from $3x + 4y - 8a$.

In Probs. 41 to 48, replace each variable by its given value and then combine the result into a single number.

41 $x + y - |z|; x = 3, y = 2, z = -1$ **42** $x + |y| - |-z|; x = 4, y = 3, z = 5$

43 $|-2x| + |-y| - |z|; x = 3, y = 4, z = 2$

44 $|3x| - |-2y| + |z|; x = 3, y = 4, z = 2$

45 $|r + s| - r + s; r = 3, s = -5$ **46** $|r| + |s| - |r - s|, r = 5, s = 1$

47 $|r - s| - |s + t| + |2s|; r = 4, s = 3, t = 5$

48 $|2r + 3s| - |2r| + |3s| - |r - 2t|; r = 2, s = 3, t = 1$

In Probs. 49 to 60, write an expression that is equal to the given one but that does not involve parentheses; then combine similar terms.

49	$(2a - 3b) + (4a + 2b)$	**50**	$(3b + 2c) + (2b - 3c)$
51	$(3c - 5d) - (2c - 6d)$	**52**	$(8c - 7d) - (6c - 5d)$
53	$2(3d - 2e) - 3(2d + 3e)$	**54**	$4(2e - 5f) - 3(3e + 4f)$
55	$-3(2f + 3g) - 2(3f - 2g)$	**56**	$-5(6g - 4h) - 4(7g + 5h)$
57	$-4(2h - 3i) + 3(3h + 6i) - 7(4h - 2i)$		
58	$-3(5i - 6j) - 4(3i + 4j) + 5(4i - 3j)$		
59	$2(5j + 4k - 6m) - 3(3j - 2k - 5m)$		
60	$3(m - 2p - 3q) - 4(-2m - 3p + q)$		

2.9 / AXIOMS AND THEOREMS OF MULTIPLICATION

In this section we shall state the axioms that determine the properties of multiplication and prove theorems that enable us to obtain the product of two signed numbers and to apply multiplication to monomials and polynomials.

Product
Multiplier
Multiplicand
Factor

The operation of multiplying b by a is indicated by $a \times b$, $a \cdot b$, or $(a)(b)$, and the *product* is denoted by ab. The number a is the *multiplier*, and b is the *multiplicand*. The numbers a and b are also called the *factors* of ab.

The first axiom given below is the closure axiom, and it states that the product of two real numbers is a real number.

Closure axiom for
multiplication

**If a and b are real numbers, there exists
a real number c such that $ab = c$**   **(2.30)**

We have seen the importance of the commutative and associative axioms in addition, and we shall presently state similar axioms for multiplication. It is readily verified that $3 \cdot 5 = 5 \cdot 3 = 15$ and that $3 \cdot 5 \cdot 7 = 3(35) = 5(21) = 7(15) = 105$. We assume that these properties hold for all real numbers and thus we may state the following axioms:

Commutative axiom
for multiplication

Associative axiom
for multiplication

$$ab = ba \qquad (2.31)$$
$$a(bc) = (ab)c \qquad (2.32)$$

These two axioms enable us to get the product $a \cdot b \cdot c$ by multiplying the product of two of the numbers by the third. Furthermore, this product is unique regardless of the choice of the two numbers for the first multiplication and the order in which the multiplication is performed. This means that

$$abc = a(bc) = (ab)c = a(cb) = (ac)b = c(ba) = (cb)a = \cdots$$

where the list of equalities can be extended to include all possible orders in which a, b, and c can be arranged and all choices of the two that are to be enclosed in parentheses. To prove this statement we must prove that each combination is equal to some of them. To illustrate the method we shall prove that $c(ba) = a(bc)$.

Proof

$$c(ba) = c(ab) \quad \text{by the commutative axiom (2.31)}$$
$$= (ab)c \quad \text{by (2.31)}$$
$$= a(bc) \quad \text{by the associative axiom (2.32)}$$

In Eq. (2.12) we stated the right-hand distributive axiom. We shall now prove that the left-hand distributive axiom

Left-hand distributive axiom

$$a(b + c) = ab + ac \tag{2.33}$$

is true.

Proof

$$a(b + c) = (b + c)a \quad \text{by the commutative axiom (2.31)}$$
$$= ba + ca \quad \text{by the right-hand distributive axiom (2.12)}$$
$$= ab + ac \quad \text{by (2.31)}$$

The numbers 1 and 0 play unique roles in multiplication. The role of 1 is established by definition. That is, we define $1 \cdot a$ to be a. Then, since $1 \cdot a = a \cdot 1$, we have

$$1 \cdot a = a \cdot 1 = a \tag{2.34}$$

Multiplicative identity

For this reason 1 is called the *multiplicative identity* element. We shall now prove that $0 \cdot a = 0$.

Proof

$$1 \cdot a + 0 \cdot a = (1 + 0)a \quad \text{by the distributive axiom}$$
$$= 1 \cdot a \quad \text{since } 1 + 0 = 1$$
$$= a \quad \text{by (2.34)}$$
Also, $\qquad 1 \cdot a + 0 = a \quad \text{by (2.34) and (2.13)}$

Hence, $0 \cdot a$ and 0 must be equal, so $0 \cdot a = 0$. Therefore by the commutative axiom we have

$$0 \cdot a = a \cdot 0 = 0 \tag{2.35}$$

The following axiom is essential for dealing with division and for computations involving fractions:

Reciprocal

If a is a real nonzero number, there exists a unique real number $1/a$, called the reciprocal of a, such that

$$a \cdot \frac{1}{a} = \frac{1}{a} \cdot a = 1 \tag{2.36}$$

We shall employ this axiom to prove the very useful *cancellation theorem* for multiplication.

$$\text{If } ab = ac \text{ and } a \neq 0, \text{ then } b = c \qquad (2.37)$$

Proof

$ab = ac$	given
$\dfrac{1}{a} \cdot ab = \dfrac{1}{a} \cdot ac$	by the multiplicativity axiom for equality (2.6)
$1 \cdot b = 1 \cdot c$	by the associative axiom for multiplication (2.32) and (2.36)
$b = c$	by (2.34)

2.10 / LAW OF SIGNS FOR MULTIPLICATION

In this section we shall develop the law of signs for multiplication and explain methods for applying it. For this purpose, we shall first prove three theorems.

THEOREM   *If $a > 0$ and $b > 0$, then $ab > 0$*

Proof

$a > 0$	given
$a \cdot b > 0 \cdot b$	by (2.25)
$a \cdot b > 0$	since $b \cdot 0 = 0$
$ab > 0$	since $a \cdot b = ab$

THEOREM   *If $a > 0$ and $b > 0$, $a(-b) = (-b)a = -ab$*

Proof

$ab + a(-b) = a[b + (-b)]$	by the left-hand distributive axiom
$= a(0)$	since $b + (-b) = 0$ by (2.14)
$= 0$	by (2.35)
Furthermore, $ab + (-ab) = 0$	by (2.14)

Hence since $ab + a(-b) = 0$ and $ab + (-ab) = 0$, we have $a(-b) = -ab$ by (2.16). Moreover, by the commutative axiom $a(-b) = (-b)a$. This completes the proof.

THEOREM   *If $a > 0$ and $b > 0$, then $(-a)(-b) = ab$*

Proof

$a(-b) + (-a)(-b) = [a + (-a)](-b)$	by the right-hand distributive axiom
$= 0(-b)$	since $a + (-a) = 0$
$= 0$	by (2.35)

Also, $a(-b) + ab = a(-b + b)$ by the left-hand
                                                      distributive axiom

                               $= a(0)$ by (2.14)
                               $= 0$ by (2.35)

Therefore, $(-a)(-b) = ab$ by (2.16)

Since a and b are positive, we have by the definition of the absolute value of a number

$$a = |a| = |-a| \qquad \text{and} \qquad b = |b| = |-b|$$

Furthermore, $|ab| = ab = |a|\,|b|$ since a, b, and ab are positive

Likewise, $|(-a)(-b)| = |ab| = |a|\,|b|$ since $(-a)(-b) = ab$

and $|a(-b)| = |(-b)a| = |-ab| = ab = |a|\,|b|$

Law of signs for multiplication

We summarize the above conclusions in the following law:

The absolute value of the product of two real numbers is equal to the product of their absolute values.

The product of two positive real numbers or of two negative real numbers is positive. The product of a positive real number and a negative real number or of a negative real number and a positive real number is negative.

EXAMPLE 1

$$4(7) = 28$$
$$6(-9) = -54$$
$$-5(8) = -40$$
$$-3(-6) = 18$$

It follows from the associative axiom (2.32) that the product of three or more positive numbers is positive. For example, $3 \cdot 5 \cdot 7 = (3 \cdot 5)7 = 15(7) = 105$, and $2 \cdot 6 \cdot 3 \cdot 8 = (2 \cdot 6)(3 \cdot 8) = 12(24) = 288$. Furthermore the product of an even number of negative numbers is positive, and the product of an odd number of negative numbers is negative. For example, $(-2)(-4)(-5)(-8) = [(-2)(-4)][(-5)(-8)] = 8(40) = 320$, and $(-4)(-6)(-3) = [(-4)(-6)](-3) = 24(-3) = -72$.

If we do not know whether certain factors in a product are positive or negative, we cannot express the product in terms of the absolute values of the factors. For example, the product of the absolute values of a and b is $|a|\,|b|$. However, if a is positive, then $ab = |a|\,|b|$ or $-|a|\,|b|$ according as b is positive or negative. If one or more of the factors are not clearly positive or clearly negative and if minus signs appear explicitly before some of the factors, the usual practice for expressing the product is as follows. First disregard the minus signs and write the product of the symbols that remain, then prefix a minus sign if the number of explicitly expressed minus signs is odd.   If the number of explicitly expressed minus signs is even, we either prefix no sign before the product or prefix a plus sign.

EXAMPLE 2

$$(3x)(2y) = 6xy$$
$$(4x)(-3y)(5z) = -60xyz$$
$$(-7a)(8b)(-3c) = 168abc$$
$$(-2m)(-5n)(-7r) = -70 \; mnr$$

2.11 / LAWS OF EXPONENTS IN MULTIPLICATION

In Sec. 2.2 we stated that a^n, where n is a positive integer, is equal to $a \cdot a \cdot a \cdots$ to n factors.   The number a is the *base*, and n is the exponent.   We now consider the product $a^m a^n$.   By definition this product is equal to the product of m a's and n a's, and this is the product of $m + n$ a's.   Hence we have

$$a^m a^n = a^{m+n} \qquad \text{where } m \text{ and } n \text{ are positive integers} \qquad \textbf{(2.38)}$$

By the definition of the nth power of a number and the use of the commutative axiom, we can prove that

$$(ab)^n = a^n b^n \qquad \text{where } n \text{ is a positive integer} \qquad \textbf{(2.39)}$$

Proof

$$(ab)^n = ab \cdot ab \cdot ab \cdots \text{ to } n \text{ factors}$$
$$= (a \cdot a \cdot a \cdots \text{ to } n \text{ factors})(b \cdot b \cdot b \cdots \text{ to } n \text{ factors})$$

by the commutative axiom

$$= a^n b^n$$

Finally we prove that

$$(a^n)^m = a^{mn} \qquad \text{where } m \text{ and } n \text{ are positive integers} \qquad \textbf{(2.40)}$$

Proof

$$(a^n)^m = a^n \cdot a^n \cdot a^n \cdots \text{ to } m \text{ factors}$$
$$= a^{n+n+n+ \cdots \text{ to } m \text{ addends}} \qquad \textbf{by (2.38)}$$
$$= a^{mn}$$

2.12 / SYMBOLS OF GROUPING

The use of parentheses has been explained in preceding sections.   Symbols of grouping in addition to parentheses are frequently needed to make the meaning of certain expressions clear and to indicate the order in which operations are to be performed.

In addition to parentheses, we use the brackets, [ ], and the braces, { },
for these purposes.

Removing symbols of grouping   It is frequently desirable to remove the symbols of grouping from an
expression, and we shall explain and illustrate the procedure. If an ex-
pression that is enclosed in parentheses is preceded or followed by a
monomial factor, as in $x - 2y(3x - y + z)$, we apply the distributive law and
replace $-2y\,(3x - y + z)$ by $-6xy + 2y^2 - 2yz$. Therefore, $x - 2y(3x - y + z) =
x - 6xy + 2y^2 - 2yz$. Similarly, $a^2 + (a^3 - ab + b^2)2a = a^2 + 2a^4 - 2a^2b + 2ab^2$.

    Since $-(n) = -1(n)$, the expression $x + y - (-2x^2 - y^2 + z^2) = x + y -
1(-2x^2 - y^2 + z^2) = x + y + 2x^2 + y^2 - z^2$ by use of the distributive law. There-
fore, if an expression enclosed in parentheses is preceded by a minus sign,
the parentheses can be removed if and only if the sign of each of the enclosed terms is changed. If an expression enclosed in parentheses is
preceded by a plus sign and no monomial factor is indicated, as in
$a + (b + c - d)$, it is understood that the factor $+1$ precedes the parentheses
but is not expressed. Since multiplying a number by 1 yields a product
equal to the number, the parentheses can be removed from $a + (b + c - d)$
with no further changes.

    Usually when braces or brackets or both appear together with paren-
theses in an expression, one or more sets of grouping symbols will be enclosed in another set. When the symbols are removed from an expression
of this type, it is advisable to remove the innermost symbols first. We shall
illustrate the procedure with the following example.

EXAMPLE $3x^2 - \{2x^2 - xy - [x(x - y) - y(2x - y)] + 4xy\} - 3y^2$

Solution   We start with the given expression, indicate the successive steps in removing the symbols of operations, and explain the purpose of each step at the
right.

$3x^2 - \{2x^2 - xy - [x(x - y) - y(2x - y)] + 4xy\} - 3y^2$
                             given expression

$= 3x^2 - \{2x^2 - xy - [x^2 - xy - 2xy + y^2] + 4xy\} - 3y^2$
                             applying the distributive law to the
                                expression in the brackets

$= 3x^2 - \{2x^2 - xy - [x^2 - 3xy + y^2] + 4xy\} - 3y^2$
                              adding similar terms in brackets

$= 3x^2 - \{2x^2 - xy - x^2 + 3xy - y^2 + 4xy\} - 3y^2$
                        since $-[x^2 - 3xy + y^2] = -x^2 + 3xy - y^2$

$= 3x^2 - \{x^2 + 6xy - y^2\} - 3y^2$ adding similar terms in braces
$= 3x^2 - x^2 - 6xy + y^2 - 3y^2$ since $-\{x^2 + 6xy - y^2\} = -x^2 - 6xy + y^2$
$= 2x^2 - 6xy - 2y^2$ adding similar terms

Inserting symbols of grouping   If a pair of grouping symbols is inserted in an expression after a plus sign, no changes in signs are necessary. If the grouping symbols are inserted after a minus sign, the signs of all enclosed terms must be changed. For example,

$$x + y - z + w = x + (y - z + w)$$

and
$$a - b + c - d = a - (b - c + d)$$

2.13 / PRODUCTS OF MONOMIALS AND POLYNOMIALS

We employ the commutative, associative, and distributive axioms together with the law of signs and the law of exponents to obtain the product of two or more monomials, of a monomial and a polynomial, and of two polynomials. We shall illustrate the method with five examples.

EXAMPLE 1 $3x^2y \cdot 4xy^2 \cdot 6x^3y^4 = 3 \cdot 4 \cdot 6 \cdot x^2 \cdot x \cdot x^3 \cdot y \cdot y^2 \cdot y^4$ by the commutative axiom (2.31)

$$= 72x^{2+1+3}y^{1+2+4}$$ by (2.38)

$$= 72x^6y^7$$

EXAMPLE 2 $-4ab^2c^3 \cdot -2a^3b^4c \cdot 6a^2bc^5 = -4 \cdot -2 \cdot 6 \cdot a \cdot a^3 \cdot a^2 \cdot b^2 \cdot b^4 \cdot b \cdot c^3 \cdot c \cdot c^5$

$$= 48a^6b^7c^9$$ by the law of signs and the law of exponents

---

EXAMPLE 3 $3ab(2a - 4b + 7a^2b) = 3ab(2a) - 3ab(4b) + 3ab(7a^2b)$

by the left-hand distributive axiom (2.33)

$$= 6a^2b - 12ab^2 + 21a^3b^2$$

EXAMPLE 4 $(3x^2y - 6xy^2 - 8y^3)(-5x^3y^2)$

$$= 3x^2y(-5x^3y^2) + (-6xy^2)(-5x^3y^2) + (-8y^3)(-5x^3y^2)$$

$$= -15x^5y^3 + 30x^4y^4 + 40x^3y^5$$ by the right-hand distributive axiom (2.12)

The method for obtaining the product of two polynomials is illustrated in Example 5.

EXAMPLE 5   To obtain the product $(-5x^2 + 2xy + 3y^2)(3x^3 - 6x^2y + 2xy^2 - 4y^3)$ we first consider the second factor as a single number, apply the right-hand distributive axiom, and then complete the computation as indicated.

$$(-5x^2 + 2xy + 3y^2)(3x^3 - 6x^2y + 2xy^2 - 4y^3)$$
$$= (-5x^2)(3x^3 - 6x^2y + 2xy^2 - 4y^3) + (2xy)(3x^3 - 6x^2y + 2xy^2 - 4y^3)$$
$$\quad + (3y^2)(3x^3 - 6x^2y + 2xy^2 - 4y^3) \qquad \text{by the right-hand distributive axiom}$$
$$= -15x^5 + 30x^4y - 10x^3y^2 + 20x^2y^3 + 6x^4y - 12x^3y^2 + 4x^2y^3 - 8xy^4$$
$$\quad + 9x^3y^2 - 18x^2y^3 + 6xy^4 - 12y^5 \qquad \text{by the right-hand distributive axiom}$$
$$= -15x^5 + 30x^4y + 6x^4y - 10x^3y^2 - 12x^3y^2 + 9x^3y^2 + 20x^2y^3 + 4x^2y^3$$
$$\quad - 18x^2y^3 - 8xy^4 + 6xy^4 - 12y^5 \qquad \text{by the commutative axiom}$$
$$= -15x^5 + 36x^4y - 13x^3y^2 + 6x^2y^3 - 2xy^4 - 12y^5$$

The above process is usually abbreviated to the method illustrated below.

$3x^3 - 6x^2y + 2xy^2 - 4y^3$	multiplicand
$- 5x^2 + 2xy + 3y^2$	multiplier
$-15x^5 + 30x^4y - 10x^3y^2 + 20x^2y^3$	multiplying by $-5x^2$
$+ 6x^4y - 12x^3y^2 + 4x^2y^3 - 8xy^4$	multiplying by $2xy$
$+ 9x^3y^2 - 18x^2y^3 + 6xy^4 - 12y^5$	multiplying by $3y^2$
$-15x^5 + 36x^4y - 13x^3y^2 + 6x^2y^3 - 2xy^4 - 12y^5$	adding coefficients

EXERCISE 2.3 Products and Grouping Symbols

Find the indicated product in each of Probs. 1 to 40.

1 $(3a^2)(4a^3)$ 2 $(2a^4)(4a^2)$ 3 $(-5a^3)(2a^5)$

4 $(-4a^5)(5a^4)$ 5 $(2a^7)(-3a)$ 6 $(3a^3)(-4a^4)$

7 $(-7a^3)(-2a^{11})$ 8 $(-5a^5)(-6a^4)$ 9 $(2x^2)^2$

10 $(3x^4)^2$ 11 $(5x^2y)^3$ 12 $(4x^3y^2)^4$

13 $(-2x^3y^2)^3$ 14 $(-3x^4y^2)^2$ 15 $(-4x)^5$

16 $(-5x^3)^4$ 17 $2x^3y^2(2x - y)$ 18 $3xy^3(3x - 2y)$

19 $-5xy^2(3x - y)$ 20 $-3x^2y^3(3x - 4y)$ 21 $-2x^2y^3(2x^2y - xy + 3xy^2)$

22 $-x^3y(4xy^3 - 3x^2y^2 + 2x^3y)$

23 $3xy(4xy^2 - 3x^2y + 5x^2y^2)$

24 $5x^2y^3(3x^3y^2 - 2x^2y^3 - 5xy^4)$

25 $2xy^2(3x - 2y) - 3x^2y(2y - x)$

26 $3xy(2x + 3y) - 2x^2y(4 - xy)$

27 $3x^3y^4z^2(2x^4y^4z^4 - 4x^2y^3z^6) - 5x^4y^5z^3(3x^3y^3z^3 - 2xy^2z^5)$

28 $3x^8y^7z^6(4x^4y^3z^5 + 5x^5yz^3) - 2x^5y^5z^8(5x^7y^5z^3 - 3xyz^3 + 7x^8y^3z)$

29 $(2x - 3y)(3x + 4y)$ 30 $(3x + 2y)(4x - 3y)$

31 $(3x - y)(2x + 3y)$ 32 $(5x - 7y)(x + 3y)$

33 $(2x - 3y)(3x^2 - 2xy + 4y^2)$ 34 $(3x + 2y)(2x^2 - 3xy - 2y^2)$

35 $(5x + 2y)(2x^2 - xy - 3y^2)$ 36 $(3x - 4y)(2x^2 - xy + y^2)$

37 $(x^2 - 3xy + 2y^2)(2x^2 + 3xy - 3y^2)$ 38 $(2x^2 - xy + 3y^2)(3x^2 - xy - 2y^2)$

39 $(5x^2 + 2xy + y^2)(2x^2 - xy + 3y^2)$ 40 $(x^2 - 3xy + 2y^2)(x^2 + 3xy - y^2)$

Remove the symbols of grouping in each of Probs. 41 to 48, and then combine similar terms.

41 $2a + [3a - (2b - a)] + 4b$ **42** $3x + [5x - (3x - y)] + 3y$

43 $2x - [2w - 3y - (2x + 4w - 3y) - 2x] - (w - y)$

44 $5a - (2b - 3c) - [3a - 4b - (a - b - 2c) + c] - (2a - b) + c$

45 $2\{2a - b[2a - c(2a - 1) + 2ac] - c\}$

46 $2x\{x^2 - x[2x - 3(x + 2) + 1] - x^2\}$

47 $4[3x - 2(x + 2y)] - 3\{x^2 - [3y + x(x - y)]\}$

48 $2x^3 - 5x\{x^2 + 3[3x - 4(x - 2) + 3] - x^2\}$

Prove the statement in each of Probs. 49 to 60 by use of the axioms and theorems of this chapter.

49   If $a \neq 0$ and $ab = 0$, then $b = 0$.

50   If $ab = cd$ and $cd = be$, then $a = e$.

51   If $a(b + c) = ab + d$, then $d = ac$.

52   If $(a + b)(c - 3) = (a + b)(d - 1)$, then $c - d = 2$.

53   If $a < b$ and $b < c$, then $a < c$.

54   If $a > b > 0$ and $c > d > 0$, then $ac > bd$.

55   If $a > 0$, $a = b$, and $ac > bd$, then $c > d$.

56   If $a > b$ and $ac < bc$, then $c < 0$.

57   Show that multiplication is a binary operation for $\{-1, 1\}$.

58   Show that multiplication is a binary operation for $\{0, 1\}$.

59   Show that multiplication is not a binary operation for $\{-1, 1, 2\}$.

60   Show that multiplication is not a binary operation for $\{0, 1, 2, 3\}$.

2.14 / DIVISION

As stated earlier, if $b \neq 0$, the quotient of a and b is expressed as $\frac{a}{b}$ or a/b. We define this quotient as the unique number x such that $bx = a$. That is,

$$\textbf{If } b \neq 0, \textbf{ then } \frac{a}{b} = x \textbf{ if and only if } bx = a \qquad (2.41)$$

Dividend   The number a is the *dividend*, b is the *divisor*, and the procedure for com-
Divisor   puting x is called *division*.

Since $a \cdot 1 = a$, by (2.34), it follows that

$$\frac{a}{a} = 1 \qquad \text{provided } a \neq 0 \qquad (2.42)$$

The requirement $b \neq 0$ in (2.41) is specified for reasons that we next discuss. If $b = 0$, then $bx = 0 \cdot x = 0$ by (2.35). Hence if $a \neq 0$, there is no replacement in the set of real numbers for x such that $bx = a$, and so the quotient does not exist. If, however, $a = 0$ and $b = 0$, we have $0 \cdot x = 0$ for any replacement for x. Consequently 0/0 does not exist as a unique number. Therefore if the divisor $b = 0$ in a/b, the quotient is not defined.

If in (2.41) $a = 0$ and $b \neq 0$, we have $0/b = 0$, since $b \cdot 0 = 0$. Hence

$$\frac{0}{b} = 0 \qquad \text{provided } b \neq 0 \tag{2.43}$$

The law of signs for division is derived from (2.42) and the law of signs for multiplication. If $bx = a$, then by the law of signs for multiplication, we have the following possibilities:

b	x	a
Positive	Positive	Positive
Negative	Positive	Negative
Negative	Negative	Positive
Positive	Negative	Negative

It follows from the above tabulation that if a and b are both positive or both negative, $x = a/b$ is positive. Furthermore, if either a or b is positive and the other is negative, $x = a/b$ is negative. Consequently we have the following law of signs for division:

Law of signs for division

The quotient of two positive or two negative numbers is positive. The quotient of a positive number and a negative number or of a negative number and a positive number is negative.

As examples, we have

$$\frac{12}{4} = 3 \qquad \frac{24}{-6} = -4 \qquad \frac{-18}{9} = -2 \qquad \frac{-9}{-3} = 3 \qquad \frac{-3}{4} = -\frac{3}{4} \qquad \frac{5}{-7} = -\frac{5}{7}$$

We shall now prove two theorems and a corollary which are essential for the simplification of quotients and for the multiplication and division of fractions. The first theorem follows directly from the definition of a quotient. Since $a \cdot 1 = a$, we have

THEOREM

$$\frac{a}{1} = a \tag{2.44}$$

The second theorem is

THEOREM

$$\frac{ac}{bd} = \frac{a}{b}\frac{c}{d} = \frac{a}{d}\frac{c}{b} \tag{2.45}$$

Proof   In order to prove that $ac/bd = a/b \cdot c/d$ we shall let $a/b = x$ and $c/d = y$, and thus have $a/b \cdot c/d = xy$. Furthermore, by the definition of a quotient (2.41), $a = bx$ and $c = dy$. Consequently $ac = bx \cdot dy = bd \cdot xy$ by the commutative and associative axioms. Hence, by (2.41), $ac/bd = xy$. Therefore $ac/bd = a/b \cdot c/d$ since each is equal to xy.

Furthermore since $bd = db$, we have

$$\frac{ac}{bd} = \frac{ac}{db} = \frac{a}{d}\frac{c}{b}$$

This theorem can be extended to cover cases where there are more than two factors in the dividend and in the divisor. For example,

$$\frac{ace}{bdf} = \frac{a(ce)}{b(df)} = \frac{a}{b}\frac{ce}{df} = \frac{a}{b}\frac{c}{d}\frac{e}{f}$$

If in (2.45) we let $b = c = 1$ and apply the commutative axiom, we have

$$\frac{a}{d} = a \cdot \frac{1}{d} = \frac{1}{d} \cdot a \qquad \text{provided } d \neq 0 \tag{2.46}$$

Our next task is to prove the following law of exponents for division:

$$\frac{a^m}{a^n} = a^{m-n} \qquad \text{if } \begin{cases} a \neq 0 \\ m \text{ and } n \text{ are positive integers} \\ m > n \end{cases} \tag{2.47}$$

Proof   By the law of exponents for multiplication (2.38) we have

$$a^n a^{m-n} = a^{n+(m-n)}$$
$$= a^m \qquad \text{since } n + (m - n) = (n - n) + m = m$$

Consequently by the definition of a quotient (2.41) we have $a^m/a^n = a^{m-n}$.

If in (2.47) $m = n$, we have $a^n/a^n = a^{n-n} = a^0$, and a^0 has no meaning according to the previous definition of exponents. However, since $a \neq 0$, we have, by (2.42), $a^n/a^n = 1$. We therefore define

$$a^0 = 1 \qquad \text{provided } a \neq 0 \tag{2.48}$$

2.15 / MONOMIAL DIVISORS

We employ one or more of the theorems from (2.42) to (2.48) to obtain and simplify a quotient when the divisor is a monomial. The method is illustrated in the following examples.

EXAMPLE 1

$$6x^8y^6z^3 \div 3x^4y^3z^2 = \frac{6x^8y^6z^3}{3x^4y^3z^2}$$

$$= \frac{6}{3}\frac{x^8}{x^4}\frac{y^6}{y^3}\frac{z^3}{z^2} \qquad \text{by (2.45)}$$

$$= 2x^{8-4}y^{6-3}z^{3-2} \qquad \text{by (2.47)}$$

$$= 2x^4y^3z$$

EXAMPLE 2

$$24a^4b^{10}c^2 \div -3a^4b^7 = \frac{24a^4b^{10}c^2}{-3a^4b^7}$$

$$= \frac{24}{-3}\frac{a^4}{a^4}\frac{b^{10}}{b^7}\frac{c^2}{1}$$

$$= -8a^0b^3c^2$$

$$= -8b^3c^2 \qquad \text{since } a^0 = 1$$

If the dividend is a polynomial, we first employ the following theorem:

THEOREM

$$\frac{a+b+c}{d} = \frac{a}{d} + \frac{b}{d} + \frac{c}{d} \qquad \text{provided } d \neq 0 \qquad \textbf{(2.49)}$$

Proof

$$\frac{a+b+c}{d} = \frac{1}{d}(a+b+c) \qquad \text{by (2.46)}$$

$$= \frac{a}{d} + \frac{b}{d} + \frac{c}{d} \qquad \text{by the left-hand distributive axiom and (2.46)}$$

EXAMPLE 3

$$(6x^6y^5 + 4x^5y^4 - 3x^4y^3 - 2x^3y^2) \div 3x^3y^2 = \frac{6x^6y^5 + 4x^5y^4 - 3x^4y^3 - 2x^3y^2}{3x^3y^2}$$

$$= \frac{6x^6y^5}{3x^3y^2} + \frac{4x^5y^4}{3x^3y^2} - \frac{3x^4y^3}{3x^3y^2} - \frac{2x^3y^2}{3x^3y^2}$$

$$= 2x^3y^3 + \tfrac{4}{3}x^2y^2 - xy - \tfrac{2}{3}$$

2.16 / THE QUOTIENT OF TWO POLYNOMIALS

Before discussing the procedure for dividing one polynomial by another, we shall consider the relation $bx = a$, with a replaced by 23 and b replaced by 5. Then we have

$$5x = 23$$

Hence $\qquad \frac{1}{5}(5x) = \frac{1}{5} \times (23) \qquad$ by Eq. (2.6) and $\qquad x = \frac{23}{5}$ $\qquad$ by Eqs. (2.36), (2.34), and (2.46)

$$= \frac{20}{5} + \frac{3}{5} = 4 + \frac{3}{5}$$

Hence, if we divide 23 by 5, we obtain $4\frac{3}{5}$ since $4 + \frac{3}{5}$ is expressed as $4\frac{3}{5}$. We call $4\frac{3}{5}$ the *complete quotient,* the integer 4 the *partial quotient,* and the numerator of $\frac{3}{5}$ the *remainder.* Likewise,

$$\frac{6x^2 + 4}{3x} = \frac{6x^2}{3x} + \frac{4}{3x} = 2x + \frac{4}{3x}$$

and here the complete quotient is $2x + 4/3x$, the partial quotient is $2x$, and the remainder is 4.

We can readily verify that in each of the above examples the following relation is satisfied:

Division
Degree of a polynomial

$$\text{Dividend} = (\text{divisor})(\text{partial quotient}) + \text{remainder} \qquad (1)$$

The *degree of a polynomial* in any variable is the greatest exponent of that variable in the polynomial. For example, the polynomial $3x^4 + 2x^3y + 5x^2y^2 + xy^3$ is of degree 4 in x and 3 in y.

In order to divide one polynomial by another, we first arrange the terms in each polynomial so that the exponents of some letter that appears in each are in descending numerical order. Then we seek the partial quotient that is a polynomial, or possibly a monomial, that satisfies the relation (1), where the degree of the remainder in the letter chosen as the basis for the arrangement of terms is less than the degree of the divisor in that letter.

Quotient

Hereafter, we shall use the word *quotient* to refer to the partial quotient. We shall illustrate the procedure with an example.

EXAMPLE 1   Find the quotient and remainder obtained by dividing $6x^2 + 5x - 1$ by $2x - 1$.

Solution   Here the dividend is $6x^2 + 5x - 1$, the divisor is $2x - 1$, and we seek the quotient that satisfies the relation

$$6x^2 + 5x - 1 = (2x - 1)(\text{quotient}) + \text{remainder} \qquad (2)$$

Since the degree of the dividend is 2, the degree of the divisor is 1, and the degree of the remainder must be less than 1, it follows that the degree of the quotient is 1. Hence we write the quotient as $ax + b$, substitute this expression in (2), and get

$$6x^2 + 5x - 1 = (2x - 1)(ax + b) + \text{remainder} \qquad (3)$$

Now we can determine the replacements for a and b by the following procedure: We first perform the indicated multiplication in (3), and get

$$6x^2 + 5x - 1 = (2x - 1)ax + (2x - 1)b + \text{remainder} \qquad (4)$$

By inspection, we see that the only terms in the left and right members of (4) that involve x^2 are $6x^2$ and $2x(ax) = 2ax^2$, respectively. Hence $2ax^2 = 6x^2$, and it follows that $a = 3$. Now we substitute 3 for a in (4) and subtract $(2x - 1)3x = 6x^2 - 3x$ from each member and get

$$8x - 1 = (2x - 1)b + \text{remainder} \qquad (5)$$

Again by inspection, we see that the only terms in (5) that involve x are $8x$ and $2bx$. Consequently, $2bx = 8x$, and therefore $b = 4$. Finally, we substitute 4 for b in (5) and subtract $(2x - 1)4 = 8x - 4$ from each member, and we get $3 = 0 + \text{remainder}$. Since the degree of 3 in x is 0, which is less than the degree of the divisor, 3 is the remainder. Hence we have

$$6x^2 + 5x - 1 = (2x - 1)(3x + 4) + 3$$

Therefore, the quotient is $3x + 4$ and the remainder is 3.

The procedure above can be condensed in the usual long-division process shown below.

$$
\begin{array}{r}
3x + 4 \qquad \textit{quotient} \\
2x - 1 \overline{)6x^2 + 5x - 1} \qquad \textit{dividend} \\
\underline{6x^2 - 3x} \qquad (2x-1)3x \\
8x - 1 \qquad \textit{subtracting} \\
\underline{8x - 4} \qquad (2x-1)4 \\
3 \qquad \textit{remainder}
\end{array}
$$

Divisor

The above example illustrates the formal steps in the process of dividing one polynomial by another. These are stated below.

Dividing one polynomial by another

1   Arrange the terms in the dividend and divisor in the order of descending powers of a letter that appears in each.

2   Divide the first term in the dividend by the first term in the divisor to get the first term in the quotient.

3   Multiply the divisor by the first term in the quotient and subtract the product from the dividend.

4   Treat the remainder obtained in step 3 as a new dividend, and repeat steps 2 and 3.

5   Continue this process until a remainder is obtained that is of lower degree than the divisor in the letter chosen in step 1 as the basis for the arrangement.

The computation can be checked by use of the relation (1). We shall further illustrate the process by another example.

EXAMPLE 2   Divide $6x^4 - 6x^2y^2 - 3y^4 + 5xy^3 - x^3y$ by $-2y^2 + 2x^2 + xy$.

Solution   We shall arrange the terms in the dividend and divisor in the order of descending powers of x and proceed as indicated below.

$$
\begin{array}{l}
\textit{Divisor} \qquad\qquad 3x^2 - 2xy + y^2 \qquad \textit{quotient} \\[2pt]
2x^2 + xy - 2y^2 \,\overline{\big)\, 6x^4 - \quad x^3y - 6x^2y^2 + 5xy^3 - 3y^4} \qquad\qquad \textit{dividend} \\[2pt]
\qquad\qquad\qquad\quad 6x^4 + 3x^3y - 6x^2y^2 \qquad\qquad\qquad\qquad (2x^2 + xy - 2y^2)\,3x^2 \\[2pt]
\qquad\qquad\qquad\quad\; -4x^3y \qquad\quad + 5xy^3 - 3y^4 \qquad\qquad\quad \textit{subtracting} \\[2pt]
\qquad\qquad\qquad\quad\; -4x^3y - 2x^2y^2 + 4xy^3 \qquad\qquad\qquad (2x^2 + xy - 2y^2)(-2xy) \\[2pt]
\qquad\qquad\qquad\qquad\qquad 2x^2y^2 + \; xy^3 - 3y^4 \qquad\qquad\;\; \textit{subtracting} \\[2pt]
\qquad\qquad\qquad\qquad\qquad 2x^2y^2 + \; xy^3 - 2y^4 \qquad\qquad\;\; (2x^2 + xy - 2y^2)\,y^2 \\[2pt]
\qquad\qquad\qquad\qquad\qquad\qquad\qquad\quad\; - \; y^4 \qquad\qquad\qquad\; \textit{remainder}
\end{array}
$$

EXERCISE 2.4 Division

Perform the division indicated in each of Probs. 1 to 20.

1 a^7/a^2	**2** d^3/d	**3** b^5/b^3	**4** c^{11}/c^5
5 a^6b^4/a^4b	**6** b^7c^5/b^6c^3	**7** c^8d^7/c^7d^2	**8** d^9e^6/d^3e^3

9 $48a^5b^4/16a^2b$ 　　　　　　　　　　**10** $75a^{11}b^9/15a^7b^2$

11 $76a^7b^6/4a^3b^4$ 　　　　　　　　　**12** $60a^8b^9/12a^7b^6$

13 $\dfrac{21a^7 - 15a^4 - 6a^3}{3a^2}$ 　　　　　**14** $\dfrac{16a^9 - 24a^7 + 10a^4}{2a^3}$

15 $\dfrac{78x^7 + 66x^6 - 54x^4}{-6x^2}$ 　　　　**16** $\dfrac{-21x^6 + 35x^5 - 14x^4}{-7x^3}$

17 $\dfrac{-28x^7y^6 + 20x^4y^7 - 16x^2y^5}{-4x^2y^3}$ 　　**18** $\dfrac{-42x^8y^9 - 28x^7y^8 + 56x^2y^3}{-14xy^2}$

19 $\dfrac{18x^9y^9 - 30x^5y^6 - 42x^8y^6}{6x^4y^5}$ 　　　**20** $\dfrac{20x^7y^4 - 35x^3y^5 - 15x^6y^5}{5x^2y^3}$

In Probs. 21 to 32, divide the first expression by the second and thereby find their quotient.

21　$2x^2 - 5x - 3,\ 2x + 1$ 　　　　　**22**　$3x^2 + 11x - 20,\ 3x - 4$

23　$4x^2 - 8x + 3,\ 2x - 3$ 　　　　　**24**　$6x^2 - 13x - 5,\ 2x - 5$

25　$4x^3 + 12x^2 + 7x - 3,\ 2x^2 + 3x - 1$

26　$3x^3 + 14x^2 - 26x + 7,\ x^2 + 5x - 7$

27　$6x^3 - x^2 - 8x + 4,\ 2x^2 + x - 2$

28　$4x^3 + 12x^2 + 5x - 6,\ 2x^2 + 3x - 2$

29　$2x^4 - x^3 + 3x^2 + 5x - 6,\ 2x^2 + x - 2$

30　$3x^4 + 4x^3 - 6x^2 + 4x - 1,\ 3x^2 - 2x + 1$

31　$4x^4 + 2x^3 - 6x^2 - 7x - 3,\ 2x^2 + 2x + 1$

32　$6x^4 + x^3 - 6x^2 - 12x - 16,\ 2x^2 - x - 4$

Find the quotient and remainder obtained by dividing the first expression in each of Probs. 33 to 40 by the second.

33 $x^3 + 4x^2 + 3x + 1, x + 2$ 34 $2x^3 + 3x^2 - 8x - 2, x + 3$

35 $2x^3 + x^2 - 5x - 4, x^2 - x - 1$ 36 $3x^3 - 8x^2 - x + 3, x^2 - 2x - 3$

37 $x^4 - 2x^2 + 10x - 6, x^2 - 2x + 3$

38 $2x^4 - 3x^3 - 14x^2 - 2, 2x^2 + 3x - 1$

39 $3x^4 + 13x^3 - 16x^2 + 7x - 2, 3x - 2$

40 $4x^4 + 14x^3 + 4x^2 - 11x - 8, 2x + 1$

2.17 / FIELDS

In the preceding sections of this chapter we developed a logical structure that is the basis for the procedures involved in the four fundamental operations in the real number system. This structure consists of definitions, axioms, and theorems, which were stated as they were needed in the development. The set of axioms stated is not a minimal set since, as we shall show later, some of them can be proved from the others. We shall list below a set of fundamental axioms from which all other axioms and theorems of this chapter can be proved. The first group are called the *axioms of equality,* and the second, for reasons to be explained later, are the *axioms of a field.* In the statements, R stands for the set of real numbers; a, b, and c stand for elements of R; $+$ stands for the operation of addition; and $\times$ stands for multiplication. We shall show later that sets exist whose elements have the properties defined by these axioms under operations that differ from the usual addition and multiplication. The number in parentheses that follows the name of the axiom is the equation number attached to the same axiom earlier in this chapter.

AXIOMS OF EQUALITY

E.1 The reflexive axiom

$$a = a \tag{2.1}$$

E.2 The symmetric axiom

$$\text{If } a = b, \text{ then } b = a \tag{2.2}$$

E.3 The transitive axiom

$$\text{If } a = b \text{ and } b = c, \text{ then } a = c \tag{2.3}$$

E.4 The replacement axiom

$$\text{If } a = b, \text{ then } a \text{ can be replaced by } b \text{ in any}$$
$$\text{mathematical statement without affecting}$$
$$\text{the truth or falsity of the statement} \tag{2.7}$$

AXIOMS OF A FIELD

A.1   Closure axiom for addition

There exists a unique number $s \in R$ such that $a + b = s$ (2.9)

A.2   Commutative axiom for addition

$$a + b = b + a \tag{2.10}$$

A.3   Associative axiom for addition

$$a + (b + c) = (a + b) + c \tag{2.11}$$

A.4   Identity element for addition

There exists an element $0 \in R$ such that $a + 0 = a$ (2.13)

A.5   Inverse element for addition

There exists a unique element $a' \in R$ such that
$$a + a' = 0 \tag{2.14}$$

Note that this statement is the same as that of Eq. (2.14) if a' is replaced by $-a$.

M.1   Closure axiom for multiplication

There exists a unique element $p \in R$ such that $a \times b = p$ (2.30)

M.2   Commutative axiom for multiplication

$$a \times b = b \times a \tag{2.31}$$

M.3   Associative axiom for multiplication

$$a \times (b \times c) = (a \times b) \times c \tag{2.32}$$

M.4   Identity element for multiplication

There exists an element $1 \in R$ such that $a \times 1 = a$ (2.34)

M.5   Inverse element for multiplication

If $a \neq 0$, there exists an element $a^{-1} \in R$ such that
$$a \times a^{-1} = 1 \tag{2.36}$$

Note that in Eq. (2.36) a^{-1} is expressed as $1/a$.

AM.1   Right-hand distributive axiom

$$(a + b) \times c = (a \times c) + (b \times c) \tag{2.12}$$

This axiom distributes the operation $\times$ over the operation $+$.

We employed the word "field" to describe the second group of axioms stated above. We now define a field as follows:

Field   A set S of elements is a field if there exist two binary operations $+$ and $\times$ (not necessarily the usual operations of addition and multiplication) such that all requirements of axioms A.1 to A.5, M.1 to M.5, and AM.1 are satisfied by the elements of S.

If $+$ and $\times$ denote the ordinary operations of addition and multiplication, then the elements of the set of real numbers R satisfy all the field axioms, and therefore R is a field. Furthermore, since a rational number is a real number, axioms A.2 to A.5, M.2 to M.5, and AM.1 hold if a, b, and c are rational numbers. Moreover, since

$$\frac{a}{b} + \frac{c}{d} = \frac{ad + bc}{ad} \qquad \text{and} \qquad \frac{a}{b} \times \frac{c}{d} = \frac{ac}{bd}$$

the sum and product of two rational numbers is rational. Hence the closure axioms A.1 and M.1 hold, and the set of rational numbers is therefore a field.

If, however, a is an integer other than 1, there is no integer k such that $a \times k = 1$. Hence the integer a has no inverse for $\times$ in the set of integers. Hence the set of integers is not a field.

We shall now discuss certain finite sets, called *residue sets*, and two operations, $\oplus$ and $\otimes$, that are not the usual operations of addition and multiplication. As an example, we shall obtain the residue set of 7. If we divide 26 by 7, we get the partial quotient 3 and the remainder 5, since $26 =$
Residue $(3 \times 7) + 5$. The remainder 5 is called the *residue* of 26 modulo 7. We use the word "modulo" for the phrase "with respect to the divisor." The divisor is called the *modulus*. Similarly, since $38 = (5 \times 7) + 3$, the residue of 38 modulo 7 is 3; and, since $5 = (0 \times 7) + 5$, the residue of 5 modulo 7 is 5. Finally, any integer can be expressed in the form $7n + r$, where n is an integer and r is an element of the set

$$S = \{0, 1, 2, 3, 4, 5, 6\}$$

Hence the residue of any specified integer modulo 7 is an element of the
Residue set   set S, and we call S the *complete residue set* modulo 7.

We shall now define the operations $\oplus$ and $\otimes$ as follows: If $a \in S$ and $b \in S$, then

$a \oplus b$ is the residue of $a + b$ modulo 7
$a \otimes b$ is the residue of $a \times b$ modulo 7

EXAMPLES

$3 \oplus 6 = 2$ since $3 + 6 = 9 = (1 \times 7) + 2$
$1 \oplus 3 = 4$ since $1 + 3 = 4 = (0 \times 7) + 4$
$5 \oplus 2 = 0$ since $5 + 2 = 7 = (1 \times 7) + 0$
$6 \oplus 0 = 6$ since $6 + 0 = 6 = (0 \times 6) + 6$
$6 \otimes 5 = 2$ since $6 \times 5 = 30 = (4 \times 7) + 2$
$5 \otimes 3 = 1$ since $5 \times 3 = 15 = (2 \times 7) + 1$
$6 \otimes 0 = 0$ since $6 \times 0 = 0 = (0 \times 7) + 0$

We shall refer to $a \oplus b$ as the *residue sum* of a and b and $a \otimes b$ as the *residue product*.

By use of the above methods, we can construct the following tables of residue sums and residue products. In the first table, the number in each rectangle is $a \oplus b$, where a and b are, respectively, the number at the left end of the line and the number at the head of the column that contains the rectangle. Similarly, in the second table, the number in each rectangle is $a \otimes b$.

RESIDUE SUMS MODULO 7

$\oplus$	0	1	2	3	4	5	6
0	0	1	2	3	4	5	6
1	1	2	3	4	5	6	0
2	2	3	4	5	6	0	1
3	3	4	5	6	0	1	2
4	4	5	6	0	1	2	3
5	5	6	0	1	2	3	4
6	6	0	1	2	3	4	5

RESIDUE PRODUCTS MODULO 7

$\otimes$	0	1	2	3	4	5	6
0	0	0	0	0	0	0	0
1	0	1	2	3	4	5	6
2	0	2	4	6	1	3	5
3	0	3	6	2	5	1	4
4	0	4	1	5	2	6	3
5	0	5	3	1	6	4	2
6	0	6	5	4	3	2	1

If we refer to the above tables, we see that 0 is the identity element for $\oplus$ and 1 is the identity element for $\otimes$, since if $a \in S$, then $a \oplus 0 = a$ and $a \otimes 1 = a$. Furthermore, since 0 appears in each line of the first table each element of S has an inverse under $\oplus$ and since 1 appears in each line of the second, except the line of zeros, each nonzero element of S has an inverse element under the operation $\otimes$. The closure axioms hold, since for each two elements in S there exist in S a unique residue sum and a unique residue product. By use of methods illustrated in the following examples, it can be verified that all the other field axioms hold, and therefore S is a field.

EXAMPLE 1 Show that $(2 \oplus 4) \oplus 6 = 2 \oplus (4 \oplus 6)$.

Solution
$$(2 \oplus 4) \oplus 6 = 6 \oplus 6 = 5$$
$$2 \oplus (4 \oplus 6) = 2 \oplus 3 \qquad \text{since } 4 \oplus 6 = 3$$
$$= 5$$

EXAMPLE 2 Show that $(2 \otimes 4) \otimes 6 = 2 \otimes (4 \otimes 6)$.

Solution
$$(2 \otimes 4) \otimes 6 = 1 \otimes 6 \qquad \text{since } 2 \otimes 4 = 1$$
$$= 6$$
$$2 \otimes (4 \otimes 6) = 2 \otimes 3 \qquad \text{since } 4 \otimes 6 = 3$$
$$= 6$$

EXAMPLE 3 Show that $(2 \oplus 4) \otimes 6 = (2 \otimes 6) \oplus (4 \oplus 6)$.

Solution
$$(2 \oplus 4) \otimes 6 = 6 \otimes 6$$
$$= 1$$
$$(2 \otimes 6) \oplus (4 \otimes 6) = 5 \oplus 3$$
$$= 1$$

It is proved in the theory of numbers that the complete set of residues modulo m is a field if and only if m is a prime number.

The complete set of residues modulo 4 is

$$T = \{0, 1, 2, 3\}$$

We show below the tables of residue sums and residue products in T.

$\oplus$	0	1	2	3
0	0	1	2	3
1	1	2	3	0
2	2	3	0	1
3	3	0	1	2

$\otimes$	0	1	2	3
0	0	0	0	0
1	0	1	2	3
2	0	2	0	2
3	0	3	2	1

By use of the above tables, it can be verified that all the field axioms hold in the set T for $\oplus$ and $\otimes$ except axiom M.5. The number 1 does not appear in the fourth line of the second table, and therefore 2 does not have an inverse under $\otimes$. Hence T is not a field.

EXERCISE 2.5 Properties of a Field

Which of the field axioms are never satisfied in the set of numbers in each of Probs. 1 to 12?

1   The set of positive integers
2   The set of nonnegative integers
3   $\{-4, -3, -2, -1, 0, 1, 2, 3, 4\}$
4   $\{x \mid x = 2n, n$ a nonnegative integer$\}$
5   $\{x \mid x = 2n + 1, n$ an integer$\}$
6   $\{x \mid x = 5n, n$ an integer$\}$
7   The set of positive prime numbers
8   $\{x \mid x = 3n + 1, n$ a nonnegative integer$\}$
9   $\{x \mid x = \dfrac{3n}{2}, n$ an integer$\}$ 10 $\{x \mid x = n + \frac{1}{2}, n$ an integer$\}$
11   $\{x \mid x = \frac{1}{2}(n + 2), n$ an integer$\}$ 12 $\{x \mid x = n^2, n$ an integer$\}$

Find the residues required in Probs. 13 to 16.

13   8 modulo 6   14   35 modulo 8
15   5 modulo 7   16   28 modulo 4

17   Write the complete set of residues modulo 3, and construct the tables of residue sums and residue products.

18   In view of the tables in Prob. 17, is the set $\{0, 1, 2\}$ a field under the operations $\oplus$ and $\otimes$?

19   Construct a table of residue products for the residue set $\{0, 1, 2, 3, 4, 5\}$ modulo 6. Which of the axioms M.1 to M.5 do not hold for this set?

20   If the operation $a * b$ is defined as below, find $b * a$ in (a) through (d). In each case, state whether or not the commutative axiom holds.
     (a) $a * b = a + b + \sqrt{ab}$ (b) $a * b = a + \sqrt{b} + ab$
     (c) $a * b = ab(a + b)$ (d) $a * b = ab^2(a^2 + b)$

EXERCISE 2.6 REVIEW

1   Prove that if $(a + b) + c = a$, then $b = -c$.
2   Prove that if $a > b$, $b > c$, and $d < 0$, then $ad < cd$.

3 Show that multiplication is a binary operation in the set of positive real numbers.

4 Show that multiplication is not a binary operation in the set of negative integers.

5 Find the sum of $2a + 3b - 4c$, $-3a + 5b - c$, and $4a + b + 5c$.

6 Find the sum of $3x - 4y - 2z$, $2x + 5y - 3z$, and $-5x - y + 5z$.

7 Subtract $2a - 5p - 4r$ from $3a - 7p + 5r$.

8 Subtract $3x - 2y - 7z$ from $3x + 2y - 7z$.

9 Subtract the sum of $2x^2 - 3x + 4$ and $3x^2 + 8x + 1$ from $6x^2 + 5x + 4$.

10 Subtract $5x^2 + 8x - 3$ from the sum of $3x^2 + 7x - 4$ and $6x^2 - x + 1$.

11 Evaluate $|x| - |y| + |x - y|$ if $x = 3$ and $y = -6$.

12 Evaluate $|x + y| - |x - y| + x - y$ if $x = 5$ and $y = -4$.

13 Remove parentheses and combine similar terms in $2(3x - 2y + 4z) - 3(2x + 3y - z)$.

14 Remove the signs of aggregation and collect similar terms in $2\{3x - 2[3x - 2(3x - 2) + 5] + 9\}$.

Find the indicated product in each of Probs. 15 to 22.

15 $(2x^3)(3x^2)$
16 $(-2x^4)(5x^3)$
17 $(4x^4)^2$
18 $(-5x^3)^3$
19 $3x^2y(2xy - y^2)$
20 $2xy^3(3x^2y - 4xy^2)$
21 $(2x - 5y)(3x + 4y)$
22 $(5x - 3y)(-2x + 7y)$

23 Show that addition and multiplication are not binary operations for $\{0, 1, 2\}$.

Find the indicated quotients in Probs. 24 to 27.

24 a^9b^7/a^7b^4
25 $36x^5y^4/12x^2y$
26 $\dfrac{8x^4y^3 - 4x^3y^3 - 10x^5y^5}{2x^2y^3}$
27 $\dfrac{9x^6y^7 - 12x^4y^5 + 18x^7y^5}{-3x^3y^4}$

Find the quotient and remainder if the first expression is divided by the second in each of Probs. 28 and 29.

28 $2x^3 - 3x^2 - 5x + 5$, $x^2 - x - 3$
29 $2x^4 + x^3 - x^2 + 14x - 9$, $x^2 + 2x - 1$

30 Find the complete residue set modulo 5 and construct tables of residue sums and residue products.

31 If $a \circ b = a^2 - ab - b$, find $b \circ a$ and state whether the commutative axiom holds.

CHAPTER THREE *Special Products and Factoring*

In this chapter we shall discuss methods that contribute to speed and accuracy in computation. Certain polynomials that occur frequently are of such a type that the computation for obtaining the product of two of them can be done mentally, and thus we can avoid the longer method discussed in Sec. 2.13.

It is frequently necessary to express a polynomial in factored form when dealing with fractions, solving equations, and simplifying complicated expressions. Hence it is important to be able to recognize factorable forms of polynomials and then obtain the factors. The procedures for factoring will be discussed in Secs. 3.3 to 3.6.

3.1 / THE PRODUCT OF TWO BINOMIALS

If the corresponding terms of two binomials are similar, as in $ax + by$ and $cx + dy$, we get the product by use of the right-hand distributive axioms shown by Eq. (2.12) and the left-hand distributive theorem shown by Eq. (2.33). The procedure is explained below.

$$(ax + by)(cx + dy)$$
$$= ax(cx + dy) + by(cx + dy) \qquad \text{by the right-hand distributive axiom, Eq. (2.12)}$$
$$= acx^2 + adxy + bcxy + bdy^2 \qquad \text{by the left-hand distributive theorem and the commutative axiom, Eqs. (2.33) and (2.31)}$$
$$= acx^2 + (ad + bc)xy + bdy^2 \qquad \text{by Eq. (2.12)}$$

Hence we have

$$(ax + by)(cx + dy) = acx^2 + (ad + bc)xy + bdy^2 \qquad \textbf{(3.1)}$$

By observing the polynomial at the right of the equality sign in Eq. (3.1), we see that we obtain the product of two binomials with similar corresponding terms by performing the following steps:

Steps in obtaining the product of two binomials with similar terms

1 Multiply the first terms in the binomials to obtain the first term in the product.

2 Add the products obtained by multiplying the first term in each binomial by the second term in the other. This yields the second term in the product.

3 Multiply the second terms in the binomials to get the third term in the product.

Ordinarily, the computation required by these three steps can be done mentally, and the result can be written with no intermediate steps. This fact is illustrated by the following example.

EXAMPLE 1   Obtain the product of $2x - 5y$ and $4x + 3y$.

Solution   The product is indicated below. To get it, we proceed as directed below the product, and then record the results in the position indicated by the flow lines.

$$(2x - 5y)(4x + 3y) = 8x^2 - 14xy - 15y^2$$

Get these products mentally:

1   $2x \cdot 4x =$ _____

2   $(2x \cdot 3y) + (-5y \cdot 4x) = 6xy - 20xy =$ _____

3   $-5y \cdot 3y =$ _____

SQUARE OF THE SUM OR DIFFERENCE OF TWO NUMBERS

The square of the sum of two numbers x and y is expressed as $(x + y)^2$. Since $(x + y)^2 = (x + y)(x + y)$, we can use Eq. (3.1) and get

$$(x + y)^2 = (x + y)(x + y)$$
$$= x^2 + (xy + xy) + y^2 \qquad \text{by Eq. (3.1)}$$
$$= x^2 + 2xy + y^2$$

Consequently, $$(x + y)^2 = x^2 + 2xy + y^2 \tag{3.2}$$

Similarly, $$(x - y)^2 = x^2 - 2xy + y^2 \tag{3.3}$$

Square of the sum or difference of two numbers

Therefore, *the square of the sum (or of the difference) of two numbers is the square of the first number, plus (or minus) twice the product of the two numbers, plus the square of the second number.*

EXAMPLE 2 By use of Eqs. (3.2) and (3.3), obtain the square of $2a + 5b$ and the square of $3x - 4y$.

Solution
$$
\begin{aligned}
(2a + 5b)^2 &= (2a)^2 + 2(2a)(5b) + (5b)^2 && \text{by Eq. (3.2)} \\
&= 4a^2 + 20ab + 25b^2 && \text{by Eqs. (2.42), (2.31), and (2.32)}
\end{aligned}
$$

$$
\begin{aligned}
(3x - 4y)^2 &= (3x)^2 - 2(3x)(4y) + (-4y)^2 && \text{by Eq. (3.3)} \\
&= 9x^2 - 24xy + 16y^2 && \text{by Eqs. (2.42), (2.31), and (2.32)}
\end{aligned}
$$

PRODUCT OF THE SUM AND DIFFERENCE OF THE SAME TWO NUMBERS

The product of the sum and the difference of the numbers a and b is expressed as $(a + b)(a - b)$. If we apply Eq. (3.1) to this product, we get

$$
\begin{aligned}
(a + b)(a - b) &= a^2 + ab - ab - b^2 \\
&= a^2 - b^2 && \text{by Eqs. (2.14) and (2.13)}
\end{aligned}
$$

Consequently, $$(a + b)(a - b) = a^2 - b^2 \tag{3.4}$$

Product of the sum and difference of the same two numbers

Therefore, *the product of the sum and the difference of the same two numbers is equal to the difference of their squares.*

We shall illustrate the application of Eq. (3.4) with an example.

EXAMPLE 3 By use of Eq. (3.4), obtain the product of $3x + 5y$ and $3x - 5y$.

Solution
$$
\begin{aligned}
(3x + 5y)(3x - 5y) &= (3x)^2 - (5y)^2 && \text{by Eq. (3.4)} \\
&= 9x^2 - 25y^2 && \text{by Eq. (2.42)}
\end{aligned}
$$

EXERCISE 3.1 Product of Binomials

Find the products indicated in the following problems by use of the methods of Sec. 3.1.

1	$(x + 3)(x - 2)$	2	$(x + 4)(x + 3)$
3	$(x - 5)(x - 4)$	4	$(x - 2)(x + 4)$
5	$(2x + 3)(4x + 5)$	6	$(3x + 5)(2x - 3)$

7    $(5x - 2)(2x - 5)$

8    $(4x - 3)(3x - 2)$

9    $(2x - 3y)(4x + 7y)$

10   $(3x + 2y)(7x - 4y)$

11   $(5x + 7y)(6x - 5y)$

12   $(6x - 7y)(7x + 5y)$

13   $(2a + b)^2$

14   $(3a + 2b)^2$

15   $(6a + 5b)^2$

16   $(7a + 4b)^2$

17   $(a^2 + 9b)^2$

18   $(a^3 + 2b)^2$

19   $(a^2 + 3b^2)^2$

20   $(2a^3 + 5b^2)^2$

21   $(3a - 2b)^2$

22   $(5a - 4b)^2$

23   $(5a - 3b)^2$

24   $(7a - 5b)^2$

25   $(a^2 - 6b)^2$

26   $(2a^2 - 3b)^2$

27   $(2a^2 - 3b^3)^2$

28   $(3a^2 - 7b^2)^2$

29   $(8)(2) = (5 + 3)(5 - 3)$

30   $(27)(21)$

31   $(63)(57)$

32   $(49)(55)$

33   $(x + 3)(x - 3)$

34   $(2x - 1)(2x + 1)$

35   $(3x + 2)(3x - 2)$

36   $(5x - 6)(5x + 6)$

37   $(2x - 5y)(2x + 5y)$

38   $(3x + 4y)(3x - 4y)$

39   $(7x + 6y)(7x - 6y)$

40   $(8x - 5y)(8x + 5y)$

41   $(2a^2 + 3b^2)(2a^2 - 3b^2)$

42   $(4a^2 + 5b^2)(4a^2 - 5b^2)$

43   $(5a^2 - 7b^3)(5a^2 + 7b^3)$

44   $(6a^3 + 7b^2)(6a^3 - 7b^2)$

45   $\left(\dfrac{x}{2} + \dfrac{y}{5}\right)\left(\dfrac{x}{2} - \dfrac{y}{5}\right)$

46   $\left(\dfrac{3x}{4} + \dfrac{4y}{3}\right)\left(\dfrac{3x}{4} - \dfrac{4y}{3}\right)$

47   $\left(\dfrac{2u}{3v} - \dfrac{5x}{2w}\right)\left(\dfrac{2u}{3v} + \dfrac{5x}{2w}\right)$

48   $\left(\dfrac{5x}{4y} + \dfrac{2y}{3x}\right)\left(\dfrac{5x}{4y} - \dfrac{2y}{3x}\right)$

49   $(x + y + z)^2 = [(x + y) + z]^2$

50   $(2x - y + z)^2$

51   $(x - y + 3z)^2$

52   $(2x + y - z)^2$

53   $(2x + y + z - 3w)^2 = [(2x + y) + (z - 3W)]^2$

54   $(x - 2y + 2z + w)^2$

55   $(x^3 + 2x^2 - 2x + 3)^2$

56   $(2x^3 + 3x^2 + 3x - 2)^2$

57   $[3(x + y) + 1][2(x + y) - 3]$

58   $[2(2a - b) - 1][3(2a - b) + 2]$

59   $[5(2x - y) + 4][3(2x - y) - 1]$

60   $[2(3x - 2y) + 5][4(3x - 2y) + 3]$

61   $[(x^2 - 3) + 2x][(x^2 - 3) - 2x]$

62   $[(x^2 - 2x) + 3][(x^2 - 2x) - 3]$

63   $[(2x^2 + y^2) + 2xy][(2x^2 + y^2) - 2xy]$

64   $[(x^2 + 3y^2) - 3xy][(x^2 + 3y^2) + 3xy]$

65   $[(x^3 + x) + (x^2 - 1)][(x^3 + x) - (x^2 - 1)]$

66   $[(x^2 + x) + (x^3 + 1)][x^2 + x) - (x^3 + 1)]$

67   $[(2x^4 + x) + (x^3 - 2x^2)][(2x^4 + x) - (x^3 - 2x^2)]$

68   $[(x^5 - 3x) + (3x^3 - 1)][(x^5 - 3x) - (3x^3 - 1)]$

69   $(97)(103) = (100 - 3)(100 + 3)$

70   $(52)(48)$ 71 $(35)(25)$ 72 $(84)(76)$

73   $1.04^2 = (1 + 0.04)^2$ 74 1.05^2 75 0.97^2 76 0.88^2

3.2 / FACTORING

A number is factored if it is expressed as the product of two or more other numbers. Several sets of factors may be possible. For example, $6 = 6 \cdot 1 = 3 \cdot 2 = 9 \cdot \frac{2}{3}$. In this sec-

Prime number tion, however, we shall consider only *prime* factors. A *prime number* is an integer greater than 1 that has no integral factors except itself and 1. Therefore, the only prime factors of 6 are 3 and 2, and we say that 6 is factored if it is expressed as $3 \cdot 2$ or $2 \cdot 3$. In the discussion of the process of factoring, we shall consider only polynomials in which the numerical coefficients are integers. We say that a polynomial with integral coeffi-

Factored cients is *factored* if it is expressed as the product of two or more irreducible
Irreducible polynomial polynomials of the same type. A polynomial is *irreducible*, or *prime*, if it cannot be expressed as the product of two polynomials of lower degree and if the coefficients have no common factor.

3.3 / COMMON FACTORS

If each term of a polynomial is di-
Common factor visible by the same monomial, the monomial is called the *common factor* of the terms of the polynomial. Such a polynomial can be factored by expressing it as the product of the common factor and the sum of the quotients obtained by dividing each term of the polynomial by the common factor. This procedure is justified by the distributive axiom. If either factor thus obtained is not prime, we continue factoring by use of one or more of the methods previously discussed. For example,

$$ab + ac - ad = a(b + c - d) \qquad \text{by Eq. (2.33)}$$
$$6a^3b + 3a^2b^2 - 18ab^3 = 3ab(2a^2 + ab - 6b^2) \qquad \text{by Eq. (2.33)}$$
$$= 3ab(2a - 3b)(a + 2b) \qquad \text{by Eq. (3.1)}$$

This method also can be applied to an expression that is the sum of two or more products which have a common factor.

EXAMPLE 1 $a(x^2 + y^2) - a(x^2 - xy - y^2)$
$$= a(x^2 + y^2 - x^2 + xy + y^2) \qquad \text{by Eqs. (2.33) and (2.19)}$$
$$= a(2y^2 + xy) \qquad \text{by Eqs. (2.10), (2.14), and (2.13)}$$
$$= ay(2y + x) \qquad \text{by Eq. (2.33)}$$

EXAMPLE 2 $(a + b)(a - b) + 2(a + b) = (a + b)(a - b + 2) \qquad \text{by Eq. (2.33)}$

EXAMPLE 3 $(x-1)(x+2) - (x-1)(2x-3)$
$$= (x-1)[(x+2) - (2x-3)] \qquad \text{by Eq. (2.33)}$$
$$= (x-1)(x+2-2x+3) \qquad \text{by Eq. (2.19)}$$
$$= (x-1)(-x+5) \qquad \text{by Eq. (2.10)}$$

3.4 / FACTORS OF A BINOMIAL

THE DIFFERENCE OF TWO SQUARES

If we interchange the members of Eq. (3.4), we obtain

$$a^2 - b^2 = (a+b)(a-b) \qquad \textbf{(3.5)}$$

Consequently, we have the following rule for factoring the difference of the squares of two numbers:

Factors of the difference of two squares

The difference of the squares of two numbers is equal to the product of the sum and the difference of the two numbers.

We shall illustrate the application of this rule with the following example.

EXAMPLE 1 Factor $49a^2 - 16b^2$, $(a+3b)^2 - 4$, and $x^2 - (y+z)^2$.

Solution 1 $\quad 49a^2 - 16b^2 = (7a)^2 - (4b)^2 \qquad \text{by Eq. (2.39)}$
$$= (7a+4b)(7a-4b) \qquad \text{by Eq. (3.5)}$$

2 $\quad (a+3b)^2 - 4 = (a+3b)^2 - 2^2$
$$= (a+3b+2)(a+3b-2) \qquad \text{by Eq. (3.5)}$$

3 $\quad x^2 - (y+z)^2 = [x + (y+z)][x - (y+z)] \qquad \text{by Eq. (3.5)}$
$$= (x+y+z)(x-y-z) \qquad \text{by Eq. (2.19)}$$

THE SUM AND DIFFERENCE OF TWO CUBES

The sum and difference of the cubes of two numbers can be expressed as $x^3 + y^3$ and $x^3 - y^3$, respectively. If we divide $x^3 + y^3$ by $x + y$ using the method of Sec. 2.16, we get $x^2 - xy + y^2$ as the quotient. Hence,

$$x^3 + y^3 = (x+y)(x^2 - xy + y^2) \qquad \textbf{(3.6)}$$

Similarly,

$$x^3 - y^3 = (x-y)(x^2 + xy + y^2) \qquad \textbf{(3.7)}$$

Consequently, we have the following two rules:

Factors of the sum or the difference of two cubes

If a binomial is expressed as the sum of the cubes of two numbers, one factor is the sum of the two numbers. The other factor is the square of the first number minus the product of the two numbers plus the square of the second number.

If a binomial is expressed as the difference of the cubes of two numbers, one factor is the difference of the two numbers. The other factor is the square of the first number plus the product of the two numbers plus the square of the second number.

EXAMPLE 2   Factor $8x^3 + 27y^3$ and $27a^3 - 64b^3$.

Solution   **1** $\begin{aligned} 8x^3 + 27y^3 &= (2x)^3 + (3y)^3 \\ &= (2x + 3y)[(2x)^2 - (2x)(3y) + (3y)^2] \\ &= (2x + 3y)(4x^2 - 6xy + 9y^2) \end{aligned}$

     by Eq. (2.39)
     by Eq. (3.6)
     by Eqs. (2.39), (2.31), and (2.32)

2 $\begin{aligned} 27a^3 - 64b^6 &= (3a)^3 - (4b^2)^3 \\ &= (3a - 4b^2)[(3a)^2 + (3a)(4b^2) + (4b^2)^2] \\ &= (3a - 4b^2)(9a^2 + 12ab^2 + 16b^4) \end{aligned}$

Frequently, the factors obtained by the use of Eqs. (3.5) to (3.7) can be further factored by a repeated application of one or more of these formulas.

EXAMPLE 3   Factor $x^6 + y^6$, $a^8 - y^8$, and $a^{12} - 8b^6$.

Solution   **1** $\begin{aligned} x^6 - y^6 &= (x^3)^2 - (y^3)^2 \\ &= (x^3 + y^3)(x^3 - y^3) \\ &= (x + y)(x^2 - xy + y^2)(x - y)(x^2 + xy + y^2) \end{aligned}$

     by Eq. (2.40)
     by Eq. (3.5)
     by Eqs. (3.6) and (3.7)

2 $\begin{aligned} x^8 - y^8 &= (x^4)^2 - (y^4)^2 \\ &= (x^4 + y^4)(x^4 - y^4) \\ &= (x^4 + y^4)(x^2 + y^2)(x^2 - y^2) \\ &= (x^4 + y^4)(x^2 + y^2)(x + y)(x - y) \end{aligned}$

     by Eq. (2.40)
     by Eq. (3.5)
     by Eq. (3.5)
     by Eq. (3.5)

3 $a^{12} - 8b^6 = (a^4)^3 - (2b^2)^3$

     by Eqs. (2.40) and (2.39)

$\begin{aligned} &= (a^4 - 2b^2)[(a^4)^2 + (a^4)(2b^2) + (2b^2)^2] \\ &= (a^4 - 2b^2)(a^8 + 2a^4b^2 + 4b^4) \end{aligned}$

     by Eq. (3.7)
     by Eqs. (2.40), (2.39), (2.31), and (2.32)

EXERCISE 3.2 Common Factors, Binomials

Factor the expression in each of Probs. 1 to 12 by the method of Sec. 3.3.

1	$2x - 10$	2	$3x + 6$	3	$5x + 15$
4	$7x - 21$	5	$x^2 + 3x$	6	$x^2 - 7x$
7	$2x^2 - 4x$	8	$3x^4 - 12x^3$	9	$2x^2 - 8x + 10$
10	$3x^2 + 6x - 9$	11	$5x^2 + 10x + 35$	12	$21x^2 - 7x + 14$

Factor the expression in each of Probs. 13 to 72.

13	$x^2 - y^2$	14	$x^2 - 4y^2$	15	$9x^2 - y^2$
16	$16x^2 - y^2$	17	$4x^2 - 9y^2$	18	$9x^2 - 25y^2$
19	$25x^2 - 49y^2$	20	$36x^2 - 121y^2$	21	$x^2 - y^4$
22	$x^6 - y^2$	23	$x^8 - y^6$	24	$x^6 - y^{14}$
25	$x^2 - 16y^8$	26	$9x^4 - y^{10}$	27	$4x^4 - 9y^6$
28	$16x^8 - 81y^4$	29	$x^3 - y^3$	30	$a^3 + b^3$
31	$m^3 + n^3$	32	$p^3 - q^3$	33	$a^3 - 8b^3$
34	$a^3 - 27b^3$	35	$8a^3 + 125b^3$	36	$27a^3 + 343b^3$
37	$27x^6 + 8y^3$	38	$729x^3 + 64y^9$	39	$125x^6 - 27y^{12}$
40	$216x^9 - 125y^6$	41	$x^6y^3 + 64$	42	$8x^9y^6 + 27$
43	$64x^{12}y^9 - 125$	44	$512x^9y^{15} - 343$	45	$(x + y)^2 - 4z^2$
46	$(x - 2y)^2 - 25z^2$	47	$4x^2 - (y - 3z)^2$	48	$81x^2 - (4y - 5z)^2$
49	$(x + y)^3 - 1$	50	$(2x - y)^3 - 8$	51	$(3x + 2y)^3 + 27$
52	$(x - 3y)^3 - 64$	53	$16x^4 - y^4$	54	$x^4 - 81y^4$
55	$81x^4 - 625y^4$	56	$16x^4 - 81y^4$	57	$x^8 - y^4$
58	$x^8 - 81$	59	$x^6 - 1$	60	$x^{12} - 1$
61	$x^6 - y^6$	62	$x^6 + 27$	63	$343x^{12} - 1$
64	$x^{18} - y^{18}$	65	$x^5 + y^5$ †	66	$x^7 + y^7$ †
67	$128x^7 + y^{14}$	68	$x^{15} - 32y^{15}$	69	$x^7 - y^7$ ‡
70	$x^5 - 32y^{10}$ §	71	$243x^{15} - 32y^{10}$	72	$125x^{12} - 27y^6$

3.5 / TRINOMIALS THAT ARE PERFECT SQUARES

If a trinomial is the square of a binomial, we know by Eqs. (3.2) and (3.3) that two of its terms are perfect squares and hence are positive and that the third term, except possibly for the sign that precedes it, is twice the product of the positive square roots of the two perfect squares. Furthermore, such a trinomial is the square of

† $x + y$ is one factor.
‡ $x - y$ is one factor.
§ $x - 2y^2$ is one factor.

a binomial composed of the positive square roots of the perfect-square terms connected by the sign that precedes the other term.

EXAMPLE   Factor $4x^2 - 12xy + 9y^2$, $9a^2 + 24ab + 16b^2$, and $(2a - 3b)^2 - 8(2a - 3b) + 16$.

Solution   1   Since $4x^2 = (2x)^2$, $9y^2 = (3y)^2$, and $12xy = 2(2x)(3y)$, we have
$$4x^2 - 12xy + 9y^2 = (2x - 3y)(2x - 3y) = (2x - 3y)^2.$$

2   $$9a^2 + 24ab + 16b^2 = (3a + 4b)^2$$

3   $$(2a - 3b)^2 - 8(2a - 3b) + 16 = [(2a - 3b) - 4]^2$$

3.6 / FACTORS OF A QUADRATIC TRINOMIAL

A trinomial of the type $ax^2 + bxy + cy^2$, where a, b, and c stand for integers, is a *quadratic trinomial with integral coefficients.* In this section we shall discuss methods for finding the two binomial factors of such a trinomial if such factors exist. Since we shall use Eq. (3.1) for this purpose, we shall rewrite it here with the members interchanged.

$$acx^2 + (ad + bc)xy + bdy^2 = (ax + by)(cx + dy) \qquad (3.8)$$

To use Eq. (3.8) in factoring $3x^2 - 10xy - 8y^2$, we must find four integral replacements for a, b, c, and d such that $ac = 3$, $bd = -8$, and $ad + bc = -10$. The only possibilities for a and c are ± 3† and ± 1, and these numbers must have the same sign since the sign preceding ac is understood to be plus. Now the possibilities for b and d are ± 4 and ∓ 2 or ± 8 and ∓ 1, where the double signs indicate that if one of the numbers is positive, the other is negative. We must now select a set of values for a, b, c, and d from the listed possibilities so that $ad + bc = -10$. If we let $a = 3$ and $c = 1$, then $ad + bc = 3d + b = -10$, and this is true if $d = -4$ and $b = 2$. Therefore, $3x^2 - 10xy - 8y^2 = (3x + 2y)(x - 4y)$.

We call attention to the fact that $(-3x - 2y)$ and $(-x + 4y)$ are also factors of $3x^2 - 10xy - 8y^2$, as the reader can verify.

It is often desirable to know whether a quadratic trinomial is factorable. We shall now state, and prove in the quadratic-equation chapter, that $ax^2 + bx + c$ is factorable rationally in terms of its coefficients (assumed to be rational) if and only if $D^2 = b^2 - 4ac$ is a nonnegative perfect square.

---

† This means $+3$ and -3.

EXAMPLE Is $12x^2 - xy - 20y^2$ factorable rationally in terms of the coefficients?

Solution Since $b^2 - 4ac = (-1)^2 - 4(12)(-20) = 31^2$ the answer is yes. The factors are readily seen to be $4x + 5y$ and $3x - 4y$.

EXERCISE 3.3 Factoring Trinomials

Factor the expression in each of Probs. 1 to 56.

1 $a^2 - 2a + 1$	2 $b^2 - 6b + 9$	3 $c^2 + 4c + 4$
4 $d^2 + 6d + 9$	5 $4a^2 - 4a + 1$	6 $9b^2 + 6b + 1$
7 $25x^2 + 10x + 1$	8 $16x^2 - 8x + 1$	9 $4x^2 - 12x + 9$
10 $9x^2 + 24x + 16$	11 $25x^2 + 60x + 36$	12 $64x^2 - 80x + 25$
13 $x^2 + x - 6$	14 $x^2 - 3x - 4$	15 $x^2 - 2x - 15$
16 $x^2 + 4x - 12$	17 $x^2 + 8x + 15$	18 $x^2 + 9x + 14$
19 $y^2 + 12y + 35$	20 $y^2 + 10y + 16$	21 $y^2 + 5y + 6$
22 $y^2 - 9y + 18$	23 $y^2 - 9y + 20$	24 $y^2 - 10y + 21$
25 $3x^2 + 7x + 2$	26 $2x^2 + 5x + 2$	27 $6x^2 + 7x - 3$
28 $4x^2 - 19x + 12$	29 $7x^2 - 10x + 3$	30 $3x^2 + 13x - 10$
31 $4x^2 + 4x - 3$	32 $5x^2 + 6x - 8$	33 $28x^2 - xy - 2y^2$
34 $24x^2 - xy - 3y^2$	35 $18x^2 - 9xy - 5y^2$	36 $12x^2 + 11xy - 15y^2$
37 $20x^2 + 3xy - 35y^2$	38 $24x^2 + 26xy - 15y^2$	39 $42x^2 + 5xy - 25y^2$
40 $45x^2 + 2xy - 15y^2$	41 $30x^2 + 7xy - 49y^2$	42 $54x^2 - 3xy - 35y^2$

43 $30x^2 + 37xy - 12y^2$ 44 $30x^2 - xy - 99y^2$

45 $48x^2 + 155xy - 77y^2$ 46 $42x^2 - 23xy - 84y^2$

47 $10x^2 + 13xy - 77y^2$ 48 $63x^2 - 124xy + 60y^2$

49 $(x + y)^2 + (x + y) - 12$ 50 $(x + y)^2 + 2(x + y) - 8$

51 $(2x + y)^2 + 2(2x + y) - 15$ 52 $(m - 2n)^2 + m - 2n - 6$

53 $10(3x - 4y)^2 + 19(3x - 4y) + 6$

54 $12x^2 - x(2y - 3z) - 6(2y - 3z)^2$

55 $15(3x - 5y)^2 + 31(3x - 5y)z - 24z^2$

56 $12(5x - 2y)^2 - (5x - 2y)z - 20z^2$

Test the expression in each of Probs. 57 to 64 to see if it is factorable rationally in terms of the coefficients, and factor those that can be factored.

57 $2x^2 + 5x - 3$	58 $2x^2 + 5x - 2$	59 $7x^2 + 6x + 5$
60 $6x^2 + 11x - 10$	61 $5x^2 + 4x + 1$	62 $4x^2 - 13x + 3$
63 $10x^2 + 11x - 6$	64 $4x^2 - 13x - 3$	

3.7 / FACTORING BY GROUPING

Frequently, the terms in a polynomial can be grouped in such a way that each group has a common factor, and then the method of Sec. 3.3 can be applied. We shall illustrate the method with two examples.

EXAMPLE 1   Factor $ax + bx - ay - by$.

Solution   We notice that the first two terms have the common factor x and that the third and fourth terms have the common factor y. Hence we group the terms in this way, $(ax + bx) - (ay + by)$, and then proceed as indicated below.

$$ax + bx - ay - by = (ax + bx) - (ay + by) \qquad \text{since by Eq. (2.19)} - (ay + by)$$
$$= -ay - by$$
$$= x(a + b) - y(a + b) \qquad \text{by Eq. (2.33)}$$
$$= (a + b)(x - y) \qquad \text{by Eqs. (2.31) and (2.33), with } a + b \text{ as the common factor}$$

EXAMPLE 2   Factor $a^2 + ab - 2b^2 + 2a - 2b$.

Solution   Since $a^2 + ab - 2b^2 = (a + 2b)(a - b)$ and $2a - 2b = 2(a - b)$, we proceed as indicated below.

$$a^2 + ab - 2b^2 + 2a - 2b = (a^2 + ab - 2b^2) + (2a - 2b)$$
$$= (a + 2b)(a - b) + 2(a - b)$$
$$= (a - b)(a + 2b + 2) \qquad \text{by Eqs. (2.31) and (2.33)}$$
$$\text{with } a - b \text{ as the}$$
$$\text{common factor}$$

Often, after the terms in a polynomial are suitably grouped, it becomes evident that the methods of Sec. 3.4 can be applied.

EXAMPLE 3   Factor $4c^2 - a^2 + 2ab - b^2$.

Solution

$$4c^2 - a^2 + 2ab - b^2 = 4c^2 - (a^2 - 2ab + b^2) \qquad \text{by Eq. (2.19)}$$
$$= 4c^2 - (a - b)^2 \qquad \text{by Eq. (3.3)}$$
$$= (2c)^2 - (a - b)^2 \qquad \text{by Eq. (2.39)}$$
$$= [2c + (a - b)][2c - (a - b)] \qquad \text{by Eq. (3.5)}$$
$$= (2c + a - b)(2c - a + b) \qquad \text{by Eq. (2.19)}$$

EXERCISE 3.4 Grouping

Factor each of the following expressions by use of grouping.

1   $xy + x + 2y + 2$ 2 $xy - 3y + 2x - 6$

3   $rs - 3r - 4s + 12$ 4 $ab - 5a + 3b - 15$

5   $2x^2 + 6xy - x - 3y$ 6 $2x^2 + 4x - xy - 2y$

7   $6x^2 + 9x - 4xy - 6y$ 8 $5x^2 - 15x - 2xy + 6y$

9   $ac + ad - 3bc - 3bd$ 10 $ac - ad + 2bc - 2bd$

11   $2ac + 2bc - ad - bd$ 12 $3ac + 2ad - 3bc - 2bd$

13   $x^3 + 3x^2 - x - 3$ 14 $2x^3 - x^2 - 8x + 4$

15   $2x^3 + x^2 - 18x - 9$ 16 $3x^3 + 2x^2 - 3x - 2$

17   $4r^3 + 10r^2 - 6r$ 18 $3s^3 - 3s^2 - 6s$

19   $10t^3 - 45t^2 - 25t$ 20 $4u^3 + 14u^2 + 12u$

21   $12a^3 - 12a^2 + 3a$ 22 $28b^3 + 28b^2 + 7b$

23   $36c^3 + 96c^2 + 64c$ 24 $20d^3 - 60d^2 + 45d$

25   $x^2 - y^2 - xz - yz$ 26 $x^2 + xy - 2y^2 + xz + 2yz$

27   $x^2 + xy - 6y^2 + xz + 2yz$ 28 $2x^2 - 7xy + 3y^2 + 2xz - 6yz$

29   $x^2 + 4xy + 4y^2 - z^2$ 30 $4x^2 + y^2 - 9z^2 - 4xy$

31   $x^2 - 9y^2 - z^2 + 6yz$ 32 $9x^2 - 4y^2 - z^2 - 4yz$

33   $x^2 - y^2 - 4y - 4$ 34 $x^2 - 4y^2 - 9z^2 + 12yz$

35   $4x^2 + 4w^2 - 9y^2 - 8xw$ 36 $9x^2 + 4w^2 - 16y^2 - 12xw$

37   $x^4 - 4x^3 + 4x^2 - 25$ 38 $x^4 - 2x^3 + x^2 - 1$

39   $y^4 + 8y^3 + 16y^2 - 9$ 40 $y^4 - 2y^3 + y^2 - 4$

41   $x^4 + 2x^2y^2 + 9y^4$ 42 $x^4 - 11x^2y^2 + 25y^4$

43   $4x^4 + 7x^2y^2 + 16y^4$ 44 $25x^4 + 16x^2y^2 + 4y^4$

45   $x^3 - y^3 + x^2 - y^2$ 46 $x^3 - x^2 + y^3 + y^2$

47   $x^3 + 8y^3 + x^2 - xy - 6y^2$ 48 $8x^3 + 2x^2 + 3xy - 2y^2 - y^3$

EXERCISE 3.5 REVIEW

Find the products in Probs. 1 to 40.

1   $(2a + 1)(a + 3)$ 2 $(3b + 1)(b + 2)$

3   $(4c + 1)(c + 3)$ 4 $(2d + 3)(3d + 1)$

5   $(a + 2b)(3a + b)$ 6 $(3b - c)(b - 2c)$

7   $(4r - s)(r + 3s)$ 8 $(2t - 5u)(3t + u)$

9   $(4r + u)(7r - 3u)$ 10 $(3s - 7t)(8t + 3s)$

11   $(5c + 7d)(4c - 3d)$ 12 $(7s - 12t)(4s + 3t)$

13   $(b + 5)(b - 5)$ 14 $(3x + 1)(3x - 1)$

15   $(5x + 1)(5x - 1)$ 16 $(3x + y)(3x - y)$

17   $(5x - 3y)(5x + 3y)$ 18 $(3x + 2y)(3x - 2y)$

19 $(3x - 4y)(3x + 4y)$

20 $(7x + 5y)(7x - 5y)$

21 $(8x + 5y)(8x - 5y)$

22 $(3x - 1)^2$

23 $(7x - 3)^2$

24 $(2x - 5)^2$

25 $(m + 3n)^2$

26 $(6p + 5q)^2$

27 $(x + 6y)^2$

28 $(5q - 4r)^2$

29 $(2x^3 - 3y^2)^2$

30 $(4p^3 + 5q^2)^2$

31 $(3a^2 + 2b^3)^2$

32 $(u + v + w)(u + v - w)$

33 $(r + s - 3t)(r + s + 3t)$

34 $(u + 3v - 2w)(u + 3v + 2w)$

35 $(a + 3b - 2c)(a + 3b + 2c)$

36 $(2a - 3b + 4c)(2a + 3b - 4c)$

37 $(a + 2b + 3c - d)^2$

38 $(3a - b - c + 2d)^2$

39 $(x^3 + 3x^2 - x - 2)^2$

40 $(r^3 - 2r^2 - 3r + 4)^2$

Factor each of the following expressions.

41 $x^2 - 16$

42 $a^2 - 9$

43 $25a^2 - 4b^2$

44 $49x^2 - 36y^2$

45 $27b^4 - 48c^2$

46 $12a^6 - 27b^4$

47 $(5x + 2y)^2 - z^2$

48 $(3a - 5b)^2 - 4c^2$

49 $45c^3d - 80cd^3$

50 $81x^5y - 49xy^7$

51 $36x^3y^2 - 25xy^6$

52 $72x^5y^4 - 98xy^2$

53 $x^2 + 8xy + 16y^2$

54 $x^2 - 10xy + 25y^2$

55 $25m^2 - 90mn + 81n^2$

56 $9p^2 - 48pq + 64q^2$

57 $49r^2 + 126rs + 81s^2$

58 $64x^2 + 80xy + 25y^2$

59 $36x^6 - 24x^3y + 4y^2$

60 $25a^4 - 30a^2b^4 + 9b^8$

61 $x^2 - 8x + 15$

62 $y^2 + 5y + 6$

63 $6x^2 - 7xy - 3y^2$

64 $10x^2 - 7xy + y^2$

65 $12r^2 + 11rs - 5s^2$

66 $10r^2 + 13rs - 3s^2$

67 $6s^2 + 10st - 4t^2$

68 $35a^2 + 34ab - 21b^2$

69 $54r^2 - 3rs - 35s^2$

70 $6a^2 - 23ab + 15b^2$

71 $30a^2 - 61ab + 30b^2$

72 $16a^4 + 24a^2b^2 + 9b^4$

73 $25x^4 + 20x^2y^2 + 4y^4$

74 $64r^4 - 112r^2s^2 + 49s^4$

75 $81r^4 + 72r^2q^2 + 16q^4$

76 $27x^3 + y^3$

77 $8x^3 - y^3$

78 $64a^3 + 8b^3$

79 $27a^6 + 1$

80 $8 + y^6$

81 $m^9 - 27n^{12}$

82 $343c^3 + 27d^3$

83 $125x^3 - 216y^3$

84 $27 - (3a + b)^3$

85 $(5c + 2d)^3 - 216$

86 $(2a - b)^3 - (a + b)^3$

87 $x^3 + x^2 - y^3 - y^2$

88 $m^2 - m - n^2 - n$

89 $4r^4 - r^2 - 4rs - 4s^2$

90 $a^3 - 9b^2 - 27b^3 + a^2$

CHAPTER FOUR *Fractions*

Numerical fractions of the type $\frac{1}{2}$, $\frac{3}{4}$, $\frac{2}{3}$, and $\frac{5}{7}$ are common in arithmetic and are used constantly in everyday living. Algebraic fractions are equally important in mathematics and in all fields to which algebra is applied. Skill in operations that involve fractions is essential for progress in any of these fields. In this chapter we shall explain the basic operations dealing with fractions.

4.1 / DEFINITIONS AND FUNDAMENTAL PRINCIPLE

In Sec. 2.16 we defined the number

Fraction
Numerator
Denominator

a/b as the quotient of a and b. The number a/b is also called a *fraction* for $b \neq 0$. The number a is the *numerator* of the fraction, and b the *denominator*. We shall frequently refer to the numerator and the denominator as *members* of the fraction.

The letters a and b in the above definition may stand for numbers or for algebraic expressions. Hence,

$$\frac{4a}{7} \qquad \frac{3x^2}{2xy} \qquad \frac{c^2 - d^2}{c^2 + d^2} \qquad \text{and} \qquad \frac{(x+y)(x^4 + xy + y^4)}{(x+y)(x^3 + x^2y + y^2)}$$

are examples of fractions.

71

Our next task will be to prove the following theorem, which states the condition under which two fractions are equal.

THEOREM   **If $ad = bc$, then $a/b = c/d$; conversely,**
                    if $a/b = c/d$, then $ad = bc$ for $b \neq 0$ and $d \neq 0$.   **(4.1)**

Proof of First
Statement

$$ad = bc$$ given

$$ad\left(\frac{1}{bd}\right) = bc\left(\frac{1}{bd}\right)$$ by Eq. (2.6), the multiplication axiom

$$\frac{ad}{bd} = \frac{bc}{bd}$$ by Eq. (2.46)

$$\frac{a}{b} \cdot \frac{d}{d} = \frac{b}{b} \cdot \frac{c}{d}$$ by Eq. (2.45)

$$\frac{a}{b} \cdot 1 = 1 \cdot \frac{c}{d}$$ since $\frac{d}{d} = \frac{b}{b} = 1$

$$\frac{a}{b} = \frac{c}{d}$$ by Eq. (2.34)

Proof of Second
Statement

$$\frac{a}{b} = \frac{c}{d}$$ given

$$\frac{a}{b} \cdot \frac{bd}{1} = \frac{c}{d} \cdot \frac{bd}{1}$$ by Eq. (2.6), the multiplication axiom

$$\frac{abd}{b} = \frac{cbd}{d}$$ by Eq. 2.45)

$$\frac{a}{1} \cdot \frac{b}{b} \cdot \frac{d}{1} = \frac{c}{1} \cdot \frac{b}{1} \cdot \frac{d}{d}$$ by Eqs. (4.2) and (2.45)

$$a \cdot 1 \cdot d = c \cdot b \cdot 1$$ by Eqs. (4.2) and (4.1)

$$ad = bc$$ by Eqs. (2.31) and (2.34)

Fundamental principle
of fractions

We shall frequently have occasion to employ the *fundamental principle of fractions*. It is:

If the numerator and denominator of a fraction are multiplied or divided by the same nonzero number and the products, or quotients, are used as the numerator and denominator, respectively, of a second fraction, then the two fractions are equal.

If we state this principle in symbols, we have

$$\frac{a}{b} = \frac{an}{bn} = \frac{a \div m}{b \div m}$$ provided $b \neq 0$, $n \neq 0$, $m \neq 0$ **(4.2)**

Proof We shall use Eq. (4.1) to prove this theorem. Since

$$abn = abn \qquad \text{by Eq. (2.1)}$$

then $\qquad a(bn) = b(an) \qquad$ by Eqs. (2.31) and (2.32), the commutative and associative axioms

Hence, $\qquad \dfrac{a}{b} = \dfrac{an}{bn} \qquad$ by Eq. (4.1)

Now replacing n by $1/m$ in the above statement, we have

$$\frac{a}{b} = \frac{a(1/m)}{b(1/m)}$$

$$= \frac{a/m}{b/m} \qquad \text{by Eq. (2.46)}$$

$$= \frac{a \div m}{b \div m} \qquad \text{since } \frac{a}{m} = a \div m \text{ and } \frac{b}{m} = b \div m$$

EXAMPLE 1

$$\frac{3}{4} = \frac{3 \cdot 5}{4 \cdot 5} = \frac{15}{20}$$

EXAMPLE 2

$$\frac{x}{y} = \frac{x \cdot 2x^2y^3}{y \cdot 2x^2y^3} = \frac{2x^3y^3}{2x^2y^4}$$

EXAMPLE 3

$$\frac{24}{36} = \frac{24 \div 12}{36 \div 12} = \frac{2}{3}$$

EXAMPLE 4

$$\frac{15x^3y^7}{25x^5y^2} = \frac{15x^3y^7 \div 5x^3y^2}{25x^5y^2 \div 5x^3y^2}$$

$$= \frac{3x^0y^5}{5x^2y^0} \qquad \text{by Eq. (2.47), the law of exponents for division}$$

$$= \frac{3y^5}{5x^2} \qquad \text{since } x^0 = y^0 = 1$$

There are three signs associated with any fraction, and they are the sign preceding the fraction, the sign preceding the numerator, and the sign preceding the denominator. We shall show that if two of these signs in a given fraction are changed, the resulting fraction is equal to the given fraction.

$$\frac{n}{d} = \frac{-1 \cdot n}{-1 \cdot d} \qquad \text{by Eq. (4.2)}$$

$$= \frac{-n}{-d}$$

Furthermore, $\qquad -\frac{-n}{d} = -1 \cdot \frac{-n}{d} \qquad$ by Eq. (2.38)

$$= \frac{-1}{1} \cdot \frac{-n}{d} \qquad \text{since } -1 = \frac{-1}{1}$$

$$= \frac{n}{d} \qquad \text{by Eq. (2.45)}$$

Similarly, $\qquad -\frac{n}{-d} = -1 \cdot \frac{n}{-d} = \frac{1}{-1} \cdot \frac{n}{-d} = \frac{n}{d}$

Therefore, $\qquad \boldsymbol{\dfrac{n}{d} = \dfrac{-n}{-d} = -\dfrac{-n}{d} = -\dfrac{n}{-d}} \qquad$ provided $d \neq 0$ $\qquad$ **(4.3)**

EXAMPLE 5

$$\frac{-x}{y - x} = \frac{-(-x)}{-(y - x)} = \frac{x}{x - y}$$

EXAMPLE 6

$$\frac{y^3 - x^3}{x - y} = -\frac{-(y^3 - x^3)}{x - y} = -\frac{x^3 - y^3}{x - y}$$

4.2 / CONVERSION OF FRACTIONS

A fraction is said to be in lowest terms if the numerator and denominator have no common factors except 1. We call such a fraction a *reduced fraction*. Consequently, to reduce a given fraction to lowest terms, we divide the numerator and denominator by each factor that is common to both and thus obtain the numerator and denominator, respectively, of the reduced fraction. If the common factors are not clearly discernible, it is advisable to factor the members of the fraction before attempting the reduction. We shall illustrate the procedure with three examples.

EXAMPLE 1   Reduce $35a^4b^2/42a^3b^3$ to lowest terms.

Solution   The common factor of the members of this fraction is $7a^3b^2$. Hence we have

$$\frac{35a^4b^2}{42a^3b^3} = \frac{35a^4b^2 \div 7a^3b^2}{42a^3b^3 \div 7a^3b^2}$$

$$= \frac{5ab^0}{6a^0b} \qquad \text{by Eq. (2.47), the law of exponents for division}$$

$$= \frac{5a}{6b} \qquad \text{by Eqs. (2.48) and (2.34), since } b^0 = a^0 = 1$$

EXAMPLE 2   Reduce $(x^3 + x^2 - 6x)/(x^3 - 3x^2 + 2x)$ to lowest terms.

Solution   We shall first factor the members of the given fraction and then proceed as indicated below.

$$\frac{x^3 + x^2 - 6x}{x^3 - 3x^2 + 2x} = \frac{x(x^2 + x - 6)}{x(x^2 - 3x + 2)} \qquad \begin{array}{l}\text{by Eq. (2.33), the left-hand distributive}\\ \text{theorem}\end{array}$$

$$= \frac{x(x-2)(x+3)}{x(x-2)(x-1)} \qquad \text{by Eq. (3.8)}$$

$$= \frac{1 \cdot 1 \cdot (x+3)}{1 \cdot 1 \cdot (x-1)} \qquad \text{dividing members by } x(x-2)$$

$$= \frac{x+3}{x-1} \qquad \text{by Eq. (2.34)}$$

EXAMPLE 3   Reduce

$$\frac{a^5 - a^4c - ab^4 + b^4c}{a^4 - a^3c - a^2b^2 + ab^2c}$$

to lowest terms.

Solution $$\frac{a^5 - a^4c - ab^4 + b^4c}{a^4 - a^3c - a^2b^2 + ab^2c} = \frac{(a^5 - a^4c) - (ab^4 - b^4c)}{(a^4 - a^3c) - (a^2b^2 - ab^2c)} \qquad \text{by Eq. (2.19)}$$

$$= \frac{a^4(a-c) - b^4(a-c)}{a^3(a-c) - ab^2(a-c)} \qquad \begin{array}{l}\text{by Eq. (2.33), the left-hand}\\ \text{distributive theorem}\end{array}$$

$$= \frac{(a^4 - b^4)(a-c)}{(a^3 - ab^2)(a-c)} \qquad \text{by Eq. (2.12)}$$

$$= \frac{(a^2 - b^2)(a^2 + b^2)(a-c)}{a(a^2 - b^2)(a-c)} \qquad \text{by Eqs. (3.5) and (2.33)}$$

$$= \frac{1 \cdot (a^2 + b^2) \cdot 1}{a \cdot 1 \cdot 1} \qquad \begin{array}{l}\text{dividing the members by}\\ (a^2 - b^2)(a-c)\end{array}$$

$$= \frac{a^2 + b^2}{a} \qquad \text{by Eq. (2.34)}$$

In many operations, it is desirable to convert a given fraction to another in which the denominator has a specified form. We accomplish this conversion by the application of Eq. (4.2), in which we multiply the members of the fraction by the number or expression necessary to produce the required denominator. This process is useful in addition and subtraction of fractions.

EXAMPLE 4   Convert $(a+b)/(a-b)$ to an equal fraction with a^2-b^2 as the denominator.

Solution   Since $(a-b)(a+b) = a^2 - b^2$, we multiply the members of the given fraction by $a+b$ and get

$$\frac{a+b}{a-b} = \frac{(a+b)(a+b)}{(a-b)(a+b)} = \frac{a^2 + 2ab + b^2}{a^2 - b^2}$$

4.3 / THE LOWEST COMMON DENOMINATOR

Lowest common denominator   The *lowest common denominator* of a set of fractions is the lowest common multiple of the denominators of the fractions in the set. It is abbreviated as lcd and is usually obtained in factored form. It must be divisible by every denominator in the set and must have no more factors than are needed to satisfy this requirement.

In obtaining the lcd, we begin by factoring each denominator in the set. Then the lcd is the product of the different factors of the denominators, each with an exponent that is equal to the greatest exponent of that factor in any denominator.

EXAMPLE 1   Find the lcd if the denominators are $(x-2)^4(x+1)$, $(x-2)(x+1)^3(x-1)$, and $(x-2)^2(x-1)^2$.

Solution   The different factors are $x-2$, $x+1$, and $x-1$ and their greatest exponents are 4, 3, and 2, respectively. Consequently, the lcd is $(x-2)^4(x+1)^3(x-1)^2$.

EXAMPLE 2   Convert each of the fractions a/xy, $3b/x^3y$, and $2c/xy^2$ to an equal fraction with the lcd as its denominator.

Solution   The lcd is x^3y^2; hence, we multiply the members of the given fractions by $x^3y^2/xy = x^2y$, $x^3y^2/x^3y = y$, and $x^3y^2/xy^2 = x^2$, respectively, and get

$$\frac{ax^2y}{x^3y^2} \qquad \frac{3by}{x^3y^2} \qquad \text{and} \qquad \frac{2cx^2}{x^3y^2}$$

EXERCISE 4.1 Conversions

Convert the fraction in each of Probs. 1 to 24 into an equal fraction with the expression to the right of the comma as denominator.

1 y/x, xy
2 b/c, c^2
3 u/v, $-uv$
4 s/t, $-t^2$

5 xy/xy^2, y
6 $4s/s^2$, s
7 $pq/2q$, 2
8 $5xy/10x^2y$, $2x$

9 $\dfrac{x-3y}{4y-x}$, $x-4y$

10 $\dfrac{x-y}{x-5y}$, $5y-x$

11 $\dfrac{2y-3x}{-x+6y}$, $x-6y$

12 $\dfrac{5x-4y}{2x-3y}$, $3y-2x$

13 $\dfrac{u+3}{u-2}$, u^2-4

14 $\dfrac{p-3}{p+4}$, p^2-16

15 $\dfrac{q+5}{q-5}$, q^2-25

16 $\dfrac{r-2}{r+3}$, r^2-9

17 $\dfrac{x+3}{x-2}$, $(x-2)(x+1)$

18 $\dfrac{y+2}{y-1}$, $(y-1)(y+3)$

19 $\dfrac{2y-7}{y+3}$, $(y+3)(y-4)$

20 $\dfrac{3w-1}{w+1}$, $(w+1)(w+2)$

21 $\dfrac{(x+3)(x-1)}{(x+3)(x+5)}$, $x+5$

22 $\dfrac{x^2-3x+2}{x^2+x-2}$, $x+2$

23 $\dfrac{x^2+x-6}{x^2-x-2}$, $x+1$

24 $\dfrac{2x^2-5x+2}{x^2+2x-8}$, $x+4$

Reduce the fraction in each of Probs. 25 to 52 to lowest terms.

25 $8x^2y^3/2xy$

26 $18x^3y^2/9x^4y$

27 $9x^5yz^3/15x^3y^2z^3$

28 $12x^4y^3z/27x^2y^4z^3$

29 $\dfrac{x^2+x-6}{x^2+2x-3}$

30 $\dfrac{x^2+x-2}{x^2-x-6}$

31 $\dfrac{2x^2+3x-2}{2x^2+5x+2}$

32 $\dfrac{3x^2-x-2}{2x^2+x-3}$

33 $\dfrac{(x+2)(x^2-4x+4)}{(x-1)(x^2+5x+6)}$

34 $\dfrac{(x-3)(x^2-x-6)}{(x+2)(x^2-4x+3)}$

35 $\dfrac{(2x-1)(x^2+2x-8)}{(x+4)(2x^2+x-1)}$

36 $\dfrac{(3x+2)(x^2-x-2)}{(x-2)(3x^2-x-2)}$

37 $\dfrac{ac-ad+bc-bd}{ac+2ad+bc+2bd}$

38 $\dfrac{ac+bc-2ad-2bd}{ac-2ad-bc+2bd}$

39 $\dfrac{ac+ad+3bc+3bd}{ac+3bc-2ad-6bd}$

40 $\dfrac{ac+ad-3bc-3bd}{3ac+3ad-bc-bd}$

41 $\dfrac{x^2-y^2}{x^3-y^3}$

42 $\dfrac{x^2-y^2}{x^3+y^3}$

43 $\dfrac{x^6-y^6}{x^4-y^4}$

44 $\dfrac{x^9+y^9}{x^6-y^6}$

45    $\dfrac{x+1}{(x+2)x+1}$ 46 $\dfrac{x-2}{(x-1)x-2}$

47    $\dfrac{x+3}{(2x+7)x+3}$ 48 $\dfrac{x+3}{(3x+10)x+3}$

49    $\dfrac{x-1}{(2x-1)x-1}$ 50 $\dfrac{x+2}{(x+3)x+2}$

51    $\dfrac{5x-3}{x(5x-3)+5x-3}$ 52 $\dfrac{3x-2}{3x(x-1)-2(x-1)}$

Convert each of the following sets of fractions to an equal set with a common denominator.

53    $\left\{\dfrac{3}{x},\dfrac{2}{xy},\dfrac{4}{x^2}\right\}$ 54 $\left\{\dfrac{1}{x^2},\dfrac{2}{xy},\dfrac{3}{y^2}\right\}$

55    $\left\{\dfrac{3}{x^2y},\dfrac{-2}{xy^2},\dfrac{1}{xy}\right\}$ 56 $\left\{\dfrac{5}{x^3y},\dfrac{-4}{x^2y^2},\dfrac{-3}{xy^3}\right\}$

57    $\left\{\dfrac{2x-y}{x-y},\dfrac{x-2y}{x+y},\dfrac{x+2y}{x^2-y^2}\right\}$ 58 $\left\{\dfrac{x-y}{x^2-xy+y^2},\dfrac{x^2+xy+y^2}{x+y}\right\}$

59    $\left\{\dfrac{x-y}{(x+y)(x-2y)},\dfrac{x+y}{(x-y)(x-2y)},\dfrac{x-2y}{x^2-y^2}\right\}$

60    $\left\{\dfrac{x-3y}{(x-2y)(x+y)},\dfrac{2x-y}{(3x-y)(x+y)},\dfrac{x-y}{(x-2y)(3x-y)}\right\}$

4.4 / ADDITION OF FRACTIONS

Equation (2.46) states that $a/b = a(1/b)$. Hence if we employ it and the left-hand distributive axiom (2.12), we have

$$\frac{a}{d}+\frac{b}{d}+\cdots+\frac{c}{d}=\frac{1}{d}(a+b+\cdots+c)=\frac{a+b+\cdots+c}{d}$$

Consequently, we conclude that:

Sum of fractions   *The sum of two or more fractions with the same denominator is a fraction whose numerator is the sum of the numerators of the separate fractions and whose denominator is the common denominator.*

EXAMPLE 1

$$\frac{3a}{2xy}+\frac{2a}{2xy}-\frac{b}{2xy}=\frac{1}{2xy}(3a+2a-b)=\frac{5a-b}{2xy}$$

If the denominators of the fractions to be added are not all equal, we begin by converting the given set of fractions to an equal set with the lcd as denominator and then proceed as above.

EXAMPLE 2 Combine $1/2x + 1/5y + (3x - 2y)/20xy$ into a single fraction.

Solution The lcd for the given fractions is $20xy$; hence, we multiply each member of the first fraction by $20xy/2x = 10y$ and each member of the second fraction by $20xy/5y = 4x$ and have

$$\frac{1}{2x} \cdot \frac{10y}{10y} + \frac{1}{5y} \cdot \frac{4x}{4x} + \frac{8x - 2y}{20xy} = \frac{1}{20xy}(10y + 4x + 8x - 2y)$$

$$= \frac{12x + 8y}{20xy} \qquad \text{combining}$$

$$= \frac{3x + 2y}{5xy} \qquad \text{reducing to lowest terms}$$

EXAMPLE 3 Combine $2/(2x - y) + 3/(2x + y) - 9xy/(4x^2 - y^2)y$ into a single fraction.

Solution The factored form of the third denominator is $(2x - y)(2x + y)y$; furthermore, each of the other denominators is a factor of this one. Therefore, $(2x - y)(2x + y)y$ is the lcd. Consequently, we multiply the members of the first fraction by $(2x - y)(2x + y)y/(2x - y) = (2x + y)y$ and the members of the second by $(2x - y)(2x + y)y/(2x + y) = (2x - y)y$ and have

$$\frac{2}{2x - y} \cdot \frac{y(2x + y)}{y(2x + y)} + \frac{3}{2x + y} \cdot \frac{y(2x - y)}{y(2x - y)} - \frac{9x}{(4x^2 - y^2)y}$$

$$= \frac{4xy + 2y^2 + 6xy - 3y^2 - 9x}{(4x^2 - y^2)y} = \frac{10xy - y^2 - 9x}{(4x^2 - y^2)y}$$

EXERCISE 4.2 Addition of Fractions

Perform the additions indicated in the following problems, and then reduce to lowest terms.

1 $\frac{3}{4} - \frac{1}{2} + \frac{1}{8}$ 2 $\frac{2}{3} + \frac{1}{2} - \frac{1}{6}$ 3 $\frac{2}{5} + \frac{1}{3} - \frac{2}{15}$ 4 $\frac{1}{6} + \frac{1}{4} - \frac{2}{3}$

5 $\dfrac{x - 1}{3} + \dfrac{2x + 3}{9} + \dfrac{x + 2}{2}$ 6 $\dfrac{2x + 3}{5} - \dfrac{3x - 2}{3} + \dfrac{x - 5}{2}$

7 $\dfrac{3x + 4}{7} - \dfrac{3x - 3}{2} + \dfrac{3x - 5}{14}$ 8 $\dfrac{5x + 1}{2} - \dfrac{5x - 2}{4} + \dfrac{7x - 3}{12}$

9 $\dfrac{3b}{2ac} - \dfrac{c}{3ab} + \dfrac{2a}{9bc}$ 10 $\dfrac{5c}{3ab} - \dfrac{3a}{2bc} - \dfrac{3b}{ac}$

11 $\dfrac{1}{3a} - \dfrac{3a + 4b}{12ab} + \dfrac{3}{4b}$ 12 $\dfrac{1}{2a} + \dfrac{1}{6b} - \dfrac{3a - 5b}{18ab}$

13 $\dfrac{4b}{21a} - \dfrac{5a}{14b} + \dfrac{9a^2 - 8b^2}{42ab}$ 14 $\dfrac{3a}{10b} - \dfrac{2b}{15a} - \dfrac{9a^2 - 4b^2}{30ab}$

15 $\dfrac{3a}{4b} - \dfrac{2b}{3a} + \dfrac{8b^2 - 8a^2}{6ab}$ 16 $\dfrac{5a}{6b} - \dfrac{2b}{3a} - \dfrac{30a^2 - 23b^2}{36ab}$

17 $\dfrac{2x-y}{3x+y}+\dfrac{5x}{2y}$

18 $\dfrac{3x+4y}{2x+y}-\dfrac{2y}{x}$

19 $\dfrac{3x}{2y}-\dfrac{3x-2y}{3x+2y}$

20 $\dfrac{3x}{5y}-\dfrac{5x+2y}{2x-y}$

21 $\dfrac{x+2y}{x-y}-\dfrac{x-2y}{x+y}$

22 $\dfrac{3x-2y}{2x-y}-\dfrac{2x+y}{3x-y}$

23 $\dfrac{5x-2y}{3x-8y}+\dfrac{3x-3y}{2x-5y}$

24 $\dfrac{2x+7y}{5x-3y}+\dfrac{5x+3y}{2x-7y}$

25 $\dfrac{r+s}{rs}-\dfrac{1}{s}+\dfrac{s}{r(r-s)}$

26 $\dfrac{s}{r+s}-\dfrac{r^2}{s(r+s)}+\dfrac{2r}{s}$

27 $\dfrac{s}{r}-\dfrac{2rs}{r(r-2s)}+\dfrac{r}{r-2s}$

28 $\dfrac{r^2-3s^2}{r(r+3s)}+\dfrac{3s}{r}+\dfrac{2s}{r+3s}$

29 $\dfrac{6x}{x^2-y^2}+\dfrac{2x}{y(x+y)}-\dfrac{3}{x-y}$

30 $\dfrac{2}{3x-2y}-\dfrac{3y}{x(3x+2y)}-\dfrac{8y}{9x^2-4y^2}$

31 $\dfrac{2x}{y(x+2y)}+\dfrac{4}{x-2y}+\dfrac{8x}{x^2-4y^2}$

32 $\dfrac{x}{x+3}-\dfrac{6x+6}{x^2-9}+\dfrac{x+1}{x-3}$

33 $\dfrac{3}{(x+2y)(x-y)}+\dfrac{2}{(x-2y)(x-y)}-\dfrac{4}{(x+2y)(x-2y)}$

34 $\dfrac{2}{(a+3b)(a+b)}-\dfrac{3}{(a+3b)(a-2b)}+\dfrac{5}{(a+b)(a-2b)}$

35 $\dfrac{4}{(a+5b)(a-4b)}+\dfrac{1}{(a-4b)(a-b)}+\dfrac{2}{(a+5b)(a-b)}$

36 $\dfrac{2}{(a+b)(a+2b)}-\dfrac{5}{(a-3b)(a+2b)}+\dfrac{4}{(a-3b)(a+b)}$

37 $\dfrac{4x+3y}{3x+2y}-\dfrac{2x+3y}{4x-3y}-\dfrac{3x-2y}{2x-3y}$

38 $\dfrac{5x+2y}{2x+5y}-\dfrac{3x+y}{x+3y}-\dfrac{2x-y}{x-2y}$

39 $\dfrac{2x-y}{x+3y}-\dfrac{x-3y}{2x+y}+\dfrac{3x+y}{3x-y}$

40 $\dfrac{x-3y}{x+2y}+\dfrac{x-2y}{4x+3y}-\dfrac{4x-3y}{x+3y}$

41 $\dfrac{x}{2x^2+xy-3y^2}+\dfrac{y}{2x^2-3xy+y^2}-\dfrac{x+y}{4x^2+4xy-3y^2}$

42 $\dfrac{2x}{x^2-2xy-3y^2}-\dfrac{y}{3x^2+4xy+y^2}+\dfrac{2x-y}{3x^2-8xy-3y^2}$

43 $\dfrac{3x}{2x^2+3xy-2y^2}+\dfrac{y}{x^2-4y^2}-\dfrac{2x+y}{2x^2-5xy+2y^2}$

44 $\dfrac{x}{2x^2+7xy+5y^2}+\dfrac{2y}{x^2-y^2}+\dfrac{3x+2y}{2x^2+3xy-5y^2}$

4.5 / MULTIPLICATION OF FRACTIONS

In Sec. 2.14 we proved that

$$\frac{c}{b}\cdot\frac{d}{a}=\frac{cd}{ba}$$

By use of the associative axiom we can extend this theorem as follows:

$$\frac{c}{b}\frac{d}{a}\frac{q}{p} = \left(\frac{c}{b}\frac{d}{a}\right)\frac{q}{p} = \frac{cd}{ba}\frac{q}{p} = \frac{cdq}{bap}$$

In fact, by using the same process, we can extend it to any integral number of fractions and have the following result:

Product of fractions

The product of two or more given fractions is a fraction whose numerator is the product of the numerators of the given fractions and whose denominator is the product of the given denominators.

EXAMPLE 1 Obtain the product of

$$\frac{x^2}{y^3} \qquad \frac{x-y}{x+y} \qquad \text{and} \qquad \frac{x^2+xy+y^2}{x^2-3xy+y^2}$$

Solution

$$\frac{x^2}{y^3} \cdot \frac{x-y}{x+y} \cdot \frac{x^2+xy+y^2}{x^2-3xy+y^2} = \frac{x^2(x-y)(x^2+xy+y^2)}{y^3(x+y)(x^2-3xy+y^2)}$$

When possible, the product should be reduced to lowest terms. For this reason, the members of the fractions should be factored, if they are reducible, before the product is formed. Then the factors that are common to the members of the product can be detected easily.

EXAMPLE 2 Obtain the product of

$$\frac{a^2-4b^2}{2a^2-7ab+3b^2} \qquad \frac{6a-3b}{2a+4b} \qquad \text{and} \qquad \frac{a^2-4ab+3b^2}{a^2-ab-2b^2}$$

Solution

$$\frac{a^2-4b^2}{2a^2-7ab+3b^2} \cdot \frac{6a-3b}{2a+4b} \cdot \frac{a^2-4ab+3b^2}{a^2-ab-2b^2}$$

$$= \frac{(a-2b)(a+2b)}{(2a-b)(a-3b)} \cdot \frac{3(2a-b)}{2(a+2b)} \cdot \frac{(a-b)(a-3b)}{(a+b)(a-2b)} \qquad \text{factoring}$$

$$= \frac{3(a-2b)(a+2b)(2a-b)(a-3b)(a-b)}{2(a-2b)(a+2b)(2a-b)(a-3b)(a+b)} \qquad \begin{array}{l}\text{by Eq. (2.45) and the}\\\text{commutative axiom}\end{array}$$

$$= \frac{3(a-b)}{2(a+b)} \qquad \begin{array}{l}\text{dividing the}\\\text{members by}\\(a-2b)(a+2b)\\(a-3b)(2a-b)\end{array}$$

EXAMPLE 3   Obtain the product of

Solution

$$\frac{x^2 - 3x + 2}{2x^2 + 3x - 2} \qquad \frac{2x^2 + 5x - 3}{x^2 - 1} \qquad \text{and} \qquad \frac{3x^2 + 6x}{2x - 4}$$

$$\frac{x^2 - 3x + 2}{2x^2 + 3x - 2} \cdot \frac{2x^2 + 5x - 3}{x^2 - 1} \cdot \frac{3x^2 + 6x}{2x - 4}$$

$$= \frac{(x - 2)(x - 1)}{(2x - 1)(x + 2)} \cdot \frac{(2x - 1)(x + 3)}{(x - 1)(x + 1)} \cdot \frac{3x(x + 2)}{2(x - 2)}$$

$$= \frac{3x(x + 3)}{2(x + 1)}$$

4.6 / DIVISION OF FRACTIONS

By the definition of a quotient, Eq. (2.41),

$$\frac{n}{d} \div \frac{r}{s} = x \qquad \text{if } \frac{r}{s} \cdot x = \frac{n}{d}$$

provided that $d \neq 0$, $r \neq 0$, and $s \neq 0$.   We now determine x as follows:

$$\frac{s}{r} \cdot \frac{r}{s} \cdot x = \frac{s}{r} \cdot \frac{n}{d} \qquad \text{multiplying each member of } \frac{r}{s} \cdot x = \frac{n}{d} \text{ by } \frac{s}{r}$$

$$\frac{sr}{rs} \cdot x = \frac{s}{r} \cdot \frac{n}{d} \qquad \text{by Eq. (4.6)}$$

$$1 \cdot x = \frac{s}{r} \cdot \frac{n}{d} \qquad \text{by Eq. (4.1), since } \frac{sr}{rs} = 1$$

$$x = \frac{n}{d} \cdot \frac{s}{r} \qquad \text{by Eq. (2.34) and the commutative axiom}$$

Therefore,

$$\frac{n}{d} \div \frac{r}{s} = \frac{n}{d} \cdot \frac{s}{r} \qquad\qquad\qquad \textbf{(4.4)}$$

Quotient of two fractions   Consequently, we have the following procedure for obtaining the quotient of two fractions:

In order to obtain the quotient of two fractions, we multiply the dividend by the reciprocal† of the divisor.

† By Eq. (2.36), the reciprocal of $\frac{r}{s}$ is the number $1 \div \frac{r}{s} = 1 \cdot \frac{s}{r} = \frac{s}{r}$.

EXAMPLE   Divide

$$\frac{x^2 - 3x + 2}{2x^2 - 7x + 3} \quad \text{by} \quad \frac{x^2 - x - 2}{2x^2 + 3x - 2}$$

Solution $\dfrac{x^2 - 3x + 2}{2x^2 - 7x + 3} \div \dfrac{x^2 - x - 2}{2x^2 + 3x - 2}$

$= \dfrac{x^2 - 3x + 2}{2x^2 - 7x + 3} \cdot \dfrac{2x^2 + 3x - 2}{x^2 - x - 2}$ by Eq. (4.4)

$= \dfrac{(x^2 - 3x + 2)(2x^2 + 3x - 2)}{(2x^2 - 7x + 3)(x^2 - x - 2)}$ by Eq. (2.45)

$= \dfrac{(x - 2)(x - 1)(2x - 1)(x + 2)}{(2x - 1)(x - 3)(x - 2)(x + 1)}$ factoring

$= \dfrac{1 \cdot (x - 1) \cdot 1 \cdot (x + 2)}{1 \cdot (x - 3) \cdot 1 \cdot (x + 1)}$ dividing the members by $(x - 2)(2x - 1)$

$= \dfrac{(x - 1)(x + 2)}{(x - 3)(x + 1)}$ by Eq. (2.34)

EXERCISE 4.3 Multiplication and Division of Fractions

Perform the indicated multiplications and divisions in the following problems. Reduce the result to lowest terms in each case.

1 $\dfrac{2x^2y^3}{3z^3w} \cdot \dfrac{9z^4w^2}{6x^4y}$

2 $\dfrac{5x^4y^3}{4xw^3} \cdot \dfrac{8x^3z}{15x^5w^2} \cdot \dfrac{z^2}{y}$

3 $\dfrac{15x^3y^4}{7w^3z^5} \cdot \dfrac{35x^2w}{3y^2z^2} \cdot \dfrac{z^4}{y}$

4 $\dfrac{9y^6w^4}{5x^3z^2} \cdot \dfrac{20x^4w^2}{12y^3z} \cdot \dfrac{2z^4}{3w^3}$

5 $\dfrac{8x^4y^2}{5x^3y^4} \div \dfrac{16x^5y^3}{10x^2w^4}$

6 $\dfrac{14x^6y^4}{5x^3z^2} \div \dfrac{21x^3y^2}{10wz^3}$

7 $\dfrac{9p^3q^0r}{24pq^2r^3} \div \dfrac{27p^4qr^0}{16pq^2r^3}$

8 $\dfrac{28y^4z^5}{15w^3x^2} \div \dfrac{35xw^2}{20y^3z^4}$

9 $\dfrac{15xy^2}{8y^3z^2} \cdot \dfrac{4x^2z^6}{3y^2z^3} \div \dfrac{10y^3z^2}{3x^4y^7}$

10 $\dfrac{34a^3b}{7b^2c} \cdot \dfrac{21c^3d}{17a^2d^3} \div \dfrac{3b^2c^3}{12ab^3}$

11 $\dfrac{68x^2y^3}{17y^2z} \cdot \dfrac{3x^2z}{4x^3y^2} \div \dfrac{12y^3z}{5xz^2}$

12 $\dfrac{72xy^0}{13y^3z^2} \cdot \dfrac{26x^3z}{9x^2y^4} \div \dfrac{4y^2z^2}{3x^0y}$

13 $\dfrac{xy + 2xz}{5x + 5y} \cdot \dfrac{xz + yz}{2xy + 4xz}$

14 $\dfrac{2x^2 + 3xy}{9x - 3y} \cdot \dfrac{3xz - yz}{2xy + 3y^2}$

15 $\dfrac{2p^2 - pq}{2q^2 - pq} \cdot \dfrac{pq - 2q^2}{pq - 2p^2}$

16 $\dfrac{2p + 6q}{2r^2 - rs} \cdot \dfrac{6rs - 3s^2}{2pq + 6q^2}$

17 $\dfrac{4x^2 - y^2}{x + 2y} \div \dfrac{4x^2 + 2xy}{xy + 2y^2}$

18 $\dfrac{x^2 - 25}{x - 4} \div \dfrac{2x - 10}{xy - 4y}$

19. $\dfrac{2x^2 + 3xy}{3xy - 2y^2} \div \dfrac{4x^2 - 9y^2}{3x^2 - 2xy}$

20. $\dfrac{x^2 - 4y^2}{x^2 + xy - 2y^2} \div \dfrac{x^2 - 2xy}{x^2 - y^2}$

21. $\dfrac{x^2 + 2xy}{2xy - 4y^2} \cdot \dfrac{y^2}{xy + 2y^2} \cdot \dfrac{x^2 - 2xy}{y}$

22. $\dfrac{x^2}{x - 5y} \cdot \dfrac{x^2 - 25y^2}{y^2} \cdot \dfrac{y}{x + 5y}$

23. $\dfrac{x^3y + x^2y^2}{x - y} \cdot \dfrac{x^2 - y^2}{xy^2 + y^3} \cdot \dfrac{y^3}{x}$

24. $\dfrac{x^2 - y^2}{x^2 y} \cdot \dfrac{y^3}{3xy + 3y^2} \cdot \dfrac{y^4}{2x^2 - 2xy}$

25. $\dfrac{w^2 + 3w}{z^2 - 2z} \cdot \dfrac{wz^2 - 2wz}{w^2 - 9} \cdot \dfrac{w - 3}{z^2}$

26. $\dfrac{p^2 + 5pq}{q^2 - q} \cdot \dfrac{q^2 - 1}{p^3q + 5p^2q^2} \cdot \dfrac{q^3}{pq + p}$

27. $\dfrac{3x^2 - 7xy}{x^2y^3 - 49y^5} \cdot \dfrac{3x + 21y}{3x^4 - 7x^3y} \cdot \dfrac{x^3yz^2}{xz - 7yz}$

28. $\dfrac{3x + y}{xy - 3xz} \cdot \dfrac{x^2yz}{x - 2y} \cdot \dfrac{y^3 - 9yz^2}{2yz + 6z^2}$

29. $\dfrac{x^2 - y^2}{x^2 y} \cdot \dfrac{xy - 2y^2}{x - y} \cdot \dfrac{xy^3}{x^2 - xy - 2y^2}$

30. $\dfrac{xy - xz + y^2 - yz}{yz} \cdot \dfrac{2yz - z^2}{xz - xy} \cdot \dfrac{x^2y}{2xy - xz - 2y^2 + yz}$

31. $\dfrac{xy + 2xz + y^2 + 2yz}{xy + 2xz} \cdot \dfrac{xy + xz}{x^2 + xy + xz + yz} \cdot \dfrac{x^2 - xy - xz + yz}{xy - xz - yz + z^2}$

32. $\dfrac{2xy - xz + 4y^2 - 2yz}{y - 2x} \cdot \dfrac{2xy - y^2}{x^2 + xy - 2y^2} \cdot \dfrac{x^2 - xy + xz - yz}{z^3 - 2yz^2}$

33. $\dfrac{x^3 + y^3}{x^2 + 3xy + 2y^2} \cdot \dfrac{x^2 - xy - 6y^2}{x^2 - 2xy - 3y^2} \div \dfrac{x^2 - xy + y^2}{2x^2 + 2xy}$

34. $\dfrac{x^2 - xy - 2y^2}{x^3 - y^3} \cdot \dfrac{x^2 + xy + y^2}{x^2 - 4y^2} \div \dfrac{x^2 + 4xy + 3y^2}{x^2 + xy - 2y^2}$

35. $\dfrac{x^3 + 8y^3}{x^2 - 4y^2} \cdot \dfrac{x^2 - xy - 2y^2}{x^2 - 2xy + 4y^2} \div \dfrac{x^2 - 2xy - 3y^2}{x^2 - 3xy}$

36. $\dfrac{27x^3 - y^3}{3x^2 - 4xy + y^2} \cdot \dfrac{x^2 + 2xy - 3y^2}{9x^2 + 3xy + y^2} \div \dfrac{x^2 - 9y^2}{xy + 3y^2}$

37. $\dfrac{(x - 2)x + 1}{(x - 1)x - 2} \cdot \dfrac{(x - 2)x + (x - 2)}{x^2 - 1} \div \dfrac{x + 6}{x + 1}$

38. $\dfrac{x(x + 2) + 2(2x + 4)}{2x(x + 2) + (x + 3)} \cdot \dfrac{x^2 - (x + 2)}{x^2 - 4} \div \dfrac{x + 4}{x}$

39. $\dfrac{(x - 3)x - 4}{(x - 9)x + 20} \cdot \dfrac{(x + 3)x - 40}{(x^2 - 9) - 8x} \div \dfrac{x + 8}{x - 9}$

40. $\dfrac{(x + 1)x - 2}{(x + 1) - 4} \cdot \dfrac{(x - 2)x + 1(x - 2)}{(x - 2)(x - 1)} \div \dfrac{x + 2}{x - 3}$

41. $\dfrac{(x + 2)x - 8}{(x + 2) + 2} \cdot \dfrac{x(x - 3) + (x - 3)}{(x - 2)(x + 1)} \div \dfrac{x - 3}{x + 2}$

42. $\dfrac{(3x - 2)x - 1}{(x + 1)(3x + 1)} \cdot \dfrac{(x + 1)3x - 1(x + 1)}{3(x + 1)(x - 1) - 8x} \div \dfrac{x + 1}{x - 3}$

43. $\dfrac{(x - 3)x - 4}{x^2 - 16} \cdot \dfrac{(x + 4)x + 1(x + 4)}{(x + 1)(x - 3)} \div \dfrac{x + 3}{x - 3}$

44. $\dfrac{x(x - 1) - 6}{x(x - 1) - 20} \cdot \dfrac{x(x - 1) + (3x - 8)}{x^2 - 9} \div \dfrac{x - 2}{x - 5}$

4.7 / COMPLEX FRACTIONS

Complex fraction
A *complex fraction* is a fraction in which at least one of the terms of one or both members is a fraction.

The following are examples of complex fractions:

$$\frac{\frac{3}{2}}{\frac{2}{3}} \qquad \frac{1 + \frac{x}{y}}{x + y} \qquad \frac{\frac{4x}{x + y} + \frac{2y}{x - y}}{3 - \frac{x^2 + y^2}{x^2 - y^2}}$$

First method for reducing a complex fraction
There are two methods for reducing a complex fraction to a simple fraction. The first method consists of multiplying the numerator and denominator of the complex fraction by the lcm of the denominators of the fractions that appear in it. This procedure is justified by the fundamental principle of fractions shown by Eq. (4.2). We shall illustrate the procedure with an example.

EXAMPLE 1   Reduce

$$\frac{y - \frac{x^2}{y}}{\frac{y^2}{x} - x}$$

to a simple fraction.

Solution   The denominators of the fractions that occur in this complex fraction are x and y. Hence the lcm is xy. Consequently, we multiply each member of the complex fraction by xy and get

$$\frac{y - \frac{x^2}{y}}{\frac{y^2}{x} - x} = \frac{\left(y - \frac{x^2}{y}\right)xy}{\left(\frac{y^2}{x} - x\right)xy} \qquad \text{by Eq. (4.2)}$$

$$= \frac{xy^2 - x^3}{y^3 - x^2 y}$$

$$= \frac{x(y^2 - x^2)}{y(y^2 - x^2)} \qquad \text{factoring}$$

$$= \frac{x}{y} \qquad \text{dividing the members by } y^2 - x^2$$

Second method for reducing a complex fraction
The second method for simplifying a complex fraction consists of reducing the numerator and denominator separately to simple fractions and then performing the indicated division.

EXAMPLE 2 Reduce the complex fraction

$$\frac{\dfrac{x-y}{x+y}-\dfrac{x+y}{x-y}}{1-\dfrac{x^2-xy-y^2}{x^2-y^2}}$$

to a simple fraction.

Solution

$$\frac{\dfrac{x-y}{x+y}-\dfrac{x+y}{x-y}}{1-\dfrac{x^2-xy-y^2}{x^2-y^2}}$$

$$=\frac{\dfrac{(x-y)^2-(x+y)^2}{(x+y)(x-y)}}{\dfrac{x^2-y^2-(x^2-xy-y^2)}{x^2-y^2}}$$ adding the fractions in the numerator and in the denominator

$$=\frac{\dfrac{x^2-2xy+y^2-x^2-2xy-y^2}{x^2-y^2}}{\dfrac{x^2-y^2-x^2+xy+y^2}{x^2-y^2}}$$ performing the indicated operations

$$=\frac{\dfrac{-4xy}{x^2-y^2}}{\dfrac{xy}{x^2-y^2}}$$ combining similar terms

$$=\frac{-4xy}{x^2-y^2}\cdot\frac{x^2-y^2}{xy}$$ multiplying the numerator by the reciprocal of the denominator

$$=\frac{-4xy(x^2-y^2)}{xy(x^2-y^2)}$$ by Eqs. (2.31) and (4.3)

$$=-4$$ dividing the members by $xy(x^2-y^2)$

If either the numerator or denominator of a complex fraction is itself a complex fraction or if both are complex fractions, each should be reduced to a simple fraction as the first step in the simplification.

EXAMPLE 3 Reduce the complex fraction

$$\frac{1+\dfrac{1}{1+\dfrac{1}{x-1}}}{1-\dfrac{1}{x+1}}$$

to a simple fraction.

Solution

$$\frac{1+\dfrac{1}{1+\dfrac{1}{x-1}}}{1-\dfrac{1}{x+1}} = \frac{1+\dfrac{x-1}{x-1+1}}{\dfrac{x+1}{x+1-1}}$$

multiplying both members of the complex fraction in the numerator by $(x-1)$ and both members of the complex fraction in the denominator by $(x+1)$

$$= \frac{1+\dfrac{x-1}{x}}{\dfrac{x+1}{x}}$$

$$= \frac{x+x-1}{x+1}$$

multiplying both members by x

$$= \frac{2x-1}{x+1}$$

EXERCISE 4.4 Complex Fractions

Reduce the following complex fractions to simple fractions.

1 $\dfrac{1-\dfrac{1}{3}}{2+\dfrac{2}{3}}$
2 $\dfrac{4+\dfrac{1}{2}}{3-\dfrac{1}{3}}$
3 $\dfrac{\dfrac{1}{2}-\dfrac{1}{3}}{\dfrac{3}{4}+\dfrac{7}{6}}$
4 $\dfrac{\dfrac{2}{5}+\dfrac{1}{3}}{\dfrac{5}{6}-\dfrac{1}{2}}$
5 $\dfrac{2-\dfrac{1}{x}}{4-\dfrac{1}{x^2}}$

6 $\dfrac{1-\dfrac{9}{x^2}}{1+\dfrac{3}{x}}$
7 $\dfrac{x-\dfrac{16}{x}}{1-\dfrac{4}{x}}$
8 $\dfrac{2+\dfrac{3}{x}}{4-\dfrac{9}{x^2}}$
9 $\dfrac{x^2-\dfrac{4}{x^2}}{x+\dfrac{2}{x}}$
10 $\dfrac{\dfrac{x}{3}-\dfrac{3}{x}}{\dfrac{1}{x}+\dfrac{2}{3x}}$

11 $\dfrac{\dfrac{x}{10}-\dfrac{1}{5}}{\dfrac{1}{2}-\dfrac{1}{x}}$
12 $\dfrac{\dfrac{4}{x}-\dfrac{1}{2}}{\dfrac{3}{2x}+\dfrac{1}{x}}$
13 $\dfrac{1-\dfrac{2}{x}-\dfrac{15}{x^2}}{1-\dfrac{1}{x}-\dfrac{12}{x^2}}$
14 $\dfrac{3-\dfrac{10}{x}+\dfrac{3}{x^2}}{3+\dfrac{5}{x}-\dfrac{2}{x^2}}$

15 $\dfrac{6+\dfrac{7}{x}+\dfrac{2}{x^2}}{2+\dfrac{7}{x}+\dfrac{3}{x^2}}$
16 $\dfrac{3-\dfrac{11}{x}+\dfrac{6}{x^2}}{4-\dfrac{13}{x}+\dfrac{3}{x^2}}$
17 $\dfrac{1+\dfrac{x}{x+y}}{1-\dfrac{3x}{x-y}}$
18 $\dfrac{2-\dfrac{x}{2x+y}}{2-\dfrac{5x}{x-y}}$

19 $\dfrac{1+\dfrac{2x}{x+y}}{1+\dfrac{x}{x+y}}$
20 $\dfrac{2-\dfrac{3x}{x-y}}{2-\dfrac{3x}{2x+y}}$
21 $\dfrac{a-2+\dfrac{a-2}{a+2}}{a-\dfrac{3a+12}{a+2}}$

22 $\dfrac{a+\dfrac{8a}{2a-1}}{a+2-\dfrac{6}{2a+3}}$
23 $\dfrac{2a-\dfrac{3a+4}{a-2}}{a-\dfrac{10a+4}{2a+3}}$
24 $\dfrac{a-\dfrac{a}{a+2}}{a+\dfrac{1}{a+2}}$

25 $\dfrac{\dfrac{p}{p+3}+\dfrac{p}{p^2-9}}{\dfrac{1}{p-3}+1}$

26 $\dfrac{\dfrac{2}{p+q}-\dfrac{1}{p-2q}}{2-\dfrac{p+q}{p-2q}}$

27 $\dfrac{\dfrac{1-2p}{1+3p}-\dfrac{1}{1-p}}{1-\dfrac{10}{1+3p}}$

28 $\dfrac{\dfrac{p+2}{p-2}-\dfrac{p}{p+2}}{3-\dfrac{4}{p+2}}$

29 $\dfrac{\dfrac{5}{x-2}+\dfrac{3}{2x+1}}{\dfrac{1+12x}{2-x}-1}$

30 $\dfrac{\dfrac{2x+3}{x+2}-\dfrac{2x}{x+1}}{\dfrac{x}{x+2}-1}$

31 $\dfrac{\dfrac{3}{2-3x}-\dfrac{2}{x+1}}{1-\dfrac{1+6x}{2-3x}}$

32 $\dfrac{\dfrac{x}{x+1}-\dfrac{x^2}{x^2-1}}{1+\dfrac{1}{x-1}}$

33 $\dfrac{\dfrac{1}{x}}{1-\dfrac{1}{1+\dfrac{2x}{y}}}$

34 $\dfrac{x-\dfrac{x}{2-\dfrac{1}{x}}}{y+\dfrac{y}{2x-1}}$

35 $\dfrac{u-\dfrac{u}{1+\dfrac{v}{u}}}{w-\dfrac{w}{u+1}{v}}$

36 $\dfrac{1+\dfrac{1}{x}}{1-\dfrac{1}{1+\dfrac{2}{x-4}}}$

EXERCISE 4.5 REVIEW

Convert the fraction in each of Probs. 1 to 10 into an equal fraction that has the expression to the right of the comma as denominator.

1 a/b, ab **2** a/b, b^2 **3** $ab/3a$, 3 **4** x/y, $-y^2$

5 $\dfrac{x-2y}{2x-y}$, $y-2x$

6 $\dfrac{x-2y}{2x-y}\cdot\dfrac{2x-y}{x+2y}$, $-x-2y$

7 $\dfrac{a+7}{a-7}$, a^2-49

8 $\dfrac{b+3}{b-1}$, $(b-1)(b+2)$

9 $\dfrac{3y-1}{y+2}$, $(y+2)(y-3)$

10 $\dfrac{(x+2)(x-3)}{(x+2)(x-4)}$, $x-4$

Reduce the fraction in each of Probs. 11 to 20 to lowest terms.

11 $6x^2y/15xy^3$ **12** $21x^3y^3z^0/35xyz^2$

13 $\dfrac{(x-2)(x+1)}{(x+4)(x-2)}$

14 $\dfrac{x^3+64y^3}{x+4y}$ **15** $\dfrac{(2x+1)(2x^2+5x+3)}{(x+1)(4x^2+4x-3)}$

16 $\dfrac{ax-ay+bx-by}{ax-2bx-ay+2by}$

17 $\dfrac{x^3-y^3}{x^2-y^2}$ **18** $\dfrac{x^6-y^6}{x^9-y^9}$

19 $\dfrac{x-3}{(x-2)x-3}$

20 $\dfrac{3x-4}{x(3x-4)-2(4-3x)}$

Convert the set of fractions in each of Probs. 21 and 22 to an equal set with a common denominator.

21 $\left\{ \dfrac{1}{x+1}, \dfrac{3}{x-3}, \dfrac{2}{2x-1} \right\}$

22 $\left\{ \dfrac{x-1}{(2x-3)(x+4)}, \dfrac{2x-5}{(x+4)(3x-1)}, \dfrac{3x+2}{(2x-3)(3x-1)} \right\}$

Perform the indicated additions in Probs. 23 to 32, and then reduce to lowest terms.

23 $\dfrac{1}{3} + \dfrac{2}{5} - \dfrac{5}{6}$

24 $\dfrac{2x-1}{3} - \dfrac{x+1}{2} + \dfrac{3x+2}{4}$

25 $\dfrac{a}{3bc} - \dfrac{b}{5ac} + \dfrac{5c}{6ab}$

26 $\dfrac{3x-y}{x-3y} + \dfrac{2x}{3y}$

27 $\dfrac{2x+y}{x-2y} - \dfrac{x-y}{x+y}$

28 $\dfrac{a+3b}{ab} - \dfrac{2}{a} + \dfrac{b}{a(a-b)}$

29 $\dfrac{3x}{x^2-4y^2} + \dfrac{x}{y(x-2y)} - \dfrac{2}{x+2y}$

30 $\dfrac{1}{(x-2)(x+1)} + \dfrac{2}{(x+2)(x+1)} - \dfrac{2}{x^2-4}$

31 $\dfrac{3x+2y}{4x+3y} - \dfrac{4x-3y}{2x+3y} - \dfrac{2x-3y}{3x-2y}$

32 $\dfrac{x}{(2x-y)(x-2y)} - \dfrac{2y}{(2x-y)(x+3y)} + \dfrac{2x-y}{(x-2y)(x+3y)}$

In each of Probs. 33 to 43, perform the indicated multiplications and divisions, and then reduce the result to lowest terms.

33 $\dfrac{3x^3y^2}{2yz^2} \cdot \dfrac{8z^2w^0}{6x^2w}$

34 $\dfrac{9w^3x^2}{7xy^4} \cdot \dfrac{14y^3z^2}{12x^2w}$

35 $\dfrac{34y^2z^3}{7z^2w} \div \dfrac{17x^2y^4}{14w^3x}$

36 $\dfrac{48z^4w^0}{13w^2x^3} \div \dfrac{32y^3z^3}{39x^2y^5}$

37 $\dfrac{xy+2xz}{3x-3y} \cdot \dfrac{x^2-y^2}{y^2+3yz+2z^2}$

38 $\dfrac{xy-y^2}{xy+xz} \cdot \dfrac{xy+xz+2yz+2y^2}{x^2+2xy-xy-2y^2}$

39 $\dfrac{x^2-2xy}{yz-2z^2} \div \dfrac{xy-2y^2}{2yz-4z^2}$

40 $\dfrac{x^2-xy-2y^2}{x^2+2xy-3y^2} \div \dfrac{x^2-2xy-3y^2}{x^2+5xy+6y^2}$

41 $\dfrac{x^3-y^3}{x^2+3xy+2y^2} \cdot \dfrac{x^2-xy-2y^2}{x^3+x^2y+xy^2} \div \dfrac{x^2-3xy+2y^2}{x^2+3xy+2y^2}$

42 $\dfrac{8x^3-27y^3}{4x^2-9y^2} \cdot \dfrac{2x^2+xy-3y^2}{4x^2y+6xy^2+9y^3} \div \dfrac{x^2-y^2}{xy+2y^2}$

43 $\dfrac{(x-3)x+2}{(x-3)x-4} \cdot \dfrac{(x-4)x+3}{(x-3)x-2(x-3)} \div \dfrac{(x+1)x-2}{x^2-3x-4}$

Simplify the following complex fractions.

44 $\dfrac{2-\dfrac{1}{5}}{3+\dfrac{3}{5}}$

45 $\dfrac{\dfrac{1}{3}+\dfrac{1}{2}}{\dfrac{1}{2}-\dfrac{1}{6}}$

46 $\dfrac{4-\dfrac{25}{x^2}}{2-\dfrac{5}{x}}$

47 $\dfrac{\dfrac{2x}{5}-\dfrac{5}{2x}}{2+\dfrac{5}{x}}$

48 $\dfrac{2+\dfrac{5}{x}-\dfrac{3}{x^2}}{\dfrac{15}{x^2}+\dfrac{11}{x}+2}$ **49** $\dfrac{2-\dfrac{5}{x}+\dfrac{2}{x^2}}{2+\dfrac{3}{x}-\dfrac{2}{x^2}}$ **50** $\dfrac{3x+\dfrac{x-5}{x-1}}{x+\dfrac{5}{3x-8}}$ **51** $\dfrac{x-\dfrac{2x+6}{2x+1}}{2x-\dfrac{7x+6}{x+3}}$

52 $\dfrac{\dfrac{2}{x-3}+\dfrac{1}{x+1}}{\dfrac{3x-1}{x-3}}$ **53** $\dfrac{\dfrac{3}{x+y}+\dfrac{1}{x-2y}}{1+\dfrac{x-y}{x-2y}}$ **54** $\dfrac{1}{1+\dfrac{1}{1+\dfrac{1}{a-1}}}$ **55** $\dfrac{1-\dfrac{3}{1-\dfrac{x}{y}}}{1+\dfrac{1}{1-\dfrac{x}{y}}}$

CHAPTER FIVE *Linear and Fractional Equations and Inequalities*

In the preceding chapters, we studied formal operations that follow prescribed rules of procedure. In this chapter, we shall study the conditions under which two algebraic expressions are equal and the conditions under which one algebraic expression is larger or smaller than another. In particular we shall find the values of the variable for which two expressions are equal or are unequal in a specified order. Some of the expressions we consider will involve absolute values.

5.1 / OPEN SENTENCES

The concept of an open sentence is illustrated as follows:

The colors in a rainbow and in the spectrum are violet, indigo, blue, green, yellow, orange, and red. Therefore, the sentence

Red is a color in a rainbow

is a true statement, while the sentence

Brown is a color in a rainbow

is false.   Now we shall consider four aspects of the sentence

$$x \text{ is a color in a rainbow} \tag{1}$$

where x is to be replaced by the name of a color.   First, the sentence (1) contains the variable x, which holds a place for the name of some color. Second, the sentence is neither true nor false as it stands.   Third, the sentence becomes a true statement if x is replaced by the name of one of the colors of the spectrum.   Fourth, the sentence is false if x is replaced by the name of a color other than one of the spectrum.

Sentences of the type shown by (1) illustrate the following definition:

Open sentence   An *open sentence* is a statement containing a variable that is neither true nor false but that becomes true or false when the variable is replaced by the name of an element chosen from an appropriate set.

In this chapter we shall consider open sentences of an algebraic nature. For example,

$$x + 2 = 5 \tag{2}$$

is an open sentence that states, "The sum of x and 2 is 5."   The sentence is true if x is replaced by 3, but it is false if x is replaced by any other number.

5.2 / DEFINITIONS

The open sentence (2) of Sec. 5.1 is an equation and illustrates the following definition:

Equation   An *equation* is an open sentence which states that two expressions, at least one of which contains one or more variables, are equal.

The following open sentences are examples of equations:

$$3x - 6 = 2x - 5 \tag{1}$$

$$3x + 2y = x - 2 \tag{2}$$

$$\frac{x + 3}{x - 4} = \frac{x}{x + 3} + 2 \tag{3}$$

The expression on each side of the equality sign in an equation is a
Members of an   *member of the equation.*
equation   In this chapter we consider equations in only one variable.   We shall explain methods for finding the replacements for which the equation is
Root   true.   Each such replacement is a *root* of the equation, and the set of all
Solution set   roots is the *solution set* for the equation.
Domain   The set of permissible replacements is called the *domain* of the equation.   The solution set is a subset of the domain.

Permissible replacement   A replacement is permissible provided each member of the equation is a number for this replacement and the replacement is consistent with the conditions of the equation or with the problem that generates the equation.

EXAMPLE 1   In (3) the left member is not a number if x is replaced by 4 and the right member is not a number if x is replaced by -3. Hence the domain of (3) is the set of all real numbers except 4 and -3.

Conditional equations   An equation that is satisfied by some numbers in its domain but not by others is called a *conditional equation,* as is an equation that is not satisfied by any number in the domain. An equation that is satisfied by each Identity   number in the domain is called an *identity.*

EXAMPLE 2   Neither member of the equation

$$\frac{3}{x-2} - \frac{2}{x-1} = \frac{x+1}{(x-2)(x-1)} \qquad (4)$$

is a number if x is replaced by 1 or 2. Therefore, the domain cannot include 1 or 2. However, if we combine the fractions in the left member, we obtain the fraction on the right. Therefore, the equation is true if x is replaced by any number other than 1 or 2. Hence the equation is an identity.

EXAMPLE 3   The equation

$$2x - 3 = x + 1 \qquad (5)$$

is true if x is replaced by 4, since $(2 \times 4) - 3 = 4 + 1$, but it is not true if x is replaced by any number other than 4. Hence (5) is a conditional equation.

EXAMPLE 4   The equation

$$\frac{x-1}{x+1} = 1 \qquad (6)$$

is true for no replacement for x, since $x - 1$ is never equal to $x + 1$. Hence (6) is a conditional equation whose solution set is $\varnothing$.

We shall now introduce the symbol $f(x)$, read "f of x," which is very important in mathematics. In this chapter we shall employ only one aspect of the meaning of $f(x)$,† and we shall use it to stand for an algebraic ex-

---

† We call attention to the fact that $f(x)$ does not mean "f times x."

pression in the variable x. This symbol will be discussed more fully in Chap. 6, where it will be given a broader interpretation.

Thus, if in a given discussion we let $f(x) = 3x^2 - 2x + 1$, then throughout the discussion, $f(x)$ stands for the trinomial $3x^2 - 2x + 1$. If the x in $f(x)$ is replaced by a specific number, then x in the expression denoted by $f(x)$ must be replaced by that number. Thus, if $f(x) = 3x^2 - 2x + 1$, then $f(2) = 3(2)^2 - 2(2) + 1 = 9$, and $f(y) = 3y^2 - 2y + 1$. Also, $f(2y) = 3(2y)^2 - 2(2y) + 1 = 12y^2 - 4y + 1$.

The notations $h(x)$, $g(x)$, and $F(x)$, or any letter followed by a second letter enclosed in parentheses, are also used to stand for algebraic expressions. The letter enclosed in parentheses denotes the variable in the expression.

Frequently, we shall use the notation $f(x)$ without indicating the expression for which it stands. When this is done, we mean that $f(x)$ represents any expression that we may choose. In this sense, then, each of the following expressions is an equation:

$$f(x) = c \qquad \text{where } c \text{ is a constant} \tag{7}$$

$$f(x) = g(x) \tag{8}$$

$$f(x) + g(x) = k(x) \tag{9}$$

Hereafter in this chapter we shall deal only with conditional equations in one variable, and we shall explain methods for solving them.

5.3 / EQUIVALENT EQUATIONS

The objective in solving an equation is to find a replacement for the variable that satisfies the equation. The simpler the equation is in form, the easier it is to solve. For example, we shall consider the equations

$$7x - 45 = 5x - 43 \tag{1}$$

and $\qquad\qquad\qquad\qquad 2x = 2 \tag{2}$

At this stage, the only way that we can find the root of (1) is to guess at a number and then substitute it for x and see if it satisfies the equation. In (2), however, it is obvious that the root is 1. Now if we substitute 1 for x in (1) and simplify, we get $-38 = -38$. Therefore, 1 is also a root of (1). We call attention to the fact that $7x - 45 = 5x - 43$ and $2x = 2$ are different statements. Each statement, however, is true if x is replaced by 1. Two equations of this type are said to be equivalent, and they illustrate the definition below.

Equivalent equations

Two equations are *equivalent* if every root of each of them is a root of the other.

If we employ the term "solution set," we can state the above definition in this way:

Two equations are *equivalent* if their solution sets are equal.

Operations that yield equivalent equations

As we shall presently see, we make extensive use of the concept of equivalent equations in the process of solving an equation. We shall therefore consider the operations that can be performed on the members of a given equation to yield an equation that is equivalent to the given one. We shall first prove the following theorem:

THEOREM

If $f(x)$, $g(x)$, and $h(x)$ are polynomials,† then $f(x) = g(x)$ and
$$f(x) + h(x) = g(x) + h(x) \text{ are equivalent equations} \qquad (5.1)$$

Before proving this theorem, we shall illustrate its meaning. If $f(x) = 3x + 1$, $g(x) = 2x + 5$, and $h(x) = -2x + 3$, then the equation $f(x) = g(x)$ is $3x + 1 = 2x + 5$. Furthermore, $f(x) + h(x) = g(x) + h(x)$ becomes $3x + 1 + (-2x + 3) = 2x + 5 + (-2x + 3)$, or $x + 4 = 8$. Now the theorem states that $3x + 1 = 2x + 5$ and $x + 4 = 8$ are equivalent. We readily see that 4 is a root of each of these equations since $4 + 4 = 8$ and $3 \cdot 4 + 1 = 2 \cdot 4 + 5 = 13$.

Proof

If the constant r is a root of $f(x) = g(x)$, then

$$f(r) = g(r)$$

Furthermore, $h(r)$ is a constant, and it follows by Eq. (2.8) that

$$f(r) + h(r) = g(r) + h(r)$$

Hence, r is a root of $f(x) + h(x) = g(x) + h(x)$.

Conversely, if r is a root of $f(x) + h(x) = g(x) + h(x)$, then

$$f(r) + h(r) = g(r) + h(r)$$

Consequently, by Eq. (2.8) we have

$$f(r) + h(r) - h(r) = g(r) + h(r) - h(r)$$

Thus, since $h(r) - h(r) = 0$, it follows that $f(r) = g(r)$. Therefore, r is a root of $f(x) = g(x)$.

We shall now illustrate the use of theorem (5.1).

---

† In most cases, the theorem is true if $f(x)$, $g(x)$, and $h(x)$ are not polynomials. The theorem as stated, however, suffices for our purposes.

EXAMPLE 1   Solve the equation

$$6x - 3 = 7 + 5x \qquad (3)$$

Solution   In this equation, $f(x) = 6x - 3$ and $g(x) = 7 + 5x$, and we shall let $h(x) = 3 - 5x$. By theorem (5.1), if we add $3 - 5x$ to each member of (3), we obtain an equation that is equivalent to (3). Thus we get

$$6x - 3 + 3 - 5x = 7 + 5x + 3 - 5x \qquad (4)$$

or, by the commutative axiom,

$$6x - 5x - 3 + 3 = 7 + 3 + 5x - 5x$$

Consequently,

$$6x - 5x = 10 \qquad \text{since } -3 + 3 = 5x - 5x = 0 \qquad (5)$$

Therefore, we have

$$x = 10 \qquad \text{since } 6x - 5x = x$$

Since the members of (4) reduce to x and 10, respectively, 10 is a root of (4); furthermore, since (3) and (4) are equivalent equations, 10 is a root of (3). Thus the solution set of (3) is $\{10\}$.

In order to check our computation, we substitute 10 for x in (3) and obtain $6(10) - 3 = 57$ for the left member and $7 + 5(10) = 57$ for the right member. Hence the two members of (3) are equal if $x = 10$.

The purpose of adding $h(x)$ to each member of $f(x) = g(x)$ is to obtain an equivalent equation in which each term of one member contains x and each term of the other is a constant. This is accomplished if $h(x)$ is the polynomial whose terms are the negatives of the terms involving x in $g(x)$ and the negatives of the constant terms in $f(x)$.

EXAMPLE 2   Obtain an equation that is equivalent to

$$4x - 3 = 2x + 9 \qquad (6)$$

such that the terms that involve x are in the left member, and the terms in the right member are constants.

Solution   If we add $-2x$ to each member of (6), we get

$$4x - 3 - 2x = 2x + 9 - 2x$$
$$= 2x - 2x + 9 \qquad \text{by the commutative axiom, Eq. (2.10)}$$
$$= 9 \qquad \text{since } 2x - 2x = 0$$

Hence we have

$$4x - 3 - 2x = 9 \qquad (7)$$

and this equation is equivalent to (6). Furthermore, the terms that involve x are in the left member. The left member, however, also has the constant -3 as a term. Hence we add 3 to each member of (7) and get

$$4x - 3 - 2x + 3 = 9 + 3$$

or $\qquad 4x - 2x = 9 + 3 \qquad$ by the commutative axiom, Eq. (2.10), and the fact that $-3 + 3 = 0$

Since the successive operations of adding $-2x$ and then 3 to each member are equivalent to the single operation of adding $-2x + 3$, we can combine the two steps thusly:

$$4x - 3 = 2x + 9 \qquad \text{given equation}$$
$$4x - 3 - 2x + 3 = 2x + 9 - 2x + 3 \qquad \text{adding } -2x + 3 \text{ to each member}$$
$$4x - 2x - 3 + 3 = 2x - 2x + 9 + 3 \qquad \text{by the commutative axiom, Eq. (2.10)}$$
$$4x - 2x = 9 + 3 \qquad \text{since } -3 + 3 = -2x + 2x = 0$$

Thus, in this case, $h(x) = -2x + 3$.

If, in the equation $f(x) = g(x)$, $f(x)$ and $g(x)$ are polynomials in which one or more of the numerical coefficients are fractions, we use the following theorem as a first step in solving the equation:

THEOREM **If k is a nonzero constant, then the equations**
$$f(x) = g(x) \text{ and } k \cdot f(x) = k \cdot g(x) \text{ are equivalent} \qquad (5.2)$$

Proof In keeping with this, $\frac{2}{3}x = \frac{5}{8}$ and $24(\frac{2}{3}x) = 24(\frac{5}{8})$ or $16x = 15$ are equivalent equations.

EXAMPLE 3 Obtain an equation that is equivalent to

$$\tfrac{1}{2}x + \tfrac{2}{3} = \tfrac{1}{4}x - \tfrac{1}{6} \qquad (8)$$

and that contains no fractions.

Solution The lcd of the denominators in the given equation is 12. Hence we multiply each member of (8) by 12 and get

$$12(\tfrac{1}{2}x + \tfrac{2}{3}) = 12(\tfrac{1}{4}x - \tfrac{1}{6}) \qquad (9)$$

and by theorem (5.2), (9) is equivalent to (8). We now simplify the members of (9) and get

$$6x + 8 = 3x - 2 \qquad \text{by the associative axiom, Eq. (2.32)}$$

5.4 / LINEAR EQUATIONS IN ONE VARIABLE

We define a linear equation in one variable as follows:

Linear equation in one variable

An equation $f(x) = g(x)$ is a *linear equation in one variable* if $f(x)$ and $g(x)$ are polynomials of degree 1 in x or if one of them is a polynomial of degree 1 and the other is a constant.

For example, $ax = b$, $3x + 4 = 7$, and $2x - 7 = 8 - 4x$ are linear equations in one variable.

The process of solving a linear equation in one variable consists of the following steps:

1　If one or more of the coefficients or constant terms in $f(x) = g(x)$ are fractions, we multiply each member of $f(x) = g(x)$ by the lcd of the denominators and equate the products. By theorem (5.2), the equation thus obtained is equivalent to $f(x) = g(x)$. If the lcd of the denominators is k, then the equation is $k \cdot f(x) = k \cdot g(x)$.

2　We next formulate the polynomial $h(x)$ so that $k \cdot f(x) + h(x)$ is a polynomial each term of which involves x and so that $k \cdot g(x) + h(x)$ involves only constants. Then we write the equation $k \cdot f(x) + h(x) = k \cdot g(x) + h(x)$. By theorems (5.2) and (5.1), this equation is equivalent to $f(x) = g(x)$.

3　By use of the commutative, associative, and distributive axioms, we combine the terms in each member of the equation obtained in step 2, and we thus obtain an equation of the type $ax = b$.

4　We now employ theorem (5.2) and multiply each member of $ax = b$ by $1/a$, and we thus obtain $x = b/a$. This number is the root of the given equation since the equation obtained in each step is equivalent to the given equation.

5　Finally, we substitute the number obtained in step 4 into the original equation in order to verify the fact that it is a root.

EXAMPLE 1　Solve the equation $6x - 7 = 2x + 1$.

Solution　Since no fractions are involved in this equation, step 1 is unnecessary; so we proceed to step 2.

The term in the right member that involves x is $2x$, and the constant term in the left member is -7; so $h(x) = -2x + 7$. Hence we add $-2x + 7$ to each member of the given equation and proceed as indicated below.

$$6x - 7 = 2x + 1 \qquad \text{given equation}$$
$$6x - 7 - 2x + 7 = 2x + 1 - 2x + 7 \qquad \text{adding } -2x + 7 \text{ to each member}$$
$$6x - 2x - 7 + 7 = 2x - 2x + 1 + 7 \qquad \text{by the commutative axiom, Eq. (2.10)}$$
$$4x = 8 \qquad \text{combining terms}$$
$$x = 2 \qquad \text{multiplying each member by } \tfrac{1}{4}$$

Check   If we substitute 2 for x in the given equation, we get $(6 \times 2) - 7 = 12 - 7 = 5$ for the left member and $(2 \times 2) + 1 = 4 + 1 = 5$ for the right member. Hence the members are equal if $x = 2$. Therefore, the root of the equation is 2.

We stated earlier that the solution set of an equation is the set of roots of the equation. Preceding Example 1, we listed steps showing that a linear equation in one variable is equivalent to an equation of the type $ax = b$. We shall now prove that $ax = b$ has only one root. We shall suppose that r and r' are roots of $ax = b$. Then we have

$$ar = b \qquad \text{and} \qquad ar' = b \qquad \text{by the definition of a root}$$

Hence, $ar = ar'$ by Eq. (2.4), since each is equal to b

Thus $r = r'$ by Eq. (2,37)

Hence a linear equation in one variable has not more than one root. Therefore, the solution set of a linear equation that is not an identity is either the empty set $\varnothing$ or a set that contains only one element. It follows, then, that the solution set of the equation in Example 1 is the set $\{2\}$. In the notation of sets, this statement can be expressed in the form

$$\{x \mid 6x - 7 = 2x + 1\} = \{2\}$$

EXAMPLE 2   Find the set indicated by $\{x \mid \tfrac{1}{2}x - \tfrac{2}{3} = \tfrac{3}{4}x + \tfrac{1}{12}\}$.

Solution   To obtain the required set, we must solve the equation

$$\tfrac{1}{2}x - \tfrac{2}{3} = \tfrac{3}{4}x + \tfrac{1}{12} \qquad\qquad (1)$$

Since the lcm of the denominators is 12, we shall employ the theorem in (5.2) and multiply each member of (1) by 12; we get

$$12\left(\tfrac{1}{2}x - \tfrac{2}{3}\right) = 12\left(\tfrac{3}{4}x + \tfrac{1}{12}\right)$$
$$6x - 8 = 9x + 1 \qquad \text{by the distributive axiom} \qquad (2)$$

Since the term that involves x in the right member is $9x$ and the constant term in the left member is -8, it follows that $h(x) = -9x + 8$. Hence we add $-9x + 8$ to each member of (2) and proceed as follows:

$$6x - 8 - 9x + 8 = 9x + 1 - 9x + 8 \qquad \text{adding } -9x + 8 \text{ to each member of (2)}$$
$$6x - 9x - 8 + 8 = 9x - 9x + 1 + 8 \qquad \text{by the commutative axiom}$$
$$-3x = 9 \qquad\qquad\qquad \text{combining terms}$$
$$x = -3 \qquad\qquad\qquad \text{multiplying each member by } -\tfrac{1}{3}$$

Check   Replacing x by -3 in the left member of (1), we get

$$\tfrac{1}{2}(-3) - \tfrac{2}{3} = -\tfrac{3}{2} - \tfrac{2}{3} = -\tfrac{9}{6} - \tfrac{4}{6} = -\tfrac{13}{6}$$

Similarly, for the right member we have

$$\tfrac{3}{4}(-3) + \tfrac{1}{12} = -\tfrac{9}{4} + \tfrac{1}{12} = -\tfrac{27}{12} + \tfrac{1}{12} = -\tfrac{26}{12} = -\tfrac{13}{6}$$

Hence -3 is a root since the members of (1) are equal if $x = -3$. Furthermore, since (1) is a linear equation, -3 is the only root. Consequently, the solution set is

$$\{x \mid \tfrac{1}{2}x - \tfrac{2}{3} = \tfrac{3}{4}x + \tfrac{1}{12}\} = \{-3\}$$

EXERCISE 5.1 Linear Equations

Find the solution set of the equation in each of Probs. 1 to 44.

1 $5x = 3x + 4$ **2** $6x = 2x + 12$ **3** $2x = 5x - 21$ **4** $4x = 7x - 15$

5 $3x = 5x + 4$ **6** $5x = 7x + 2$ **7** $7x = 3x + 12$ **8** $3x = 5x - 18$

9 $4(x + 2) + 3(x - 6) = -3$

10 $7(x - 2) - 4(x + 1) = -21$

11 $5(x + 4) - 2(3x + 2) = 19$

12 $9(x - 3) + 6(2x - 1) = 9$

13 $\tfrac{2}{3}x - 2 = \tfrac{1}{6}x + 3$

14 $\tfrac{3}{8}x + 2 = \tfrac{1}{4}x + 3$

15 $\tfrac{5}{12}x + 1 = \tfrac{3}{4}x - 3$

16 $\tfrac{5}{6}x + 7 = \tfrac{3}{4}x + 6$

17 $\tfrac{1}{2}x + \tfrac{1}{3}x + \tfrac{1}{4} = \tfrac{3}{4}x + \tfrac{3}{4}$

18 $\tfrac{1}{6}x + \tfrac{3}{8}x - \tfrac{1}{2} = \tfrac{3}{4}x - 3$

19 $\tfrac{5}{6}x - \tfrac{2}{9} - \tfrac{2}{9}x = \tfrac{5}{9}x + \tfrac{7}{9}$

20 $\tfrac{2}{3}x + \tfrac{1}{2}x + \tfrac{1}{2} = \tfrac{5}{6}x + 1$

21 $9(\tfrac{1}{3}x + \tfrac{1}{5}) - 4(\tfrac{1}{2}x + \tfrac{1}{4}) = 1$

22 $3(\tfrac{1}{3}x + \tfrac{1}{6}) - (\tfrac{1}{2}x - \tfrac{1}{3}) = 1$

23 $2(\tfrac{7}{18}x - \tfrac{5}{18} - \tfrac{1}{12}x) = 3(\tfrac{7}{36}x + \tfrac{4}{27})$

24 $3(\tfrac{1}{6}x + \tfrac{1}{9}x - \tfrac{1}{12}) = 2(\tfrac{3}{2}x + \tfrac{3}{2})$

25 $\dfrac{2x - 3}{3} = x - 3$

26 $\dfrac{3x + 4}{2} = 3x - 4$

27 $\dfrac{3x + 5}{4} = 2x - 5$

28 $\dfrac{5x - 7}{3} = 2x - 5$

29 $\dfrac{3x - 2}{5} + 3 = \dfrac{4x - 1}{3}$

30 $\dfrac{2x - 3}{3} + 1 = \dfrac{3x + 2}{5}$

31 $\dfrac{2x - 9}{3} - 2 = \dfrac{x - 3}{3}$

32 $\dfrac{5x + 1}{8} + 3 = \dfrac{3x + 1}{2}$

33 $\dfrac{3x + 7}{2} + 3x - 7 = \dfrac{2x + 3}{5}$

34 $\dfrac{3x + 10}{2} - x - 4 = \dfrac{3x + 6}{4}$

35 $\dfrac{5x + 7}{2} = \dfrac{3x + 5}{4} + 2x + 3$

36 $\dfrac{4x + 5}{5} = \dfrac{3x - 15}{2} + 2x - 5$

37 $\quad a(x+b) = a(b-x) + 2a + 2b$ 38 $\quad a(x+2) + b(x+3) = a + 2b$

39 $\quad (a+b)(x+1) = x(2a+b)$ 40 $\quad a(x-2) + b(x+2) = a - 3b$

41 $\quad \dfrac{ax}{b} - \dfrac{bx}{a} = \dfrac{(a+b)^2}{ab}$ 42 $\quad \dfrac{x-2a}{b} - \dfrac{3}{2} = \dfrac{b-x}{2a} + \dfrac{1}{2}$

43 $\quad \dfrac{b(bx-1)}{a} - \dfrac{a(1+ax)}{b} = 1$ 44 $\quad \dfrac{bx}{a} - \dfrac{9ax}{b} = 3a + b$

5.5 / LINEAR INEQUALITIES

Inequality

Linear inequality

An inequality in one variable is an open sentence of the type $f(x) > g(x)$ or $f(x) < g(x)$. If either $f(x)$ or $g(x)$ is a polynomial of degree 1 in x and the other is a polynomial of degree 1 in x or is a constant, then the inequality is *linear*.

For example, $2x + 2 > x - 3$, $5x + 3 < 2x + 1$, and $x + 4 > 0$ are linear inequalities.

Solution set

The *solution set* of an inequality is the set of replacements for x for which the inequality is a true statement. For example, it can be verified that $x - 1 > 3$ is true if x is replaced by a number greater than 4. Hence the solution set of the inequality is $\{x \mid x > 4\}$, or stated in another way, $\{x \mid x - 1 > 3\} = \{x \mid x > 4\}$.

The procedure for finding the solution set is called *solving the inequality*. If the inequality is linear, this procedure is similar to that for solving a linear equation and involves the concept of the equivalence of inequalities.

Equivalent inequalities

Two inequalities are *equivalent* if they have the same solution set.

We shall now give several theorems that state conditions under which two inequalities are equivalent.

Theorems on equivalent inequalities

If $f(x)$, $g(x)$, and $h(x)$ are expressions, constants included, then for all values of x for which the three expressions are real numbers, $f(x) > g(x)$ is equivalent to each of the following:

$$f(x) + h(x) > g(x) + h(x) \tag{5.3}$$

$$f(x) \cdot h(x) > g(x) \cdot h(x) \qquad \text{for } x \in \{x \mid h(x) > 0\} \tag{5.4}$$

$$f(x) \cdot h(x) < g(x) \cdot h(x) \qquad \text{for } x \in \{x \mid h(x) < 0\} \tag{5.5}$$

Similar statements can be made if $f(x) < g(x)$.

If k is a positive constant, then $f(x) > g(x)$ and $k \cdot f(x) > k \cdot g(x)$ are equivalent as are $f(x) < g(x)$ and $k \cdot f(x) < k \cdot g(x)$ $\tag{5.4a}$

If c is a negative constant, then $f(x) > g(x)$ and $c \cdot f(x) < c \cdot g(x)$ are equivalent as are $f(x) < g(x)$ and $c \cdot f(x) > c \cdot g(x)$ $\tag{5.5a}$

The proofs of these theorems are similar to those of the theorems shown by Eqs. (5.1) and (5.2) and are based on the axioms shown by Eqs.

(2.24) and (2.25). The details of the proofs are left as an exercise for the student.

EXAMPLE 1 Find the solution set of $5x - 9 > 2x + 3$.

Solution

$$5x - 9 > 2x + 3$$ given

$$5x - 9 - 2x + 9 > 2x + 3 - 2x + 9$$ by theorem (5.3), with
                 $h(x) = -2x + 9$

$$3x > 12$$ combining terms

$$x > 4$$ by theorem (5.4a), with $k = \frac{1}{3}$

Hence the solution set is $\{x \mid x > 4\}$.

EXAMPLE 2 Find the solution set of $\frac{1}{6}x - \frac{3}{4} < \frac{3}{8}x + \frac{1}{2}$.

Solution

$$\frac{1}{6}x - \frac{3}{4} < \frac{3}{8}x + \frac{1}{2}$$ given

$$24\left(\frac{1}{6}x - \frac{3}{4}\right) < 24\left(\frac{3}{8}x + \frac{1}{2}\right)$$ by theorem (5.4a), with $k = 24$

$$4x - 18 < 9x + 12$$ by the left-hand distributive theorem

$$4x - 18 - 9x + 18 < 9x + 12 - 9x + 18$$ by theorem (5.3), with
                 $h(x) = -9x + 18$

$$-5x < 30$$ combining terms

$$x > -6$$ by theorem (5.5a), with $c = -\frac{1}{5}$

It should be noticed that except for special cases, the solution set of a conditional linear equation contains only one element, whereas the solution set of a linear inequality may contain an unlimited number of elements. For example,

$$\{x \mid 2x - 1 = x + 3\} = 4$$
$$\{x \mid 2x - 1 > x + 3\} = \{x \mid x > 4\}$$

5.6 / NONLINEAR INEQUALITIES

There are many types of nonlinear inequalities, just as there are many types of nonlinear equations. We shall deal with those nonlinear inequalities whose left members can be factored into linear factors and those that are fractions with the unknown in the denominator and probably also in the numerator. In solving such inequalities, we make use of the fact that the product or quotient of any number of positive numbers is positive, the product or quotient of an even number of negative numbers is positive, and the product or quotient of an odd number of negative numbers is negative. We shall illustrate the procedure to be followed by solving several examples.

EXAMPLE 1 Find the solution set of $(x + 1)/(x - 2) > 0$.

Solution The fraction $(x + 1)/(x - 2)$ is positive, i.e., greater than zero, if both the numerator and denominator are positive and if both are negative. Hence, we seek the set of numbers that satisfies the two inequalities

$$x + 1 > 0 \tag{1}$$

$$x - 2 > 0 \tag{2}$$

simultaneously, and also those which satisfy

$$x + 1 < 0 \tag{3}$$

$$x - 2 < 0 \tag{4}$$

simultaneously.

If we let S_1, S_2, S_3, and S_4 stand for the solution sets of (1), (2), (3), and (4), respectively, then

$$S_1 = \{x \mid x > -1\}$$
$$S_2 = \{x \mid x > 2\}$$
$$S_3 = \{x \mid x < -1\}$$
$$S_4 = \{x \mid x < 2\}$$

The sets S_1, S_2, S_3, and S_4 are indicated on the number line in Fig. 5.1.

Now the set of numbers that satisfies *both* (1) and (2) must belong to S_1 *and* to S_2, and it is therefore the intersection, $S_1 \cap S_2$, of the two sets.

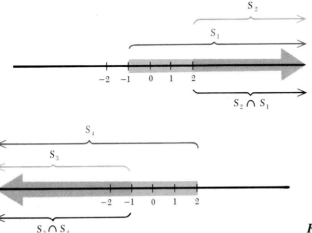

Figure 5.1

From Fig. 5.1 we see that

$$S_1 \cap S_2 = \{x \,|\, x > 2\}$$

Likewise, the set of numbers that satisfies *both* (3) and (4) is the intersection $S_3 \cap S_4$. Again, from Fig. 5.1, we see that

$$S_3 \cap S_4 = \{x \,|\, x < -1\}$$

Therefore, the solution set of

$$\frac{x+1}{x-2} > 0$$

is the set of those x in $S_1 \cap S_2$ and the set of those x in $S_3 \cap S_4$. It is therefore the *union* of the two intersections. Consequently the desired solution set is

$$(S_1 \cap S_2) \cup (S_3 \cap S_4) = \{x \,|\, x > 2\} \cup \{x \,|\, x < -1\}$$

EXAMPLE 2   Find the solution set of

$$\frac{x^2 + x + 2}{x + 1} < 2 + x \tag{5}$$

Solution   We add $-x - 2$ to each member of (5) and obtain the equivalent inequalities

$$\frac{x^2 + x + 2}{x + 1} - x - 2 < 0 \tag{6}$$

$$\frac{x^2 + x + 2 - x^2 - 3x - 2}{x + 1} < 0 \qquad \text{adding expressions in the left member}$$

$$\frac{-2x}{x + 1} < 0 \qquad \text{combining similar terms}$$

Now the fraction on the left is less than zero if

$$-2x > 0 \qquad \text{and} \qquad x + 1 < 0 \tag{7}$$

and also if $-2x < 0 \qquad \text{and} \qquad x + 1 > 0 \tag{8}$

The set of numbers that satisfies the inequalities in (7) simultaneously is

$$\{x \,|\, x < 0\} \cap \{x \,|\, x < -1\} = \{x \,|\, x < -1\}$$

and the set that satisfies the inequalities in (8) is

$$\{x \,|\, x > 0\} \cap \{x \,|\, x > -1\} = \{x \,|\, x > 0\}$$

Hence the solution set of (5) is

$$\{x \,|\, x < -1\} \cup \{x \,|\, x > 0\}$$

EXAMPLE 3   Find the solution set of $(x-1)(x+2)(x-3) < 0$.

Solution   We shall make use of the fact that the product is negative if all three factors are negative or if any one of the factors is negative and the other two are positive. Consequently, the solution set is the union of these four sets. We shall begin by finding the four sets.

If S_1 is the set for which each factor is negative, then

$$S_1 = \{x \mid x-1 < 0\} \cap \{x \mid x+2 < 0\} \cap \{x \mid x-3 < 0\}$$
$$= \{x \mid x < 1\} \cap \{x \mid x < -2\} \cap \{x \mid x < 3\}$$
$$= \{x \mid x < -2\}$$

since, if a number is smaller than the smallest of three numbers, it is smaller than the other two.

If S_2 is the set for which $x-1 < 0$ and the other two factors are positive, then

$$S_2 = \{x \mid x-1 < 0\} \cap \{x \mid x+2 > 0\} \cap \{x \mid x-3 > 0\}$$
$$= \{x \mid x < 1\} \cap \{x \mid x > -2\} \cap \{x \mid x > 3\}$$
$$= \varnothing$$

since x cannot be less than 1 and greater than 3 simultaneously.

If S_3 is the set for which the second factor is negative and the other two factors positive, then

$$S_3 = \{x \mid x-1 > 0\} \cap \{x \mid x+2 < 0\} \cap \{x \mid x-3 > 0\}$$
$$= \{x \mid x > 1\} \cap \{x \mid x < -2\} \cap \{x \mid x > 3\}$$
$$= \varnothing$$

since x cannot be less than -2 and greater than 3 simultaneously.

Finally, if S_4 is the set for which the third factor is negative and the other two positive, then

$$S_4 = \{x \mid x-1 > 0\} \cap \{x \mid x+2 > 0\} \cap \{x \mid x-3 < 0\}$$
$$= \{x \mid x-1 > 0\} \cap \{x \mid x-3 < 0\}$$

since if $x-1 > 0$, it follows that $x+2 > 0$. Therefore,

$$S_4 = \{x \mid x > 1\} \cap \{x \mid x < 3\}$$
$$= \{x \mid 1 < x < 3\}$$

Consequently, the solution set of the given inequality is

$$S = S_1 \cup S_2 \cup S_3 \cup S_4$$
$$= \{x \mid x < -2\} \cup \varnothing \cup \varnothing \cup \{x \mid 1 < x < 3\}$$
$$= \{x \mid x < -2\} \cup \{x \mid 1 < x < 3\}$$

This can be put in words by saying that the solution set consists of all numbers less than -2 and all numbers between 1 and 3.

EXERCISE 5.2 Linear and Nonlinear Inequalities

Find the solution set of each of the following inequalities.

1 $3x > 12$ 2 $5x < 25$ 3 $-2x < 10$ 4 $-7x > -21$

5 $3x - 1 > x + 3$ 6 $6x - 5 > 3x + 4$

7 $7x + 3 > 4x + 6$ 8 $5x - 7 > 3x + 1$

9 $6x - 7 < 3x + 2$ 10 $8x - 9 < 2x + 3$

11 $7x - 20 < 2x + 5$ 12 $9x - 13 < 2x + 8$

13 $3x + 2 > 5x - 4$ 14 $2x + 7 > 5x + 4$

15 $8x + 7 > 6x + 3$ 16 $x + 6 > 3x + 12$

17 $\frac{3}{4}x + \frac{5}{6} < \frac{1}{2}x - \frac{2}{3}$ 18 $\frac{3}{10}x + \frac{7}{20} < \frac{2}{5}x + \frac{1}{4}$

19 $\frac{1}{8}x + \frac{5}{6} < \frac{7}{8}x + \frac{1}{12}$ 20 $\frac{5}{6}x + \frac{3}{4} < \frac{1}{4}x - \frac{2}{3}$

21 $\frac{1}{2}x - \frac{1}{3} > \frac{3}{4}x + \frac{7}{6}$ 22 $\frac{1}{6}x + \frac{1}{3} > \frac{1}{2}x + \frac{2}{3}$

23 $\frac{4}{7}x + \frac{3}{4} > \frac{1}{2}x + \frac{1}{4}$ 24 $\frac{5}{9}x + \frac{5}{6} > \frac{1}{2}x + \frac{1}{2}$

25 $(x - 3)(x + 1) > 0$ 26 $(x - 4)(x - 2) > 0$

27 $(x + 2)(x + 3) > 0$ 28 $(x - 2)(x - 5) > 0$

29 $(2x + 3)(5x - 2) < 0$ 30 $(2x - 7)(7x + 2) < 0$

31 $(5x - 3)(2x + 7) < 0$ 32 $(3x + 5)(2x - 3) < 0$

33 $\dfrac{x + 2}{2x - 1} > 0$ 34 $\dfrac{3x - 5}{x + 1} > 0$

35 $\dfrac{3x - 5}{2x + 6} > 0$ 36 $\dfrac{4x + 7}{2x - 3} > 0$

37 $\dfrac{x^2 - x + 2}{x - 1} < x + 3$ 38 $\dfrac{x^2 + 2x + 3}{x + 1} < x + 2$

39 $\dfrac{2x^2 + 5x - 1}{2x - 1} < x + 2$ 40 $\dfrac{3x^2 + 8x - 2}{3x - 4} < x - 2$

41 $(x + 1)(2x - 1)(3x + 4) > 0$ 42 $(x - 3)(3x + 1)(2x + 3) > 0$

43 $(3x - 7)(7x + 3)(x + 2) > 0$ 44 $(2x + 5)(x - 3)(2x - 3) > 0$

45 $(x + 2)(2x + 1)(3x - 5) < 0$ 46 $(3x - 8)(x + 2)(5x - 1) < 0$

47 $(2x - 7)(7x + 2)(x - 4) < 0$ 48 $(4x - 7)(x + 1)(2x - 7) < 0$

5.7 / FRACTIONAL EQUATIONS

Fractional equation   If at least one algebraic fraction with the variable in the denominator appears in an equation, then the equation is a *fractional equation*. For example,

$$\frac{x}{x + 1} + \frac{5}{8} = \frac{5}{2(x + 1)} - \frac{3}{4}$$

is a fractional equation, but

$$\frac{x + 1}{2} - 5x = 7$$

is not.

We employ the following theorem in solving a fractional equation:

THEOREM *If k(x) is a polynomial, then each root of f(x) = g(x) is also a root of*
$$k(x) \cdot f(x) = k(x) \cdot g(x) \qquad (5.6)$$

Proof If r is a root of $f(x) = g(x)$, then $f(r) = g(r)$. Furthermore $k(r)$ is a constant. Hence by the multiplicativity axiom for equalities, we have $k(r) \cdot f(r) = k(r) \cdot g(r)$. Hence r is a root of $k(x) \cdot f(x) = k(x) \cdot g(x)$.

As an example, consider $f(x) = 3x - 2$, $g(x) = 5x + 8$, and $h(x) = x - 3$. Then $f(x) = g(x)$ becomes

$$3x - 2 = 5x + 8 \qquad (1)$$

and $h(x) \cdot f(x) = h(x) \cdot g(x)$ is

$$(x - 3)(3x - 2) = (x - 3)(5x + 8) \qquad (2)$$

Now if x is replaced by -5, the members of (1) are equal since each is -17, and the members of (2) are both $(-8)(-17)$. Hence -5 is a root of each equation.

The converse of theorem (5.6) is not true as seen from the fact that 3 is not a root of $3x - 2 = 5x + 8$ since $7 \neq 23$, but 3 is a root of $(x - 3)(3x - 2) = (x - 3)(5x + 8)$, since if x is replaced by 3, each member of the latter equation is equal to 0.

If $f(x) = g(x)$ is a fractional equation and $k(x)$ is the lcm of the denominators, then $k(x) \cdot f(x) = k(x) \cdot g(x)$ will contain no fractions. If the latter equation is linear, we can solve it by the methods of Sec. 5.4.

EXAMPLE 1 Solve the equation

$$\frac{x}{x + 1} + \frac{5}{8} = \frac{5}{2(x + 1)} + \frac{3}{4} \qquad (3)$$

Solution We shall employ theorem (5.6) as a first step in solving Eq. (3). Since the lcm of the denominators is $8(x + 1)$, we let $k(x) = 8(x + 1)$; then

$$8(x + 1)\left(\frac{x}{x + 1} + \frac{5}{8}\right) = 8(x + 1)\left[\frac{5}{2(x + 1)} + \frac{3}{4}\right] \qquad (4)$$

$$8x + 5(x + 1) = 4(5) + 6(x + 1) \qquad \text{multiplying by the lcm}$$
$$8x + 5x + 5 = 20 + 6x + 6 \qquad \text{by the right-hand distributive axiom}$$

$$8x + 5x + 5 - 6x - 5 = 20 + 6x + 6 - 6x - 5 \qquad \text{adding } -6x - 5 \text{ to each member}$$

$$7x = 21 \qquad \text{combining similar terms}$$
$$x = 3 \qquad \text{multiplying each member by } \tfrac{1}{7}$$

By theorem (5.6) we know that each root of (3) is a root of (4), but we do not know that the converse is true. Hence, we must replace x by 3 in (3) and see if it is a root. When this is done, we see that each member is equal to $\frac{11}{8}$. Hence the root of (3) is 3, and we have

$$\left\{ x \,\middle|\, \frac{x}{x+1} + \frac{5}{8} = \frac{5}{2(x+1)} + \frac{3}{4} \right\} = \{3\}$$

EXAMPLE 2 Find

$$\left\{ x \,\middle|\, \frac{2}{x+1} - 3 = \frac{4x+6}{x+1} \right\}$$

Solution The required numbers are the roots of

$$\frac{2}{x+1} - 3 = \frac{4x+6}{x+1} \tag{5}$$

We first employ theorem (5.6), multiply each member by $x + 1$, and get

$$2 - 3x - 3 = 4x + 6 \tag{6}$$

$$2 - 3x - 3 - 4x - 2 + 3 = 4x + 6 - 4x - 2 + 3 \qquad \text{by (5.3) with } h(x) =$$
$$\qquad\qquad -4x - 2 + 3$$
$$-7x = 7 \qquad\qquad \text{combining similar terms}$$
$$x = -1 \qquad\qquad \text{multiplying each member}$$
$$\text{by } -\tfrac{1}{7}$$

If we now replace x by -1 in (5), the left member becomes $\frac{2}{0} - 3$, which is not a number. Furthermore, the right member becomes $\frac{2}{0}$, which also is not a number. Hence, since neither member of Eq. (5) is defined when $x = -1$, we cannot accept -1 as a root. Furthermore, by theorem (5.6) each root of Eq. (5) is a root of Eq. (6), and -1 is the only root of Eq. (6). Hence we conclude that Eq. (5) has no roots. Therefore,

$$\left\{ x \,\middle|\, \frac{2}{x+1} - 3 = \frac{4x+6}{x+1} \right\} = \varnothing$$

where $\varnothing$ is the empty set.

EXERCISE 5.3 Fractional Equations

Find the solution set of the equation in each of Probs. 1 to 12.

1 $\quad \dfrac{x+2}{x-1} = \dfrac{x+7}{x+1}$

2 $\quad \dfrac{x+3}{x} = \dfrac{x+8}{x+2}$

3 $\dfrac{3x+4}{2x-1} = \dfrac{6x+1}{4x-3}$

4 $\dfrac{8x+3}{4x+7} = \dfrac{2x-3}{x-7}$

5 $\dfrac{4}{x-1} - \dfrac{3}{x+2} = \dfrac{18}{(x+2)(x-1)}$

6 $\dfrac{5}{x+4} + \dfrac{1}{x-2} = \dfrac{30}{(x+4)(x-2)}$

7 $\dfrac{6}{x+3} - \dfrac{2}{x-1} = \dfrac{8}{(x+3)(x-1)}$

8 $\dfrac{6}{x+6} - \dfrac{2}{x+3} = \dfrac{18}{(x+6)(x+3)}$

9 $\dfrac{5}{x+3} - \dfrac{3}{2x-5} = \dfrac{15}{2x^2+x-15}$

10 $\dfrac{2}{x+2} - \dfrac{5}{2x+1} = \dfrac{-20}{2x^2+5x+2}$

11 $\dfrac{3}{x+1} - \dfrac{4}{3x-4} = \dfrac{2x}{2x^2-x}$

12 $\dfrac{2}{x-5} - \dfrac{7}{2x+5} = \dfrac{21}{2x^2-5x-25}$

13 Find $\left\{ x \middle| \dfrac{2}{x+2} + \dfrac{3}{x+6} = \dfrac{5}{x+4} \right\}$.

14 Find $\left\{ x \middle| \dfrac{2}{x+6} + \dfrac{2}{2x-9} = \dfrac{3}{x-1} \right\}$.

15 Find $\left\{ x \middle| \dfrac{1}{x+5} + \dfrac{3}{3x-1} = \dfrac{2}{x+1} \right\}$.

16 Find $\left\{ x \middle| \dfrac{2}{x+5} + \dfrac{6}{3x+5} = \dfrac{4}{x+3} \right\}$.

17 Find $\left\{ x \middle| \dfrac{3}{3x+7} - \dfrac{1}{2x-2} = \dfrac{.5}{x+9} \right\}$.

18 Find $\left\{ x \middle| \dfrac{6}{3x-4} - \dfrac{1}{x-3} = \dfrac{1}{x+2} \right\}$.

19 Find $\left\{ x \middle| \dfrac{6}{2x-1} - \dfrac{1}{x-3} = \dfrac{2}{x+2} \right\}$.

20 Find $\left\{ x \middle| \dfrac{1}{x+2} + \dfrac{3}{3x-8} = \dfrac{2}{x-1} \right\}$.

Show that the solution set of the equation in each of Probs. 21 to 28 is $\varnothing$.

21 $\dfrac{x-6}{x-7} - 2 = \dfrac{1}{x-7}$

22 $\dfrac{x+2}{x-9} = 5 + \dfrac{11}{x-9}$

23 $\dfrac{2x-3}{x-3} - 1 = \dfrac{3}{x-3}$

24 $\dfrac{3x+1}{x-4} - 2 = \dfrac{13}{x-4}$

25 $\dfrac{5}{x+3} - \dfrac{2}{x-2} = \dfrac{x-12}{x^2+x-6}$

26 $\dfrac{6}{x-5} - \dfrac{1}{x+5} = \dfrac{2x+20}{x^2-25}$

27 $\dfrac{x+2}{x-2} + \dfrac{2x}{x+1} = \dfrac{x+4}{x-2} + \dfrac{2x^2-7x-3}{(x-2)(x+1)}$

28 $\dfrac{x+2}{x+6} + \dfrac{x-1}{x-3} = \dfrac{x+5}{x+6} + \dfrac{x^2+x+9}{(x+6)(x-3)}$

Find the solution set of the equation in each of Probs. 29 to 36, for the letter given at the right of the comma.

29 $C = \dfrac{5}{9}(F-32),\ F$

30 $C = \dfrac{Ak}{4\pi d},\ d$

31 $\dfrac{p}{q} = \dfrac{f}{q-f},\ f$

32 $\dfrac{1}{p} + \dfrac{1}{q} = \dfrac{2}{R},\ q$

33 $m = \dfrac{c(1-p)}{1-d},\ p$

34 $I = \dfrac{Ne}{R+Nr},\ r$

35 $S = \dfrac{a-ar^n}{1-r},\ a$

36 $M = \dfrac{L}{F}\left(\dfrac{25}{f} + 1\right),\ f$

5.8 / EQUALITIES AND INEQUALITIES THAT INVOLVE ABSOLUTE VALUES

In order to solve an equation that involves absolute values, we make use of the definition of $|a|$ which states that $|a| = a$ for $a \geq 0$ and $|a| = -a$ for $a < 0$. We shall illustrate the procedure for solving equations that involve absolute values with three examples.

EXAMPLE 1   Solve $|x| = 4$.

Solution   If we make use of the absolute value of x, the given equation becomes $x = 4$ for $x > 0$ and $-x = 4$ for $x < 0$. Consequently, the solution set is $\{-4, 4\}$.

EXAMPLE 2   Solve $|2x - 1| = 5$.

Solution   By use of the definition of absolute value, the given equation becomes

$$2x - 1 = 5 \quad \text{for} \quad 2x - 1 \geq 0 \tag{2}$$

and

$$-2x + 1 = 5 \quad \text{for} \quad 2x - 1 < 0 \tag{3}$$

From (2), we have $x = 3$ for $x \geq \frac{1}{2}$, and from (3) we get $x = -2$ for $x < \frac{1}{2}$. Consequently, the solution set is $\{-2, 3\}$.

EXAMPLE 3   Solve $|3x - 2| = |x + 2|$.

Solution   We know that the two absolute values are equal if the two numbers are equal and also if they are numerically equal but of opposite sign. We shall solve accordingly.

$$\begin{array}{lcl} 3x - 2 = x + 2 & \quad \text{or} \quad & 3x - 2 = -x - 2 \\ 2x = 4 & & 4x = 0 \\ x = 2 & & x = 0 \end{array}$$

Consequently, the solution set of the given equation is $\{0, 2\}$.

Essentially the same procedure is used in solving inequalities that involve absolute values as in solving equations that involve absolute values. We shall illustrate the procedure after a bit of preliminary discussion.

If we use the definition of the absolute value of a number, we see that an inequality of the type

$$|ax + b| < c \quad c > 0 \tag{4}$$

requires that $ax + b$ be between c and $-c$; hence, if a replacement for x

satisfies *both ax + b < c and ax + b > −c*, it will satisfy (4). Therefore, the solution set of (4) is the intersection of

$$\{x\,|\,ax + b < c\} \qquad \text{and} \qquad \{x\,|\,ax + b > −c\}$$

EXAMPLE 4 Solve $|3x − 4| < 5$.

Solution This inequality is satisfied if x satisfies both $3x − 4 < 5$ and $3x − 4 > −5$. Thus, the solution is $\{x\,|\,3x − 4 < 5\} \cap \{x\,|\,3x − 4 > −5\}$. Now, adding 4 to each member of each of these inequalities and dividing by 3, we see that

$$\{x\,|\,x < 3\} \cap \{x\,|\,x > −1/3\} = \{x\,|\,−1/3 < x < 3\}$$

is the desired solution.

EXAMPLE 5 Solve $|−2x + 7| < 9$.

Solution This inequality is satisfied if both $−2x + 7 < 9$ and $−2x + 7 > −9$ are satisfied; and hence if both $−2x < 2$ and $−2x > −16$ are satisfied. Now, dividing by $−2$ in each inequality and changing the direction in keeping with $(5.5a)$, we find that $\{x\,|\,−1 < x < 8\}$ is the desired solution set.

If we apply the definition of absolute value to

$$|ax + b| > c \qquad c > 0 \tag{5}$$

we find that (5) is satisfied if *either ax + b > c or ax + b < −c*. Consequently, the solution set of (5) is the union of

$$\{x\,|\,ax + b > c\} \qquad \text{and} \qquad \{x\,|\,ax + b < −c\}$$

EXAMPLE 6 Solve $|3x + 2| > 4$.

Solution If we apply the definition of the absolute value of a number to the given inequality, we see that it is satisfied if either $3x + 2 > 4$ or $3x + 2 < −4$; now solving this pair of inequalities, we find that $x > \frac{2}{3}$ and $x < −2$. Therefore, the desired solution set is

$$\{x\,|\,x > \tfrac{2}{3}\} \cup \{x\,|\,x < −2\}$$

EXAMPLE 7 Solve $|3x − 2| > |x + 2|$.

Solution We must consider the same three cases as we did in Example 3.
If both $3x − 2$ and $x + 2$ are nonnegative, then (6) becomes $3x − 2 >$

$x + 2$, that is, $x > 2$ for $\{x \mid 3x - 2 \geq 0\} \cap \{x \mid x + 2 \geq 0\}$; hence, $x > 2$ for $\{x \mid x \geq \frac{2}{3}\} \cap \{x \mid x \geq -2\}$. Therefore $x > 2$ is a part of the solution set.

If $3x - 2 \geq 0$ and $x + 2 < 0$, then (6) becomes $3x - 2 > -x - 2$, that is, $x > 0$ for $\{x \mid 3x - 2 \geq 0\} \cap \{x \mid x + 2 < 0\} = \{x \mid x \geq \frac{2}{3}\} \cap \{x \mid x < -2\} = \emptyset$. Therefore, nothing is added to the solution set.

If $3x - 2 < 0$ and $x + 2 \geq 0$, then (6) becomes $-3x + 2 > x + 2$, that is, $x < 0$ for $\{x \mid x < \frac{2}{3}\} \cap \{x \mid x \geq -2\} = \{x \mid -2 \leq x < \frac{2}{3}\}$; hence, $x < 0$ is added to the solution set.

Consequently, the solution set is $\{x \mid x > 2\} \cup \{x \mid x < 0\}$.

EXERCISE 5.4 Equations and Inequalities that Involve Absolute Values

Find the solution set of each of the following equations and inequalities.

1	$\lvert x \rvert = 2$	2	$\lvert x \rvert = 0$	3	$\lvert -x \rvert = 3$	4	$\lvert -x \rvert = 5$
5	$\lvert x - 1 \rvert = 1$	6	$\lvert x + 3 \rvert = 7$	7	$\lvert x + 4 \rvert = 6$	8	$\lvert x - 2 \rvert = 3$
9	$\lvert 2x - 1 \rvert = 5$	10	$\lvert 3x + 4 \rvert = 10$	11	$\lvert 5x + 3 \rvert = 12$	12	$\lvert 2x - 3 \rvert = 3$

13	$\lvert -2x + 7 \rvert = 3$	14	$\lvert -3x + 1 \rvert = 7$	15	$\lvert -4x - 5 \rvert = 5$
16	$\lvert -6x + 7 \rvert = 13$	17	$\lvert x + 2 \rvert = \lvert x - 1 \rvert$	18	$\lvert x - 3 \rvert = \lvert -x + 1 \rvert$
19	$\lvert x - 3 \rvert = \lvert -x + 3 \rvert$	20	$\lvert x + 4 \rvert = \lvert x - 6 \rvert$	21	$\lvert 2x - 5 \rvert = \lvert x + 4 \rvert$
22	$\lvert 3x + 7 \rvert = \lvert 2x - 3 \rvert$	23	$\lvert 4x + 3 \rvert = \lvert 3x + 4 \rvert$	24	$\lvert 5x - 2 \rvert = \lvert 3x + 4 \rvert$
25	$\lvert x \rvert < 2$	26	$\lvert x \rvert < 5$	27	$\lvert x \rvert > 3$
28	$\lvert x \rvert > 7$	29	$\lvert x + 3 \rvert < 4$	30	$\lvert x - 2 \rvert < 6$
31	$\lvert x - 5 \rvert < 3$	32	$\lvert x + 7 \rvert < 5$	33	$\lvert 2x - 1 \rvert > 5$
34	$\lvert 3x - 5 \rvert > 1$	35	$\lvert 3x + 4 \rvert > 2$	36	$\lvert 5x - 7 \rvert > -3$
37	$\lvert -x + 2 \rvert < 3$	38	$\lvert -2x - 1 \rvert < 5$	39	$\lvert -3x + 5 \rvert > 2$
40	$\lvert -5x + 3 \rvert > 2$	41	$\lvert x + 2 \rvert < \lvert x - 4 \rvert$	42	$\lvert x - 1 \rvert < \lvert x - 3 \rvert$
43	$\lvert 2x - 3 \rvert < \lvert x + 1 \rvert$	44	$\lvert 3x + 2 \rvert < \lvert 2x + 3 \rvert$	45	$\lvert 3x - 4 \rvert > \lvert x + 6 \rvert$
46	$\lvert 2x - 1 \rvert > \lvert -2x + 1 \rvert$	47	$\lvert 4x + 1 \rvert > \lvert 2x - 3 \rvert$	48	$\lvert 5x - 2 \rvert > \lvert 3x + 4 \rvert$

5.9 / SOLVING STATED PROBLEMS

A stated problem is a description of a situation that involves both known and unknown quantities and that also involves certain relations between these quantities. If the problem is solvable by means of one equation, it must be possible to find two combinations of the quantities in the problem that are equal. Furthermore, at least one of the combinations must involve the variable.

The process of solving a stated problem by means of an equation is not always simple, and considerable practice is necessary before one becomes adept at problem solving.   The following approach is suggested:

Solving stated problems   1   Read the problem carefully and study it until the situation is thoroughly understood.

2   Identify the quantities, both known and unknown, that are involved in the problem.

3   Select one of the unknowns and represent it by a letter, usually x, and then express the other unknowns in terms of this letter.

4   Search the problem for the information that tells which quantities, or what combinations of them, are equal.

5   When the desired combinations are found, set them equal to each other, thus obtaining an equation.

6   Solve the equation thus obtained and check the number in the solution set in the problem.

It is usually helpful to tabulate the data given in the problem, as is done in the following illustrative examples.

We shall present below several examples of the various types of problems that can be solved by means of equations.   The general procedure explained in the examples should be applied to similar problems that occur in the next exercise and in other exercises that involve stated problems.

PROBLEMS INVOLVING MOTION AT A UNIFORM VELOCITY

Problems that involve motion usually state a relation between the distances traveled, between the velocities (or speeds), or between the periods of time involved.   The fundamental formula for use in solving such problems is

$$d = vt$$

where d represents the distance, v represents the velocity (or speed), and t represents the period of time.   When this formula is used, d and v must be expressed in terms of the same linear unit and v and t must be expressed in the same unit of time.   The formula can be solved for v and t in order to get the two additional formulas

$$v = \frac{d}{t} \quad \text{and} \quad t = \frac{d}{v}$$

EXAMPLE 1   A party of hunters made a trip of 380 miles to a hunting lodge in 7 hours. They traveled 4 hours on a paved highway and the remainder of the time on a pasture road.   If the average velocity through the pasture was 25 mi/ hour less than that on the highway, find the average velocity and the distance traveled on each part of the trip.

Solution   We shall first tabulate the data:

	Time, hours	Velocity, mi/hour	Distance, mi
On paved highway	4	x	$4x$
On pasture road	$7 - 4 = 3$	$x - 25$	$3(x - 25)$
Total	7		380

Quantities that are equal:

$$\text{(Distance on highway)} + \text{(distance through pasture)} = 380$$
$$4x \qquad + \qquad 3(x - 25) \qquad = 380$$

In the above table, the unknown quantities are expressed in terms of x and are printed in color, and these unknown quantities are the two velocities and the distance on each part of the trip.   The known quantities are 380 mi, the total distance; 7 hours, the total time; 4 hours, the time spent on the highway; and 25 mi/hour, the amount by which the velocity on the highway exceeds that through the pasture.   The time spent on the pasture road was 7 hours − 4 hours = 3 hours, and the total distance is equal to the sum of the distances traveled on each of the two parts.

If we let

$$x = \text{the speed on the highway, in miles per hour}$$

then $x - 25 = $ the speed through the pasture

Furthermore, $4x = $ the distance traveled on the highway, in miles
$$3(x - 25) = \text{the distance traveled through the pasture}$$

Hence, $4x + 3(x - 25) = 380$, the total distance in miles

This is the desired equation, and we solve it below.

$4x + 3x - 75 = 380$	by the distributive axiom
$4x + 3x = 380 + 75$	adding 75 to each member
$7x = 455$	combining terms
$x = 65$	multiplying each member by $\frac{1}{7}$

Thus 65 mi/hour is the velocity on the highway, and 40 mi/hour is the velocity through the pasture since $65 - 25 = 40$.

$$4 \times 65 = 260 \text{ mi traveled on the highway}$$
$$3 \times 40 = 120 \text{ mi traveled through the pasture}$$

Check $260 + 120 = 380.$

EXAMPLE 2 Three airports, A, B, and C, are located on a north-south line. B is 645 mi north of A, and C is 540 mi north of B. A pilot flew from A to B, delayed 2 hours, and continued to C. The wind was blowing from the south at 15 mi/hour during the first part of the trip, but during the delay it changed to the north with a velocity of 20 mi/hour. If each flight required the same period of time, find the airspeed (i.e., the speed delivered by the propeller) of the plane.

Solution We proceed as follows:

Let $\qquad x =$ the airspeed, in miles per hour

Then

$\qquad x + 15 =$ the speed of the plane from A to B, in miles per hour

and

$\qquad x - 20 =$ the speed of the plane from B to C, in miles per hour

Furthermore, $\quad \dfrac{645}{x + 15} =$ number of hours required for the first flight

$\dfrac{540}{x - 20} =$ number of hours required for the second flight

Hence, since these two periods of time are equal, we have

$$\frac{645}{x + 15} = \frac{540}{x - 20}$$

This is the required equation, and we solve it as follows:

$$(x - 20)(x + 15)\frac{645}{x + 15} = (x - 20)(x + 15)\frac{540}{x - 20}$$

multiplying each member by the lcm of the denominators

$$(x - 20)645 = (x + 15)540$$

performing the indicated multiplication

$$645x - 12{,}900 = 540x + 8{,}100$$

by the distributive axiom

$$645x - 12{,}900 - 540x + 12{,}900 = 540x + 8{,}100 - 540x + 12{,}900$$

adding $-540x + 12{,}900$ to each member

$$105x = 21{,}000 \qquad \text{combining terms}$$

$$x = 200 \qquad \text{multiplying each member by } \tfrac{1}{105}$$

Check $x + 15 = 200 + 15 = 215$, and this is the velocity of the plane during the first flight; $\frac{645}{215} = 3$, so the first flight required 3 hours. Furthermore, $x - 20 = 200 - 20 = 180$ and $\frac{540}{180} = 3$, so the second flight required 3 hours. Consequently, the airspeed of 200 mi/hour satisfies the conditions of the problem.

WORK PROBLEMS

Problems that involve the rate of doing certain things can often be solved by first finding the fractional part of the task done by each individual, or by each agent, in one unit of time and then finding a relation between these fractional parts. If this method is used, the unit 1 represents the entire job that is to be done.

EXAMPLE 3   A farmer can plow a field in 4 days by using a tractor. His hired hand can plow the same field in 6 days by using a smaller tractor. How many days will be required for the plowing if they work together?

Solution   We let

$x =$ the number of days required to plow the field if they work together

Then $\dfrac{1}{x} =$ the part of the field plowed in 1 day by the two

Furthermore,

$\frac{1}{4} =$ the part of the field plowed in 1 day by the farmer
$\frac{1}{6} =$ the part of the field plowed in 1 day by the hired hand

Consequently, $$\frac{1}{4} + \frac{1}{6} = \frac{1}{x}$$

This is the required equation, and we solve it as follows:

$$12x\left(\frac{1}{4} + \frac{1}{6}\right) = 12x\frac{1}{x} \qquad \text{multiplying each member by the lcm, } 12x, \text{ of the denominators}$$

$$3x + 2x = 12 \qquad \text{by the distributive axiom}$$

$$5x = 12$$

$$x = 2\tfrac{2}{5} \qquad \text{days}$$

Check   If they plow the field in $2\frac{2}{5}$ days, then they complete $1/2\frac{2}{5} = \frac{5}{12}$ of it in 1 day. Furthermore, one plows one-sixth of it in 1 day, and the other plows one-fourth; thus:

$$\frac{1}{6} + \frac{1}{4} = \frac{2 + 3}{12} = \frac{5}{12}$$

EXAMPLE 4 If, in Example 3, the hired hand worked 1 day with the smaller machine and then was joined by his employer with the larger one, how many days were required for them to finish the plowing?

Solution The hired hand plowed one-sixth of the field in 1 day, and hence five-sixths of it remained unplowed.
We let

$$x = \text{the number of days required for the two to finish the job}$$

Then $\dfrac{x}{4} =$ the part plowed by the farmer, since he plows one-fourth of the field in 1 day

$\dfrac{x}{6} =$ the part plowed by the hired hand

Therefore, $\dfrac{x}{4} + \dfrac{x}{6} = \dfrac{5}{6}$

This is the required equation, and it is solved as follows:

$$12\left(\frac{x}{4} + \frac{x}{6}\right) = 12 \times \tfrac{5}{6} \qquad \text{multiplying each member by the lcm of the denominators}$$

$$3x + 2x = 10 \qquad \text{by the distributive axiom}$$

$$5x = 10$$

$$x = 2$$

Check In 2 days, the farmer plowed $\tfrac{2}{4} = \tfrac{1}{2}$ of the field and the hired hand plowed $\tfrac{2}{6} = \tfrac{1}{3}$ of it, and

$$\frac{1}{2} + \frac{1}{3} = \frac{3+2}{6} = \frac{5}{6}$$

MIXTURE PROBLEMS

Many problems involve the combination of certain substances of known strengths, usually expressed in percentages, into a mixture of required strength in one of the substances. Others involve the mixing of certain commodities of specified prices. In such problems, it should be remembered that the total amount of any given element in a mixture is equal to the sum of the amounts of that element in the substances combined and that the value of any mixture is the sum of the values of the substances that are put together.

EXAMPLE 5 How many gallons of a liquid that is 74 percent alcohol must be combined with 5 gal of one that is 90 percent alcohol in order to obtain a mixture that is 84 percent alcohol?

Solution   If we let x represent the number of gallons of the first liquid and remember that 74 percent of x is $0.74x$, then the table below, showing the data in the problem, is self-explanatory.

Since (the number of gallons of alcohol in the mixture) = (the number of gallons in the first liquid) + (the number of gallons in the second liquid), we have

$$0.74x + 4.5 = 0.84(x + 5)$$

the solution to which is given directly below the table.

	Number of gallons	Percentage of alcohol	Number of gallons of alcohol
First liquid	x	74	$0.74x$
Second liquid	5	90	$0.90 \times 5 = 4.5$
Mixture	$x + 5$	84	$0.84(x + 5)$

$$0.74x + 4.5 = 0.84x + 4.2 \qquad \text{by the distributive axiom}$$

$$0.74x + 4.5 - 4.5 - 0.84x = 0.84x + 4.2 - 4.5 - 0.84x \qquad \text{adding } -4.5 - 0.84x \text{ to each member}$$

$$-0.10x = -0.3 \qquad \text{combining terms}$$

$$x = 3 \qquad \text{multiplying each member by } -10$$

Hence the required number of gallons of the first mixture is 3.

Check
$$(0.74 \times 3) + 4.5 = 2.22 + 4.5 = 6.72$$
$$0.84 \times (3 + 5) = 0.84 \times 8 = 6.72$$

MISCELLANEOUS PROBLEMS

In addition to the three types of problems discussed above, there is a wide variety of problems that can be solved by means of equations. The fundamental approach to all of them is the same. It consists of finding two equal quantities, one or both of which involve the unknown. We shall discuss three other kinds of problems, giving the general principle or formula to be used in solving each.

Many problems in physics and mechanics involve the lever. A lever is a rigid bar supported at a point, called the *fulcrum*, that is usually between the two ends of the bar. If two weights W_1 and W_2 at distances L_1 and L_2, respectively, from the fulcrum are balanced on a lever, then

$$W_1L_1 = W_2L_2$$

Furthermore, if a force F at a distance D from the fulcrum will just raise a weight R that is a distance d from the fulcrum, then

$$FD = Rd$$

In solving problems dealing with investments, the formula usually employed is

$$I = Prt$$

where P is the principal, or sum invested; I is the interest earned on the investment; r, expressed as a percent, is the rate of interest, or earning per unit of time; and t is the total time the principal is invested.

Problems involving the digits in a number depend upon the place value of our number system. For example, if h is the hundreds digit in a three-place number, t is the tens digit, and u is the units digit, then $100h + 10t + u$ is the number. If the hundreds digit and the units digit are interchanged, then the number is $100u + 10t + h$.

EXERCISE 5.5 **Stated Problems**

1 Find three consecutive integers whose sum is 75.

2 One fall a family spent $112 outfitting their two children for school. If the clothes for the older child cost $1\frac{1}{3}$ the cost of those for the younger, how much did they spend for each child?

3 According to the 1970 census, the population of Mattville was 41,209. If this population was 5,015 less than twice the population of Mattville in the 1960 census, what was the population increase in the 10 years?

4 Mr. Lee jogged a total of 6,600 yd in three nights. If each night he increased his distance 440 yd, how far did he jog on the first night?

5 The Gardner family spent $625 buying a band instrument for each of their two children. If one instrument cost $195 more than the other, how much did each instrument cost?

6 The winning candidate for president of the freshman class received 2,898 votes. If that was 210 more than half the votes cast, how many freshmen voted?

7 Fred noticed he had worked one-third of the problems in his math assignment and that when he had worked eight more problems he would be halfway through the assignment. How many problems were in the assignment?

8 John agreed to work on his uncle's ranch 3 months one summer for $650 and a used car. At the end of 2 months he was needed at home,

so his uncle paid him $200 and the car. What was the value of the car?

9   Scott did yard work for two families, one of whom paid him $1.50/hour while the other paid him $1.25/hour. If he earned $20.75 in 15 hours, how long did he work at each rate?

10   Todd, Tim, and Rick worked a total of 18 hours clearing weeds from a vacant lot so they could use it as a baseball diamond. If Todd and Tim worked a total of 12 hours and Rick worked 1 hour longer than Tim, how long did each boy work?

11   A family making a long car trip drove one-fourth of the distance the first day, 575 miles on the second day, and 125 miles more than the first day on the third day. How long was the trip?

12   The associated student body collected $1,399 from selling 879 tickets to a campus dance. If tickets cost $1.75 each with a $.025 reduction to those students having an associated student body card, how many of the students had associated student body cards?

13   On a trip Ralph noticed that his car averaged 21 mi/gal of gas except for the days he used the air conditioning, and then it averaged only 17 mi/gal. If he used 91 gal of gas to drive 1,751 mi, on how many of those miles did he use the air conditioning?

14   A portion of $31,750 was invested at 5 percent, and the remainder at 6 percent. If the total income from the money is $1,750, how much was invested at each rate?

15   The Thompsons spent $1,488 on carpeting for their new home. The carpeting used in the living room and the hall cost $13/sq yd, and that used in the bedrooms cost $10/sq yd. If the bedroom area used 20 sq yd more than the living room, how much did the Thompsons spend on each type of carpet?

16   At the beginning of the summer, Brian and Bruce each earned $13.20/day from their summer jobs. After a time, Brian was assigned more responsibility and then earned $14.80/day. If the boys each worked 65 days and together earned a total of $1,764, how long did Brian work at the higher rate?

17   The owner of a duplex collected $4,100 in 1 year from renting his two apartments. Find the rent charged for each if one rented for $25/month more than the other and if the more expensive one was vacant for 2 months.

18   A mountain resort that featured skiing in the winter was partially staffed by college students in the summer. One summer there were three times as many students employed as there were year-round employees. When September came, 40 of the students went back to

school and 30 other people were hired for the winter. If there were then twice as many nonstudents as students, how many people staffed the resort in the winter?

19 Fred is 3 years older than his sister Mary. In 7 years she will be six-sevenths of his age. How old are they?

20 The petty-cash drawer of a small office contained $16.25. If there were twice as many nickels as quarters and as many dimes as nickels and quarters combined, how many coins of each type were there?

21 A man who drove 30 mi to work each day picked up a friend on the way. If he was able to average 40 mi/hour on the trip and drove 15 min longer with the friend than without him, how far did he live from the friend's house?

22 A boy rode his motorbike 20 min to a friend's home, and then the two drove in a car 30 min to a beach 35 mi from the first boy's home. If the car speed was 10 mi/hour faster than that of the motorbike, how fast did the car travel?

23 A bicycle club left the campus to ride to a park 24 mi away for an outing. One member of the club left from the same place by car with picnic supplies $1\frac{1}{2}$ hours later, traveled at a speed four times as fast, and arrived at the park at the same time as the cyclists. How fast did he drive?

24 A large plane left an airport a few seconds ahead of a small private plane that followed the same flight plan during the first hour of its flight. The speed of the large plane was five times that of the small one, and at the end of the hour, it was 500 mi ahead. What was the speed of the large plane?

25 A group of horseback riders left a stable for a ride to a mountain lookout. On the way up, they averaged 3 mi/hour. On the way back along the same trail, they averaged $5\frac{1}{4}$ mi/hour, and the return trip took $\frac{5}{7}$ hour less time than the outward trip. How long was the entire horseback ride?

26 A student left his college town on a bus that traveled 60 mi/hour. Three hours later, his father left home at a speed of 50 mi/hour to meet the bus. If the college was 345 mi from the student's home and the father met the bus as it arrived at the station, how long did the father drive to meet the bus?

27 Two brothers took turns washing the family car on weekends. John could usually wash the car in 45 min, whereas Jim took 30 min for the job. One weekend they were in a hurry to go to a football game, so they worked together. How long did it take them?

28 Mrs. James spent 1 hour addressing one-third of the family Christmas cards, and Mr. James spent $1\frac{1}{4}$ hours addressing another third. Continuing the addressing together, how long did they take to finish the rest of the cards?

29   A three-man maintenance crew could clean a certain building in 4 hours, whereas a four-man crew could do the job in 3 hours. If one man of the four-man crew was an hour late, how long did the job take?

30   Three secretaries worked together typing a group of form letters. Miss Jones could have done them alone in 2 hours, Miss Brown in 3 hours, and Miss Smith in 2 hours. How long did the secretaries need to type the letters working together?

31   A chemical mixing tank can be filled by two hoses. One requires 42 min to fill the tank, and the other 30 min. If both hoses are used, how much time is needed to fill the tank?

32   Jean, Carol, and Linda were on a committee to compile and staple the pages of their club newsletter. Each girl could have done the job alone in 4 hours. Jean started at 3:30 P.M., Carol came at 3:45, and Linda joined the work at 4. What time did they finish?

33   Dave, Joe, and Mike were assigned the job of cataloging the music for their school band. Dave could have completed the assignment in 2 hours alone, Joe in 3 hours, and Mike in 4 hours. The boys started work together, but Joe left at the end of a half-hour and Mike at the end of an hour. How long did Dave work alone to finish the cataloging?

34   A swimming pool can be filled in 6 hours and requires 9 hours to drain. If the drain was accidently left open for 6 hours while the pool was being filled, how long did filling the pool require?

35   The intake pipe to a reservoir is controlled by an automatic valve that closes when the reservoir is full and opens again when three-fourths of the water has been drained. The intake pipe can fill the reservoir in 6 hours, and the outlet can drain it in 16 hours. If the outlet is open continuously, how much time elapses between the two instants the reservoir is full?

36   How many pounds of chocolates costing $0.80/lb may be mixed with 6 lb of chocolates costing $1/lb to produce a mixture that can be sold for $0.85/lb?

37   Phil paid $21.43 for a collection of seven records. Some were on sale for $3.94, and some for $1.89. How many of each type did he buy?

38   A contractor mixed two batches of concrete that were 9.3 and 11.3 percent cement to obtain 4,500 lb of concrete that was 10.8 percent cement. How many pounds of each type of concrete was used?

39   The specifications for a shipment of gravel set as a minimum standard

that 85 percent of the gravel should pass through a screen of a certain size. One load of 6 cu yd tested at only 65 percent. How much gravel testing at 90 percent must be added to the 65 percent load to make it acceptable?

40 Only 5 percent of the area of a city could be developed into parks, whereas in the unincorporated area outside the city limits 25 percent of the area could be developed into parks. If the city covered an area of 300 sq mi, how many square miles of suburbs had to be annexed so the city could develop 10 percent of its area as parks?

41 A chemist mixed 40 milliliters of 8 percent hydrochloric acid with 60 milliliters of 12 percent hydrochloric acid solution. He used a portion of this solution and replaced it with distilled water. If the new solution tested 5.2 percent hydrochloric acid, how much of the original mixture did he use?

42 A small plane was scheduled to fly from Los Angeles to San Francisco. The flight was against a head wind of 10 mi/hour. Threat of mechanical failure forced the plane to turn back, and it returned to Los Angeles with a tail wind of 10 mi/hour, landing $1\frac{1}{2}$ hours after it had taken off. If the plane had a uniform airspeed of 150 mi/hour, how far had it gone before turning back?

43 A group of tourists took a sightseeing bus trip of 240 mi and then boarded a plane which took them to their next stop 550 mi away. The average speed of the plane was 16.5 times that of the bus, and their travel time was 6 hours and 50 min. Find the average bus speed and the average plane speed.

44 A boy rode his bicycle 15 mi with a tail wind of 8 mi/hour, but in the same time rode only 3 mi of the return trip with the same wind now against him. How fast would he have traveled with no wind?

45 A rancher drove 40 mi/hour on a gravel road to the main highway, on which he traveled at 60 mi/hour until he reached a city that was 110 mi from his home. If his trip took 2 hours, how far was his ranch from the main highway?

46 A businessman traveled 870 mi to attend a company conference. He drove his car 30 mi to an airport and flew the rest of the way. If his plane speed was 12 times that of the car and he flew 48 min longer than he drove, how long did he fly?

47 A man invested $2,400 in the common stock of one company and $1,280 in the stock of another. The price per share of the second was four-fifths the price per share of the first. The next day the price of the more expensive stock advanced $1.50/share, the price of the other declined $0.75/share, and as a result, the value of his investment increased $31. Find the price per share of the more expensive stock.

48   Mrs. Johnson planned to spend $78 for fabric to make draperies. She found her fabric on sale at 20 percent less per yard than she expected and was able to buy her drapery fabric plus 4 extra yards for a bed-spread for $83.20. How much fabric had she planned to buy, and what was the original cost per yard?

EXERCISE 5.6 REVIEW

Find the solution set of the equation or inequality in each of Probs. 1 to 32.

1   $2x + 5 = 3x + 2$

2   $3x - 4 = 8x + 6$

3   $3(x - 3) = 5(x - 1) - 8$

4   $3(2x - 3) = 4(x + 1) - 3$

5   $\frac{2}{3}x + 1 = \frac{5}{6}x$

6   $\frac{i}{8}x - 2 = \frac{3}{4}x - 1$

7   $5(\frac{1}{3}x + \frac{2}{5}) = 2(\frac{4}{5}x + \frac{3}{2})$

8   $2(\frac{11}{16}x - \frac{5}{18}) = 3(\frac{4}{3} + \frac{7}{4}x)$

9   $\dfrac{2x + 11}{5} = \dfrac{x + 17}{10} + x + 4$

10 $\dfrac{3x - 4}{4} = \dfrac{x - 6}{3} + x - 4$

11 $3x - 5 > x + 1$

12 $5x + 2 > 8x - 4$

13 $4x + 3 < x + 6$

14 $7x - 9 < 2x + 11$

15 $(2x - 1)(3x + 7) < 0$

16 $(5x - 8)(3x + 1) > 0$

17 $\dfrac{5x + 9}{6x - 5} > 0$

18 $\dfrac{3x - 7}{7x + 3} < 0$

19 $(x + 2)(x - 4)(2x + 9) < 0$

20 $(3x + 8)(2x - 5)(5x + 3) > 0$

21 $\dfrac{x + 2}{x - 1} = \dfrac{x + 8}{x + 2}$

22 $\dfrac{3x + 6}{2x - 3} = \dfrac{3x + 1}{2x - 4}$

23 $\dfrac{10}{x - 1} - \dfrac{7}{x + 1} = \dfrac{35}{x^2 - 1}$

24 $\dfrac{9}{x - 3} - \dfrac{4}{x - 6} = \dfrac{18}{x^2 - 9x + 18}$

25 $|x + 3| = 4$

26 $|-2x + 1| = 7$

27 $|x + 2| = |2x + 1|$

28 $|5x - 4| = |2x + 8|$

29 $|x + 3| < 4$

30 $|-2x + 1| > 7$

31 $|x + 2| > |2x + 1|$

32 $|5x - 4| < |2x + 11|$

33 Solve $F = 32 + 9C/5$ for C.

34 Solve $S = \dfrac{n}{2}(a + l)$ for l.

35 The correct formula for converting from Centigrade to Fahrenheit temperature is $F = 1.8C + 32$, but a much easier mental calculation is obtained by use of $f = 2C + 30$. Show that $|F - f| < 5°$ if $5° < F < 95°$.

CHAPTER SIX *Functions, Relations, and Graphs*

In Chap. 2 we discussed the concept of one-to-one correspondence, which related each element of one set of numbers to one and only one element of another, and we used the concept in the discussion of the set of natural numbers. In this chapter we shall discuss sets of pairs of related numbers and shall explain concepts and methods that will be used extensively in the remainder of this book.

6.1 / ORDERED PAIRS OF NUMBERS

If a man earns $50/day and works 5 days, then the following tabulation shows the correspondence between the number of days worked and the number of dollars earned:

Days worked	1	2	3	4	5
Dollars earned	50	100	150	200	250

This correspondence can also be indicated by the following sets of pairs:

$$\{(1, 50), (2, 100), (3, 150), (4, 200), (5, 250)\} \tag{1}$$

Ordered pair   Each pair of numbers in (1) is an *ordered pair* and illustrates the following definition:

A pair of numbers (x, y) is an *ordered pair* if the interpretation of each number depends on its position in the pair.

An ordered pair can also be defined in this way:

If S and T are two sets of numbers, then (x, y) is an ordered pair if $x \in S$ and $y \in T$.

Note that the first number in each pair in (1) belongs to $\{1, 2, 3, 4, 5\}$ and the second belongs to $\{50, 100, 150, 200, 250\}$.

The two numbers in an ordered pair are called the *components of the pair.*

Equality of   Two ordered pairs (a, b) and (c, d) are equal if and only if $a = c$ and
ordered pairs $b = d$.

6.2 / RELATIONS

We define a relation as follows:

If D is a set of numbers and if there exists a rule that associates each element x of D with one or more numbers y, and if R is the set of all numbers y, then the set of ordered pairs (x, y) such that $x \in D$ and $y \in R$ is
Relation   called a *relation* with *domain D* and *range R*. Hence, $D \times R$ is a relation.

The rule that establishes the correspondence between the elements of D and R may be an equation in two variables, a table of statistics composed of two columns, tables such as A.1 to A.4 in the Appendix of this text, or any agreement that clearly pairs each number in one set with a number in another.

EXAMPLE 1   Write the relation established by the equation $x - y^2 = 0$, where x belongs to $D = \{0, 1, 4, 9, 16\}$.

Solution   We first solve the given equation for y^2 and get

$$y^2 = x$$

Then $y = \pm\sqrt{x}$ since $(\pm\sqrt{x})^2 = x$. Now we assign the number 0 to x, get $y = 0$, and we have the ordered pair $(0, 0)$. Similarly, if $x = 1$, $y = \pm 1$, and we get the ordered pairs $(1, 1)$ and $(1, -1)$. We continue this procedure by assigning each of the other numbers in D to x, calculating each corresponding value of y, and then arranging the resulting ordered pairs so that the second components appear in order of magnitude. Thus we obtain the relation

$$\{(16, -4), (9, -3), (4, -2), (1, -1), (0, 0), (1, 1), (4, 2), (9, 3), (16, 4)\}$$

Figure 6.1

This relation can be expressed in a more compact form as

$$\{(x, y) \mid x - y^2 = 0 \text{ and } x \in \{0, 1, 4, 9, 16\}\}$$

Note that the range of the relation is $\{-4, -3, -2, -1, 0, 1, 2, 3, 4\}$.

EXAMPLE 2   By use of Table A.3 in the Appendix, write the relation

$$\{(x, \sqrt[3]{x}) \mid x \in \{-3, -2, -1, 0, 1, 2, 3\}\}$$

where the second component in each pair is correct to four digits.

Solution   By referring to Table A.3 we see that $\sqrt[3]{3}$ correct to four digits is 1.442. Hence $\sqrt[3]{-3}$ is approximately equal to -1.442, and we therefore have the ordered pairs $(-3, -1.442)$ and $(3, 1.442)$. Using a similar procedure for the other numbers in D, we obtain

$$\{(x, \sqrt[3]{x}) \mid x \in \{-3, -2, -1, 0, 1, 2, 3\}\} =$$
$$\{(-3, -1.442), (-2, -1.260), (-1, -1), (0, 0), (1, 1), (2, 1.260), (3, 1.442)\}$$

EXAMPLE 3   If $D = \{1, 2, 3, 4\}$ and $R = \{5, 6, 7, 8, 9\}$, write the relation obtained by matching the nth number in D with both the nth number in R and the $(n + 1)$th number in R.

Solution   The components of each ordered pair in the required relation are the numbers at the ends of the arrows in the diagram in Fig. 6.1. Hence the relation is

$$\{(1, 5), (1, 6), (2, 6), (2, 7), (3, 7), (3, 8), (4, 8), (4, 9)\}$$

6.3 / FUNCTIONS

By referring to Examples 1 and 3 of Sec. 6.2, we see that couples of ordered pairs appear that have equal first components but different second components. For example, in Example 1, we have $(16, -4)$ and $(16, 4)$; $(9, -3)$ and $(9, 3)$; $(4, -2)$ and $(4, 2)$; and $(1, -1)$ and $(1, 1)$. In Example 2, however, no two pairs have the same first component and different second components. That relation is an example of a function. Functions are of utmost importance in mathematics, and we shall discuss some of their basic properties in the remainder of this section. We start with the following definition:

Function

If D is a set of numbers, if there exists a rule such that for each element x of D exactly one number y is determined, and if R is the set of all such numbers y, then the set of ordered pairs $\{(x, y) \mid x \in D \text{ and } y \in R\}$ is a *function* with the *domain* D and *range* R.

As in the case of relations, the rule that establishes the correspondence may be an equation in two variables or any other agreement or device that associates each element of D with only one element of R. In this chapter we shall confine our discussion to functions in which the rule is an algebraic equation that is usually in the form $y = f(x)$.

EXAMPLE 1   Find the set of ordered pairs $\{(x, y)\}$ if $y = x^2 - 2x - 3$ and $D = \{x \mid x$ is an integer and $1 \leq x \leq 4\}$.

Solution   We first note that $D = \{1, 2, 3, 4\}$; furthermore, the corresponding number pairs (x, y) are

$$x = 1, \; y = 1^2 - (2 \times 1) - 3 = 1 - 2 - 3 = -4$$
$$x = 2, \; y = 2^2 - (2 \times 2) - 3 = 4 - 4 - 3 = -3$$
$$x = 3, \; y = 3^2 - (2 \times 3) - 3 = 9 - 6 - 3 = 0$$
$$x = 4, \; y = 4^2 - (2 \times 4) - 3 = 16 - 8 - 3 = 5$$

Hence $\{(x, y)\} = \{(1, -4), (2, -3), (3, 0), (4, 5)\}$.

Independent variable

Dependent variable

In the function $\{(x, y) \mid y = f(x)\}$ defined by $y = f(x)$, x is called the *independent* variable and y is the *dependent* variable. In other words, the independent variable in a function is the variable whose replacement set is the set of first numbers in the ordered pairs in the function, and the dependent variable is the variable whose replacement set is the set of second numbers in these ordered pairs. If the domain of the function is not specified, it is understood to be the set of real numbers.

EXAMPLE 2 $\{(x, y) \mid x^2 + y^2 = 1\}$ is a relation but not a function since there are two values of y for some (actually all but 0) values of x.

6.4 / FUNCTIONAL NOTATION

It has been customary for some time to designate the function defined by $y = f(x)$ by the letter f, and either of the following notations may be used for this purpose:

$$f = \{(x, y) \mid y = f(x)\}$$
$$f = \{(x, f(x))\}$$

The second expression for f illustrates the fact that it is not necessary to introduce the letter y for the dependent variable, since $f(x)$ stands for the second number in the ordered pair with the first number x. Thus in the function $f = \{(x, f(x))\}$, f designates the function; $f(x)$ is the value of f for a specified replacement for x and is called a *function value*. For example, if $y = 3x^2 - 2x + 4$, we may designate the function f in either of the following ways:

Function value

$$f = \{(x, y) \mid y = 3x^2 - 2x + 4\}$$
$$f = \{(x, f(x)) \mid f(x) = 3x^2 - 2x + 4\}$$
$$f = \{(x, 3x^2 - 2x + 4)\}$$

Furthermore, the value of f for $x = 5$ is $f(5) = (3 \times 5^2) - (2 \times 5) + 4 = 75 - 10 + 4 = 69$.

If more than one function is involved in a particular discussion, it is customary to designate one of them by f and the others by letters other than f. Thus the functions defined by $y = h(x)$, $y = g(x)$, and $y = k(x)$ are $h = \{(x, h(x))\}$, $g = \{(x, g(x))\}$, and $k = \{(x, k(x))\}$, respectively.

EXAMPLE 1 If $f(x) = (x-2)/(x+1)$, find the function values $f(2)$ and $f(\frac{1}{2})$.

Solution

$$f(2) = \frac{2-2}{2+1} = \frac{0}{3} = 0 \qquad f(\tfrac{1}{2}) = \frac{\frac{1}{2}-2}{\frac{1}{2}+1} = \frac{2(\frac{1}{2}-2)}{2(\frac{1}{2}+1)}$$

$$= \frac{1-4}{1+2} = -1$$

EXAMPLE 2 If $f(x) = x^2 - x - 3$, $g(x) = (x^2 - 1)/(x+2)$, and $F(x) = f(x) + g(x)$, find $F(2)$.

Solution

$$F(2) = f(2) + g(2)$$

$$= 2^2 - 2 - 3 + \frac{2^2 - 1}{2 + 2}$$

$$= 4 - 2 - 3 + \tfrac{3}{4}$$

$$= -1 + \tfrac{3}{4}$$

$$= -\tfrac{1}{4}$$

EXAMPLE 3 If $D = \{x \mid x \text{ is an integer and } -2 \le x \le 1\}$, find the function $\{[x, f(x)] \mid f(x) = x^3 - 3 \text{ and } x \in D\}$.

Solution

$$D = \{-2, -1, 0, 1\}$$
$$f(-2) = (-2)^3 - 3 = -8 - 3 = -11$$
$$f(-1) = (-1)^3 - 3 = -1 - 3 = -4$$
$$f(0) = 0^3 - 3 = -3$$
$$f(1) = 1^3 - 3 = -2$$

Consequently,

$$f = \{(x, f(x)) \mid f(x) = x^3 - 1, x \text{ is an integer, and } -2 \leq x \leq 1\}$$
$$= \{(-2, -11), (-1, -4), (0, -3), (1, -2)\}$$

EXAMPLE 4   If $f(x) = 3x + 4$ and $D = \{x \mid -1 \leq x \leq 3\}$, find the range R of $f(x)$.

Solution   We shall first prove that the function value of $3x + 4$ increases as x increases. If $X > x$, then

$$3X > 3x \qquad \text{by Eq. (2.25)}$$

and it follows that

$$3X + 4 > 3x + 4 \qquad \text{by Eq. (2.24)}$$

Consequently, if x belongs to D, the function value $f(x) = 3x + 4$ is least when $x = -1$ and greatest when $x = 3$. Hence, since $f(-1) = -3 + 4 = 1$ and $f(3) = 9 + 4 = 13$, $R = \{y \mid 1 \leq y \leq 13\}$.

Occasionally, the independent variable in an equation in two variables is expressed in terms of a third variable. For example, suppose $y = f(x) = x^2 + 1$ and $x = g(t) = t - 1$. We can express y in terms of t by replacing x in $f(x)$ by $g(t) = t - 1$. If this is done, we obtain

$$y = F(t) = f[g(t)] = (t - 1)^2 + 1 = t^2 - 2t + 2$$

Now if $D = \{0, 1, 2, 3, 4\}$ and $t \in D$, we have the three functions

$$g = \{(t, x) \mid x = g(t) = t - 1, t \in D\}$$
$$= \{(0, -1), (1, 0), (2, 1), (3, 2), (4, 3)\}$$

$$f = \{(x, y) \mid y = f(x) = x^2 + 1, x \in \{-1, 0, 1, 2, 3\}\}$$
$$= \{(-1, 2), (0, 1), (1, 2), (2, 5), (3, 10)\}$$

$$F = \{(t, y) \mid y = F(t) = f[g(t)] = t^2 - 2t + 2, t \in D\}$$
$$= \{(0, 2), (1, 1), (2, 2), (3, 5), (4, 10)\}$$

Note that the domain of g and F is D, the domain of f is the range of g, and the range of F is the range of f. This example illustrates the following definition:

If $f = \{(x, y) \mid y = f(x)\}$, $g = \{(t, x) \mid x = g(t)\}$, and $F = \{(t, y) \mid y = F(t) = f[g(t)]\}$, then F is a composite function whose components are f and g.

Composite function F is often represented by $f \circ g$. In order to be able to find F under the above conditions, we must have x in the domain of g and $g(x)$ in the domain of f. Consequently, the domain of $f \circ g$ consists of those elements x in the domain of g such that $g(x)$ belongs to the domain of f.

EXAMPLE 5 If $y = f(x) = (x^2 - 2)/(x^2 + 4)$ and $x = g(t) = t + 1$, obtain the equation $y = F(t) = f[g(t)]$.

Solution

$$y = F(t) = f[g(t)]$$
$$= \frac{(t+1)^2 - 2}{(t+1)^2 + 4} = \frac{t^2 + 2t - 1}{t^2 + 2t + 5}$$

EXERCISE 6.1 Relations, Functions, Functional Notation

1 Find the set of ordered pairs obtained by pairing the $(n+1)$st element of $S = \{1, 2, 3, 4, 5\}$ with the nth element of $T = \{1, 3, 5, 7\}$.

2 If $S = \{3, 5, 7, 9\}$ and $T = \{2, 4, 6, 8\}$, find the set of ordered pairs obtained by pairing the nth element of S with the $(5-n)$th element of T.

3 Find the set of ordered pairs obtained by pairing the nth element of $S = \{m, a, t, h\}$ with the $(5 - n)$th element of $T = \{l, o, m, l\}$.

4 Write out the set of ordered pairs obtained by pairing each of $1, 3, 5, 4$, 8, and 12 with the number that is diametrically opposite it on the face of a clock.

5 Is the set of ordered pairs $\{(x, x^2) \mid x = 1, 2, 3, 5\}$ a function or merely a relation? Why?

6 Is the set of ordered pairs $\{(x, y) \mid y^2 = x, x \in \{0, 1, 9, 25\}\}$ a function or merely a relation? Why?

7 Is the set of ordered pairs $\{(x, y) \mid y^2 = 2x - 1, x \in \{1, 5, 13\}\}$ a function or merely a relation? Why?

8 Is the set of ordered pairs $\{(x, y) \mid y = x^2 - 1, x \in R\}$ a function or merely a relation? Why?

Write out each set of ordered pairs described in Probs. 9 to 16, and state whether it is or is not a function.

9 $\{(x, 3x - 2) \mid x$ is an integer and $2 \leq x \leq 6\}$
10 $\{(x, x^2 - 1) \mid x$ is an integer and $-2 \leq x \leq 2\}$
11 $\{(x, y) \mid y = x^2 - 3x, x \in \{3, 4, 5, 6\}\}$
12 $\{(x, y^2) \mid y = 2x - 1, x$ is an integer and $-1 \leq x \leq 3\}$
13 $\{(x, y) \mid y^2 = x - 3, x$ is an integer and $3 \leq x \leq 6\}$
14 $\{(x, y) \mid 2y^2 = x + 1, x \in \{1, 7, 15, 31\}\}$
15 $\{(x, y) \mid y^2 = 16 - x^2, x \in \{0, \sqrt{7}, 4\}\}$
16 $\{(x, y^2) \mid y^2 = 16 - x^2, x \in \{0, \sqrt{7}, 4\}\}$

17   If $f(x) = 3x - 1$, find $f(-2)$ and $f(1)$.
18   If $g(x) = x - 3$, find $g(0)$ and $g(5)$.
19   If $h(x) = 2x + 5$, find $h(-3)$ and $h(2)$.
20   If $x(h) = 4h + 1$, find $x(-1)$ and $x(3)$.
21   If $s(x) = 2x^2 - 3x + 1$, find $s(3)$ and $s(b)$.
22   If $q(x) = x^2 + x - 3$, find $q(a)$ and $q(3)$.
23   If $p(q) = q^2 + q + 2$, find $p(-1)$ and $p(b)$.
24   If $r(u) = -3u^2 + 2u - 1$, find $r(0)$ and $r(x)$.
25   If $f(x) = 3x + 1$, find $f(2)$, $f(h)$, and $f(2 + h)$.
26   If $f(x) = 2x - 3$, find $f(3)$, $f(h)$, and $f(3 + h)$.
27   If $g(x) = x^2 - x + 1$, find $g(h)$, $g(1)$, and $g(1 + h)$.
28   If $g(x) = 2x^2 + 3x - 4$, find $g(4)$, $g(h)$, and $g(4 + h)$.
29   If $h(x) = 3x + 4$, find $h(x + 1) - h(x)$.
30   If $f(x) = x^2 - x + 2$, find $f(x + h) - f(x)$.
31   If $s(x) = 2x^2 + 3x + 1$, find $s(x + h) - s(x)$.
32   If $t(x) = 4x - 3$, find $t(x + h) - t(x)$.
33   If $f(x) = x + 1$ and $x = g(t) = t^2 - 1$, find $F(t) = f[g(t)]$ and $F(4)$.
34   If $f(x) = 3x + 9$, and $x = g(t) = 2t - 3$, find $F(t) = f[g(t)]$ and $F(0)$.
35   If $f(x) = x^2 - 2x + 2$, and $x = g(t) = t + 1$, find $F(t) = f[g(t)]$ and $F(1)$.
36   If $f(x) = 2x^2 - 12x + 18$ and $x = g(t) = t + 3$, find $F(t) = f[g(t)]$ and $F(1)$.

Find the set of ordered number pairs that make up $F = \{(t, F(t)) \mid F(t) = f[g(t)]\}$ in each of Probs. 37 to 40.

37   $x = g(t) = 2t - 3$, $f(x) = 3x + 1$, $x \in D$, $D = \{1, 3, 5, 7\}$
38   $x = g(t) = t - 13$, $f(x) = 5x + 2$, $x \in D$, $D = \{-2, 0, 2, 4\}$
39   $x = g(t) = 2t + 1$, $f(x) = x^2 - 4x + 1$, $x \in D$, $D = \{-1, 0, 1, 2\}$
40   $x = g(t) = t + 2$, $f(x) = x^2 + x - 1$, $x \in D$, $D = \{1, 4, 9, 16\}$

41   Find the range of $f = \{(x, 2x - 1)\}$ if $D = \{1, 2, 3, 5\}$.
42   Find the range of $f = \{(x, y) \mid y = 3x - 2\}$ if $D = \{0, 2, 3, 6\}$.
43   Find the range of $f = \{(x, y^2) \mid y = x + 3\}$ if $D = \{-2, -1, 0, 1, 2\}$.
44   Find the range of $f = \{(x, x^2 + 2)\}$ if $D = \{x \mid x$ is an integer strictly between -1 and $4\}$.

Find $f \cap g$ in each of Probs. 45 to 48.

45   $f = \{(x, x - 1)\}$, $g = \{(x, 2x - 3)\}$, $D = \{0, 2, 4, 7\}$
46   $f = \{(x, 2x - 5)\}$, $g = \{(x, 2x + 1)\}$, $D = R$
47   $f = \{(x, 2x^2 - 4x - 1)\}$, $g = \{(x, x - 4)\}$, $D = \{1, 1.5, 2, 3\}$
48   $f = \{(x, 2x^2 - 5x - 3)\}$, $g = \{(x, x^2 - 2x - 5)\}$, $D = \{0, 1, 2, 3\}$

6.5 / THE RECTANGULAR COORDINATE SYSTEM

In this section we shall introduce a device for associating an ordered pair of numbers with a point in a plane. Invented by the French mathematician and philosopher René Descartes (1596–1650), it is called the *rectangular* or *cartesian* coordinate system.

In order to set up this system, we construct two perpendicular number lines in the plane and choose a suitable scale on each. For convenience these lines are horizontal and vertical, and the unit length on each is the same (see Fig. 6.2), although neither restriction is necessary. The two lines are called the *coordinate axes,* the horizontal line being the *X* axis and the vertical line the *Y* axis. The intersection of the two lines is the *origin,* designated by the letter *O.* The coordinate axes divide the plane into four sections called *quadrants.* These quadrants are numbered I, II, III, and IV counterclockwise, as indicated in Fig. 6.2*a.*

Next it is agreed that horizontal distances measured to the right from the *Y* axis are positive, and horizontal distances measured to the left are negative. Similarly, vertical distances measured upward from the *X* axis are positive, and vertical distances measured downward are negative. These distances, because of their signs, are called *directed distances.* Finally, we agree that the first number in an ordered pair of numbers represents the directed distance from the *Y* axis to a point and the second number in the pair represents the directed distance from the *X* axis to the point. It follows then that an ordered pair of numbers uniquely determines the position of a point in the plane. For example, (4, 1) determines the point that is 4 units to the right of the *Y* axis and 1 unit above the *X* axis. This point is designated by *Q* in Fig. 6.2*b.* Similarly, the ordered pair (−5, −1) determines the point *S* in Fig. 6.2*b* that is 5 units to the left of the *Y* axis and 1 unit below the *X* axis. Conversely, each point in the plane determines a unique ordered pair of numbers. For example, the point *P* in Fig. 6.2*b* is

Coordinate axes

X axis, Y axis, Origin

Directed distance

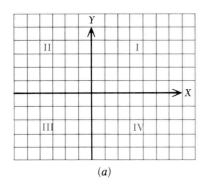

(a)

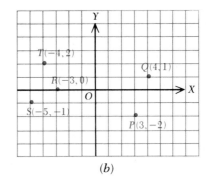

(b)

Figure 6.2

Cartesian plane

3 units to the right of the Y axis and 2 units below the X axis, and so P determines the ordered pair $(3, -2)$. A plane in which the coordinate axes have been constructed is called a *cartesian plane.*

Coordinates

Abscissa

Ordinate

The two numbers in an ordered pair that is associated with a point in the cartesian plane are called the *coordinates* of the point. The first number is called the *abscissa* of the point, and it is the directed distance from the Y axis to the point. The second number in the pair is the *ordinate* of the point, and it represents the directed distance from the X axis to the point.

Plotting a point

The procedure for locating a point in the plane by means of its coordinates is called *plotting* the point. The notation $P(a, b)$ means that P is the point whose coordinates are (a, b). In order to plot the point $T(-4, 2)$, we count 4 units to the left of the origin on the X axis and then upward 2 units and thus arrive at the point. Similarly, the point $R(-3, 0)$ is 3 units to the left of the origin and on the X axis. The general point and its coordinates are written $P(x, y)$.

6.6 / THE GRAPH OF A FUNCTION

By use of the rectangular coordinate system, we can obtain a geometric representation, or a geometric "picture," of a function. In order to obtain this representation, we require that each ordered pair of numbers (x, y) of a function be the coordinates of a point in the cartesian plane, with x as the abscissa and y as the ordinate. Now we define the graph of a function as follows:

Graph of a function

The graph of a function is the totality of points (x, y) whose coordinates constitute the set of ordered pairs of the function with x a number in the domain D and y the corresponding number in the range R.

The graphs of most of the functions that we shall discuss in this chapter are smooth continuous curves.† When we say that the graph of a function is a curve, what we mean is that the point determined by each ordered pair of numbers in the function is on the curve and that the coordinates of each point on the curve are an ordered pair of numbers in the function.

We shall illustrate the procedure for obtaining the graph of a function by explaining the steps in the construction of the graph of the function defined by

$$y = x^2 - 2x - 2 \tag{1}$$

The first step is to assign several values to x and then calculate each corre-

---

† We are not in position to give a rigorous definition of a smooth continuous curve. For our present purposes, however, the following *description* will suffice: A smooth continuous curve contains no breaks or gaps, and there are no sudden or abrupt changes in its direction.

sponding value of y. Before doing this, however, it is advisable to make a table like the one below in which to record the corresponding values of x and y.

x	
y	

The values selected for x, in most cases, should be small, usually integers less than 10. In this case, we start with the integer 0 and then successively assign the integers 1, 2, 3, 4, -1, and -2 to x. These integers were selected arbitrarily and, as we shall see, will be used to determine a portion of the graph. We next calculate the value of y corresponding to each of the selected integers by replacing x in (1) by each of them. Thus we get the corresponding number pairs

$$x = 0,\ y = 0^2 - 2(0) - 2 = -2$$
$$x = 1,\ y = 1^2 - 2(1) - 2 = -3$$
$$x = 2,\ y = 2^2 - 2(2) - 2 = -2$$

and so on until all the integers listed above have been used. When a number is assigned to x, it should be recorded in the table and the corresponding value of y entered below it. The numbers assigned to x should be entered in order of magnitude from left to right. When the value of y for each of the values of x has been calculated and the results entered in the table, we have

x	-2	-1	0	1	2	3	4
y	6	1	-2	-3	-2	1	6

Now we plot the points (x, y) thus determined, as shown in Fig. 6.3, and connect them with a smooth curve. This curve is a portion of the graph defined by $y = x^2 - 2x - 2$, and it indicates the following facts about the behavior of the function:

1   The least value of y is -3 and occurs if $x = 1$.

2   The value of y increases rapidly as x increases or decreases from 1.

3   The value of y is zero when x is equal to the abscissas of the points where the graph crosses the X axis or at $x = 2.7$ (approximately) and at -0.7 (approximately).

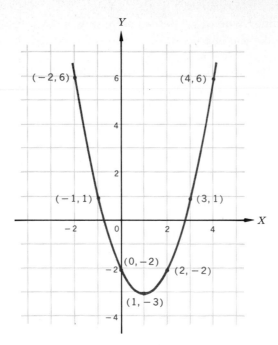

Figure 6.3

Zero of a function

    A *zero* of the function defined by $y = f(x)$ is the value of the independent variable x for which $y = 0$. In other words, if the second number in one of the ordered pairs in a function is zero, then the first number in the pair is a zero of the function. For example, the zero of the function defined by $f(x) = 2x - 5$ is $2\frac{1}{2}$ since $(2 \times 2\frac{1}{2}) - 5 = 0$. The zeros of many classes of functions can be determined by algebraic methods, but we must depend upon graphical methods for others. In the latter method, we construct the graph of the function and estimate the abscissa of each point where the graph crosses the X axis. This estimate can be made as accurate as necessary by choosing a sufficiently large scale.

    If either the domain or the range or both is the entire set of real numbers, then the graph of the function is unlimited in extent and only a portion of the graph can be constructed. In many cases, the domain and range are limited by the equation that defines the function. For example, the domain of the function determined by $y = \sqrt{25 - x^2}$ is $D = \{x \mid -5 \leq x \leq 5\}$, and the range is $R = \{y \mid 0 \leq y \leq 5\}$. In this chapter we shall be concerned with portions of graphs that are in regions relatively near the origin. Consequently, in assigning values to the independent variable, we generally start with 0, then assign several consecutive positive integral values, and next assign several negative integral values. We must continue to assign values to x until we have a sufficient number of points to determine the general trend of the curve.

In some cases, the equation that defines a function is such that the points obtained by assigning consecutive integers to x are so far apart that we cannot tell how to draw the curve that connects them. For example, if we assign $-1, 0$, and 1 to x in $y = 9x^2 - 1$, we obtain the values $8, -1$, and 8 for y and thus have the points $(-1, 8)$, $(0, -1)$, and $(1, 8)$, or the points A, B, and C in Fig. 6.4. These points are not sufficient to determine the shape of the curve. If, however, we assign the numbers $-\frac{2}{3}, -\frac{1}{3}, \frac{1}{3}$, and $\frac{2}{3}$ to x, calculate each corresponding value of y, and plot the points thus determined, we obtain the additional points D, E, F, and G. Using these, together with A, B, and C, we can draw the curve.

On the other hand, an equation $y = f(x)$ that determines a function may be of such nature that the points determined by assigning consecutive integers to x are clustered so closely together in the plane that only a small portion of the curve is determined. In such cases, we assign values to x that are more widely separated. For example, if in the equation

$$y = \frac{x^2}{100}$$

we assign the consecutive integers from 0 to 10, inclusive, to x, the corresponding values of y are less than or equal to 1, and the points determined by the ordered pairs of numbers thus obtained give us no idea of the nature of the curve. If, however, we assign the numbers 0, 10, 20, 30, and 40 to x, we obtain the points $(0, 0)$, $(10, 1)$, $(20, 4)$, $(30, 9)$, and $(40, 16)$, and these points determine the portion of the curve shown in Fig. 6.5.

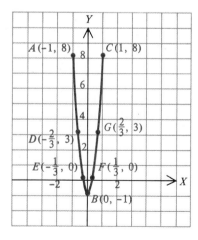

Figure 6.4

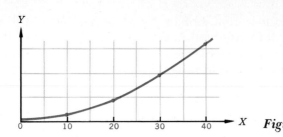

Figure 6.5

6.7 / LINEAR FUNCTIONS

Linear function

It is proved in analytic geometry that the graph of the function defined by $y = ax + b$ is a straight line, and for this reason the function $\{(x, y) \mid y = ax + b\}$ is called a *linear function*. Furthermore, it is proved in analytic geometry that the graph of the function defined by $y = \sqrt{r^2 - x^2}$ is the upper half of a circle of radius r with the center at the origin.

Since the graph of a linear function is a straight line, it is fully determined by two points whose coordinates are ordered pairs in the function. If the graph does not pass through the origin and is not parallel to either axis, the coordinates of the points where it crosses the coordinate axes are readily determined. The abscissa of the point of intersection of the graph

x intercept
y intercept

and the X axis is called the *x intercept*, and the ordinate of the point of intersection of the graph and the Y axis is the *y intercept*. We obtain the x intercept by replacing y by 0 in $y = ax + b$ and solving for x. Similarly, we obtain the y intercept by replacing x by 0 and solving for y. Consequently, we con-

Procedure for constructing the graph of a linear function

struct the graph of the function defined by $y = ax + b$ by finding the two intercepts, plotting the points that correspond to the two intercepts, and then drawing the straight line through the two points thus determined. As a check, it is advisable to find a third point whose coordinates satisfy the equation. If this point is not on the line, then an error has been made.

EXAMPLE 1   Construct the graph of the function defined by $y = 3x - 9$.

Solution   We find the intercepts by assigning 0 to x and solving for y and by assigning 0 to y and solving for x. Then we find the coordinates of a third point by assigning 4 to x and solving for y. Thus we obtain the following table of corresponding values:

x	0	3	4
y	−9	0	3

We now plot the points determined by these pairs of corresponding numbers, draw a straight line through them, and thus obtain the graph in Fig. 6.6.

If $b = 0$ in $y = ax + b$, the graph of the function passes through the origin, so each intercept is zero. Consequently, we must obtain the coordinates of points other than those of the intersections with the axes in order to determine the graph. We illustrate this situation in Example 2.

EXAMPLE 2   Construct the graph of the function defined by $y = 3x$.

Solution   We assign the numbers -3, 0, and 3 to x, calculate each corresponding value of y, and thus get the following table:

x	-3	0	3
y	-9	0	9

We now plot the points determined by the pairs of corresponding numbers, draw a straight line through them, and thus get the graph shown in Fig. 6.7. Notice that this graph is just the graph in Fig. 6.6 moved up 9 units.

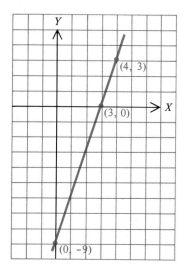

Figure 6.6

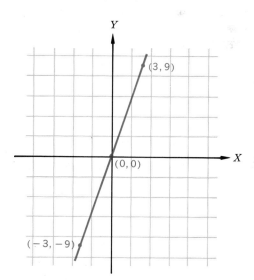

Figure 6.7

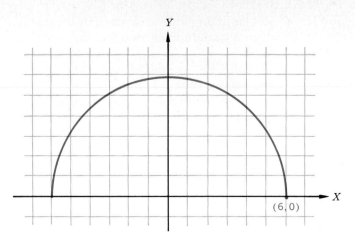

Figure 6.8

EXAMPLE 3 Construct the graph of the function defined by $y = \sqrt{36 - x^2}$.

Solution The graph of the function defined by this equation is the semicircle above the X axis with its center at $(0, 0)$ and with radius equal to 6. It is shown in Fig. 6.8.

EXERCISE 6.2 Graphs

1 Plot the points determined by the following ordered pairs of numbers: $(2, 3)$, $(3, -1)$, $(-2, 4)$, $(0, 5)$, $(-2, -3)$, $(-4, 2)$, $(-3, -5)$, $(-3, 0)$. $(5, 1)$, $(0, -3)$.

2 Plot the points $P(x, y)$ if $y = 3x$ and $x \in \{-2, -1, 2, 4\}$.

3 Plot the points $P(x, y)$ if $y = -x$ and $x \in \{-1, 0, 2, 3\}$.

4 Plot the points $P(x, y)$ if $y = -2x$ and $x \in \{-3, 1, 2, 5\}$.

Describe the lines or rays determined by the conditions in each of Probs. 5 to 12.

5 The ordinate of each point is zero.

6 The two coordinates of each point are equal and positive.

7 The two coordinates of each point are numerically equal but of opposite sign.

8 The abscissa of each point is 3.

9 The ray on which $P(x, y)$ lies if $x = 1$ and $y \geq 2$

10 The ray on which $P(x, y)$ lies if $x \leq 1$ and $y = 3$

11 The ray on which $P(x, y)$ lies if $x = -2$ and $y \leq 0$

12 The ray on which $P(x, y)$ lies if $y = -4$ and $x \geq 2$

Construct the graph of the function defined by the equation in each of Probs. 13 to 36.

13 $y = x + 1$	14 $y = x - 3$	15 $y = -x + 4$
16 $y = -x - 2$	17 $y = -2x + 3$	18 $y = -3x - 7$
19 $y = 4x - 1$	20 $y = 2x + 4$	21 $y = x^2$
22 $y = x^2 + 3$	23 $y = x^2 - x$	24 $y = x^2 + 3x - 2$
25 $y = 2x^2 + x$	26 $y = 2x^2 + 1$	27 $y = 3x^2 - 3$
28 $y = 3x^2 - 2x$	29 $y = 2x^2 - 4x - 1$	30 $y = 3x^2 + 4x - 1$
31 $y = 2x^2 + 5x + 3$	32 $y = 4x^2 - 5x + 2$	33 $y = -x^2 - 2x + 1$
34 $y = -3x^2 + 2x - 5$	35 $y = -x^2 - 2x - 1$	36 $y = -2x^2 + 5x + 1$

In each of Probs. 37 to 44, find the intercepts of the indicated function and then construct the graph for the specified domain.

37   $\{(x, y) \mid y = 3x - 4\}, D = \{-3 \le x \le 5\}$

38   $\{(x, y) \mid y = -2x - 3\}, D = \{-4 \le x \le 4\}$

39   $\{(x, y) \mid y = -3x + 5\}, D = \{-2 \le x \le 6\}$

40   $\{(x, y) \mid y = 4x + 7, D = \{-5 \le x \le 3\}$

41   $\{(x, y) \mid y = \sqrt{36 - x^2}\}, D = \{x \mid -6 \le x \le 6\}$

42   $\{(x, y) \mid y = \sqrt{9 - x^2}\}, D = \{x \mid 0 \le x \le 3\}$

43   $\{(x, y) \mid y = -\sqrt{16 - x^2}\}, D = \{x \mid -4 \le x \le 0\}$

44   $\{(x, y) \mid y = -\sqrt{49 - x^2}\}, D = \{x \mid -7 \le x \le 7\}$

6.8 / SPECIAL FUNCTIONS

In this section we shall discuss examples of functions whose graphs are not continuous curves. As a first example, we shall consider the function

$$f = \begin{cases} f(x) = x + 1 & \text{if } x \le 2 \\ f(x) = -x + 4 & \text{if } x > 2 \end{cases}$$

This function is defined by the equations

$$y = x + 1 \qquad \text{if } x \le 2 \qquad (1)$$

$$y = -x + 4 \qquad \text{if } x > 2 \qquad (2)$$

As we pointed out in Sec. 6.6, the graph of $y = x + 1$ is a straight line, and the x and y intercepts of this line are -1 and 1, respectively. Furthermore, if $x = 2$, $y = 3$, so the line passes through the point $(2, 3)$. Since, however, the domain of x in this case is $D = \{x \mid x \le 2\}$, the graph does not extend to the right of the point $(2, 3)$. Hence the graph of the function defined by (1) is a ray whose right extremity is $(2, 3)$ and which extends

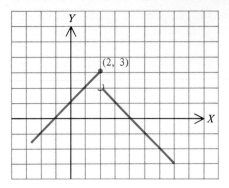

Figure 6.9

downward to the left through the points $(0, 1)$ and $(-1, 0)$, as indicated in Fig. 6.9.

We now consider (2). The graph of the function defined by $y = -x + 4$ is the straight line that passes through $(2, 2)$ and $(4, 0)$ since if $x = 2, y = 2$, and if $x = 4, y = 0$. If, however, we impose the restriction $x > 2$, the point $(2, 2)$ is not on the graph of $y = -x + 4$ and the graph does not extend to the left of $(2, 2)$. Hence the graph consists of all points on the half line through $(4, 0)$ in Fig. 6.9 with the exception of the point $(2, 2)$.

Bracket function   An important function in the theory of numbers is the bracket function $[x]$. By the notation $[x]$, we mean the *greatest integer* that is less than or equal to x. Hence in the equation

$$y = [x] \tag{3}$$

if $x = \frac{1}{2}$, then $y = [\frac{1}{2}] = 0$ since 0 is the greatest integer less than $\frac{1}{2}$. Similarly, if $x = 2\frac{1}{2}$, $y = [2\frac{1}{2}] = 2$ since 2 is the greatest integer less than $2\frac{1}{2}$. Since $[x]$ is the greatest integer less than or *equal* to x, then $[n] = n$ if n is an integer. In general, if x is in the interval $0 \leq x < 1$, then $y = 0$; and for x such that $1 \leq x < 2$, $y = 1$. Hence, by the above and by similar arguments, we have the following corresponding values of x and y:

$$0 \leq x < 1, y = 0$$
$$1 \leq x < 2, y = 1$$
$$2 \leq x < 3, y = 2$$
$$\cdots \cdots \cdots \cdots$$
$$n \leq x < n + 1, y = n$$

Consequently, the function defined by (3) is a set of horizontal line segments, four of which are shown in Fig. 6.10.

The cartesian product of two sets of integers is a relation whose graph is a set of isolated points. For example, if $T = \{0, 1, 2, 3, 4\}$ and $S = \{1, 2, 3, 4, 5\}$, then the cartesian product

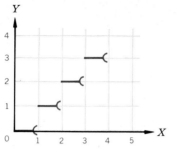

Figure 6.10

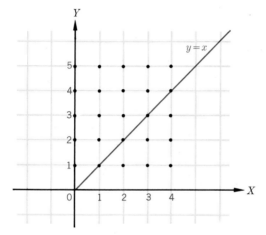

Figure 6.11

$T \times S = \{(0, 1), (0, 2), (0, 3), (0, 4), (0, 5), (1, 1), (1, 2), (1, 3), (1, 4),$
$(1, 5), (2, 1), (2, 2), (2, 3), (2, 4), (2, 5), (3, 1), (3, 2), (3, 3),$
$(3, 4), (3, 5), (4, 1), (4, 2), (4, 3), (4, 4), (4, 5)\}$

The graph of this relation is the set of points shown in Fig. 6.11.

The graph of $\{(x, y) \mid (x, y) \in T \times S, \text{ and } y = x\}$ is the set of points on the graph of $y = x$ in the figure.

The graphs of $\{(x, y) \mid (x, y) \in T \times S, \text{ and } y > x\}$ and of $\{(x, y) \mid (x, y) \in T \times S, \text{ and } y < x\}$ are, respectively, the set of points above the line $y = x$ in the figure and the set of points below the line.

6.9 / INVERSE FUNCTIONS

We studied relations in Sec. 6.2 and functions in Sec. 6.3 and shall now consider the two functions

$$f = \{(1, 2), (3, 4), (5, 6)\} \qquad \text{and} \qquad g = \{(1, 2), (3, 4), (5, 4)\}$$

Both have $\{1, 3, 5\}$ as the domain, whereas the range of f is $\{2, 4, 6\}$ and that of g is $\{2, 4\}$.

If we interchange the first and second elements in f, we obtain

$$F = \{(2, 1), (4, 3), (6, 5)\}$$

and they constitute a function with domain $\{2, 4, 6\}$ and range $\{1, 3, 5\}$. If, however, we interchange the first and second elements in g, we get

$$\{(2, 1), (4, 3), (4, 5)\}$$

and this is a relation but not a function since two pairs have the same first element 4 and different second elements 3 and 5.

The set of ordered pairs obtained by interchanging the elements in each ordered pair of a function is called the *inverse relation*. Furthermore, if the inverse relation is a function, it is called the *inverse function*. If the function is designated by f, then we use f^{-1} to indicate the inverse. Consequently, if $f = \{(2, 3), (4, 5), (6, 7)\}$, then $f^{-1} = \{(3, 2), (5, 4), (7, 6)\}$.

It follows from the definition of a function that *if a function f is such that whenever two of its ordered pairs with different first elements also have different second elements, then the inverse function f^{-1} exists.* The inverse function is obtained by interchanging first and second elements of each ordered pair in f.

If we designate a function by

$$f = \{(x, y) \,|\, y = f(x)\}$$

then $y = f(x)$ defines the function, and $x = f(y)$ defines its inverse. If the latter equation is solvable for y, the solution is ordinarily put in the form $y = f^{-1}(x)$, and we designate the inverse relation by

$$f^{-1} = \{(x, y) \,|\, y = f^{-1}(x)\}$$

and it is the inverse function provided $y = f^{-1}(x)$ defines a function.

Inverse relation
Inverse function

EXAMPLE 1   Find the inverse of

$$f = \{(x, y) \,|\, y = f(x) = 2x + 4\}$$

and sketch the graphs of the function and its inverse.

Solution   In order to find the defining equation of the inverse of the function defined by $y = 2x + 4$, we must solve the equation for x and then interchange x and y. Thus, solving for x, we get $x = \frac{1}{2}y - 2$, and interchanging x and y, we get $y = \frac{1}{2}x - 2$. Consequently,

$$f^{-1} = \{(x, y) \,|\, y = f^{-1}(x) = \tfrac{1}{2}x - 2\}$$

is the inverse of f, and it is a function. Both graphs are shown in Fig. 6.12.

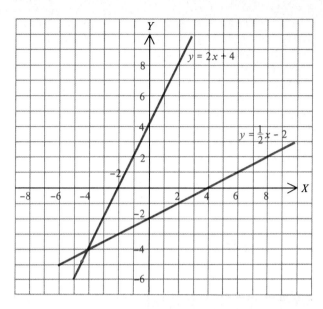

Figure 6.12

EXAMPLE 2 If $F = \{(x, y) \mid y = F(x) = \sqrt{x^2 + 16}\}$, find F^{-1}.

Solution The defining equation for F is

$$y = \sqrt{x^2 + 16} \tag{1}$$

and we solve it for x in terms of y as follows:

$$\begin{aligned} y^2 &= x^2 + 16 && \text{equating the squares of the members of (1)} \\ x &= \sqrt{y^2 - 16} && \text{solving for } x \end{aligned} \tag{2}$$

Note that since $x \geq 0$, we are concerned only with the positive square root of $y^2 - 16$. Furthermore, since x is real, $y \geq 4$ in (2). Now we interchange x and y in (2) and obtain the defining equation for F^{-1}. It is

$$y = \sqrt{x^2 - 16} \qquad \text{for } x \geq 4 \tag{3}$$

Hence, $\qquad F^{-1} = \{(x, y) \mid y = F^{-1}(x) = \sqrt{x^2 - 16} \text{ and } x \geq 4\}$

Note that the domain and range of F are $\{x \mid x \geq 0\}$ and $\{y \mid y \geq 4\}$, respectively. Furthermore the domain and range of F^{-1} are $\{x \mid x \geq 4\}$ and $\{y \mid y \geq 0\}$.

Now if x is an element of the intersection $\{x \mid x \geq 0\} \cap \{x \mid x \geq 4\}$ of the domains of F and F^{-1}, then $F(x)$ is an element of the domain of F^{-1}, and $F^{-1}(x)$ is an element in the domain of F. Furthermore, for each x in this intersection,

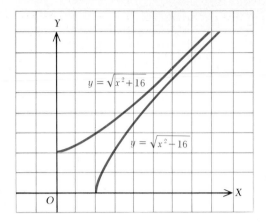

Figure 6.13

$$F[F^{-1}(x)] = \sqrt{(\sqrt{x^2 - 16})^2 + 16}$$
$$= \sqrt{x^2 - 16 + 16}$$
$$= \sqrt{x^2} = x$$

and

$$F^{-1}[F(x)] = \sqrt{(\sqrt{x^2 + 16})^2 - 16}$$
$$= \sqrt{x^2 + 16 - 16}$$
$$= \sqrt{x^2} = x$$

Hence, in the specified domain, $F[F^{-1}(x)] = F^{-1}[F(x)]$.

The graphs of the functions defined by Eqs. (1) and (3) are shown in Fig. 6.13. In this figure, it can be seen that the graph of the inverse function is situated in the same position relative to the X axis as the graph of the function is situated with respect to the Y axis. Either is the reflection of the other in the line $y = x$.

6.10 / VARIATION

The term *variation* is often used in describing some of the simpler functions. We shall see how it is used in stated problems.

Direct variation   If a number y *varies directly* as another number x, where $x \neq 0$, then $y = kx$. The word "directly" is often omitted when discussing direct variation. If the weight w of a piece of pipe varies directly as its length L, then $w = kL$.

Inverse variation   If one number y *varies inversely* as another x, then $y = k/x$. Thus, if the volume V of a confined mass of gas at a constant temperature varies inversely as the pressure P, then $V = k/P$. This is known as Boyle's law.

Joint variation   If one number *varies jointly* as two or more others, then it varies as their product. Thus, if the volume V of a box varies jointly as its length L, width W, and height H, then $V = kLWH$.

Combined variation   If one number varies jointly as several others and inversely as still others, then the variation is referred to as a *combined variation*. Thus, if y varies jointly as x and z and inversely as w, then $y = kxz/w$.

Constant of variation   In each of the four types of variations defined above, k is called the *constant of variation*. The constant can be determined if a set of values for the variables is known. Thus if in the example given for direct variation, $w = 90$ for $L = 15$, then $90 = k(15)$ and $k = 6$; hence, $w = 6L$.

A typical problem in variation involves a set of values for all the variables and a second set for all but one of the variables. After the variation has been expressed as an equation, the value of the constant of variation can be found by making use of the complete set of values of the variables as in the previous example. Finally, the value of the variable not included in the incomplete set of values can be determined by use of the incomplete set. If we want to find the value of w for $L = 17$, we need only substitute 17 for L in $w = 6L$ and thereby get $w = 6(17) = 102$.

EXAMPLE 1   The horsepower required to propel a ship varies as the cube of the speed. If the horsepower required for a speed of 15 knots is 10,125, find the horsepower required for a speed of 20 knots.

Solution   If we let

$$P = \text{required horsepower}$$
$$s = \text{speed, in knots}$$

then, since P varies as s^3, we have

$$P = ks^3 \qquad (1)$$

We are given that $P = 10{,}125$ for $s = 15$ knots. By substituting these values in (1), we get

$$10{,}125 = k(15^3)$$

and

$$k = \frac{10{,}125}{15^3} = \frac{10{,}125}{3{,}375} = 3$$

Now we substitute $k = 3$ and $s = 20$ in (1) and have

$$P = 3(20^3) = 3(8{,}000)$$
$$= 24{,}000 \text{ horsepower}$$

EXAMPLE 2 The weight of a rectangular block of metal varies jointly as the length, the width, and the thickness. If the weight of a 12- by 8- by 6-in. block of aluminum is 18.7 lb, find the weight of a 16- by 10- by 4-in. block.

Solution 1 We let

$$W = \text{weight, in pounds}$$
$$l = \text{length, in inches}$$
$$w = \text{width, in inches}$$
$$t = \text{thickness, in inches}$$

Then, since the weight varies jointly as the length, width, and thickness, we have

$$W = klwt$$

2 When $l = 12$ in., $w = 8$ in., and $t = 6$ in., $W = 18.7$ lb. Therefore,

$$18.7 = k(12)(8)(6)$$
$$= 576k$$

and

$$k = \frac{18.7}{576}$$

3 On substituting $k = 18.7/576$, $l = 16$, $w = 10$, and $t = 4$ in the equation $W = klwt$, we obtain

$$W = \frac{18.7}{576}(16)(10)(4)$$
$$= 20.8 \text{ lb}$$

as the weight of the 16- by 10- by 4-in. block. The reader should note that in the example k is the weight of 1 in.^{3} of aluminum.

EXAMPLE 3 The safe load of a beam with a rectangular cross section that is supported at each end varies jointly as the product of the width and the square of the depth and inversely as the length of the beam between supports. If the safe load of a beam 3 in. wide and 6 in. deep with supports 8 ft apart is 2,700 lb, find the safe load of a beam of the same material that is 4 in. wide and 10 in. deep with supports 12 ft apart.

Solution 1 We let

$$w = \text{width of beam, in inches}$$
$$d = \text{depth of beam, in inches}$$
$$l = \text{length between supports, in feet}$$
$$L = \text{safe load, in pounds}$$

Then

$$L = \frac{kwd^2}{l}$$

2 According to the first set of data, when $w = 3$, $d = 6$, and $l = 8$, then $L = 2,700$. Therefore,

$$2,700 = \frac{k(3)(6^2)}{8}$$

$$21,600 = 108l$$

and

$$k = 200$$

3 Consequently, if $w = 4$, $d = 10$, $l = 12$, and $k = 200$, we have

$$L = \frac{200(4)(10^2)}{12}$$

$$= 6,666\tfrac{2}{3}$$

EXERCISE 6.3 Special Functions, Inverses, Variation

1 Plot the graph of $T \times S$ if $T = \{1, 2, 3\}$ and $S = \{2, 4, 6\}$.
2 Plot the graph of $\{(x, y) \mid y > x\} \cap T \times S$ if $T = S = \{1, 3, 5\}$.
3 Plot the graph of $\{(x, y) \mid y = x\} \cap T \times S$ if $T = \{-1, 0, 1\}$ and $S = \{1, 2, 3\}$.
4 Plot the graph of $\{(x, y) \mid y < x\} \cap T \times S$ if $T = \{0, 1, 2\}$ and $S = \{1, 2, 5\}$.

Construct the graph of the function in each of Probs. 5 to 16.

5 $y = \begin{cases} x & \text{for } x < 0 \\ x - 1 & \text{for } x \geq 0 \end{cases}$
6 $y = \begin{cases} x + 2 & \text{for } x \geq -1 \\ x + 1 & \text{for } x < -1 \end{cases}$

7 $y = \begin{cases} -x + 3 & \text{for } x > 3 \\ \sqrt{9 - x^2} & \text{for } x \leq 3 \end{cases}$
8 $y = \begin{cases} 2x - 1 & \text{for } |x| \geq 3 \\ x^2 + 2 & \text{for } |x| < 3 \end{cases}$

9 $y = |x|$
10 $y = |x| + x$

11 $y = |x| + 2$
12 $y = |x + 2|$

13 $y = [x] + 1, D = \{-2 < x \leq 3\}$
14 $y = [x + 1], D = \{-2 < x \leq 3\}$

15 $y = 2[x], D = \{-1 < x \leq 3\}$
16 $y = [2x], D = \{-1 < x \leq 3\}$

Find the inverse of the relation or function defined by the equation in each of Probs. 17 to 28. State whether the inverse is a function or merely a relation, and give its domain.

17 $y^2 = 4 - x^2, D = \{x \mid 0 \leq x \leq 2\}$
18 $y^2 = 18 - 2x^2, D = \{x \mid 0 \leq x \leq 3\}$

19 $y^2 = x^2 - 16, D = \{x \mid 4 \leq x < 6\}$
20 $y^2 = 4(1 - x^2), D = \{x \mid 0 \leq x \leq 1\}$

21 $f = \{(x, x - 1) \mid x \geq 2\}$
22 $f = \{(x, 2x + 3) \mid x \geq -1\}$

23 $f = \{(x, -x + 2) \mid x \geq 1\}$
24 $f = \{(x, 3x - 1) \mid x \geq 0\}$

25 $f = \left\{(x, y) \mid y = \dfrac{x + 3}{x}, x \geq 1\right\}$
26 $f = \left\{(x, y) \mid y = \dfrac{x - 2}{x}, x > 4\right\}$

27 $f = \left\{ (x, y) \mid y = \dfrac{x}{x-2}, x \ge 4 \right\}$ **28** $f = \left\{ (x, y) \mid y = \dfrac{x-1}{x+4}, x > 1 \right\}$

29 If W varies directly as x and is 12 for $x = 4$, find the value of W for $x = 2$.

30 The horsepower required to propel a ship varies as the cube of its speed. Find the ratio of the power required at 14 knots to that required at 7 knots.

31 The kinetic energy of a body varies as the square of its velocity. Find the ratio of the kinetic energy of a car at 20 mi/hour to that of the same car at 50 mi/hour.

32 As a body falls from rest, its velocity varies as the time in flight. If the velocity of a body at the end of 2 sec is 64.4 ft/sec, find the velocity at the end of 5 sec.

33 If a body is above the surface of the earth, its weight varies inversely as the square of the distance of the body from the center of the earth. If a man weighs 160 lb on the surface of the earth, how much will he weigh 200 mi above the surface? Assume the radius of the earth is 4,000 mi.

34 The current I varies as the electromotive force E and inversely as the resistance R. If in a system a current of 20 amperes flows through a resistance of 20 ohms with an electromotive force of 100 volts, find the current that 150 volts will send through the system.

35 The illumination produced on a surface by a source of light varies directly as the candlepower of the source and inversely as the square of the distance between the source and the surface. Compare the illumination produced by a 512-candela lamp that is 8 ft from a surface with that of a 72-candela lamp 2 ft from the surface.

36 If other factors are equal, the centrifugal force on a circular curve varies inversely as the radius of the turn. If the centrifugal force is 18,000 lb for a curve of radius 50 ft, find the centrifugal force on a curve of radius 150 ft.

37 If y varies jointly as x and w and is 72 for $x = 9$ and $w = 4$, find the value of y for $x = 18$ and $w = 1.5$.

38 The simple interest earned in a given time varies jointly as the principal and the rate. If $300 earned $45 at a 6 percent rate, how much would be earned by $500 at 5 percent in the same time?

39 The volume of a regular pyramid varies jointly as the altitude and the area of the base. If the volume of a regular pyramid of altitude 5 in. and base 9 sq in. is 15 in.³, find the volume of another of altitude 4 in. and base area 6 sq in.

40 The crushing load of a circular pillar varies as the fourth power of the diameter and inversely as the square of the height of the pillar. If 256 tons is needed to crush a pillar 8 in. in diameter and 20 ft high, find the load needed to crush a pillar 10 in. in diameter and 15 ft high.

EXERCISE 6.4 **REVIEW**

1   Is the set of ordered pairs $\{(x, y) \mid y = x^2\}$ a function? Why?
2   Is the set of ordered pairs $\{(x, y) \mid y^2 = x\}$ a function? Why?

Write out the sets of ordered pairs described in Probs. 3 to 5, and state whether each is a function.

3   $\{(x, 2x + 1) \mid x$ is an integer and $-2 \le x \le 3\}$
4   $\{(x, y) \mid y^2 = x + 3$ and $-1 \le x \le 3\}$
5   $\{(x, y) \mid y^2 = 25 - x^2, x \in \{0, 3, 4, 5\}$

6   If $f(x) = 2x - 3$, find $f(-2), f(0)$, and $f(3)$.
7   If $f(x) = x^2 - 2x + 5$, find $f(-1), f(1)$, and $f(4)$.
8   If $f(x) = x^2 - x - 3$, find $f(x + h) - f(x)$.
9   If $f(x) = 3x + 4$, find $f(2 + h) - f(2)$.
10   If $f(x) = 2x - 3$ and $x = f(t) = 2t - 3$, find $F(t) = f[f(t)]$ and $F(2)$.
11   If $f(x) = x^2 + 3x - 1$ and $x = g(t) = t + 2$, find $F(t) = f[g(t)]$ and $F(4)$.
12   Find the range of $f = \{(x, y) \mid y = 3x - 4\}$ if $D \in \{-3, -1, 1, 3, 5\}$.
13   Find the range of $f = \{(x, y^2) \mid y = x + 2\}$ if $D \in \{-2, -1, 0, 4, 7\}$.
14   If $f = \{(x, 3x - 5)\}$ and $g = \{(x, 2x + 3)\}$, find $f \cap g$.
15   If $f = \{(x, x^2 - 4x + 7)\}$ and $g = \{(x, 2x^2 - 3x + 5)\}$ find $f \cap g$.
16   Describe the ray on which $P(x, y)$ lies if $x = -1$ and $y \le 3$.
17   Describe the ray on which $P(x, y)$ lies if $x \ge -2$ and $y = 2$.

Construct the graph of the function defined by the equation in each of Probs. 18 to 25.

18   $y = 2x + 7$ 19 $y = -3x + 1$
20   $y = x^2 + 2x - 1$ 21 $y = -2x^2 + 3x - 4$
22   $y = \sqrt{49 - x^2}, -7 \le x \le 7$ 23 $y = |x| - 2, -4 \le x \le 2$
24   $y = [x^2 - 1], D = \{-2 < x \le 1\}$ 25 $f = \{(x, -x + 3)\} \mid x \ge 3\}$

26   The mechanical advantage of a jackscrew varies directly as the length of the lever arm. If the mechanical advantage of a jackscrew is 192 when a 3-ft lever arm is used, what is the mechanical advantage for a 2-ft arm?

27   If the other factors are fixed, the lift on a wing of a plane varies with the density of the air. If the density of air at sea level is 0.08 lb/ft^3 and the lift on a wing is 2,500 lb, find the lift at such an altitude that the density is 0.06 lb/ft^3.

28 For a given load, the amount a wire stretches varies inversely as the square of the diameter. If a wire with a diameter of 0.6 in. is stretched 0.006 in. by a given load, how much will a wire of the same material with a diameter of 0.2 in. be stretched by the same load?

29 Under the same load, the sag of beams of the same material, length, and width varies inversely as the cube of the thickness. If a beam 4 in. thick sags 1/64 in. when a load is placed on it, find the sag of a beam 2 in. thick under the same load.

30 The pressure on the bottom of a container holding a liquid varies jointly as the depth and specific gravity of the liquid. The pressure on the bottom of a beaker filled to a depth of 10 in. with mercury is 4.90 lb/sq in. The specific gravity of mercury is 13.6. Find the pressure on the bottom of an upright drum filled to a depth of 3 ft with oil of specific gravity 0.8.

31 The kinetic energy of a moving body varies jointly as its mass and the square of its velocity. If the kinetic energy of a body of mass 10 g moving at 5 cm/second is 125 ergs, find the kinetic energy of a mass of 6 g moving at 10 cm/second.

CHAPTER SEVEN *Exponents and Radicals*

In Chap. 2 we stated the meaning of the notation a^n for n a positive integer, and we defined the number a^0 for $a \neq 0$. Furthermore, we derived four rules that deal with positive integral exponents, and we used the definitions and rules to a limited extent. In areas where mathematics is used, a broader concept of an exponent is needed. Consequently, in this chapter we shall extend the definition of an exponent to include negative numbers and fractions. The extensions will be made so that the laws developed in Chap. 2 will hold. We also shall develop additional laws as needed and shall explain how to use the extended concept of an exponent and how to apply the former laws and the new ones in more complicated situations than those which occurred in Chap. 2.

7.1 / NONNEGATIVE INTEGRAL EXPONENTS

For the convenience of the reader, we shall repeat the definitions and laws of exponents stated and derived in Chap. 2.

If n is a positive integer, then

$$a^n = \underbrace{a \cdot a \cdot a \cdot \cdots \cdot a}_{n \text{ factors}}$$

153

$$a^m a^n = a^{m+n} \qquad m \text{ and } n \text{ positive integers} \qquad (2.38)$$

$$(ab)^n = a^n b^n \qquad n \text{ a positive integer} \qquad (2.39)$$

$$\frac{a^m}{a^n} = a^{m-n} \qquad m \text{ and } n \text{ integers, } m > n > 0, \text{ and } a \neq 0 \quad (2.47)$$

$$a^0 = 1 \qquad a \neq 0 \qquad (2.48)$$

We shall now derive the laws for obtaining the power of a quotient and the power of a power.

$$\left(\frac{a}{b}\right)^n = \frac{a}{b} \cdot \frac{a}{b} \cdot \frac{a}{b} \cdots \text{ to } n \text{ factors each equal to } \frac{a}{b}$$

$$= \frac{a \cdot a \cdot a \cdots \text{ to } n \text{ factors each equal to } a}{b \cdot b \cdot b \cdots \text{ to } n \text{ factors each equal to } b}$$

$$= \frac{a^n}{b^n}. \qquad \text{by the definition of the product of two or more fractions}$$
$$\qquad\qquad \text{by the definition of an } n\text{th power}$$

Power of a quotient Hence we have
$$\left(\frac{a}{b}\right)^n = \frac{a^n}{b^n} \qquad (7.1)$$

Also, $\qquad (a^n)^m = a^n \cdot a^n \cdot a^n \cdots$ to m factors each equal to a^n
$$= a^{n+n+n+\cdots\text{to } m \; n's} \qquad \text{by (2.38)}$$
$$= a^{mn}$$

Power of a power Therefore, we have
$$(a^n)^m = a^{mn} \qquad (7.2)$$

We shall illustrate the application of the above laws with several examples.

EXAMPLE 1 $\qquad\qquad\qquad a^3 a^5 a^7 = a^{3+5+7} = a^{15} \qquad$ by (2.38)

EXAMPLE 2 $\qquad\qquad\qquad \dfrac{a^9}{a^4} = a^{9-4} = a^5 \qquad$ by (2.47)

EXAMPLE 3 $\qquad\qquad\qquad \dfrac{a^4 a^7}{a^3} = \dfrac{a^{11}}{a^3} = a^8 \qquad$ by (2.38) and (2.47)

EXAMPLE 4 $\qquad\qquad\qquad (a^6)^4 = a^{24} \qquad$ by (7.2)

EXAMPLE 5 $\qquad\qquad\qquad (a^3 b^4)^5 = a^{15} b^{20} \qquad$ by (2.39) and (7.2)

EXAMPLE 6 $\qquad\qquad\qquad \left(\dfrac{a^7}{b^4}\right)^3 = \dfrac{a^{21}}{b^{12}} \qquad$ by (7.1) and (7.2)

The methods for obtaining the product and quotient of two similar monomials are illustrated in the following examples:

EXAMPLE 7

$$(6a^3b^2)(8a^5b^4) = 6 \cdot 8 \cdot a^3a^5b^2b^4 \qquad \text{by the commutative law}$$
$$= 48a^8b^6 \qquad \text{by (2.38)}$$

EXAMPLE 8

$$\frac{48x^7y^9}{4x^2y^5} = \frac{48}{4} \cdot \frac{x^7}{x^2} \cdot \frac{y^9}{y^5}$$
$$= 12x^5y^4 \qquad \text{by (2.47)}$$

To simplify an expression involving positive integral exponents, we make all possible applications of the above laws. The following examples illustrate the procedure.

EXAMPLE 9

$$(4x^2y^3z^7)^4 = 4^4(x^2)^4(y^3)^4(z^7)^4 \qquad \text{by (2.39)}$$
$$= 256x^8y^{12}z^{28} \qquad \text{by (7.2)}$$

EXAMPLE 10

$$\left(\frac{5x^5y^8}{3z^4w^2}\right)^3 = \frac{5^3(x^5)^3(y^8)^3}{3^3(z^4)^3(w^2)^3} \qquad \text{by (7.1) and (2.39)}$$

$$= \frac{125x^{15}y^{24}}{27z^{12}w^6} \qquad \text{by (7.2)}$$

EXAMPLE 11 $\left(\dfrac{4a^3b^4}{3c^2d^3}\right)^4\left(\dfrac{9c^5d}{2a^2b^5}\right)^2 = \dfrac{4^4(a^3)^4(b^4)^4}{3^4(c^2)^4(d^3)^4} \cdot \dfrac{9^2(c^5)^2d^2}{2^2(a^2)^2(b^5)^2} \qquad$ by (7.1) and (2.39)

$$= \frac{256a^{12}b^{16}}{81c^8d^{12}} \cdot \frac{81c^{10}d^2}{4a^4b^{10}} \qquad \text{by (7.2)}$$

$$= 64 \cdot \frac{a^{12}}{a^4} \cdot \frac{b^{16}}{b^{10}} \cdot \frac{c^{10}}{c^8} \cdot \frac{d^2}{d^{12}}$$

$$= \frac{64a^8b^6c^2}{d^{10}} \qquad \text{by (2.47)}$$

EXERCISE 7.1 Simplification of Exponential Expressions

Perform the indicated operations and simplify.

1	2^32^4	2	3^23^3	3	5^25^4	4	4^14^2
5	$5^4/5^3$	6	$3^5/3^2$	7	$6^7/6^3$	8	$4^5/4^2$
9	$(2^2)^3$	10	$(4^3)^2$	11	$(3^2)^4$	12	$(5^3)^3$
13	$(3/4)^4$	14	$(2/5)^3$	15	$(2/7)^3$	16	$(2/3)^5$
17	$(3^24^1)^3$	18	$(2^33^2)^4$	19	$(5^22^3)^3$	20	$(3^55^2)^2$

21 $(2x^3y^2)(3x^2y)$ $\qquad$ 22 $(-5x^2y^4)(2x^0y^2)$ $\qquad$ 23 $(4x^0y^2z)(3x^3y^3z^2)$

24 $(-2x^3y^3z)(-3x^2yz^4)$ $\qquad$ 25 $\dfrac{8x^3y^5}{4x^2y}$ $\qquad$ 26 $\dfrac{12x^5y^4}{6x^3y^3}$

27   $\dfrac{18x^4y^3}{3x^3y^2}$ 28 $\dfrac{24x^5y^6}{6x^3y^4}$ 29 $(2a^2bc^3)^3$

30   $(3a^3b^2c^4)^2$ 31 $(5ab^2c^4)^4$ 32 $(7a^3b^2c^4)^3$

33   $\left(\dfrac{a^3b^2}{2c^3d^4}\right)^2$ 34 $\left(\dfrac{a^4b^3c}{2d^2}\right)^3$ 35 $\left(\dfrac{2x^4y^3}{3w^2z}\right)^2$

36   $\left(\dfrac{3x^3y^5}{2p^2q^3}\right)^4$ 37 $(2a^2b^3)^2(3ab^0)^4$ 38 $(3a^3b)^3(2a^2b^3)^2$

39   $(5cd^2)^2(2c^2d^3)^3$ 40 $(3x^2y^3)^3(5x^2y)^4$ 41 $\dfrac{2u^2v^3}{3w^4}\cdot\dfrac{6w^2u}{8u^2v^2}$

42   $\dfrac{15b^2c^5}{16d^3}\cdot\dfrac{4b^4d^3}{5c^6}$ 43 $\dfrac{3x^2y^3}{7x^4z^2}\cdot\dfrac{28y^3}{33x^4z^3}$ 44 $\dfrac{7b^4c^3}{8d^2}\cdot\dfrac{16b^3d^3}{21c^4d^5}$

45   $\left(\dfrac{c^4d^3}{a^4}\right)^2\left(\dfrac{a^3}{c^2d}\right)^3$ 46 $\left(\dfrac{3a^2}{b^3c^0}\right)^4\left(\dfrac{bc^2}{9a^4}\right)^2$ 47 $\left(\dfrac{4a^2b^5}{c^3}\right)^2\left(\dfrac{c^4}{12a^3b}\right)^3$

48   $\left(\dfrac{6a^4}{b^2c^5}\right)^3\left(\dfrac{bc^2}{2a^3}\right)^4$ 49 $\left(\dfrac{6a^2b^3}{c^3d^4}\right)^3\div\left(\dfrac{3a^3b^4}{c^4d^5}\right)^2$ 50 $\left(\dfrac{14c}{15a^2b}\right)^2\div\left(\dfrac{7c^2}{30a^5b^3}\right)^2$

51   $\left(\dfrac{6x^3y^2}{z^0}\right)^3\div\left(\dfrac{3x^4y^2}{z^3}\right)^4$ 52 $\left(\dfrac{2d^3a^2}{3b^4c^3}\right)^4\div\left(\dfrac{4d^2a^4}{6b^3c^2}\right)^2$ 53 $\dfrac{(a^2b^3c^4)^3}{(a^3bc^3)^2(ab^0c^2)^4}$

54   $\dfrac{(a^6b^5c^3)^3}{(a^2b^3c^2)^3(a^3b^2c)^4}$ 55 $\dfrac{(a^2c^3)^4(b^3d^4c^5)^2}{(a^2d^3c^4)^5}$ 56 $\dfrac{(a^2b^2c^4)^2(a^3bc^2)^3}{(a^3b^2c)^5}$

57   $\dfrac{x^{3a+1}y^{a+3}}{x^{a+2}y^{a-1}}$ 58 $\dfrac{x^{2a-3}y^{b-4}}{x^{a+2}y^{b+3}}$ 59 $\dfrac{x^{a+3}y^{b-4}}{x^{a+2}y^{2-b}}$

60   $\dfrac{x^{a+2}y^{b+4}}{x^{2-a}y^{b-1}}$ 61 $\dfrac{(a^{b+3}d^{b+2})^3}{a^{3b}d^6}$ 62 $\dfrac{(a^{n+2}b^{n-3})^2}{a^{2n}b^2}$

63   $\dfrac{(a^{2n-3}b^{n+2})^3}{a^{n-9}b^{3n-1}}$ 64 $\dfrac{(a^{3n+1}b^{2n-1})^4}{a^{6n+4}b^{2n-5}}$ 65 $\left(\dfrac{a^{b-3}}{a^{2b-3}}\right)^c$

66   $\left(\dfrac{a^b}{a^{b-2}}\right)^c$ 67 $\dfrac{(a^{n+2}b^{2n+1})^p}{(a^{p+1}b^{2p-2})^n}$ 68 $\dfrac{(a^{2-n}b^{n+3})^p}{(a^{2-p}b^{p+3})^n}$

7.2 / NEGATIVE INTEGRAL EXPONENTS

As stated in the introduction, we shall extend the concept of exponents in this chapter and shall make the extensions so that the laws in Sec. 7.1 still hold. The first extension was made when we defined a^0 as 1, provided $a \neq 0$. It will be remembered that it is necessary to define a^0 in this way if Eq. (2.47) is to hold true when $m = n$. We shall now investigate the situation if we disregard the restriction $m > n$ in Eq. (2.47), and we shall let $m = t$ and $n = 2t$, where t is a positive integer. These replacements for m and n yield

$$\frac{a^t}{a^{2t}} = a^{t-2t} = a^{-t}$$

We know, however, by the fundamental principle of fractions shown by Eq. (4.4), that

$$\frac{a^t}{a^{2t}} = \frac{a^t \div a^t}{a^{2t} \div a^t} = \frac{a^{t-t}}{a^{2t-t}} \qquad \text{by Eqs. (4.4) and (2.44)}$$

$$= \frac{a^0}{a^t} = \frac{1}{a^t} \qquad \text{since } a^0 = 1$$

Consequently, if the law shown by Eq. (2.47) is to hold true for $m < n$, it is necessary to define a^{-t} as follows:

Definition of a
negative exponent

$$a^{-t} = \frac{1}{a^t} \qquad a \neq 0 \qquad \qquad \textbf{(7.3)}$$

We shall now prove that the other laws in Sec. 7.1 hold if a^{-t} is defined as above. Since each of these laws was derived from either Eq. (2.38) or Eq. (2.47), it follows that each of them will be true for the above definition of a negative exponent provided that Eqs. (2.38) and (2.47) are true. The definition was made so that Eq. (2.47) is true. Hence it remains only to prove that Eq. (2.38) is true. Consequently, we shall prove that

$$a^{-t} \cdot a^{-s} = a^{-t-s} \qquad a \neq 0, \ t \text{ and } s \text{ positive integers}$$

Proof

$$a^{-t} \cdot a^{-s} = \frac{1}{a^t} \cdot \frac{1}{a^s} \qquad \text{by Eq. (7.3)}$$

$$= \frac{1}{a^t a^s} \qquad \text{by Eq. (2.46)}$$

$$= \frac{1}{a^{t+s}} \qquad \text{by Eq. (2.38)}$$

$$= a^{-(t+s)} \qquad \text{by Eq. (7.3)}$$

$$= a^{-t-s} \qquad \text{by Eq. (2.19)}$$

Therefore, we may remove all restrictions from the laws in Sec. 7.1 except that m and n be integers and $a \neq 0$ in Eqs. (2.47) and (2.48) and that $b \neq 0$ in Eq. (7.1).

It is frequently desirable to convert a fraction with negative exponents in either or both members to an equal fraction in which all exponents are positive. We employ the fundamental principle of fractions shown by Eq. (4.2) for this purpose and illustrate the procedure with several examples.

EXAMPLE 1 Convert the fraction $a^x b^{-y}/c^z d^{-w}$ to an equal fraction in which all exponents are positive.

Solution The terms with negative exponents in this fraction are b^{-y} and d^{-w}. Hence if we multiply each member by $b^y d^w$, we obtain an equal fraction with no exponents negative. Thus,

$$\frac{a^x b^{-y}}{c^z d^{-w}} = \frac{a^x b^{-y} \cdot b^y d^w}{c^z d^{-w} \cdot b^y d^w} \qquad \text{by Eq. (4.2)}$$

$$= \frac{a^x b^0 d^w}{c^z d^0 b^y} \qquad \text{by Eqs. (2.31) and (2.38)}$$

$$= \frac{a^x d^w}{c^z b^y} \qquad \text{since } b^0 = d^0 = 1$$

We call attention to the fact that in the above example we multiplied both members of the fraction by the product of positive integral powers of b and d. A similar procedure applied to any fraction will yield an equal fraction in which all exponents are positive provided that the product by which the members of the fraction are multiplied is properly chosen. The following examples illustrate the method for choosing this product and the subsequent steps in the conversion of the fraction.

EXAMPLE 2   Convert $2^{-3}a^{-2}bc^{-1}/4^{-2}xy^{-3}z^4$ to an equal fraction in which all exponents are positive.

Solution 1   The terms with negative exponents which occur in this fraction are 2^{-3}, a^{-2}, c^{-1}, 4^{-2}, and y^{-3}. We now form the product of the numbers obtained by changing the sign of the exponent in each of the above powers, and we get $2^3 4^2 a^2 c^1 y^3$. Then we multiply the members of the given fraction by this product and get

$$\frac{2^{-3}a^{-2}bc^{-1}}{4^{-2}xy^{-3}z^4} = \frac{(2^{-3}a^{-2}bc^{-1})(2^3 4^2 a^2 c^1 y^3)}{(4^{-2}xy^{-3}z^4)(2^3 4^2 a^2 c^1 y^3)} \qquad \text{by Eq. (4.2)}$$

$$= \frac{2^{-3+3}4^2 a^{-2+2}bc^{-1+1}y^3}{2^3 4^{-2+2}a^2 c^1 x y^{-3+3}z^4} \qquad \text{by Eqs. (2.31) and (2.38)}$$

$$= \frac{2^0 4^2 a^0 bc^0 y^3}{2^3 4^0 a^2 cxy^0 z^4}$$

$$= \frac{16by^3}{8a^2 cxz^4} \qquad \text{since } 2^0 = a^0 = c^0 = 4^0 = y^0 = 1$$

$$= \frac{2by^3}{a^2 cxz^4} \qquad \text{dividing the members by 8}$$

Solution 2 $$\frac{2^{-3}a^{-2}bc^{-1}}{4^{-2}xy^{-3}z^4} = \frac{\left(\dfrac{b}{2^3 a^2 c}\right)}{\left(\dfrac{xz^4}{4^2 y^3}\right)} \qquad \text{by (7.2)}$$

$$= \frac{b}{2^3 a^2 c} \cdot \frac{4^2 y^3}{xz^4}$$

$$= \frac{2by^3}{a^2 cxz^4}$$

EXAMPLE 3   Convert $2c^{-1}d^{-4}e^2/3c^{-3}d^{-2}e^{-1}$ to an equal fraction with only positive exponents.

Solution   In this fraction, we have c^{-1} in the numerator and c^{-3} in the denominator, and c^{-3} has the exponent with the greater absolute value. Also d^{-4} is in the numerator, and d^{-2} is in the denominator; here d^{-4} has the exponent with the greater absolute value. Furthermore, e^{-1} appears in the denominator. Hence if we multiply each member of the fraction by $c^3d^4e^1$, we will obtain an equal fraction in which all exponents are positive. This procedure yields

$$\frac{2c^{-1}d^{-4}e^2}{3c^{-3}d^{-2}e^{-1}} = \frac{(2c^{-1}d^{-4}e^2)(c^3d^4e^1)}{(3c^{-3}d^{-2}e^{-1})(c^3d^4e^1)} \qquad \text{by Eq. (4.2)}$$

$$= \frac{2c^{-1+3}d^{-4+4}e^{2+1}}{3c^{-3+3}d^{-2+4}e^{-1+1}} \qquad \text{by Eqs. (2.31) and (2.38)}$$

$$= \frac{2c^2d^0e^3}{3c^0d^2e^0}$$

$$= \frac{2c^2e^3}{3d^2} \qquad \text{since } d^0 = c^0 = e^0 = 1$$

EXAMPLE 4   Perform the operations indicated in $(a^{-2}b^3/2^{-2}a^{-1}b^{-2})^{-3}$ and express the result without negative exponents

Solution   In this problem, we have two alternatives for the first step. We can first convert the fraction in the parentheses to an equal fraction in which all exponents are positive and then apply Eq. (7.3); or we can apply Eqs. (7.1) and (2.39) first, then use Eq. (2.40), and finally proceed as in the above examples.

Method 1   The terms with negative exponents that occur in the fraction in parentheses are 2^{-2}, a^{-1}, a^{-2}, and b^{-2}. However, in the case of the negative exponents of a, a^{-2} has the exponent with the greater absolute value. Consequently, we multiply the members of the fraction by $2^2a^2b^2$ and get

$$\left(\frac{a^{-2}b^3}{2^{-2}a^{-1}b^{-2}}\right)^{-3} = \left(\frac{a^{-2}b^3}{2^{-2}a^{-1}b^{-2}} \cdot \frac{2^2a^2b^2}{2^2a^2b^2}\right)^{-3} \qquad \text{by Eq. (4.4)}$$

$$= \left(\frac{2^2a^0b^5}{2^0ab^0}\right)^{-3} \qquad \text{by Eqs. (2.31) and (2.38)}$$

$$= \left(\frac{4b^5}{a}\right)^{-3} \qquad \text{since } a^0 = 2^0 = b^0 = 1$$

$$= \frac{1}{\left(\dfrac{4b^5}{a}\right)^3} \qquad \text{by Eq. (7.3)}$$

$$= \frac{1}{\dfrac{64b^{15}}{a^3}} \qquad \text{by Eqs. (7.1) and (2.40)}$$

$$= \frac{a^3}{64b^{15}} \qquad \text{multiplying the members of the complex fraction by } a^3$$

Method 2   If we apply Eqs. (7.1) and (2.39) first, we get

$$\left(\frac{a^{-2}b^3}{2^{-2}a^{-1}b^{-2}}\right)^{-3} = \frac{(a^{-2})^{-3}(b^3)^{-3}}{(2^{-2})^{-3}(a^{-1})^{-3}(b^{-2})^{-3}} \qquad \text{by Eqs. (7.1) and (2.39)}$$

$$= \frac{a^6 b^{-9}}{2^6 a^3 b^6} \qquad \text{by Eq. (2.40)}$$

$$= \frac{(a^6 b^{-9})(b^9)}{(2^6 a^3 b^6)(b^9)} \qquad \text{by Eq. (4.2)}$$

$$= \frac{a^6 b^0}{64 a^3 b^{15}} \qquad \text{by Eq. (2.38)}$$

$$= \frac{a^3}{64 b^{15}} \qquad \text{dividing the members by } a^3 \text{ and replacing } b^0 \text{ by 1}$$

The methods employed in the above examples can also be applied to fractions in which either or both of the members are the sum of two or more numbers.

EXAMPLE 5   Convert the fraction $(4x^{-2} - 9y^{-2})/(3x + 2y)$ to an equal fraction in which all exponents are positive.

Solution   The terms with negative exponents that appear in this fraction are x^{-2} and y^{-2}. Consequently, we multiply each member of the fraction by $x^2 y^2$ and proceed as indicated below.

$$\frac{4x^{-2} - 9y^{-2}}{3x + 2y} = \frac{(4x^{-2} - 9y^{-2})(x^2 y^2)}{(3x + 2y)(x^2 y^2)} \qquad \text{by Eq. (4.2)}$$

$$= \frac{4x^{-2+2}y^2 - 9x^2 y^{-2+2}}{3x^3 y^2 + 2x^2 y^3} \qquad \text{by Eqs. (2.12), (2.31), and (2.38)}$$

$$= \frac{4x^0 y^2 - 9x^2 y^0}{3x^3 y^2 + 2x^2 y^3}$$

$$= \frac{4y^2 - 9x^2}{3x^3 y^2 + 2x^2 y^3} \qquad \text{by Eqs. (2.48) and (2.34)}$$

$$= \frac{(2y + 3x)(2y - 3x)}{x^2 y^2 (3x + 2y)} \qquad \text{factoring}$$

$$= \frac{2y - 3x}{x^2 y^2} \qquad \text{dividing the members by } (3x + 2y)$$

EXERCISE 7.2 **Operations with Negative Exponents**

Find the value of the expression in each of Probs. 1 to 28.

1	3^{-2}	**2**	4^{-1}	**3**	2^{-4}	**4**	1^{-9}
5	$4^{-1}4^{-2}$	**6**	$4^{-3}4^0$	**7**	$3^{-2}3$	**8**	$2^{-5}2^{-2}$
9	$8^{-2}/8^{-3}$	**10**	$3^{-5}/3^{-7}$	**11**	$2^{-4}/2^{-1}$	**12**	$5^{-1}/5^{-3}$
13	$(2^{-3})^1$	**14**	$(2^{-1})^3$	**15**	$(3^{-2})^{-2}$	**16**	$(4^0)^{-5}$
17	$(2^{-3})^2(3^2)^{-2}$	**18**	$(3^{-1})^2(4^2)^{-1}$	**19**	$(3^{-2})^{-2}(4^{-1})^1$	**20**	$(2^{-3})^3(3^{-2})^1$
21	$(2^{-1}3^{-2})^3$	**22**	$(3^{-2}2)^{-1}$	**23**	$(4^{-2}2^2)^2$	**24**	$(3^{-1}4^2)^{-2}$
25	$(3^{-1}/2^{-2})^2$	**26**	$(2^{-1}/3^{-2})^3$	**27**	$(2^{-3}/3^2)^{-2}$	**28**	$(3^3/2^{-3})^{-2}$

Write the expression in each of Probs. 29 to 36 without a denominator by using negative exponents if necessary.

29	a^2/b^{-2}	**30**	a^3/b	**31**	a^2/b^3
32	a^3/a^5	**33**	$a^2b^{-3}/a^3b^{-4}c^2$	**34**	$a^{-1}b^{-3}/a^2b^3c^6$
35	$3a^0b^3/a^{-1}b^{-2}c^3$	**36**	$2^3x^{-4}y^{-3}/4^2x^{-5}y^{-2}z^0$		

In Probs. 37 to 80, combine wherever possible by use of Eqs. (2.42) to (2.47), (7.1), and (7.2), and express the results without zero or negative exponents.

37	$a^{-1}b$	**38**	$c^{-2}d^{-1}$	**39**	a^0b^{-3}	**40**	$a^{-3}b^{-2}$
41	a^{-1}/b^{-2}	**42**	a^{-3}/b^2	**43**	a^2/b^{-3}	**44**	a^{-3}/b^{-4}
45	$\dfrac{3p^{-3}q^{-4}}{6^{-1}p^2q^{-2}}$	**46**	$\dfrac{9^{-2}p^{-4}q^0}{3^{-3}p^{-3}q^{-2}}$	**47**	$\dfrac{2a^{-2}b^{-1}}{3^{-1}a^3b^{-4}}$	**48**	$\dfrac{3^{-2}a^{-2}b^{-3}}{6^{-1}a^{-4}b^{-2}}$
49	$(a^{-2}/b^{-3})^2$	**50**	$(a^{-4}/b)^{-3}$	**51**	$(a^3/b^{-1})^{-2}$	**52**	$(a^3/b^2)^{-4}$
53	$(a^{-1}b^{-2})^2$	**54**	$(a^{-2}b)^{-2}$	**55**	$(a^3b^{-1})^{-3}$	**56**	$(a^{-4}b^3)^3$
57	$\left(\dfrac{a^2n^0t^{-2}}{a^{-2}nt^{-5}}\right)^{-1}$	**58**	$\left(\dfrac{b^{-2}u^3g^{-4}}{bu^{-2}g^{-1}}\right)^2$	**59**	$\left(\dfrac{s^{-2}a^2t^{-1}}{s^2a^{-1}t^2}\right)^2$	**60**	$\left(\dfrac{a^2p^{-3}e^{-1}}{a^{-1}p^{-4}e^0}\right)^{-2}$
61	$b^2+\dfrac{1}{b^{-2}}$	**62**	$3a^{-2}-\dfrac{2}{a^2}$	**63**	$c^3-\dfrac{2}{c^{-3}}$	**64**	$2a^{-1}-\dfrac{2}{a}$
65	$\dfrac{x^{-3}-y^{-3}}{x^{-2}y^{-2}}$			**66**	$\dfrac{x^{-2}y^{-2}}{x^{-2}-y^{-2}}$		
67	$\dfrac{x^{-2}y^{-1}-x^{-1}y^{-2}}{x^{-2}-y^{-2}}$			**68**	$\dfrac{x^{-3}y^{-2}+x^{-2}y^{-3}}{x^{-3}+y^{-3}}$		
69	$\dfrac{x^{-2}-3x^{-1}y^{-1}-4y^{-2}}{x^{-2}y^{-1}-4x^{-1}y^{-2}}$			**70**	$\dfrac{y^{-2}+2x^{-1}y^{-1}+x^{-2}}{x^{-1}y^{-2}+x^{-2}y^{-1}}$		
71	$\dfrac{x^{-1}y^{-2}+2x^{-2}y^{-1}}{y^{-2}+4x^{-1}y^{-1}+4x^{-2}}$			**72**	$\dfrac{x^{-2}y^{-1}+x^{-1}y^{-2}}{x^{-2}-y^{-2}}$		

73 $(x-1)^{-2}-2(x-1)^{-3}(x+1)$

74 $-3(x-1)^2(x-2)^{-4}+(x-1)(x-2)^{-3}$

75 $2(4x-1)^{-1}(2x+1)^{-2}+4(4x-1)^{-2}(2x+1)^{-1}$

76 $3(2x-1)^{-2}(3x+2)^{-2}+4(2x-1)^{-3}(3x+2)^{-1}$

77 $3(2x+3)^{-1}(3x-2)^{-3}+(3x-2)^{-2}(2x+3)^{-2}$

78   $2(3x-1)^{-2}(1-2x)^{-2} - 6(3x-1)^{-3}(1-2x)^{-1}$
79   $5(3x-4)^2(1-5x)^{-3} + 3(3x-4)(1-5x)^{-2}$
80   $3(4x+1)^{-1}(2-3x)^{-3} - 2(2-3x)^{-2}(4x+1)^{-2}$

7.3 / RATIONAL EXPONENTS AND RADICALS

In this section we further extend the definition of a^n to include the case in which n is the quotient of two integers, in such a way that the laws stated in Sec. 7.1 are valid for rational exponents.

If law (7.2) is valid for $n = 1/k$ and $m = k$, then $(a^{1/k})^k = a^{k/k} = a$. Hence we define $a^{1/k}$ as a number whose kth power is a. Without further restrictions, however, $a^{1/k}$ may have more than one value. For example, $16^{1/2}$ is both 4 and -4, since $4^2 = (-4)^2 = 16$. We shall remove this ambiguity by means of the following definitions.

kth root of a number, principal root

The number b is a *kth root* of a if $b^k = a$.

If a positive kth root of a constant a exists, then it is the *principal kth root of a*. If no positive kth root of a exists but there is a negative kth root, then that is the principal kth root of a.

Radical of order k
Radicand
Index

We use the radical $\sqrt[k]{a}$ to designate the principal kth root of a. This symbol is called a *radical of order k*. The number a is the *radicand*, and k is the *index* of the radical. If the index of the radical is not written, it is understood to be 2. For example,

$$\sqrt{16} = 4 \qquad \text{and} \qquad \sqrt[3]{-27} = -3$$

We now make use of the above definitions to define $a^{1/k}$ in terms of a root. If a is a constant then

Definition of a rational exponent

$$a^{1/k} = \sqrt[k]{a} \qquad\qquad (7.4)$$

If a is a positive constant, then there is a positive kth root of a, and therefore $a^{1/k}$ is positive. For example, $4^{1/2} = 2$, $64^{1/3} = 4$, and $32^{1/5} = 2$. If a is a negative constant and k is an odd integer, it can be proved by the methods of Chap. 12 that only one real kth root of a exists and that this root is negative. For example, $(-8)^{1/3} = -2$, and there is no other real number whose third power is -8. Furthermore, if a is negative and k is even, no real kth root of a exists, since any even power of a real number is positive. For example, there is no real number whose square is -4, and there is no real number whose fourth power is -256, since both the square and the fourth power of a real number are positive. Since we shall deal only with real numbers in this chapter, we shall exclude the case in which the radicand is negative and the index of the radical is even. To be able to handle that case, a further extension of the number system, which will be discussed in Chaps. 8 and 12, is required.

If in (7.4) we replace a by b^j, we have $(b^j)^{1/k} = \sqrt[k]{b^j}$. Therefore, $(b^j)^{1/k}$ is a kth root of b^j. Furthermore, if we raise $(b^{1/k})^j$ to the kth power, we get

$$\begin{aligned}[(b^{1/k})^j]^k &= (b^{1/k})^{jk} \qquad \text{by (7.2)}\\ &= (b^{1/k})^{kj} \qquad \text{by the commutative law}\\ &= b^j \qquad \text{by the definition of } b^{1/k}\end{aligned}$$

Hence, $(b^j)^{1/k}$ and $(b^{1/k})^j$ are both kth roots of b^j, and it can be proved† that, except for the excluded case, the two roots have the same sign and are therefore equal. Consequently, we have

$$b^{j/k} = (b^j)^{1/k} = (b^{1/k})^j = \sqrt[k]{b^j} = (\sqrt[k]{b})^j \qquad \text{provided } \sqrt[k]{b} \text{ is real} \quad \textbf{(7.5)}$$

The proofs that the laws of exponents stated in Sec. 7.1 hold for rational exponents as defined in (7.4) can be established by use of (7.5) and arguments similar to the above.‡

The processes involved in working with fractional exponents are illustrated in the following examples.

EXAMPLE 1 Evaluate $4^{1/2}$, $8^{2/3}$, $(-32)^{3/5}$.

Solution

$$\begin{aligned}4^{1/2} &= \sqrt{4} \qquad\qquad\quad \text{by (7.4)}\\ &= 2\\ 8^{2/3} &= (\sqrt[3]{8})^2 \qquad\qquad \text{by (7.5)}\\ &= 2^2 = 4\\ (-32)^{3/5} &= (\sqrt[5]{-32})^3 \qquad \text{by (7.5)}\\ &= (-2)^3 = -8\end{aligned}$$

EXAMPLE 2 Express $3a^{1/2}b^{5/2}$ and $5a^{3/4}/b^{5/4}$ in radical form.

Solution

$$\begin{aligned}3a^{1/2}b^{5/2} &= 3(ab^5)^{1/2} \qquad \text{by (2.39)}\\ &= 3\sqrt{ab^5} \qquad\quad \text{by (7.4)}\\ \frac{5a^{3/4}}{b^{5/4}} &= 5\left(\frac{a^3}{b^5}\right)^{1/4} \qquad \text{by (7.1) with } n = \tfrac{1}{4}\\ &= 5\sqrt[4]{\frac{a^3}{b^5}} \qquad\quad\ \text{by (7.4)}\end{aligned}$$

---

† To prove this statement, we first assume that b is positive. Then b^j and $b^{1/k}$ are each positive. Therefore, $(b^j)^{1/k}$ and $(b^{1/k})^j$ are positive and hence are equal. If b is negative, k is odd, and j is odd, b^j and $b^{1/k}$ are negative. Therefore, $(b^j)^{1/k}$ and $(b^{1/k})^j$ are negative and therefore are equal. The argument for the case b negative, k odd, and j even can be made in a similar way.
‡ As an exercise the reader may prove that $a^{1/n}a^{1/m}$ and $a^{1/n+1/m}$ are both mnth roots of a^ma^n and that, except for the excluded case, they have the same sign, thus proving that $a^{1/m}a^{1/n} = a^{1/m+1/n}$.

EXAMPLE 3   Find the product of $3a^{1/2}$ and $2a^{2/3}$.

Solution
$$(3a^{1/2})(2a^{2/3}) = 6a^{1/2+2/3}$$
$$= 6a^{(3+4)/6}$$
$$= 6a^{7/6}$$

EXAMPLE 4   Find the quotient of $8x^{1/2}y^{5/6}$ and $5x^{1/4}y^{1/3}$.

Solution
$$8x^{1/2}y^{5/6} \div 5x^{1/4}y^{1/3} = \tfrac{8}{5}x^{1/2-1/4}y^{5/6-1/3} = \tfrac{8}{5}x^{1/4}y^{3/6}$$
$$= \tfrac{8}{5}x^{1/4}y^{1/2}$$

EXAMPLE 5   Combine $[(4x^4y^{3/4}z^2)/(9x^2y^{1/4}z)]^{1/2}$ wherever possible using the laws of exponents, and express the result without zero or negative exponents.

Solution
$$\left(\frac{4x^4y^{3/4}z^2}{9x^2y^{1/4}z}\right)^{1/2} = \left(\frac{4}{9}\frac{x^4}{x^2}\frac{y^{3/4}}{y^{1/4}}\frac{z^2}{z}\right)^{1/2}$$
$$= \left(\tfrac{4}{9}x^2y^{2/4}z\right)^{1/2}$$
$$= \tfrac{2}{3}xy^{1/4}z^{1/2}$$

EXAMPLE 6   Combine $[(4x^2y^{3/4}z^{1/6})/(32x^{-1}y^0z^{-5/6})]^{-1/3}$ wherever possible, and express the result without zero or negative exponents.

Solution
$$\left(\frac{4x^2y^{3/4}z^{1/6}}{32x^{-1}y^0z^{-5/6}}\right)^{-1/3} = \left(\frac{4}{32}\frac{x^2}{x^{-1}}\frac{y^{3/4}}{y^0}\frac{z^{1/6}}{z^{-5/6}}\right)^{-1/3}$$
$$= \left(\tfrac{1}{8}x^3y^{3/4}z\right)^{-1/3}$$
$$= \frac{2}{xy^{1/4}z^{1/3}}$$

EXERCISE 7.3 Radical and Exponential Expressions

Convert each number or expression in Probs. 1 to 36 to a form without fractional exponents or radicals.

1 $64^{1/2}$	2 $64^{1/3}$	3 $64^{1/6}$	4 $243^{1/5}$
5 $0.0625^{1/4}$	6 $0.07776^{1/5}$	7 $0.008^{1/3}$	8 $0.09^{1/2}$
9 $8^{2/3}$	10 $16^{3/4}$	11 $243^{3/5}$	12 $343^{2/3}$
13 $0.04^{3/2}$	14 $0.027^{2/3}$	15 $0.0081^{3/4}$	16 $0.00032^{2/5}$
17 $64^{-2/3}$	18 $(-8)^{-2/3}$	19 $32^{-3/5}$	20 $625^{-3/4}$
21 $\left(\dfrac{36}{25}\right)^{3/2}$	22 $\left(\dfrac{8}{27}\right)^{2/3}$	23 $\left(\dfrac{81}{625}\right)^{3/4}$	24 $\left(\dfrac{32}{243}\right)^{3/5}$

25 $\left(\dfrac{1,296}{625}\right)^{-3/4}$ **26** $\left(\dfrac{32}{243}\right)^{-2/5}$ **27** $\left(\dfrac{64}{27}\right)^{-2/3}$ **28** $\left(\dfrac{729}{64}\right)^{-5/6}$

29 $\sqrt{49}$ **30** $\sqrt[3]{125}$ **31** $\sqrt[4]{16}$ **32** $\sqrt[3]{343}$

33 $\sqrt[3]{64^2}$ **34** $\sqrt{64^3}$ **35** $\sqrt[5]{32^4}$ **36** $\sqrt[4]{81^3}$

Simplify the expression in each of Probs. 37 to 44.

37 $\sqrt{16a^2b^4}$ **38** $\sqrt[3]{8a^6b^9}$ **39** $\sqrt[4]{16a^8b^{12}}$ **40** $\sqrt[5]{243a^{10}b^{20}}$

41 $\sqrt[4]{\dfrac{16a^8}{b^{12}}}$ **42** $\sqrt[5]{\dfrac{32a^{10}}{b^{25}}}$ **43** $\sqrt[3]{\dfrac{27a^6}{b^9}}$ **44** $\sqrt{\dfrac{729a^8}{b^6}}$

Put the expression in each of Probs. 45 to 52 in radical form.

45 $r^{1/3}s^{2/5}$ **46** $a^{3/4}b^{1/3}$ **47** $a^{2/7}b^{3/7}$ **48** $r^{1/5}s^{2/5}$

49 $x^{3/5}y^{-2/5}$ **50** $x^{-1/4}y^{3/4}$ **51** $x^{2/3}y^{-1/3}$ **52** $x^{3/7}y^{-5/7}$

Combine the following expressions wherever possible by use of the laws of exponents, and express each result without zero or negative exponents.

53 $a^{1/4}a^{5/6}$ **54** $b^{2/3}b^{1/2}$ **55** $c^{3/5}c^{1/4}$ **56** $d^{2/5}d^{1/3}$

57 $\dfrac{p^{3/7}}{p^{1/3}}$ **58** $\dfrac{q^{5/9}}{q^{1/3}}$ **59** $\dfrac{r^{3/5}}{r^{1/2}}$ **60** $\dfrac{s^{5/6}}{s^{1/4}}$

61 $(2x^{1/2})(5x^{1/3})$ **62** $(2x^{1/4})(3x^{1/3})$ **63** $(3x^{1/2})(4x^{1/5})$

64 $(5x^{1/3})(4x^{1/5})$ **65** $\dfrac{6x^{1/2}y^{2/3}}{2x^{3/4}y^{3/5}}$ **66** $\dfrac{27x^{5/9}y^{-1/2}}{9x^{4/9}y^{-3/2}}$

67 $\dfrac{16^{3/4}x^{1/6}y^{-1/2}}{2x^{-1/6}y}$ **68** $\dfrac{32^{4/5}x^{3/2}y^{-1}}{8^{4/3}x^{2/3}y^0}$ **69** $(9x^4y^6)^{1/2}$

70 $(16x^0y^8)^{1/2}$ **71** $(32a^{15})^{1/5}$ **72** $(64a^{-12}b^6)^{1/6}$

73 $(125a^{3/7}b^{-3})^{1/3}$ **74** $(81a^{-8}b^{4/5})^{-1/4}$ **75** $(243^{-1}a^{15})^{1/5}$

76 $(25x^{-4}y^{4/7})^{-1/2}$ **77** $\left(\dfrac{27x^{-6}y}{8x^3y^{-1/3}}\right)^{-2/3}$ **78** $\left(\dfrac{234x^0y^{5/3}}{x^{5/6}y^{5/4}}\right)^{-1/5}$

79 $\left(\dfrac{729^{-1}x^6y^{-1}}{x^{3/2}y^0}\right)^{-1/6}$ **80** $\left(\dfrac{4a^{-1}b^{2/3}}{25a^3b^0}\right)^{-1/2}$ **81** $(9x^4y^6)^{1/2}(8x^6y^9)^{-1/3}$

82 $(16x^{-4}y^2)^{1/2}(125x^6y^{-3})^{1/3}$ **83** $(81a^8b^4)^{-3/4}(32a^{-10}b^5)^{1/5}$

84 $(8a^6b^9)^{-2/3}(4a^4b^6)^{1/2}$ **85** $(a^{3/4}+b^{1/2})(a^{3/4}-b^{1/2})$

86 $(2a^{1/2}-b^{1/5})(2a^{1/2}+b^{1/5})$ **87** $(x^{1/3}-y^{1/3})(x^{2/3}+x^{1/3}y^{1/3}+y^{2/3})$

88 $(a^{2/3}+b^{2/3})(a^{4/3}-a^{2/3}b^{2/3}+b^{4/3})$

89 $(x+1)(2x+3)^{-1/2}+(2x+3)^{1/2}$

90 $-(x+3)(5-2x)^{-1/2}+(5-2x)^{1/2}$

91 $3(3x+2)(2x-5)^{-1/4}+6(2x-5)^{3/4}$

92 $3(2x-5)(3x+4)^{-3/4}+8(3x+4)^{1/4}$

93 $3(2x-1)^{2/3}(x+1)^{-1/2}+8(2x-1)^{-1/3}(x+1)^{1/2}$

94 $3(3x+2)^{2/3}(x+1)^{-1/2}+4(3x+2)^{-1/3}(x+1)^{1/2}$

95 $3(2x+3)^{1/4}(3x-1)^{-1/2}+(3x-1)^{1/2}(2x+3)^{-3/4}$

96 $3(3x+5)^{2/3}(4x-3)^{-1/4}+2(3x+5)^{-1/3}(4x-3)^{3/4}$

97 $\left(\dfrac{x^{1/(b-2)}}{x^{1/(b+2)}}\right)^{(b^2-4)/b}$

98 $\left(\dfrac{x^{a+5b}}{x^{3b}}\right)^{a/(a+2b)}$

99 $\left(\dfrac{x^{a+3b}}{x^a}\right)^a\left(\dfrac{x^{a-2b}}{x^{-2b}}\right)^b$

100 $\left(\dfrac{x^{a-5b}}{x^{-4b}}\right)^{b/(a^2-b^2)}$

101 $\left[(x^{1/(a+2b)})^{(a^2-4b^2)b}\right]^{b/(a-2b)}$

102 $\left[(x^{(a+b)/(a-b)})^{(a-b)/b}\right]^{b/(a+b)}$

103 $\left[(x^{a^2-b^2})^{1/(a-b)}\right]^{2/(a+b)}$

104 $\left[(x^{a^3-b^3})^{1/(a^2+ab+b^2)}\right]^{(a+b)(a-b)}$

7.4 / LAWS OF RADICALS

In this section we develop three useful laws of radicals, and in the remainder of the chapter we shall explain their use.

Since laws (2.39) and (7.1) are valid for $n = 1/k$, we have

$$(ab)^{1/k} = a^{1/k}b^{1/k} \qquad \text{and} \qquad \left(\frac{a}{b}\right)^{1/k} = \frac{a^{1/k}}{b^{1/k}}$$

If we convert these two equalities to radical form, we get

$$\sqrt[k]{ab} = \sqrt[k]{a}\,\sqrt[k]{b} \qquad\qquad (7.6)$$

and

$$\sqrt[k]{\frac{a}{b}} = \frac{\sqrt[k]{a}}{\sqrt[k]{b}} \qquad\qquad (7.7)$$

We use law (7.6) to remove rational factors from the radicand, to insert factors into the radicand, and to obtain the product of two radicals of the same order. We use law (7.7) to obtain the quotient of two radicals of the same order and to rationalize monomial denominators.

REMOVAL OF RATIONAL FACTORS FROM THE RADICAND
If the radicand of a radical of order k has factors that are kth powers, we can remove the kth roots of these powers from the radicand. For example,

$$\sqrt{125} = \sqrt{25(5)} = \sqrt{5^2}\,\sqrt{5} = 5\sqrt{5}$$

$$\sqrt[3]{256} = \sqrt[3]{64(4)} = \sqrt[3]{4^3 4} = 4\sqrt[3]{4}$$

$$\sqrt[4]{128a^6b^9} = \sqrt[4]{16a^4b^8(8a^2b)} = \sqrt[4]{(2ab^2)^4 8a^2b} = 2ab^2\sqrt[4]{8a^2b}$$

MULTIPLICATION OF RADICALS OF THE SAME ORDER
To obtain the product of two radicals of the same order, we use law (7.6) read from right to left. For example, $\sqrt[3]{c}\,\sqrt[3]{d} = \sqrt[3]{cd}$. Usually, products obtained in this way should be simplified by removing all possible rational factors from the radical. The procedure is illustrated in the following two examples.

$$\sqrt{8x^3y}\,\sqrt{6x^2y^5} = \sqrt{48x^5y^6} = \sqrt{16x^4y^6(3x)} = 4x^2y^3\sqrt{3x}$$

$$\sqrt[3]{9a^5b^2}\ \sqrt[3]{81a^2b^7} = \sqrt[3]{729a^7b^9} = \sqrt[3]{3^6a^6b^9(a)} = 3^2a^2b^3\sqrt[3]{a} = 9a^2b^3\sqrt[3]{a}$$

INSERTION OF RATIONAL FACTORS INTO THE RADICAND

Frequently, it is desirable to write a radical expression that has a rational coefficient as a single radical in which the rational coefficient is absorbed in the radicand; this is done by use of the inverse of the process described in the preceding subsection. For example, in $2a\sqrt{4ab}$, we may insert $2a$ into the radicand in this way:

$$2a\sqrt{4ab} = \sqrt{4a^2}\ \sqrt{4ab} = \sqrt{16a^3b}$$

The following example further illustrates the procedure:

$$4x\sqrt[3]{2y} = \sqrt[3]{(4x)^3}\ \sqrt[3]{2y} = \sqrt[3]{64x^3(2y)} = \sqrt[3]{128x^3y}$$

THE QUOTIENT OF TWO RADICALS OF THE SAME ORDER

To get the quotient of two radicals of the same order, we use law (7.7) read from right to left. For example, $\sqrt[n]{a}/\sqrt[n]{b} = \sqrt[n]{a/b}$. The following two examples illustrate the method to be used.

$$\frac{\sqrt{128a^3b^5}}{\sqrt{2ab^2}} = \sqrt{\frac{128a^3b^5}{2ab^2}} = \sqrt{64a^2b^3} = \sqrt{(64a^2b^2)b} = 8ab\sqrt{b}$$

$$\frac{\sqrt[3]{625x^{10}y^7z^{11}}}{\sqrt[3]{5x^2yz^4}} = \sqrt[3]{\frac{625x^{10}y^7z^{11}}{5x^2yz^4}} = \sqrt[3]{125x^8y^6z^7} = \sqrt[3]{(5x^2y^2z^2)^3x^2z}$$

$$= 5x^2y^2z^2\sqrt[3]{x^2z}$$

RATIONALIZATION OF DENOMINATORS

Rationalizing
the denominator

It is frequently desirable to convert a radical with a fractional radicand to a form in which no radical appears in the denominator. This process is called *rationalizing the denominator*. If the denominator of the radicand is a monomial, we use law (7.7) for the purpose. For example, to rationalize the denominator of $\sqrt{8a/3bc^3}$, we multiply the numerator and denominator of the radicand by the expression of lowest power that will convert the denominator into a perfect square. In this case, the expression is $3bc$; hence, we have

$$\sqrt{\frac{8a}{3bc^3}} = \sqrt{\frac{8a(3bc)}{3bc^3(3bc)}} = \sqrt{\frac{2^26abc}{(3bc^2)^2}} = \frac{2\sqrt{6abc}}{3bc^2}$$

As a second example,

$$\sqrt[3]{\frac{4xy^2}{5x^5y^7}} = \sqrt[3]{\frac{4xy^2(25xy^2)}{5x^5y^7(25xy^2)}} = \sqrt[3]{\frac{100x^2yy^3}{(5x^2y^3)^3}} = \frac{y\sqrt[3]{100x^2y}}{5x^2y^3} = \frac{\sqrt[3]{100x^2y}}{5x^2y^2}$$

RATIONALIZING BINOMIAL DENOMINATORS

If the denominator of a fraction is the sum or the difference of two terms at least one of which contains a radical expression of the second order, we rationalize the denominator by the method illustrated below.

EXAMPLE 1   Rationalize the denominator in $4/(\sqrt{5} - 1)$.

Solution   Since by Eq. (3.4), $(\sqrt{5} - 1)(\sqrt{5} + 1) = 5 - 1 = 4$, we multiply the members of the given fraction by $(\sqrt{5} + 1)$ and complete the problem as below.

$$\frac{4}{\sqrt{5} - 1} = \frac{4(\sqrt{5} + 1)}{(\sqrt{5} - 1)(\sqrt{5} + 1)} \qquad \text{by Eq. (4.4)}$$

$$= \frac{4(\sqrt{5} + 1)}{5 - 1}$$

$$= \frac{4(\sqrt{5} + 1)}{4}$$

$$= \sqrt{5} + 1 \qquad \text{by Eq. (4.2)}$$

EXAMPLE 2   Rationalize the denominator in $(a + b + 2\sqrt{ab})/(\sqrt{a} + \sqrt{b})$.

Solution   We note that if we multiply the denominator of the fraction by $\sqrt{a} - \sqrt{b}$, we obtain $a - b$ by Eq. (3.4). Consequently, we proceed as below.

$$\frac{a + b + 2\sqrt{ab}}{\sqrt{a} + \sqrt{b}} = \frac{(a + b + 2\sqrt{ab})(\sqrt{a} - \sqrt{b})}{(\sqrt{a} + \sqrt{b})(\sqrt{a} - \sqrt{b})} \qquad \text{by Eq. (4.2)}$$

$$= \frac{a\sqrt{a} - a\sqrt{b} + b\sqrt{a} - b\sqrt{b} + 2a\sqrt{b} - 2b\sqrt{a}}{a - b}$$

$$= \frac{a\sqrt{a} + b\sqrt{a} - 2b\sqrt{a} - a\sqrt{b} - b\sqrt{b} + 2a\sqrt{b}}{a - b} \qquad \text{by Eq. (2.10)}$$

$$= \frac{(a + b - 2b)\sqrt{a} + (-a - b + 2a)\sqrt{b}}{a - b} \qquad \text{by Eq. (2.12)}$$

$$= \frac{(a - b)\sqrt{a} + (a - b)\sqrt{b}}{a - b}$$

$$= \frac{(a - b)(\sqrt{a} + \sqrt{b})}{a - b} \qquad \text{by Eq. (2.33)}$$

$$= \sqrt{a} + \sqrt{b} \qquad \text{by Eq. (4.2)}$$

EXERCISE 7.4 Radical Expressions

Remove all possible factors from the radicand in Probs. 1 to 36.

1 $\sqrt{18}$ **2** $\sqrt{50}$ **3** $\sqrt{75}$ **4** $\sqrt{294}$

5 $\sqrt[3]{48}$	6 $\sqrt[3]{250}$	7 $\sqrt[4]{80}$	8 $\sqrt[5]{160}$
9 $\sqrt{9a^2b^5}$	10 $\sqrt{8a^3b^7}$	11 $\sqrt{96ab^3}$	12 $\sqrt{125a^9b^3}$
13 $\sqrt{20a^7b^8}$	14 $\sqrt{112a^{11}b^9}$	15 $\sqrt{108a^9b^9}$	16 $\sqrt{175a^6b^7}$
17 $\sqrt[3]{16a^4b}$	18 $\sqrt[3]{108a^6b^4}$	19 $\sqrt[3]{375a^7b^8}$	20 $\sqrt[3]{686a^9b^4}$
21 $\sqrt[4]{48a^8b^5}$	22 $\sqrt[4]{162a^4b^7}$	23 $\sqrt[5]{96a^6b^7}$	24 $\sqrt{486a^{11}b^6}$
25 $\sqrt{\dfrac{6a^7}{4b^4}}$	26 $\sqrt{\dfrac{12r^5}{9s^6}}$	27 $\sqrt{\dfrac{80p^3}{25q^4}}$	28 $\sqrt{\dfrac{18r^5}{16s^8}}$
29 $\sqrt{\dfrac{7b^2}{12a^3}}$	30 $\sqrt{\dfrac{12p^4}{25r^5}}$	31 $\sqrt[3]{\dfrac{27a^7}{12b^5}}$	32 $\sqrt[3]{\dfrac{10d^8}{27c^7}}$
33 $\sqrt[3]{r^{-3a}s^{4a}}$	34 $\sqrt[4]{r^{4a}s^{7a}}$	35 $\sqrt{p^{4a}q^{6a}}$	36 $\sqrt{t^{6a}u^{10a}}$

Incorporate the number that appears before the radical as a part of the radicand in Probs. 37 to 44.

37 $3\sqrt{5}$	38 $2\sqrt{6}$	39 $3\sqrt[3]{3}$	40 $5\sqrt[4]{2}$
41 $2x\sqrt[3]{3xy}$	42 $3x\sqrt[4]{2x^2y}$	43 $2xy^2\sqrt{3x^2y}$	44 $3xy\sqrt[5]{3x^2y}$

In each of Probs. 45 to 64, combine into a single radical and simplify.

45 $\sqrt{3}\,\sqrt{27}$	46 $\sqrt{5}\,\sqrt{125}$	47 $\sqrt{5}\,\sqrt{20}$	48 $\sqrt{24}\,\sqrt{6}$
49 $\sqrt[4]{2}\,\sqrt[4]{8}$	50 $\sqrt[3]{4}\,\sqrt[3]{54}$	51 $\sqrt[5]{27}\,\sqrt[5]{45}$	52 $\sqrt[4]{686}\,\sqrt[4]{168}$
53 $\sqrt{3xy^3}\,\sqrt{147x^3y^5}$	54 $\sqrt{10x^3y^2}\,\sqrt{15xy^4}$		55 $\sqrt{2xy^5}\,\sqrt{200x^3y}$
56 $\sqrt{6x^3y}\,\sqrt{18xy^2}$	57 $\sqrt[3]{9xy^2}\,\sqrt[3]{6x^2y^3}$		58 $\sqrt[3]{16x^4y^2}\,\sqrt[3]{20x^2y^6}$
59 $\sqrt[4]{8x^3y}\,\sqrt[4]{14x^5y^2}$	60 $\sqrt[4]{54x^2y^7}\,\sqrt[4]{120x^6y^2}$		61 $\sqrt[5]{16x^4y^3}\,\sqrt[5]{18x^6y}$
62 $\sqrt[5]{162x^7y}\,\sqrt[5]{96x^3y^6}$	63 $\sqrt[6]{32x^7y}\,\sqrt{54xy^5}$		64 $\sqrt[7]{16x^4y^6}\,\sqrt[7]{40x^3y^2}$

In each of Probs. 65 to 80, rationalize the denominator and simplify.

65 $\sqrt{\dfrac{3}{2}}$	66 $\sqrt[3]{\dfrac{5}{4}}$	67 $\sqrt[4]{\dfrac{2}{27}}$	68 $\sqrt{\dfrac{7}{3}}$
69 $\sqrt{\dfrac{3x^3}{2y}}$	70 $\sqrt{\dfrac{2x}{7y^3}}$	71 $\sqrt{\dfrac{4x^5}{25y^7}}$	72 $\sqrt{\dfrac{27x^7}{8y^5}}$
73 $\sqrt{\dfrac{54x^9y^5}{24x^3y}}$	74 $\sqrt{\dfrac{243x^7y^0}{147x^2y^{-3}}}$	75 $\sqrt{\dfrac{24r^5y^3}{6r^{-3}y^{-4}}}$	76 $\sqrt{\dfrac{64x^7y^3}{98x^3y}}$
77 $\sqrt[3]{\dfrac{9a^5b^2}{16a^{-1}b^{-3}}}$	78 $\sqrt[3]{\dfrac{6a^{-1}b^2}{125a^{-5}b^{-2}}}$	79 $\sqrt[4]{\dfrac{80a^3b^5}{162a^{-1}b^{-3}}}$	80 $\sqrt[5]{\dfrac{96a^3b^6}{486a^{-4}b^0}}$

Rationalize the denominator in each of Probs. 81 to 88.

81 $\dfrac{1}{1+\sqrt{2}}$	82 $\dfrac{3}{\sqrt{5}+1}$	83 $\dfrac{2+\sqrt{3}}{1-\sqrt{3}}$	84 $\dfrac{5+\sqrt{7}}{1-\sqrt{7}}$

85 $\dfrac{\sqrt{3} + 2\sqrt{2}}{\sqrt{2} - \sqrt{3}}$ **86** $\dfrac{1 + 2\sqrt{5}}{\sqrt{3} + \sqrt{5}}$ **87** $\dfrac{\sqrt{14} - 2}{\sqrt{7} - \sqrt{2}}$ **88** $\dfrac{13 - \sqrt{42}}{\sqrt{6} - \sqrt{7}}$

7.5 / CHANGING THE ORDER OF A RADICAL

If it is possible to convert a given radical expression to an equal one of lower order, it is usually advisable to do so. This conversion is possible if the radicand can be expressed as a power in which the exponent is one of the factors of the index of the radical expression. For example, in $\sqrt[9]{64a^6b^{12}}$, the index is $9 = 3 \cdot 3$ and the radicand is $64a^6b^{12} = (4a^2b^4)^3$. Hence,

$$
\begin{aligned}
\sqrt[9]{64a^6b^{12}} &= \sqrt[9]{(4a^2b^4)^3} \\
&= (4a^2b^4)^{3/9} & \text{by Eq. (7.5)} \\
&= (4a^2b^4)^{1/3} & \text{since } \tfrac{3}{9} = \tfrac{1}{3} \\
&= \sqrt[3]{4a^2b^4} & \text{by Eq. (7.4)}
\end{aligned}
$$

Similarly, in the radical $\sqrt[12]{a^{10}b^4c^8}$, the radicand is equal to $(a^5b^2c^4)^2$. Consequently,

$$
\begin{aligned}
\sqrt[12]{a^{10}b^4c^8} &= \sqrt[12]{(a^5b^2c^4)^2} \\
&= (a^5b^2c^4)^{2/12} & \text{by Eq. (7.4)} \\
&= (a^5b^2c^4)^{1/6} & \text{since } \tfrac{2}{12} = \tfrac{1}{6} \\
&= \sqrt[6]{a^5b^2c^4}
\end{aligned}
$$

The same procedure can be employed to convert a given radical expression to an equal radical expression in which the new index is a multiple of the given index. The method is illustrated in Examples 1 and 2.

EXAMPLE 1   Convert the radical expression $\sqrt[3]{4a^2b^5c^7}$ to an equal radical expression of order 9.

Solution
$$
\begin{aligned}
\sqrt[3]{4a^2b^5c^7} &= \sqrt[3]{2^2a^2b^5c^7} & \text{since } 4 = 2^2 \\
&= 2^{2/3}a^{2/3}b^{5/3}c^{7/3} & \text{by Eqs. (7.4) and (2.39)} \\
&= 2^{6/9}a^{6/9}b^{15/9}c^{21/9} & \text{since } \tfrac{2}{3} = \tfrac{6}{9},\ \tfrac{5}{3} = \tfrac{15}{9},\ \text{and } \tfrac{7}{3} = \tfrac{21}{9} \\
&= (2^6a^6b^{15}c^{21})^{1/9} & \text{by Eq. (2.39)} \\
&= \sqrt[9]{2^6a^6b^{15}c^{21}} & \text{by Eq. (7.4)}
\end{aligned}
$$

EXAMPLE 2   Convert $\sqrt[3]{2a^2b}$, $\sqrt{3ab}$, and $\sqrt[4]{4a^3b^2}$ to radical expressions of the same order.

Solution   Since the lcm of the indices of the radical expressions is 12, we shall convert each radical expression to one of order 12.

$$\sqrt[3]{2a^2b} = (2a^2b)^{1/3} \qquad \text{by Eq. (7.4)}$$
$$= (2a^2b)^{4/12}$$
$$= \sqrt[12]{(2a^2b)^4} \qquad \text{by Eq. (7.5)}$$
$$= \sqrt[12]{16a^8b^4} \qquad \text{by Eq. (2.39)}$$

Similarly,
$$\sqrt{3ab} = (3ab)^{1/2}$$
$$= (3ab)^{6/12}$$
$$= \sqrt[12]{(3ab)^6}$$
$$= \sqrt[12]{729a^6b^6}$$

$$\sqrt[4]{4a^3b^2} = (4a^3b^2)^{1/4}$$
$$= (4a^3b^2)^{3/12}$$
$$= \sqrt[12]{(4a^3b^2)^3}$$
$$= \sqrt[12]{64a^9b^6}$$

We occasionally encounter a radical expression in which the radicand is also a radical expression, as in $\sqrt[r]{\sqrt[u]{a}}$. We can convert this expression to a single radical expression by use of Eqs. (7.4) and (2.40) in this way:

$$\sqrt[r]{\sqrt[u]{a}} = (a^{1/u})^{1/r} \qquad \text{by Eq. (7.4)}$$
$$= a^{1/ru} \qquad \text{by Eq. (2.40)}$$
$$= \sqrt[ru]{a} \qquad \text{by Eq. (7.4)}$$

Hence we have
$$\sqrt[r]{\sqrt[u]{a}} = \sqrt[ru]{a} \qquad\qquad (7.8)$$

EXAMPLE 3 Convert $\sqrt[3]{\sqrt[4]{a^{15}b^{12}c^9}}$ to a single radical expression, simplify, and then convert the simplified radical expression to the lowest possible order.

Solution
$$\sqrt[3]{\sqrt[4]{a^{15}b^{12}c^9}} = \sqrt[12]{a^{15}b^{12}c^9} \qquad \text{by Eq. (7.8)}$$
$$= \sqrt[12]{(ab)^{12}a^3c^9} \qquad \text{factoring}$$
$$= ab\sqrt[12]{a^3c^9} \qquad \text{by Eq. (7.6)}$$
$$= ab(a^{3/12}c^{9/12}) \qquad \text{by Eqs. (7.5) and (2.39)}$$
$$= ab(a^{1/4}c^{3/4})$$
$$= ab(ac^3)^{1/4} \qquad \text{by Eq. (2.39)}$$
$$= ab\sqrt[4]{ac^3} \qquad \text{by Eq. (7.4)}$$

7.6 / ADDITION AND MULTIPLICATION OF RADICAL EXPRESSIONS

If the distributive axiom is to be used in adding two or more radical expressions, the radical expressions must be of the same order and have identical radicands. For example, the radicands in $\sqrt{a} + \sqrt{b}$ are not the same, and the orders in $\sqrt{a} + \sqrt[3]{a}$ are different; hence neither sum can be expressed as a single radical expression. Frequently, however, the radical expressions to be added can be converted to others to which the distributive axiom may be applied. The procedure is illustrated in the following example.

EXAMPLE 1   Combine $\sqrt{108} + \sqrt{48} - \sqrt{3}$ into a single radical expression.   (7.6)

Solution

$$\sqrt{108} + \sqrt{48} - \sqrt{3} = \sqrt{36 \cdot 3} + \sqrt{16 \cdot 3} - \sqrt{3} \qquad \text{factoring}$$
$$= 6\sqrt{3} + 4\sqrt{3} - \sqrt{3} \qquad \text{by Eq. (7.6)}$$
$$= (6 + 4 - 1)\sqrt{3} \qquad \text{by Eq. (2.12)}$$
$$= 9\sqrt{3}$$

If the radical expressions to be added have different orders or if the radicands are not identical, the following procedure is suggested:

1   Simplify each radical expression and rationalize all denominators.

2   If it is possible to convert one or more of the radical expressions to a lower order, do so.

3   If possible, assemble the resulting radical expressions into one or more groups composed of radical expressions of the same order and with identical radicands; then apply the distributive axiom to each group.

EXAMPLE 2   Perform the addition indicated in $\sqrt{8a^3b^3} + \sqrt[3]{ab} - \sqrt{2/ab} - \sqrt[3]{8a^4b^4} - \sqrt[4]{4a^2b^2}$.

Solution   In this problem, no two radicands are identical and we have radical expressions of order 2, 3, and 4. The radical expression $\sqrt[4]{4a^2b^2}$, however, can be converted to a lower order in this way: $\sqrt[4]{4a^2b^2} = \sqrt[4]{(2ab)^2} = (2ab)^{2/4} = (2ab)^{1/2} = \sqrt{2ab}$. We therefore replace $\sqrt[4]{4a^2b^2}$ by $\sqrt{2ab}$ and then complete the solution as indicated below.

$$\sqrt{8a^3b^3} + \sqrt[3]{ab} - \sqrt{\frac{2}{ab}} - \sqrt[3]{8a^4b^4} \ - \sqrt[4]{4a^2b^2}$$
$$= \sqrt{(2ab)^2 2ab} + \sqrt[3]{ab} - \sqrt{\frac{2ab}{a^2b^2}} - \sqrt[3]{(2ab)^3 ab} - \sqrt{2ab}$$

$\qquad\qquad\qquad$ factoring radicands and replacing $\dfrac{2}{ab}$ by $\dfrac{2ab}{a^2b^2}$

$$= 2ab\sqrt{2ab} + \sqrt[3]{ab} - \frac{1}{ab}\sqrt{2ab} - 2ab\sqrt[3]{ab} - \sqrt{2ab}$$

$\qquad\qquad\qquad$ by Eqs. (7.6) and (7.7)

$$= 2ab\sqrt{2ab} - \frac{1}{ab}\sqrt{2ab} - \sqrt{2ab} + \sqrt[3]{ab} - 2ab\sqrt[3]{ab}$$

$\qquad\qquad\qquad$ by the commutative axiom, Eq. (2.10)

$$= \left(2ab - \frac{1}{ab} - 1\right)\sqrt{2ab} + (1 - 2ab)\sqrt[3]{ab}$$

$\qquad\qquad\qquad$ by the distributive axiom, Eq. (2.12)

We obtain the product of two multinomials whose terms involve radical expressions by the method of Sec. 2.13 and the use of the law shown in Sec. 7.6. The method is illustrated in Example 3.

EXAMPLE 3 Find the product of $\sqrt{x} - 2\sqrt{xy} + 2\sqrt{y}$ and $\sqrt{x} + 2\sqrt{xy} - \sqrt{y}$.

Solution

$\sqrt{x} - 2\sqrt{xy} + 2\sqrt{y}$	multiplicand
$\sqrt{x} + 2\sqrt{xy} - \sqrt{y}$	multiplier
$x - 2x\sqrt{y} + 2\sqrt{xy}$	multiplying the multiplicand by $\sqrt{x}$
$2x\sqrt{y} \qquad -4xy + 4y\sqrt{x}$	multiplying by $2\sqrt{xy}$
$\qquad -\sqrt{xy} \qquad + 2y\sqrt{x} - 2y$	multiplying by $-\sqrt{y}$
$x \qquad + \sqrt{xy} - 4xy + 6y\sqrt{x} - 2y$	product

We call attention to the fact that the above procedure, as in Sec. 2.13, is a repeated application of the distributive axiom.

INSERTING A FACTOR INTO THE RADICAND
It is sometimes desirable to convert an expression such as $ab\sqrt[n]{c}$ to another in which no factors appear at the left of the radical sign. We accomplish this by use of Eqs. (7.5) and (7.6), as follows:

$$ab\sqrt[n]{c} = \sqrt[n]{a^n b^n}\ \sqrt[n]{c} \qquad \text{by Eq. (7.5)}$$
$$= \sqrt[n]{a^n b^n c} \qquad \text{by Eq. (7.6)}$$

EXAMPLE 4 Insert the factor $a^2 b$ of $a^2 b\sqrt[3]{ab}$ into the radicand.

Solution

$$a^2 b\sqrt[3]{ab} = \sqrt[3]{(a^2 b)^3}\ \sqrt[3]{ab} \qquad \text{by Eq. (7.5)}$$
$$= \sqrt[3]{(a^2 b)^3 (ab)} \qquad \text{by Eq. (7.6)}$$
$$= \sqrt[3]{a^6 b^3 (ab)} \qquad \text{by Eq. (2.39)}$$
$$= \sqrt[3]{a^7 b^4} \qquad \text{by Eq. (2.38)}$$

EXAMPLE 5 In each of the following expressions, insert the integral factor at the left of the radical sign into the radicand and then arrange the given expressions in order of magnitude: $3\sqrt{3}, 2\sqrt{6}, 4\sqrt{2}$.

Solution

$$3\sqrt{3} = \sqrt{3^2 \cdot 3} = \sqrt{3^3} = \sqrt{27}$$
$$2\sqrt{6} = \sqrt{2^2 \cdot 6} = \sqrt{4 \cdot 6} = \sqrt{24}$$
$$4\sqrt{2} = \sqrt{4^2 \cdot 2} = \sqrt{16 \cdot 2} = \sqrt{32}$$

Since $24 < 27 < 32$, the desired arrangement is $2\sqrt{6}, 3\sqrt{3}, 4\sqrt{2}$.

EXERCISE 7.5 **Order, Sum, Product of Radicals**

Express each of Probs. 1 to 8 as a single radical.

1 $\sqrt{\sqrt{3}}$ 2 $\sqrt[3]{\sqrt{7}}$ 3 $\sqrt[4]{\sqrt[3]{5}}$ 4 $\sqrt[5]{\sqrt[4]{6}}$

5 $\sqrt{\sqrt[3]{2a}}$ 6 $\sqrt[3]{\sqrt[3]{6b}}$ 7 $\sqrt[3]{\sqrt[5]{3c}}$ 8 $\sqrt[4]{\sqrt{13b}}$

In each of Probs. 9 to 16, convert the radical to an equal one with the number to the right as index.

9 $\sqrt{2x}$, 4   10 $\sqrt[3]{5y}$, 6   11 $\sqrt[3]{5y}$, 9   12 $\sqrt[4]{9x}$, 8

13 $\sqrt[3]{3x^2y}$, 12   14 $\sqrt{2x^3y^2}$, 8   15 $\sqrt[4]{2xy^3}$, 8   16 $\sqrt[5]{3x^2y^4}$, 20

Change the radicals in each of Probs. 17 to 20 to radicals of the same order.

17 $\sqrt{a}$, $\sqrt[3]{a}$, $\sqrt[6]{a}$ 18 $\sqrt{a}$, $\sqrt[3]{a}$, $\sqrt[4]{a}$

19 $\sqrt[3]{2ac}$, $\sqrt[4]{2ac^2}$, $\sqrt[6]{2a^2c}$ 20 $\sqrt{ab}$, $\sqrt[3]{a^2b}$, $\sqrt[5]{ab^2}$

Reduce the order of the radicals in Probs. 21 to 28.

21 $\sqrt[4]{16x^2y^4}$ 22 $\sqrt[6]{8x^6y^3}$ 23 $\sqrt[6]{36x^6y^4}$ 24 $\sqrt[10]{32x^5y^{10}}$

25 $\sqrt[8]{81x^8y^4}$ 26 $\sqrt[12]{64x^6y^9}$ 27 $\sqrt[9]{8x^6y^{12}}$ 28 $\sqrt[15]{216x^9y^6}$

In Probs. 29 to 56, make all combinations that can be made by addition and subtraction after, if necessary, rationalizing each denominator and removing all possible factors from the radicand.

29 $\sqrt{3} - \sqrt{12} + \sqrt{75}$

30 $\sqrt{2} - \sqrt{98} + \sqrt{242}$

31 $\sqrt{8} - \sqrt{50} + \sqrt{98} - \sqrt{18}$

32 $\sqrt{5} + \sqrt{20} - \sqrt{45} + \sqrt{80}$

33 $\sqrt[3]{3} + \sqrt[3]{24} - \sqrt[3]{81} + \sqrt[3]{192}$

34 $\sqrt[3]{6} + \sqrt[3]{48} + \sqrt[3]{162} - \sqrt[3]{750}$

35 $\sqrt[4]{2} + \sqrt[4]{162} + \sqrt[4]{512}$

36 $\sqrt[5]{3} + \sqrt[5]{96} - \sqrt[5]{729}$

37 $\sqrt{18} + \sqrt[3]{54} - \sqrt{50} + \sqrt[3]{250}$

38 $\sqrt[3]{375} + \sqrt{72} - \sqrt[3]{192} + \sqrt{98}$

39 $\sqrt{75} + \sqrt{50} - \sqrt{48} + \sqrt{128}$

40 $\sqrt{72} + \sqrt{12} + \sqrt{162} - \sqrt{147}$

41 $\sqrt[3]{ab^3} + 2\sqrt{a^3b} - a\sqrt{ab^3}$

42 $2\sqrt{st^2} + 3\sqrt{s^3} + 5\sqrt{s^3t^4}$

43 $\sqrt{12x^3} - \sqrt{75x^3y^2} + \sqrt{27xy^4z^2}$

44 $\sqrt{8u^3v^2} - \sqrt{18u^5} + \sqrt{8uv^4}$

45 $\sqrt{\dfrac{3}{x}} + \dfrac{\sqrt{12x^2}}{x} + \sqrt{\dfrac{48}{x}}$

46 $\dfrac{\sqrt{9x}}{x} + \dfrac{\sqrt{16x^3}}{x^2} - \dfrac{3}{\sqrt{x}}$

47 $\dfrac{c\sqrt{18c}}{d} + \dfrac{\sqrt{8c^3}}{d} - \dfrac{6c^2}{d\sqrt{2c}}$

48 $\dfrac{x^2\sqrt{3x}}{y} - \dfrac{\sqrt{75x}}{5y} + \dfrac{3x\sqrt{12x}}{2y}$

49 $\sqrt{25a^2b} - \sqrt{9ab^2} - \sqrt{16a^2b} + \sqrt{36ab^2}$

50 $\sqrt{5ab^3} + \sqrt{a^3b^2} + \sqrt{20ab^5} + \sqrt{ab^4}$

51 $\sqrt[3]{27a^4b^5} - \sqrt[3]{8a^7b^8} + \sqrt[3]{64a^5b^7} - \sqrt[3]{27a^2b^4}$

52 $\sqrt[3]{64x^8y^4} + \sqrt[3]{8x^4y^5} - \sqrt[3]{8x^5y^4} - \sqrt[3]{27x^{10}y^2}$

53 $\sqrt{2ab} + \sqrt[4]{4a^2b^2} + \sqrt[6]{8a^3b^3}$ 54 $\sqrt[3]{27a^4b} - \sqrt[6]{a^8b^2} + \sqrt[9]{a^{12}b^3}$

55 $\sqrt{4a^2b} - \sqrt[4]{16a^4b^6} - \sqrt[8]{256a^8b^4}$ 56 $\sqrt[5]{ab} - \sqrt[10]{a^2b^2} + \sqrt[15]{a^3b^3}$

EXERCISE 7.6 **REVIEW**

Perform the indicated operations and simplify in Probs. 1 to 46.

1 $3^2 3^4$ 2 $6^5/6^3$ 3 $(2/3)^4$ 4 $(2^2)^3$

5 $(2^2 3^3)^3$ 6 $(5x^2y^3)(3x^0y^4)$ 7 $12x^4y^5/3x^2y^2$ 8 $(a^2b^4)^2$

9 $(2a^3b^2)^3$ 10 $(a^2b^0)^4/(a^3b^2)^2$ 11 $\left(\dfrac{a^3b^2}{c^3}\right)^2\left(\dfrac{a^2c^4}{b}\right)^3$

12 $\left(\dfrac{8a^3b^2}{c^4d^3}\right)^3 \div \left(\dfrac{4a^4b^3}{c^5d^4}\right)^2$ 13 $\dfrac{(a^{2n-1}\,b^{n+3})^2}{a^{n-2}\,b^{n+1}}$ 14 3^{-2}

15 $3^{-1}/3^{-3}$ 16 $(3^{-2})^2$ 17 $2^{-4}2^3$

18 a^3/b^{-2} 19 $a^{-1}b^{-2}/a^2b^2c^3$ 20 $\dfrac{3a^{-1}b^{-2}}{2^{-1}a^3b^{-5}}$

21 $\left(\dfrac{a^{-2}}{b^3}\right)^{-3}$ 22 $125^{1/3}$ 23 $0.64^{1/2}$

24 $\sqrt[6]{64}$ 25 $\sqrt[4]{81}$ 26 $\sqrt{25a^2b^6}$

27 $\sqrt[3]{\dfrac{8a^3}{b^6}}$ 28 $\sqrt{\dfrac{676a^4}{b^2}}$ 29 $(3x^{1/4})(2x^{1/3})$

30 $(5x^{1/3})(3x^{1/5})$ 31 $(25x^2y^4)^{1/2}$ 32 $(64a^{3/5}b^{-3})^{1/3}$

33 $(8x^3y^6)^{1/3}(9x^2y^6)^{-1/2}$ 34 $\left(\dfrac{x^{2a+b}}{x^{a+b}}\right)^{1/a}$

35 $4(3x+2)^{1/2}(2x-3)^{-2/3} + 9(2x-3)^{1/3}(3x+2)^{-1/2}$

36 $\sqrt{27}$ 37 $\sqrt{80}$ 38 $\sqrt[3]{40}$ 39 $\sqrt[5]{96}$

40 $\sqrt{8}\sqrt{2}$ 41 $\sqrt[3]{18}\sqrt[3]{12}$ 42 $\sqrt[5]{a^2}\sqrt[5]{a^4}$ 43 $\sqrt[3]{20x^2y^4}\sqrt[3]{50xy^2}$

44 $\sqrt{15x^2y}\sqrt{6x^4y^3}$ 45 $\sqrt[3]{\dfrac{135x^{10}y^{-1}}{320x^{-2}y^5}}$ 46 $\dfrac{3-\sqrt{2}}{1+\sqrt{2}}$

Express the radicals in each of Probs. 47 to 58 as a single radical.

47 $\sqrt[3]{\sqrt{9}}$ 48 $\sqrt{\sqrt[3]{16}}$ 49 $\sqrt[4]{\sqrt[3]{x^8}}$ 50 $\sqrt[3]{\sqrt{64}}$

51 $\sqrt{2} + \sqrt{50} - \sqrt{72}$ 52 $\sqrt{9a} + \sqrt{4b} - \sqrt[3]{27a}$

53 $3\sqrt{ab^3} + \sqrt{4a^3b} + \sqrt{9a^3b^3}$ 54 $\sqrt{3ab} - \sqrt[4]{9a^2b^2} + \sqrt[6]{27a^3b^3}$

55 $(\sqrt{3} + \sqrt{7})(\sqrt{3} - \sqrt{7})$ 56 $(2\sqrt{2} + 3\sqrt{5})(3\sqrt{2} - \sqrt{5})$

57 $(\sqrt{u} + \sqrt{v})(u - \sqrt{uv} + v)$

58 $(\sqrt{a} + \sqrt{b} + \sqrt{c})(\sqrt{a} - \sqrt{b} + \sqrt{c})$

CHAPTER EIGHT

Quadratic Equations

In previous chapters we have dealt with problems that could be solved by means of equations involving only the first power of the variable. Practical problems and theoretical considerations, however, frequently lead to more complicated equations. For example, if the resistance of the air is neglected, the distance s that a compact body falls through the air in t sec is given by the formula $s = 16.1t^2 + v_0t$, where s is expressed in feet and v_0 is the initial vertical velocity in feet per second. Thus, if a body is released from an airplane that is 5,000 ft high and is given an initial downward velocity of 20 ft/sec, then the time required for the body to reach the ground is a root of the equation $5,000 = 16.1t^2 + 20t$. This equation involves the square of the variable t, and it is called a *quadratic equation*, after the Latin word for "square." In this chapter we shall develop methods for solving such equations.

8.1 / DEFINITIONS

We define a quadratic equation as follows:

Quadratic equation   The equation $f(x) = g(x)$ is a *quadratic equation* in x if (1) one of $f(x)$ and $g(x)$ is a polynomial of second degree in x and the other is a poly-

176

nomial of first degree in x or is a constant, or (2) both $f(x)$ and $g(x)$ are polynomials of the second degree in x with the coefficients of x^2 not equal.

Examples of quadratic equations are

$$3x^2 - 5x + 1 = x^2 + 3x - 2$$
$$2x^2 - 4x + 2 = 6x + 3$$
$$3x^2 - 2x + 1 = 4x$$
$$5x^2 - 2x = 3$$

Before discussing methods for solving the general quadratic equation, we shall obtain the solution set of an equation of the type

$$ax^2 - b = 0 \qquad\qquad (1)$$

We proceed as follows:

$$x^2 = \frac{b}{a} \qquad \text{solving for } x^2$$

$$x = \pm\sqrt{\frac{b}{a}} \qquad \text{since } \left(\pm\sqrt{\frac{b}{a}}\right)^2 = \frac{b}{a}$$

$$= \pm\frac{\sqrt{ab}}{a} \qquad \text{by Eqs. (4.2) and (7.7)}$$

Hence the solution set of (1) is $\{\sqrt{ab}/a, -\sqrt{ab}/a\}$.

We shall prove in Sec. 8.5 that any given quadratic equation is equivalent to an equation of the type

$$(x + d)^2 = k \qquad\qquad (2)$$

Consequently, by the definition of a square root, an element in the solution set of (2) must satisfy the equation $x + d = \pm\sqrt{k}$ since $(\pm\sqrt{k})^2 = k$. Therefore,

$$x = -d \pm \sqrt{k}$$

and the solution set of (2) is $\{-d + \sqrt{k}, -d - \sqrt{k}\}$. Hence we have the following theorem:

THEOREM **If the equation $(x + d)^2 = k$ is given, then $x + d = \pm\sqrt{k}$ and the solution set of the equation is $\{-d + \sqrt{k}, -d - \sqrt{k}\}$ (8.1)**

Thus there are two elements in the solution set of a quadratic equation. In other words, a quadratic equation has two roots. It may happen, however, that the two elements in the solution set are equal. For example, if $k = 0$ in (2), then the solution set is $\{-d + 0, -d - 0\} = \{-d, -d\}$.

Before discussing the general procedure for solving a quadratic equation, we shall explain a method that is applicable to a wide variety of equations.

8.2 / SOLUTION BY FACTORING

If a quadratic equation is equivalent to an equation of the type

$$(ax + b)(cx + d) = 0 \qquad (1)$$

then by Eq. (2.35) the equation is satisfied by any replacement for x that will make either factor of the left member of (1) equal to zero. Hence we may obtain the roots of (1) by setting each factor of the left number separately equal to zero and solving the resulting equation for x. Thus we get

$ax + b = 0$	setting the first factor in (1) equal to 0
$x = -\dfrac{b}{a}$	solving for x
$cx + d = 0$	setting the second factor in (1) equal to 0
$x = -\dfrac{d}{c}$	solving for x

Consequently, the solution set of (1) is $\{-b/a, -d/c\}$.

As the above discussion illustrates, the method of solving a quadratic equation by factoring consists of the following steps:

1 From the given equation, obtain an equivalent equation in which the right member is zero.

2 Factor the left member of the equation obtained in step 1.

3 Set each factor obtained in step 2 equal to zero, and solve the resulting equations for x.

4 The elements of the solution set are the two roots obtained in step 3.

EXAMPLE 1   Solve the equation $2x^2 = x + 6$ by the factoring method.

Solution

$2x^2 = x + 6$	given equation
$2x^2 - x - 6 = 0$	adding $-x - 6$ to each member
$(2x + 3)(x - 2) = 0$	factoring the left member
$2x + 3 = 0$	setting the first factor equal to 0
$2x = -3$	
$x = -\frac{3}{2}$	
$x - 2 = 0$	setting the second factor equal to 0
$x = 2$	

Consequently, the solution set is $\{-\frac{3}{2}, 2\}$. This can be verified by substituting each element of the set for x in the given equation.

EXAMPLE 2   Solve the equation $5y^2 = 6y$ by the factoring method.

Solution

$$5y^2 = 6y \qquad \text{given equation}$$
$$5y^2 - 6y = 0 \qquad \text{adding } -6y \text{ to each member}$$
$$y(5y - 6) = 0 \qquad \text{factoring the left member}$$
$$y = 0 \qquad \text{setting the first factor equal to 0}$$
$$5y - 6 = 0 \qquad \text{setting the second factor equal to 0}$$
$$5y = 6$$
$$y = \tfrac{6}{5}$$

Therefore the solution set is $\{0, \tfrac{6}{5}\}$.

Check   If, in the given equation, we replace y by 0, each member is equal to 0. If we replace y by $\tfrac{6}{5}$, we have $5 \times \tfrac{36}{25} = \tfrac{36}{5}$. This verifies the fact that $\{0, \tfrac{6}{5}\}$ is the required solution set.

NOTE: We wish to impress the reader with the fact that this method is applicable *only when the right member of the equation is zero.* If one of the factors in the left member is zero, their product is zero regardless of the value of the other factor. If, however, the right member of the equation is not zero, as in $(x - 1)(x - 2) = 6$, we cannot assign a value arbitrarily to one of the factors without at the same time fixing the value of the other. For example, if in the above equation we let $x - 1 = 2$, then $x - 2$ must be 3 since the product of the two factors is 6, but $x - 2 = (x - 1) - 1 = 1$, not 3, if $x - 1 = 2$.

EXERCISE 8.1 Solution by Factoring

Find the solution set of the pure quadratics in each of Probs. 1 to 12.

1	$x^2 - 1 = 0$	2	$x^2 - 9 = 0$	3	$x^2 - 16 = 0$
4	$x^2 - 36 = 0$	5	$4x^2 - 1 = 0$	6	$9x^2 - 1 = 0$
7	$25x^2 - 1 = 0$	8	$49x^2 - 1 = 0$	9	$4x^2 - 9 = 0$
10	$25x^2 - 4 = 0$	11	$16x^2 - 81 = 0$	12	$36x^2 - 49 = 0$

Find the solution set of each of the following equations by factoring.

13	$x^2 - 3x + 2 = 0$	14	$x^2 - 5x + 6 = 0$	15	$x^2 - 4x + 3 = 0$
16	$x^2 - 7x + 10 = 0$	17	$x^2 + x - 6 = 0$	18	$x^2 + x - 2 = 0$
19	$x^2 + 7x + 10 = 0$	20	$x^2 + 4x + 3 = 0$	21	$2x^2 - 7x + 3 = 0$
22	$3x^2 - 7x + 2 = 0$	23	$5x^2 - 21x + 4 = 0$	24	$4x^2 - 13x + 3 = 0$
25	$2x^2 - 3x = 2$	26	$3x^2 - 8x = 3$	27	$5x^2 - 24x = 5$
28	$6x^2 - 35x = 6$	29	$3x^2 + 8 = 14x$	30	$5x^2 + 6 = 17x$

31 $5x^2 + 12 = 32x$	**32** $4x^2 + 12 = 19x$	**33** $7x^2 - 33x - 10 = 0$
34 $4x^2 - 9x - 9 = 0$	**35** $3x^2 - 4x - 4 = 0$	**36** $5x^2 - 7x - 6 = 0$
37 $5x^2 + 13x - 6 = 0$	**38** $5x^2 + 7x - 6 = 0$	**39** $3x^2 + x - 2 = 0$
40 $5x^2 + x - 4 = 0$	**41** $6x^2 - 13x + 6 = 0$	**42** $6x^2 + 5x - 6 = 0$
43 $12x^2 - 25x + 12 = 0$	**44** $12x^2 + 7x - 12 = 0$	**45** $15x^2 - 19x + 6 = 0$
46 $35x^2 - 24x + 4 = 0$	**47** $16x^2 - 34x + 15 = 0$	**48** $14x^2 + 43x - 21 = 0$
49 $15x^2 = 2x + 8$	**50** $10x^2 = 19x - 6$	**51** $72x^2 = -13x + 15$
52 $63x^2 = x + 12$	**53** $x^2 - ax = 2x - 2a$	**54** $x^2 + bx = 3x + 3b$
55 $3x^2 - x = 2a - 6ax$	**56** $5x^2 + 2x = -15ax - 6a$	
57 $10x^2 - 19ax + 6a^2 = 0$	**58** $6x^2 - 15bx - 4cx + 10bc = 0$	
59 $pqx^2 - 2(p^2 + q^2)x + 4pq = 0$	**60** $6x^2 - (9a + 4c)x + 6ac = 0$	

8.3 / COMPLEX NUMBERS

In Sec. 8.1, we stated that every quadratic equation is equivalent to an equation of the type $(x + d)^2 = k$ and that its roots are $x = -d \pm \sqrt{k}$. If $k < 0$, then $\sqrt{k}$ is not a real number since the square of any nonzero real number is positive. Hence, we must ignore numbers represented by the square root of a negative number or define an additional set or system of numbers. We choose the latter course.

The early mathematicians did not understand the numbers we are about to define and called them *imaginary*. In the eighteenth century, however, the mathematicians Gauss and Argand each devised a geometrical interpretation for numbers of this type, and since that time, they have become very useful and important not only in mathematics but also in physics, electrical engineering, and electronics.

We shall discuss the gaussian interpretation more fully in Chap. 12. Here we shall introduce the gaussian notation so that we can use it when necessary in solving the equations in this chapter. We shall let the letter i stand for $\sqrt{-1}$. Then $i^2 = -1$, $i^3 = i^2(i) = -1(i) = -i$, and $i^4 = i^2(i^2) = -1 \times (-1) = 1$. Therefore, the successive powers of i are the recurring numbers i, -1, $-i$, and 1. In terms of this notation, we have $\sqrt{-4} = 2\sqrt{-1} = 2i$, $\sqrt{-9} = 3\sqrt{-1} = 3i$, and if $n > 0$, $\sqrt{-n} = \sqrt{n(-1)} = \sqrt{n}\sqrt{-1} = \sqrt{n}\, i$. We describe such numbers in the following definition.

Complex number   A number of the type $a + bi$, where a and b are real numbers and $i = \sqrt{-1}$, is called a *complex number*.

Imaginary number   If neither a nor b is zero, then $a + bi$ is also called an *imaginary number*. If $a = 0$ and $b \neq 0$, then $a + bi = bi$, and this number is a *pure* imaginary number. If $b = 0$, then $a + bi = a$, and this is a real number.

The imaginary numbers $2 + 3i$ and $\frac{3}{4} - \frac{1}{4}i$, the pure imaginary number $2i$, and the real number 3 are all examples of complex numbers.

8.4 / SOLUTION BY COMPLETING THE SQUARE

As we pointed out in Sec. 8.1, the solution set of an equation of the type

$$(x + d)^2 = k \tag{1}$$

is readily obtained. Hence, if we can obtain an equation of the type (1) that is equivalent to the given equation, we can obtain the solution set without difficulty. We shall therefore examine the procedure for accomplishing this purpose. Since we shall employ theorems (5.1) and (5.2) repeatedly in the application of this method, the reader is advised to review those theorems at this point.

We shall first examine the conditions under which a quadratic trinomial of the type $x^2 + bx + c$ is a perfect square. By Eq. (3.2),

$$(x + d)^2 = x^2 + 2dx + d^2$$

and from this identity, we see that the constant term in the quadratic trinomial at the right of the equality sign is the *square of one-half the coefficient of x.* Consequently:

A quadratic trinomial in x with the coefficient of x^2 equal to 1 is a perfect square if the *constant term is the square of one-half the coefficient of x.*

EXAMPLE 1 $x^2 - 6x + 9$ is a perfect square since $9 = (-\frac{6}{2})^2$. Furthermore, $x^2 - 6x + 9 = (x - 3)^2$.

EXAMPLE 2 $x^2 + 5x + \frac{25}{4}$ is a perfect square since $\frac{25}{4} = (\frac{5}{2})^2$. Also, $x^2 + 5x + \frac{25}{4} = (x + \frac{5}{2})^2$.

EXAMPLE 3 $x^2 - \frac{1}{3}x + \frac{1}{36}$ is a perfect square since $\frac{1}{36} = (\frac{1}{2} \times \frac{1}{3})^2$ and $x^2 - \frac{1}{3}x + \frac{1}{36} = (x - \frac{1}{6})^2$.

Example 4 illustrates the use of the above information in solving a quadratic equation.

EXAMPLE 4 Solve the equation

$$x = 3 - 2x^2 \tag{2}$$

by completing the square.

Solution We first obtain an equation that is equivalent to (2) and of the type $x^2 + bx = c$. The procedure is to employ theorem (5.1), add $2x^2$ to each member of (2), and obtain

$$2x^2 + x = 3 \tag{3}$$

then employ theorem (5.2), multiply each member of (3) by $\frac{1}{2}$, and get

$$x^2 + \tfrac{1}{2}x = \tfrac{3}{2} \tag{4}$$

The next step is to *complete the square* by adding the square of one-half the coefficient of x, or $(\frac{1}{2} \times \frac{1}{2})^2 = \frac{1}{16}$, to each member of (4). Thus, we get

$$x^2 + \tfrac{1}{2}x + \tfrac{1}{16} = \tfrac{3}{2} + \tfrac{1}{16}$$
$$= \tfrac{25}{16}$$
$$(x + \tfrac{1}{4})^2 = (\tfrac{5}{4})^2 \qquad \text{since } x^2 + \tfrac{1}{2}x + \tfrac{1}{16} = (x + \tfrac{1}{4})^2 \text{ and } \tfrac{25}{16} = (\tfrac{5}{4})^2$$

We now employ (8.1), equate the square roots of the members of the last equation, prefix the square root of $(\frac{5}{4})^2$ by the plus and minus signs, and thus obtain

$$x + \tfrac{1}{4} = \pm\tfrac{5}{4} \tag{5}$$

We now complete the solution process as below.

$$x = \pm\tfrac{5}{4} - \tfrac{1}{4} \qquad \text{adding } -\tfrac{1}{4} \text{ to each member of (5)}$$
$$= \tfrac{5}{4} - \tfrac{1}{4} \text{ and } -\tfrac{5}{4} - \tfrac{1}{4}$$
$$= 1 \text{ and} -\tfrac{3}{2}$$

Hence the solution set is $\{1, -\tfrac{3}{2}\}$.

Check If we replace x by 1 and then by $-\frac{3}{2}$ in (2), we have

	LEFT MEMBER	RIGHT MEMBER
$x = 1$	1	$3 - (2 \times 1^2) = 3 - 2 = 1$
$x = -\tfrac{3}{2}$	$-\tfrac{3}{2}$	$3 - [2 \times (-\tfrac{3}{2})^2] = 3 - \tfrac{9}{2}$
		$= \dfrac{6-9}{2} = -\dfrac{3}{2}$

The above example illustrates the following formal steps that are used in the process of solving a quadratic equation by the method of completing the square:

Completing the square

1 Rewrite the equation with the x^2 and the x terms on the left and the constant on the right.

2 Divide each term of the given equation by the coefficient of x^2.

3 Complete the square of the left member.

4 Simplify the right member obtained in step 3.

5 Take the square root of each member. Be sure to use both $+$ and $-$ with the right member.

6 Solve the two linear equations obtained in step 5.

EXAMPLE 5   Solve the equation $4x^2 = 4x + 11$ by completing the square.

Solution

$$4x^2 = 4x + 11 \qquad \text{given equation}$$
$$4x^2 - 4x = 11 \qquad \text{adding } -4x \text{ to each member}$$
$$x^2 - x = \tfrac{11}{4} \qquad \text{multiplying each member by } \tfrac{1}{4}$$
$$x^2 - x + (-\tfrac{1}{2})^2 = \tfrac{11}{4} + (-\tfrac{1}{2})^2 \qquad \text{adding } [\tfrac{1}{2} \times (-1)]^2 \text{ to each member}$$
$$= \tfrac{11}{4} + \tfrac{1}{4}$$
$$= \tfrac{12}{4}$$
$$(x - \tfrac{1}{2})^2 = 3$$
$$x - \tfrac{1}{2} = \pm\sqrt{3} \qquad \text{equating the square roots of the members}$$
$$x = \tfrac{1}{2} \pm \sqrt{3} \qquad \text{solving for } x$$

Since $\sqrt{3}$ is an irrational number, the above roots cannot be simplified further. Hence the solution set is

$$\{\tfrac{1}{2} + \sqrt{3},\ \tfrac{1}{2} - \sqrt{3}\}$$

Check   We replace x in the given equation by $\tfrac{1}{2} \pm \sqrt{3}$ and get for the left member

$$4(\tfrac{1}{2} \pm \sqrt{3})^2 = 4[\tfrac{1}{4} \pm 2(\tfrac{1}{2}\sqrt{3}) + (\sqrt{3})^2] = 4(\tfrac{1}{4} \pm \sqrt{3} + 3) = 13 \pm 4\sqrt{3}$$

Similarly, for the right member,

$$4(\tfrac{1}{2} \pm \sqrt{3}) + 11 = 2 \pm 4\sqrt{3} + 11 = 13 \pm 4\sqrt{3}$$

If an approximate value of each root is wanted, we obtain the value of $\sqrt{3}$ to as many decimal places as desired and then calculate the value of $\tfrac{1}{2} \pm \sqrt{3}$. To three decimal places, $\sqrt{3} = 1.732$. Consequently, $\tfrac{1}{2} \pm \sqrt{3} = 0.5 \pm 1.732 = 2.232$ and -1.232.

EXAMPLE 6   Solve the equation $x^2 + 8 = 4x$ by completing the square.

Solution

$$x^2 + 8 = 4x \qquad \text{given equation}$$
$$x^2 - 4x = -8 \qquad \text{adding } -4x - 8 \text{ to each member;}$$
$$\qquad\qquad\qquad \text{note that step 2 is unnecessary}$$
$$x^2 - 4x + (-2)^2 = -8 + (-2)^2 \qquad \text{adding } [\tfrac{1}{2} \times (-4)]^2 \text{ to each member}$$
$$= -4$$
$$(x - 2)^2 = -4$$
$$x - 2 = \pm\sqrt{-4} \qquad \text{equating the square roots of the members}$$
$$= \pm\sqrt{4 \times (-1)}$$
$$= \pm 2\sqrt{-1} = \pm 2i$$
$$x = 2 \pm 2i$$

Therefore the solution set is $\{2 + 2i,\ 2 - 2i\}$.

These solutions can be checked in the usual manner.

EXERCISE 8.2 **Completing the Square**

Find the solution set of the equation in each of Probs. 1 to 48 by completing the square.

1 $x^2 - 3x + 2 = 0$ 2 $x^2 - 5x + 6 = 0$
3 $x^2 - 8x + 15 = 0$ 4 $x^2 - 5x + 4 = 0$
5 $x^2 - x - 6 = 0$ 6 $x^2 + x - 2 = 0$
7 $x^2 + 6x + 8 = 0$ 8 $x^2 + 8x + 15 = 0$
9 $2x^2 - 5x + 2 = 0$ 10 $3x^2 - 11x + 6 = 0$
11 $4x^2 - 23x + 15 = 0$ 12 $5x^2 - 8x + 3 = 0$
13 $3x^2 + x - 2 = 0$ 14 $3x^2 - 5x - 2 = 0$
15 $7x^2 - 33x - 10 = 0$ 16 $5x^2 + 17x + 6 = 0$
17 $15x^2 + 10 = 31x$ 18 $20x^2 + 15 = 37x$
19 $14x^2 = 53x - 14$ 20 $10x^2 = 12 + 7x$
21 $6x^2 - 5x = 21$ 22 $15x^2 + 19x = -6$
23 $30x^2 = 30 - 11x$ 24 $10x^2 = -23x - 12$
25 $x^2 - 2x - 1 = 0$ 26 $x^2 - 4x + 1 = 0$
27 $x^2 - 6x + 4 = 0$ 28 $x^2 - 10x + 22 = 0$
29 $x^2 - 8x + 14 = 0$ 30 $x^2 - 4x - 1 = 0$
31 $x^2 + 6x + 6 = 0$ 32 $x^2 + 2x - 6 = 0$
33 $x^2 - 8x + 20 = 0$ 34 $x^2 - 6x + 10 = 0$
35 $x^2 - 10x + 34 = 0$ 36 $x^2 - 6x + 34 = 0$
37 $x^2 + 4x + 13 = 0$ 38 $x^2 + 2x + 17 = 0$
39 $x^2 + 14x + 53 = 0$ 40 $x^2 + 4x + 53 = 0$
41 $x^2 - ax - bx + ab = 0$ 42 $x^2 - (2a + 3b)x + 6ab = 0$
43 $x^2 - 3ax + 2a^2 = 0$ 44 $x^2 - bx - 2b^2 = 0$
45 $2a^2x^2 - 3abx + b^2 = 0$ 46 $c^2x^2 + acx - 2a^2 = 0$
47 $4x^2 - 4ax + a^2 + b^2 = 0$ 48 $9x^2 - 6ax + a^2 + 4b^2 = 0$

8.5 / THE QUADRATIC FORMULA

In this section we shall derive a formula for finding the solution set of the quadratic equation

$$ax^2 + bx + c = 0 \qquad (8.2)$$

Standard form
which is said to be in *standard form* and which is equivalent to the quadratic equation $f(x) = g(x)$.

EXAMPLE 1   Obtain the quadratic equation in standard form that is equivalent to $4x - 3 = 5x^2$, and identify a, b, and c.

Solution

$$4x - 3 = 5x^2 \qquad \text{given equation}$$
$$4x - 3 - 5x^2 = 5x^2 - 5x^2 \qquad \text{adding } -5x^2 \text{ to each member}$$
$$-5x^2 + 4x - 3 = 0 \qquad \text{by the commutative and associative axioms}$$

This is the required equation; it has $a = -5$, $b = 4$, and $c = -3$.

We shall now solve Eq. (8.2) by the method of completing the square.

$$ax^2 + bx + c = 0 \qquad \text{given equation}$$
$$ax^2 + bx = -c \qquad \text{adding } -c \text{ to each member}$$
$$x^2 + \frac{b}{a}x = \frac{-c}{a} \qquad \text{multiplying each member by } \frac{1}{a}$$
$$x^2 + \frac{b}{a}x + \left(\frac{b}{2a}\right)^2 = \frac{-c}{a} + \frac{b^2}{4a^2} \qquad \text{adding } \left(\frac{1}{2}\frac{b}{a}\right)^2 \text{ to each member}$$
$$\left(x + \frac{b}{2a}\right)^2 = \frac{b^2 - 4ac}{4a^2}† \qquad \text{simplifying}$$
$$x + \frac{b}{2a} = \pm\frac{\sqrt{b^2 - 4ac}}{2a} \qquad \text{equating the square roots of the members}$$
$$x = -\frac{b}{2a} \pm \frac{\sqrt{b^2 - 4ac}}{2a} \qquad \text{solving for } x$$

Now, since the denominators of the right member of the last equation are the same, we have

The quadratic formula

$$x = \frac{-b \pm \sqrt{b^2 - 4ac}}{2a} \qquad (8.3)$$

Consequently, the solution set of Eq. (8.2) is

$$\left\{\frac{-b + \sqrt{b^2 - 4ac}}{2a}, \frac{-b - \sqrt{b^2 - 4ac}}{2a}\right\}$$

Equation (8.3) is known as the *quadratic formula,* and it can be used to obtain the solution set of any quadratic equation.

EXAMPLE 2   Obtain the solution set of $6x^2 = 12 + x$ by use of the quadratic formula.

Solution   We shall first obtain a quadratic equation in standard form that is equivalent to the given equation, and then we shall apply the formula. Note that in the standard form, the term involving x^2 is first, the term involving x is second, the constant term is third, and the right member is zero. The procedure for solving follows.

$$6x^2 = 12 + x \qquad \text{given equation}$$
$$6x^2 - x - 12 = 0 \qquad \text{adding } -x - 12 \text{ to each member}$$

† This proves that any given quadratic equation is equivalent to an equation of the form $(x+d)^2 = k$. Here, $d = b/2a$ and $k = (b^2 - 4ac)/4a^2$.

The above equation is in standard form, and if we compare it with Eq. (8.2), we see that $a = 6$, $b = -1$, and $c = -12$. If we replace a, b, and c in Eq. (8.3) by these numbers, we get

$$x = \frac{-(-1) \pm \sqrt{(-1)^2 - [4 \times 6 \times (-12)]}}{2 \times 6}$$

$$= \frac{1 \pm \sqrt{1 + 288}}{12}$$

$$= \frac{1 \pm \sqrt{289}}{12} = \frac{1 \pm 17}{12}$$

$$= \tfrac{3}{2} \text{ and } -\tfrac{4}{3}$$

Hence the solution set is $\{\tfrac{3}{2}, -\tfrac{4}{3}\}$.

Check   Replacing x by $\tfrac{3}{2}$ and then by $-\tfrac{4}{3}$ in the given equation, we have

	LEFT MEMBER	RIGHT MEMBER
$x = \tfrac{3}{2}$	$6 \times (\tfrac{3}{2})^2 = 6 \times \tfrac{9}{4} = \tfrac{27}{2}$	$12 + \tfrac{3}{2} = \dfrac{24 + 3}{2} = \tfrac{27}{2}$
$x = -\tfrac{4}{3}$	$6 \times (-\tfrac{4}{3})^2 = 6 \times \tfrac{16}{9} = \tfrac{32}{3}$	$12 + (-\tfrac{4}{3}) = \dfrac{36 - 4}{3} = \tfrac{32}{3}$

EXAMPLE 3   Find the solution set of $8x - 13 = 4x^2$ by use of the quadratic formula.

Solution

$$8x - 13 = 4x^2 \qquad \text{given equation}$$
$$-4x^2 + 8x - 13 = 0 \qquad \text{adding } -4x^2 \text{ to each number}$$

$$x = \frac{-8 \pm \sqrt{8^2 - [4 \times (-4) \times (-13)]}}{2 \times (-4)} \qquad \text{replacing } a \text{ by } -4, b \text{ by } 8, \text{ and } c \text{ by } -13 \text{ in Eq. (8.3)}$$

$$= \frac{-8 \pm \sqrt{64 - 208}}{-8} = \frac{-8 \pm \sqrt{-144}}{-8} = \frac{-8 \pm 12i}{-8}$$

$$= \frac{2 \mp 3i}{2} \qquad \text{dividing each member of the fraction by } -4$$

Hence the solution set is $\{(2 - 3i)/2,\ (2 + 3i)/2\}$.

Check   If we replace x by $(2 \pm 3i)/2$ in the given equation, we have for the left member

$$\left(8 \times \frac{2 \pm 3i}{2}\right) - 13 = 8 \pm 12i - 13 = -5 \pm 12i$$

and for the right member

$$4 \times \left(\frac{2 \pm 3i}{2}\right)^2 = 4 \times \frac{4 \pm 12i + 9i^2}{4}$$

$$= 4 \times \frac{4 \pm 12i - 9}{4} \qquad \text{since } i^2 = -1$$

$$= -5 \pm 12i$$

EXAMPLE 4 Use the quadratic formula to solve $y^2 + xy - 4y - 2x^2 - 5x + 3 = 0$ for y in terms of x.

Solution

$$y^2 + xy - 4y - 2x^2 - 5x + 3 = 0 \qquad \text{given equation}$$

$$y^2 + (x - 4)y - 2x^2 - 5x + 3 = 0 \qquad \text{by the right-hand distributive axiom,}$$
$$\text{Eq. (2.12)}$$

The coefficient of y^2 is 1, the coefficient of y is $(x - 4)$, and the terms not involving y are $-2x^2 - 5x + 3$. Consequently, $a = 1$, $b = x - 4$, and $c = -2x^2 - 5x + 3$. We now substitute in the quadratic formula and get

$$y = \frac{-x + 4 \pm \sqrt{(x - 4)^2 - 4(-2x^2 - 5x + 3)}}{2}$$

$$= \frac{-x + 4 \pm \sqrt{x^2 - 8x + 16 + 8x^2 + 20x - 12}}{2}$$

$$= \frac{-x + 4 \pm \sqrt{9x^2 + 12x + 4}}{2}$$

$$= \frac{-x + 4 \pm (3x + 2)}{2}$$

Hence,

$$y = \frac{-x + 4 + 3x + 2}{2} \qquad \text{using the positive sign before the parentheses}$$

$$= x + 3$$

and

$$y = \frac{-x + 4 - 3x - 2}{2} \qquad \text{using the negative sign before the parentheses}$$

$$= -2x + 1$$

Therefore, the solution set is $\{x + 3, -2x + 1\}$. It can be checked by the method used in the previous examples.

EXERCISE 8.3 **The Quadratic Formula**

Find the solution set of the equation in each of Probs. 1 to 52 by use of the quadratic formula.

1 $x^2 - 4x + 3 = 0$ 2 $x^2 - 7x + 10 = 0$ 3 $x^2 - 6x + 8 = 0$

4	$x^2 + 2x - 15 = 0$	5	$x^2 + 5x - 6 = 0$	6	$x^2 - 5x - 14 = 0$
7	$x^2 + 7x + 12 = 0$	8	$x^2 + 7x + 10 = 0$	9	$5x^2 - 17x + 6 = 0$
10	$3x^2 - 7x + 2 = 0$	11	$7x^2 - 17x + 6 = 0$	12	$2x^2 - 9x + 4 = 0$
13	$2x^2 - 9x - 5 = 0$	14	$3x^2 - 16x - 12 = 0$	15	$3x^2 - 5x - 12 = 0$
16	$7x^2 - 11x - 6 = 0$	17	$12x^2 - 17x + 6 = 0$	18	$20x^2 - 31x + 12 = 0$
19	$35x^2 - 43x + 12 = 0$	20	$21x^2 - 23x + 6 = 0$	21	$20x^2 - 7x - 6 = 0$
22	$12x^2 + x - 6 = 0$	23	$6x^2 + 7x + 2 = 0$	24	$25x^2 + 25x + 6 = 0$
25	$x^2 - 2x - 2 = 0$	26	$x^2 - 4x - 1 = 0$	27	$x^2 - 6x + 7 = 0$
28	$x^2 - 10x + 18 = 0$	29	$4x^2 - 8x + 1 = 0$	30	$9x^2 - 18x + 7 = 0$
31	$9x^2 - 12x - 1 = 0$	32	$4x^2 - 20x + 22 = 0$	33	$x^2 - 4x + 13 = 0$
34	$x^2 - 6x + 13 = 0$	35	$x^2 - 8x + 20 = 0$	36	$x^2 - 10x + 34 = 0$
37	$4x^2 - 12x + 10 = 0$	38	$9x^2 - 12x + 5 = 0$	39	$25x^2 - 30x + 13 = 0$
40	$4x^2 - 20x + 41 = 0$	41	$x^2 - x + 1 = 0$	42	$3x^2 - 4x + 3 = 0$
43	$9x^2 - 18x + 11 = 0$	44	$4x^2 - 16x + 23 = 0$	45	$x^2 - 3ax + 2a^2 = 0$
46	$x^2 + 2ax - 3a^2 = 0$		47	$x^2 - (a+b)x + ab = 0$	
48	$x^2 + (a-2b)x - 2ab = 0$		49	$4x^2 - 4ax + a^2 + b^2 = 0$	
50	$9x^2 - 12ax + 4a^2 + b^2 = 0$		51	$16x^2 - 24ax + 9a^2 + 4b^2 = 0$	
52	$9x^2 - 24ax + 16a^2 + 4b^2 = 0$				

Solve each of Probs. 53 to 60 for y in terms of x.

53	$y^2 + (3 - 2x)y + x^2 - 3x + 2 = 0$	54	$y^2 + (2 - 2x)y + x^2 - 2x - 3 = 0$
55	$y^2 + (1 - 2x)y + x^2 - x - 6 = 0$	56	$y^2 + (1 - 2x)y + x^2 - x - 2 = 0$
57	$y^2 - (3x + 1)y + 2x^2 + 3x - 2 = 0$	58	$y^2 - (1 + 4x)y + 3x^2 - x - 2 = 0$
59	$y^2 + (1 - 4x)y + 3x^2 + x - 2 = 0$	60	$y^2 + (2 - 3x)y + 2x^2 - 9x - 35 = 0$

Use the quadratic formula and Table A.3 in the Appendix to find the solution of each equation to three decimal places.

61	$x^2 - 4x + 1 = 0$	62	$x^2 - 6x + 7 = 0$
63	$x^2 - 10x + 20 = 0$	64	$x^2 - 4x - 3 = 0$
65	$4x^2 - 12x + 7 = 0$	66	$9x^2 - 12x + 1 = 0$
67	$8x^2 - 12x + 1 = 0$	68	$9x^2 - 30x + 17 = 0$

8.6 / EQUATIONS IN QUADRATIC FORM

The unknown in a quadratic equation may be any quantity. This unknown quantity must, however, enter to the second power and may enter to the first power. Thus, if $f(x)$ is the unknown, then

$$a[f(x)]^2 + b[f(x)] + c = 0 \qquad (8.4)$$

where a, b, and c are constants is a quadratic equation. Such an equation

Equation in
quadratic form

is ordinarily said to be in *quadratic form.* If the unknown $f(x)$ in an equation in quadratic form is linear or quadratic, we can find the values of x
which satisfy the given equation after solving it for $f(x)$.

EXAMPLE 1   Solve the equation $(x^2 - 3x)^2 - 2(x^2 - 3x) - 8 = 0$ for x.

Solution   The given equation is a quadratic provided we think of $x^2 - 3x$ as the unknown. We shall do that and solve for $x^2 - 3x$ by use of the quadratic
formula. Thus,

$$x^2 - 3x = \frac{-(-2) \pm \sqrt{(-2)^2 - 4(1)(-8)}}{2(1)}$$

$$= \frac{2 \pm 6}{2} = 4, -2$$

We now find the desired values of x by solving $x^2 - 3x = 4$ and $x^2 - 3x = -2$.
From the former, we get $x = 4$ and -1, and from the latter, we find that
$x = 2$ and 1. Therefore, $\{4, -1, 2, 1\}$ is the solution set of the given
equation.

EXAMPLE 2   Solve the equation $x^4 - 5x^2 - 36 = 0$ for x.

Solution   The given equation is a quadratic provided we consider x^2 as the unknown
since the equation is then

$$(x^2)^2 - 5(x^2) - 36 = 0$$

Consequently, $(x^2 - 9)(x^2 + 4) = 0$ factoring
and $x^2 = 9, -4$
Therefore, $x = \pm 3, \pm 2i$

and the solution set of the given equation is $\{3, -3, 2i, -2i\}$.

EXAMPLE 3   Find the solution set of

$$\left(\frac{x-2}{2x-1}\right)^2 - 3\left(\frac{x-2}{2x-1}\right) - 4 = 0$$

Solution   If we think of $(x-2)/(2x-1)$ as the unknown, we have a quadratic with
$a = 1$, $b = -3$, and $c = -4$. Its solution is

$$\frac{x-2}{2x-1} = \frac{-(3) \pm \sqrt{(-3)^2 - 4(1)(-4)}}{2(1)}$$ using the quadratic
                                                                  formula

$$= \frac{3 \pm \sqrt{9 + 16}}{2} = \frac{3 \pm 5}{2}$$

$$= 4, -1$$

Setting $(x - 2)/(2x - 1)$ equal to 4 and solving gives

$$\frac{x - 2}{2x - 1} = 4$$
$$x - 2 = 8x - 4 \qquad \text{clearing of fractions}$$
$$x = \tfrac{2}{7}$$

Similarly,

$$\frac{x - 2}{2x - 1} = -1$$
$$x - 2 = -2x + 1 \qquad \text{clearing of fractions}$$
$$x = 1$$

Consequently, the solution set of the given equation is $\{\tfrac{2}{7}, 1\}$. It can be checked by substituting each of these values for x in the given equation.

EXERCISE 8.4 Equations in Quadratic Form

Think of each of the following equations as a quadratic, solve it as such, and then find the solution set for x.

1 $x^4 - 5x^2 + 4 = 0$
2 $x^4 - 13x^2 + 36 = 0$
3 $x^4 - 17x^2 + 16 = 0$
4 $x^4 - 29x^2 + 100 = 0$
5 $x^4 + 12x^2 - 64 = 0$
6 $x^4 + 3x^2 - 4 = 0$
7 $x^4 + 16x^2 - 225 = 0$
8 $x^4 + 8x^2 - 9 = 0$
9 $36x^4 - 13x^2 + 1 = 0$
10 $100x^4 - 29x^2 + 1 = 0$
11 $36x^4 - 7x^2 - 4 = 0$
12 $64x^4 - 364x^2 - 225 = 0$
13 $x^6 - 9x^3 + 8 = 0$
14 $x^6 - 35x^3 + 216 = 0$
15 $x^6 + 26x^3 - 27 = 0$
16 $x^6 + 9x^3 + 8 = 0$
17 $x^8 - 17x^4 + 16 = 0$
18 $x^8 - 82x^4 + 81 = 0$
19 $16x^8 - 17x^4 + 1 = 0$
20 $81x^8 - 97x^4 + 16 = 0$
21 $36x^{-4} - 25x^{-2} + 4 = 0$
22 $400x^{-4} - 289x^{-2} + 36 = 0$
23 $16x^{-4} + 96x^{-2} - 25 = 0$
24 $36x^{-4} - 17x^{-2} - 144 = 0$
25 $x^{4/3} - 5x^{2/3} + 4 = 0$
26 $x^{4/3} - 13x^{2/3} + 36 = 0$
27 $x^{4/3} - 15x^{2/3} - 16 = 0$
28 $x^{4/3} + 3x^{2/3} - 4 = 0$
29 $x^{4/5} + 5x^{2/5} - 36 = 0$
30 $x^{4/5} - 15x^{2/5} - 16 = 0$
31 $x^{4/5} - 21x^{2/5} - 100 = 0$
32 $x^{4/5} + 15x^{2/5} - 16 = 0$
33 $36x^{-4/3} - 13x^{-2/3} + 1 = 0$
34 $144x^{-4/3} - 73x^{-2/3} + 4 = 0$
35 $36x^{-4/3} - 241x^{-2/3} + 100 = 0$
36 $36x^{-4/3} - 73x^{-2/3} + 16 = 0$
37 $(x^2 - 5)^2 - 3(x^2 - 5) - 4 = 0$
38 $(x^2 + 1)^2 - 4(x^2 + 1) - 5 = 0$
39 $(x^2 - 7)^2 + (x^2 - 7) - 6 = 0$
40 $(x^2 + 4)^2 - 12(x^2 + 4) + 32 = 0$
41 $(x^2 - 3x)^2 - 2(x^2 - 3x) - 8 = 0$
42 $(x^2 - 4x)^2 + 7(x^2 - 4x) + 12 = 0$
43 $(x^2 + 5x)^2 + 10(x^2 + 5x) + 24 = 0$
44 $(x^2 + 2x)^2 - 8(x^2 + 2x) + 15 = 0$

45 $\left(\dfrac{2x-1}{x+3}\right)^2 - 4\left(\dfrac{2x-1}{x+3}\right) + 3 = 0$ 46 $\left(\dfrac{3x+2}{2x+1}\right)^2 - 3\left(\dfrac{3x+2}{2x+1}\right) + 2 = 0$

47 $\left(\dfrac{3x-1}{x+5}\right)^2 - 3\left(\dfrac{3x-1}{x+5}\right) - 28 = 0$ 48 $2\left(\dfrac{3x+4}{2x-3}\right)^2 + 7\left(\dfrac{3x+4}{2x-3}\right) + 6 = 0$

49 $\dfrac{x-3}{x+1} - 1 - 2\left(\dfrac{x+1}{x-3}\right) = 0$ HINT: Multiply through by $(x-3)/(x+1)$ or its reciprocal.

50 $3\left(\dfrac{2x+3}{3x-9}\right) - 4 + \dfrac{3x-9}{2x+3} = 0$

51 $\dfrac{2x-7}{x+4} - 10\dfrac{x+4}{(2x-7)} - 3 = 0$

52 $\dfrac{2x-3}{2x+1} - 1 - 2\left(\dfrac{2x+1}{2x-3}\right) = 0$

8.7 / RADICAL EQUATIONS

By the theorem shown by Eq. (5.4), *any root of $f(x) = g(x)$ is also a root of $f^2(x) = g^2(x)$. In terms of sets, the solution set of $f(x) = g(x)$ is a subset of the solution set of $f^2(x) = g^2(x)$.* The converse of this theorem is not true, as is seen in the example below.

EXAMPLE 1   If $x - 1 = 2$ then $x = 3$, while $(x - 1)^2 = 2^2$ has for its solution set $\{3, -1\}$. Thus $(x - 1)^2 = 2^2$ has roots that $x - 1 = 2$ does not have.

The procedure for solving an equation that contains three or fewer radicals, all of which are of second order, consists of the following steps.

Solution of radical equations

1   Obtain an equation that is equivalent to the given equation and that has one radical and no other terms in one member. This is called *isolating a radical.*

2   Equate the squares of the members of the equation obtained in step 1.

3   If the equation obtained in step 2 contains one or more radicals, repeat the process until an equation free of radicals is obtained. The radical-free equation is called the *rationalized equation.*

Rationalized equation

4   Solve the rationalized equation.

5   Replace x in the given equation by each solution of the rationalized equation to determine which are and are not solutions of the given equation.

We illustrate the procedure in the following examples.

EXAMPLE 2   Find the solution set of the equation $\sqrt{2x^2 - 2x + 1} - 2x + 3 = 0$.

Solution   In order to solve the equation

$$\sqrt{2x^2 - 2x + 1} - 2x + 3 = 0 \qquad (1)$$

we first isolate the radical (step 1) by adding $2x - 3$ to each member, and we obtain

$$\sqrt{2x^2 - 2x + 1} = 2x - 3 \qquad (2)$$

Next (step 2) we equate the squares of the members of (2), and we get

$$2x^2 - 2x + 1 = 4x^2 - 12x + 9 \qquad (3)$$

Equation (3) involves no radical expressions, so we solve it for x (step 4) by the method indicated below.

$2x^2 - 4x^2 - 2x + 12x + 1 - 9 = 0$	adding $-4x^2 + 12x - 9$ to each member of (3) and rearranging terms
$-2x^2 + 10x - 8 = 0$	combining terms
$x^2 - 5x + 4 = 0$	multiplying each member by $-\frac{1}{2}$
$(x - 4)(x - 1) = 0$	factoring left member
$x - 4 = 0$	
$x = 4$	
$x - 1 = 0$	
$x = 1$	

Therefore the solution set of (3) is $\{4, 1\}$.

Now we replace x in (1) by 4 (step 5) and get

$$\sqrt{(2 \times 4^2) - (2 \times 4) + 1} - (2 \times 4) + 3 = \sqrt{32 - 8 + 1} - 8 + 3$$
$$= \sqrt{25} - 5 = 5 - 5 = 0$$

Consequently, 4 is a root of (1).

When, however, x is replaced by 1 in the left member of (1), we get

$$\sqrt{(2 \times 1^2) - (2 \times 1) + 1} - (2 \times 1) + 3 = \sqrt{2 - 2 + 1} - 2 + 3$$
$$= 1 - 2 + 3$$
$$= 2$$

Consequently, 1 is not a root of (1) since the right member is 0. Hence the only root of (1) is 4, and the solution set is $\{4\}$.

EXAMPLE 3   Find the solution set of

$$\sqrt{11x - 6} = \sqrt{4x + 5} - \sqrt{x - 1} \qquad (4)$$

Solution Since we have one radical expression isolated on the left of the equality sign, we proceed at once to step 2 and equate the squares of the members of (4). We then proceed as below.

$$11x - 6 = 4x + 5 - 2\sqrt{(4x+5)(x-1)} + x - 1$$

$$11x - 6 - 4x - 5 - x + 1 = -2\sqrt{(4x+5)(x-1)} \quad \text{adding}$$
$$-4x - 5 - x + 1$$
to each member

$$6x - 10 = -2\sqrt{(4x+5)(x-1)} \quad \text{combining terms}$$

$$3x - 5 = -\sqrt{4x^2 + x - 5} \quad \text{multiplying each member}$$
by $\frac{1}{2}$ and simplifying
the radicand

$$9x^2 - 30x + 25 = 4x^2 + x - 5 \quad \text{equating the squares}$$
of the members

$$5x^2 - 31x + 30 = 0 \quad \text{adding} \quad (5)$$
$$-4x^2 - x + 5$$
to each member

$$x = \frac{31 \pm \sqrt{31^2 - (4 \times 5 \times 30)}}{10} \quad \text{by the quadratic formula}$$

$$= \frac{31 \pm \sqrt{961 - 600}}{10}$$

$$= \frac{31 \pm \sqrt{361}}{10}$$

$$= \frac{31 \pm 19}{10}$$

$$= \tfrac{50}{10} \text{ and } \tfrac{12}{10}$$

$$= 5 \text{ and } \tfrac{6}{5}$$

Hence the solution set of (5) is $\{5, \tfrac{6}{5}\}$.

Now (step 5) we replace x in (4) by 5 and get $\sqrt{(11 \times 5) - 6} = \sqrt{55 - 6} = \sqrt{49} = 7$ for the left member and $\sqrt{(4 \times 5) + 5} - \sqrt{5 - 1} = \sqrt{25} - \sqrt{4} = 5 - 2 = 3$ for the right member. Since the members of (4) are not equal when x is replaced by 5, 5 is not a root of (4).

If we replace x by $\tfrac{6}{5}$ in (4), we get

$$\sqrt{\left(11 \times \frac{6}{5}\right) - 6} = \sqrt{\frac{66}{5} - 6} = \sqrt{\frac{66 - 30}{5}} = \frac{6}{\sqrt{5}}$$

for the left member and

$$\sqrt{\left(4 \times \frac{6}{5}\right) + 5} - \sqrt{\frac{6}{5} - 1} = \sqrt{\frac{49}{5}} - \sqrt{\frac{1}{5}} = \frac{7}{\sqrt{5}} - \frac{1}{\sqrt{5}} = \frac{6}{\sqrt{5}}$$

for the right member. Consequently, since the members are equal if $x = \frac{6}{5}$, $\frac{6}{5}$ is a root of (4). Therefore, the solution set of (4) is $\{\frac{6}{5}\}$.

EXERCISE 8.5 Radical Equations

Find the solution set of each of the following equations.

1 $\sqrt{2x + 5} = 3$

2 $\sqrt{5x - 1} = 2$

3 $\sqrt{3x - 2} = -1$

4 $\sqrt{4x - 3} = 3$

5 $\sqrt{x + 1}\sqrt{x - 2} = 2$

6 $\sqrt{5x + 4}\sqrt{2 - x} = 3$

7 $\sqrt{x + 5}\sqrt{2x - 4} = 6$

8 $\sqrt{2x + 3}\sqrt{x + 5} = 2$

9 $\sqrt{10 + 3x} = x + 4$

10 $\sqrt{9 + 4x} = 3 + 2x$

11 $\sqrt{x + 4} = x - 2$

12 $\sqrt{x + 3} = x - 3$

13 $\sqrt{x + 12} = 2\sqrt{x}$

14 $\sqrt{x + 6} = \sqrt{2x + 3}$

15 $\sqrt{3x + 5} = -\sqrt{5x + 3}$

16 $\sqrt{3x + 10} = 2\sqrt{x + 2}$

17 $\sqrt{5x + 1} - \sqrt{6x - 2} = 0$

18 $\sqrt{x + 3} - \sqrt{3x + 7} = 0$

19 $\sqrt{7x + 2} - \sqrt{4x + 5} = 0$

20 $\sqrt{x - 1} - \sqrt{3x - 11} = 0$

21 $\sqrt{3x - 2}/\sqrt{x - 1} = 2$

22 $\sqrt{4x - 7}/\sqrt{x - 3} = 3$

23 $\sqrt{2 - x}/\sqrt{x + 3} = 2$

24 $\sqrt{5x + 9}/\sqrt{3 - x} = 1$

25 $\sqrt{x^2 + 2x - 2} = x + 4$

26 $\sqrt{x^2 + 2x + 9} = x + 5$

27 $\sqrt{x^2 - 2x + 5} = x + 1$

28 $\sqrt{x^2 + 7x + 4} = x + 2$

29 $\sqrt{x^2 - 3x + 5} = \sqrt{x + 5}$

30 $\sqrt{x^2 - 5x + 2} = \sqrt{x + 9}$

31 $\sqrt{x^2 + 3x + 9} = \sqrt{8x + 9}$

32 $\sqrt{x^2 - 3x + 1} = \sqrt{x - 2}$

33 $\sqrt{x + 2} = 3x - 4$

34 $\sqrt{5x + 9} = 4x + 2$

35 $\sqrt{3x + 10} = 2x + 6$

36 $\sqrt{2x + 3} = 2x - 3$

37 $\sqrt{x^2 + 3x} = 3x - 1$

38 $\sqrt{x^2 + 2x + 8} = 3x - 2$

39 $\sqrt{2x^2 - 7x + 5} = 2x - 5$

40 $\sqrt{2x^2 + 5x + 1} = x + 5$

41 $\sqrt{x + 2} + \sqrt{2x + 5} = 5$

42 $\sqrt{x - 1} - \sqrt{6 - x} = 1$

43 $\sqrt{4x - 3} - \sqrt{x + 1} = 1$

44 $\sqrt{x + 6} + \sqrt{2x + 11} = 2$

45 $\dfrac{\sqrt{2x + 7} + 1}{\sqrt{5x - 1} + 2} = 1$

46 $\dfrac{\sqrt{3x - 5} + 1}{\sqrt{x + 1} - 1} = 3$

47 $\dfrac{\sqrt{3x + 4} + 2}{\sqrt{4x + 1} + 1} = 2$

48 $\dfrac{\sqrt{3x + 4} + 4}{\sqrt{2x + 1} + 1} = 2$

49 $\sqrt{3 + \sqrt{x}} = \sqrt{x} + 1$

50 $\sqrt{2 + \sqrt{x}} = 3\sqrt{x} - 4$

51 $\sqrt{2 - \sqrt{x - 1}} = \sqrt{x - 1}$

52 $\sqrt{7 + \sqrt{x + 1}} = \sqrt{x + 1} + 1$

53 $\sqrt{x + 2} + \sqrt{3x + 3} = \sqrt{12x + 1}$

54 $\sqrt{x + 9} - \sqrt{x + 4} = \sqrt{x + 1}$

55 $\sqrt{2x - 1} - \sqrt{3x - 2} = \sqrt{x - 1}$

56 $\sqrt{2x + 3} + \sqrt{x + 10} = \sqrt{3 - 13x}$

57 $\sqrt{3x-2}+\sqrt{x-1}=\sqrt{4x+1}$ 58 $\sqrt{3x-5}-\sqrt{x-2}=\sqrt{3x-8}$

59 $\sqrt{2x+3}+\sqrt{4x+5}=\sqrt{3-x}$ 60 $\sqrt{2x+4}+\sqrt{3x+1}=\sqrt{x+9}$

61 $\sqrt{x^2+2x-8}-\sqrt{x^2-3x}=2$ 62 $\sqrt{x^2+3x-2}-\sqrt{x^2+3x-9}=1$

63 $\sqrt{2x^2-5x+2}+\sqrt{2x^2-3x-1}=3$

64 $\sqrt{2x^2+7x+1}+\sqrt{2x^2+3x+4}=3$

65 $\sqrt{5bx-b^2}+b=\sqrt{7bx+2b^2}$ 66 $\sqrt{ax+8a^2}-\sqrt{5a^2-ax}=a$

67 $\sqrt{3tx-2t^2}-t=\sqrt{2tx-3t^2}$ 68 $\sqrt{ax+3a^2}-\sqrt{2a^2-ax}=a$

8.8 / NATURE OF THE ROOTS

In the remainder of this chapter we shall discuss methods that enable us to obtain information about the roots of a quadratic equation without solving the equation. We shall use the standard form of the quadratic equation

$$ax^2 + bx + c = 0 \qquad (8.2)$$

and the quadratic formula

$$x = \frac{-b \pm \sqrt{b^2 - 4ac}}{2a} \qquad (8.3)$$

for this purpose. We shall let the letter D represent the radicand in Eq. (8.3); thus

$$D = b^2 - 4ac \qquad (8.5)$$

As we shall see, the value of D enables us to determine the nature of the roots of a quadratic equation. For this reason D, or $b^2 - 4ac$, is called

Discriminant — the *discriminant* of the equation.

We shall let r represent the root of Eq. (8.2) obtained by using the plus sign preceding the radical in Eq. (8.3) and s represent the root obtained by use of the minus sign. Thus we have

$$r = \frac{-b + \sqrt{D}}{2a} \qquad s = \frac{-b - \sqrt{D}}{2a} \qquad (8.6)$$

We shall first assume that a, b, and c in Eq. (8.2) are rational numbers. Under this assumption, we have the following possibilities:

1 If $D = 0$, then $r = s = -b/2a$. Hence the roots of Eq. (8.2) are rational and equal.

2 If $D < 0$, then $\sqrt{D}$ is a pure imaginary number, and consequently r and s are imaginary.

3 If $D > 0$, two situations may exist: First, if D is a perfect square, then

$\sqrt{D}$ is a rational number, and it follows that r and s are rational. Second, if D is not a perfect square, $\sqrt{D}$ is an irrational number, and r and s are therefore irrational. In either case, r is not equal to s, since $(-b + \sqrt{D})/2a \neq (-b - \sqrt{D})/2a$.

We summarize these conclusions below:

Discriminant	Roots
$D = 0$	Rational and equal
$D > 0$ and a perfect square	Rational and unequal
$D > 0$ and not a perfect square	Irrational and unequal
$D < 0$	Imaginary

We illustrate the use of the above information in the following examples.

EXAMPLE 1 $9x^2 - 24x + 16 = 0$ $D = (-24)^2 - (4 \times 9 \times 16)$ roots are rational
$= 576 - 576 = 0$ and equal

EXAMPLE 2 $2x^2 + 3x - 20 = 0$ $D = 3^2 - [4 \times 2 \times (-20)]$ roots are rational
$= 9 + 160 = 169 = 13^2$ and unequal

EXAMPLE 3 $3x^2 - 2x - \frac{7}{5} = 0$ $D = (-2)^2 - (4 \times 3 \times -\frac{7}{5})$ roots are irrational
$= 4 + \frac{84}{5} = \frac{104}{5}$ and unequal

EXAMPLE 4 $5x^2 - 6x + 8 = 0$ $D = (-6)^2 - (4 \times 5 \times 8)$ roots are imaginary
$= 36 - 160 = -124$

If we assume that a, b, and c are real but not necessarily rational, then the information we obtain about r and s is less specific. Under this assumption, we have the following conclusions:

1 If $D = 0$, the roots are real and equal, since $r = s = -b/2a$ and we only know that a and b are real.
2 If $D > 0$, the roots are real and unequal.
3 If $D < 0$, the roots are imaginary.

EXAMPLE 5 $2x^2 - 2\sqrt{10}x + 5 = 0$ $D = (-2\sqrt{10})^2 - (4 \times 2 \times 5)$ roots are real
$= 40 - 40 = 0$ and equal

EXAMPLE 6	$\sqrt{3}x^2 - 5x + \sqrt{12} = 0$	$D = (-5)^2 - 4(\sqrt{3})(\sqrt{12})$ $= 25 - 24 = 1$	roots are real† and unequal

EXAMPLE 7	$\sqrt{5}x^2 - \sqrt{3}x + \sqrt{2} = 0$	$D = (-\sqrt{3})^2 - 4(\sqrt{5})(\sqrt{2})$ $= 3 - 4\sqrt{10} < 0$	roots are imaginary

8.9 / THE SUM AND PRODUCT OF THE ROOTS

By use of Eq. (8.6), we can show that the sum and the product of the roots of a quadratic equation are simple combinations of the coefficients in the equation. For example, the sum of the two roots is

$$r + s = \frac{-b + \sqrt{D}}{2a} + \frac{-b - \sqrt{D}}{2a} = \frac{-2b}{2a} = -\frac{b}{a}$$

Consequently, we have

Sum of roots

$$r + s = -\frac{b}{a} \tag{8.7}$$

Similarly,

$$rs = \left(\frac{-b + \sqrt{D}}{2a}\right)\left(\frac{-b - \sqrt{D}}{2a}\right)$$

$$= \frac{b^2 - D}{4a^2} = \frac{b^2 - b^2 + 4ac}{4a^2} \qquad \text{since } D = b^2 - 4ac$$

$$= \frac{4ac}{4a^2} = \frac{c}{a}$$

Product of two roots   Therefore,

$$rs = \frac{c}{a} \tag{8.8}$$

Rule for finding the sum and product of roots

Since r and s are the two roots of the equation $ax^2 + bx + c = 0$, it follows that

The sum of the two roots of a quadratic equation is equal to the negative of the quotient of the coefficients of x and x^2, and the product of the two roots is the quotient of the constant term and the coefficient of x^2.

Equations (8.7) and (8.8) are useful as a rapid check on the roots of a quadratic equation. In Example 2 of Sec. 8.5, we found that the solution set of $6x^2 - x - 12 = 0$ is $\{\frac{3}{2}, -\frac{4}{3}\}$. By Eqs. (8.7) and (8.8), we see that the

† In this case, D is a perfect square. Since, however, $r = (5 + 1)/2\sqrt{3} = 6/2\sqrt{3} = \sqrt{3}$ and $= (5 - 1)/2\sqrt{3} = 2/\sqrt{3} = 2\sqrt{3}/3$, r and s are not rational.

sum and product of the two roots should be $\frac{1}{6}$ and $-\frac{12}{6} = -2$, respectively. Then, since $\frac{3}{2} + (-\frac{4}{3}) = \frac{9}{6} - \frac{8}{6} = \frac{1}{6}$ and $\frac{3}{2} \times (-\frac{4}{3}) = -\frac{12}{6} = -2$, the two roots satisfy the required conditions.

We shall now discuss two examples that illustrate methods for using the principles of this and the preceding section to obtain information about the coefficients in equations whose roots satisfy predetermined conditions.

EXAMPLE 1   In the equation $2x^2 + (k-3)x + 3k - 5 = 0$, determine k so that the sum and the product of the roots are equal.

Solution   In this problem, $a = 2$, $b = k - 3$, and $c = 3k - 5$. Consequently, by Eq. (8.7) the sum of the roots is

$$-\frac{b}{a} = -\frac{k-3}{2}$$

and by Eq. (8.8) the product is

$$\frac{c}{a} = \frac{3k-5}{2}$$

Consequently, if the sum is equal to the product, we have

$$-\frac{k-3}{2} = \frac{3k-5}{2}$$

and we solve this equation for k as follows:

$$-k + 3 = 3k - 5 \qquad \text{multiplying each member by 2}$$
$$-4k = -8 \qquad \text{adding } -3k - 3 \text{ to each member}$$
$$k = 2 \qquad \text{multiplying each member by } -\frac{1}{4}$$

EXAMPLE 2   Given the equation $ax^2 + bx + c = 0$, (1) show that if one root is the negative of the other, then $b = 0$; and (2) show that if one root is zero, then $c = 0$.

Solution   1   If one root is the negative of the other, we have by Eq. (8.6)

$$\frac{-b + \sqrt{D}}{2a} = -\frac{-b - \sqrt{D}}{2a}$$

If we multiply each of the members of the above equation by $2a$, we have

$$-b + \sqrt{D} = b + \sqrt{D}$$

Now, by adding $-b - \sqrt{D}$ to each member of the latter equation, we get

$$-2b = 0$$

Hence, if we multiply each member by $-\frac{1}{2}$, we have

$$b = 0$$

2 If one root is zero, we shall assume that this root is s. Then by Eq. (8.8) we have

$$\frac{c}{a} = r(0) = 0 \qquad \text{by Eq. (2.35)}$$
$$c = a(0) \qquad \text{multiplying by } a$$
$$= 0 \qquad \text{by Eq. (2.35)}$$

8.10 / FACTORS OF A QUADRATIC TRINOMIAL

We shall now prove that if r and s are roots of the equation

$$ax^2 + bx + c = 0 \qquad (8.2)$$

then $x - r$ and $x - s$ are factors of the quadratic trinomial $ax^2 + bx + c$.

Proof By Eqs. (8.7) and (8.8), $-b/a = r + s$ and $c/a = rs$. Consequently, it follows that $b = -a(r + s)$ and $c = ars$, and if we replace b and c with these expressions, we have

$$ax^2 + bx + c = ax^2 - a(r + s)x + ars$$
$$= a[x^2 - (r + s)x + rs]$$
$$\mathbf{ax^2 + bx + c = a(x - r)(x - s)} \qquad (8.9)$$

If a, b, and c are rational and $b^2 - 4ac$ is a perfect square, then r and s are rational numbers. Therefore, the quadratic trinomial $ax^2 + bx + c$ can be expressed as the product of two rational linear factors if a, b, and c are rational numbers and if $b^2 - 4ac$ is a perfect square.

If the coefficients in a quadratic trinomial satisfy the above conditions, it can be factored by the usual methods. However, if the coefficients are large numbers, considerable time may be required to find the proper combinations. In such cases, it may be easier to use Eqs. (8.2) and (8.9) in the method illustrated below.

EXAMPLE 1 Factor the trinomial $72x^2 + 95x - 1,000$.

Solution We first set the given trinomial equal to zero and obtain the quadratic equation

$$72x^2 + 95x - 1,000 = 0$$

Then by use of the quadratic formula, we get

$$x = \frac{-95 \pm \sqrt{9,025 + 288,000}}{144} = \frac{-95 \pm \sqrt{297,025}}{144}$$

$$= \frac{-95 \pm 545}{144} = \frac{450}{144} \text{ and } -\frac{640}{144}$$

$$= \tfrac{25}{8} \text{ and } -\tfrac{40}{9}$$

Consequently, $r = \tfrac{25}{8}$ and $s = -\tfrac{40}{9}$.

Therefore, by Eq. (8.9),

$$72x^2 + 95x - 1,000 = 72(x - \tfrac{25}{8})(x + \tfrac{40}{9}) \qquad \text{since } a = 72$$

$$= 72\left(\frac{8x - 25}{8}\right)\left(\frac{9x + 40}{9}\right)$$

$$= (8x - 25)(9x + 40)$$

At times it is desirable to formulate a quadratic equation whose roots are specified in advance. If the roots of Eq. (8.2) are r and s, then by Eq. (8.9) the equation can be expressed in the form

$$a(x - r)(x - s) = 0 \tag{8.10}$$

where the letter a can be replaced by any chosen nonzero number. If the specified roots, r and s, are integers, we replace a by the number 1. If one or both of r and s are fractions, we replace a by the denominator or the product of the denominators. The method is illustrated in the following examples.

EXAMPLE 2   Obtain a quadratic equation whose roots are 3 and -2.

Solution   Since 3 and -2 are integers, we replace a by 1, r by 3, and s by -2 in Eq. (8.10) and obtain

$$(x - 3)(x + 2) = 0$$
$$x^2 - x - 6 = 0 \qquad \text{performing the indicated multiplication}$$

EXAMPLE 3   Obtain a quadratic equation whose roots are -4 and $\tfrac{2}{3}$.

Solution   Since only one of the roots is a fraction and the denominator is 3, replace a by 3, r by -4, and s by $\tfrac{2}{3}$ in Eq. (8.10) and simplify the result as indicated below.

$$3(x + 4)(x - \tfrac{2}{3}) = 0$$
$$3x^2 + 10x - 8 = 0 \qquad \text{multiplying the two binomials}$$

EXAMPLE 4   Obtain a quadratic equation whose roots are $-\frac{3}{4}$ and $\frac{5}{6}$.

Solution   Here, r and s are fractions and the product of the denominators is 24. Consequently, in Eq. (8.10), we replace a by 24, r by $-\frac{3}{4}$, and s by $\frac{5}{6}$, and we simplify as indicated.

$$24\left(x + \tfrac{3}{4}\right)\left(x - \tfrac{5}{6}\right) = 0$$
$$24x^2 - 2x - 15 = 0$$

EXERCISE 8.6 Use of the Discriminant

Determine the nature of the roots of the equation in each of Probs. 1 to 20.

1    $2x^2 - 3x - 2 = 0$ 2 $4x^2 + 11x - 3 = 0$
3    $6x^2 + 7x + 2 = 0$ 4 $15x^2 + 14x + 3 = 0$
5    $2x^2 - 3x - 3 = 0$ 6 $4x^2 + 11x - 2 = 0$
7    $6x^2 + 10x + 3 = 0$ 8 $15x^2 + 14x + 2 = 0$
9    $2x^2 - 3x - \sqrt{2} = 0$ 10 $4x^2 + 11x - \sqrt{3} = 0$
11    $\sqrt{6}x^2 + 7x + 2 = 0$ 12 $\sqrt{15}x^2 + 14x + 3 = 0$
13    $2x^2 - 3x + 2 = 0$ 14 $4x^2 + 11x + 9 = 0$
15    $6x^2 + 7x + 3 = 0$ 16 $15x^2 + 14x + 4 = 0$
17    $x^2 - 6x + 9 = 0$ 18 $x^2 + 8x + 16 = 0$
19    $4x^2 + 12x + 9 = 0$ 20 $9x^2 - 24x + 16 = 0$

Find the sum and the product of the roots in each of Probs. 21 to 32 without solving the equation.

21    $3x^2 + 5x + 1 = 0$ 22 $5x^2 + x + 2 = 0$
23    $7x^2 - 3x + 4 = 0$ 24 $3x^2 + 6x - 5 = 0$
25    $2x^2 + 3 = 4x$ 26 $4x^2 + 5x = 1$
27    $6x^2 = 2x - 5$ 28 $2x^2 - 6 = 3x$
29    $\sqrt{3}x^2 + \sqrt{3}x + 1 = 0$ 30 $\sqrt{2}x^2 + 3x + 2\sqrt{2} = 0$
31    $(1 + \sqrt{2})x^2 + \sqrt{2}x + 3 = 0$ 32 $(4 + \sqrt{3})x^2 + 2\sqrt{3}x - 4 = 0$

Determine k in each of Probs. 33 to 36 so that the roots of the equation are equal.

33    $2x^2 + kx + k = 0$ 34 $x^2 + (k - 3)x + 1 = 0$
35    $(k + 1)x^2 - 12x + 3k = 0$ 36 $(k - 5)x^2 - (k - 2)x - 1 = 0$

Determine k in each of Probs. 37 to 40 so that the sum and product of the roots are equal.

37    $x^2 + 3kx + k - 1 = 0$ 38 $7x^2 + (2k - 1)x - 3k + 2 = 0$
39    $5x^2 + (2k - 3)x - 2k + 3 = 0$ 40 $3x^2 + (k - 5)x + k - 3 = 0$

Find the values of k in each of Probs. 41 to 44 so that one root is zero.

41 $2x^2 + 3kx + k^2 = 4$ **42** $3x^2 + (k-1)x + k^2 + 2k = 0$

43 $kx^2 + 4x + 2k - 1 = 0$ **44** $(k+1)x^2 - 3x + k^2 + 3k + 2 = 0$

Form a quadratic equation that has the numbers listed in each of Probs. 45 to 56 as roots.

45 $3, 4$ **46** $2, -\frac{1}{3}$ **47** $-3, -\frac{5}{4}$

48 $7, 7$ **49** $2 + 3i, 2 - 3i$ **50** $3 + 5i, 3 - 5i$

51 $4 + 5i, 4 - 5i$ **52** $2 + i\sqrt{3}, 2 - i\sqrt{3}$ **53** $4 + \sqrt{5}, 4 - \sqrt{5}$

54 $2 + \sqrt{3}, 2 - \sqrt{3}$ **55** $1 + \sqrt{7}, 1 - \sqrt{7}$ **56** $3 + \sqrt{2}, 3 - \sqrt{2}$

Find the factors of the quadratic given in each of Probs. 57 to 68.

57 $6x^2 + 19x - 7$ **58** $15x^2 - 14x - 8$ **59** $14x^2 - 29x - 15$

60 $18x^2 + 21x - 4$ **61** $4x^2 + 4x - 5$ **62** $x^2 - 4x - 1$

63 $9x^2 + 12x - 1$ **64** $25x^2 + 40x + 11$ **65** $16x^2 + 8x + 37$

66 $49x^2 + 28x + 5$ **67** $9x^2 + 24x + 20$ **68** $25x^2 - 30x + 34$

8.11 / PROBLEMS THAT LEAD TO QUADRATIC EQUATIONS

Many stated problems, especially those which deal with products or quotients that involve the variable, lead to quadratic equations. The method for obtaining the equation for solving such a problem is the same as that in Sec. 5.9, and the reader should review that section at this time. It should be noted here that often a problem that can be solved by use of a quadratic equation has only one solution, while the equation has two roots. In such cases, the root that does not satisfy the conditions of the problem is discarded.

EXAMPLE 1 A rectangular building whose depth is twice its frontage is divided into two parts by a partition that is 30 ft from and parallel to the front wall. If the rear portion of the building contains 3,500 sq ft, find the dimensions of the building.

Solution In problems of this type, it is advisable to draw a diagram such as Fig. 8.1. In doing so, we find it best to let

$$x = \text{the frontage of the building, in feet}$$

Then $2x = \text{the depth of the building, in feet}$

and $2x - 30 = \text{the length of the rear portion, in feet}$

Since the area of a rectangle is equal to the product of the length and the

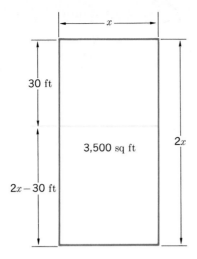

Figure 8.1

width, the area of the rear portion of the building is $x(2x - 30)$ sq ft. Furthermore, since we know that this area is 3,500 sq ft, we have

$$x(2x - 30) = 3,500 \qquad (1)$$

We solve (1) as follows:

$$2x^2 - 30x = 3,500 \qquad \text{performing the indicated multiplication}$$
$$2x^2 - 30x - 3,500 = 0 \qquad \text{adding } -3,500 \text{ to each member}$$
$$(x - 50)(2x + 70) = 0 \qquad \text{factoring the left member}$$
$$x - 50 = 0$$
$$x = 50$$
$$2x + 70 = 0$$
$$x = -35$$

Hence the solution set of (1) is $\{50, -35\}$. Since, however, the dimensions of the building are positive, we reject -35; and we have

$$x = 50 \qquad \text{frontage in feet}$$
$$2x = 100 \qquad \text{depth in feet}$$

Check $2x - 30 = 100 - 30 = 70$ and 50 ft $\times$ 70 ft $= 3,500$ sq ft.

EXAMPLE 2 The periods of time required for two painters to paint a square yard of floor differ by 1 min. Together, they can paint 27 sq yd in 1 hour. How long does it take each to paint 1 sq yd?

Solution In this problem, the numbers in our equation will be expressed in units of time. Therefore, we let

$x =$ the number of minutes required by the faster painter to paint 1 sq yd

Then $x + 1 =$ the number of minutes required by the other

Consequently,

$\dfrac{1}{x} =$ the fractional part of a square yard painted by the first man in 1 min

and

$\dfrac{1}{x+1} =$ the fractional part of a square yard painted by the other in 1 min

Consequently,

$\dfrac{1}{x} + \dfrac{1}{x+1} =$ the fractional part of a square yard painted by both men in 1 min

Since, however, together they painted 27 sq yd in 60 min, they covered $\frac{27}{60} = \frac{9}{20}$ sq yd in 1 min.   Therefore,

$$\frac{1}{x} + \frac{1}{x+1} = \frac{9}{20} \tag{2}$$

Equation (2) is the desired equation, and we solve it as follows:

$20x + 20 + 20x = 9x^2 + 9x$ multiplying each member of (2) by $20x(x+1)$

$-9x^2 + 31x + 20 = 0$ adding $-9x^2 - 9x$ to each member

The solution set of the latter equation is $\{-\frac{5}{9}, 4\}$.   We reject $-\frac{5}{9}$, however, since a negative time has no meaning in this problem.   Therefore, we have

$x = 4$ the number of minutes required by the faster painter to paint 1 sq yd

$x + 1 = 5$ the number of minutes required by the slower painter

Check   5 min $-$ 4 min $=$ 1 min.

EXERCISE 8.7 Problems Solvable by Use of Quadratic Equations

1 Find two consecutive integers whose product exceeds their sum by 19.

2 The difference between the square of a positive number and five times the number is 36.   Find the number.

3 The tens digit of a certain number is 4 more than the units digit.   The sum of the squares of the two digits is 26.   Find the number.

4 Find a negative number such that the sum of its square and six times the number is 27.

5 Find two numbers that differ by 8 and whose product is 273.

6 Find two numbers that differ by 11 and whose product is 1,230.

7 Divide 67 into two parts whose product is 1,120.

8 Divide 83 into two parts whose product is 600.

9 If the length of the side of a square is increased by 6 units, the area is multiplied by 4. Find the original side length.

10 The product of a positive even integer and the reciprocal of the next larger positive even integer equals the reciprocal of the first integer. Find that integer.

11 The sum of a number and its reciprocal is $\frac{25}{12}$. Find the number.

12 Two numbers differ by 9, and the sum of their reciprocals is $\frac{5}{12}$. Find the numbers.

13 The area of a parallelogram is 77 sq ft. Find the base and altitude if the former exceeds the latter by 4 ft.

14 What are the dimensions of a rectangular tablecloth if $6\frac{2}{3}$ yd of lace were required to edge it and $2\frac{2}{3}$ yd of material 1 yd wide were required to make it?

15 To make a rectangular concrete patio, a man used 70 ft of forming, into which he poured 150 ft³ of concrete to form a slab 6 in. thick. What were the dimensions of the patio?

16 The cost of carpeting a living room with carpeting that cost \$12/sq yd was \$576. What were the dimensions of the room if the length exceeded the width by 6 ft?

17 A country gentleman spent \$6,732 to install fencing, which cost \$0.85/ft, around his rectangular estate. If the area of his estate was 80 acres, what were the lengths of its sides? Note: 1 acre contains 43,560 sq ft.

18 A man bought 60 lb of fertilizer to spread over his rectangular backyard. If he spread it at a rate of 20 lb to 1,000 sq ft and the length of his yard exceeded the width by 10 ft, what were the dimensions of the yard?

19 Residents in a new housing development used a 125-ft-long path cut diagonally across a vacant corner lot until construction started on the lot. They then had to walk a total of 175 ft along two sides of the lot to go around the corner. What were the dimensions of the lot?

20 To tile a family-room floor, 432 square asphalt tiles, 10 in. on a side, were required. If the length of the room exceeded the width by 5 ft, what were the dimensions of the room?

21 John drove 325 miles from the campus to his home. While there, he had a new motor put in his car, so that his driving speed on the return trip was 15 mi/hour less than on his trip home. If his return trip took $1\frac{1}{2}$ hours longer than his trip home, how fast did he drive each way?

22 A man who lived on an island went to the mainland 36 mi away by boat and returned home by seaplane. His total traveling time was $1\frac{4}{5}$ hours, and his flying speed averaged five times as fast as his boat speed. How fast did he travel each way?

23 A plane flew 1,560 mi and back in 5 hours. Find the speed of the plane in still air if the wind velocity was 25 mi/hour.

24 An executive who had a radiophone in his car started from his home to visit one of his branch offices. He had driven 10 mi when he received an urgent call from his home office and drove 15 mi directly there at a speed 15 mi/hour faster than he had been driving. If he arrived at his home office 40 min after he left home, what was his faster speed?

25 Two brothers washed the family car in 24 min. Previously, when they each had washed the car alone, it had been found the younger boy took 20 min longer to do the job than the older boy. How long did the older boy take to wash the car?

26 One summer Jack and Henry got jobs feeding the animals in the zoo. The job usually took them 1 hour a day. On Jack's day off, Henry took $\frac{5}{6}$ hour longer to feed the animals than Jack did on Henry's day off. How long did Jack take to feed the animals alone?

27 On each of two consecutive years, a band boosters' club collected $150 from the sale of boosters' buttons. The second year, they had increased their price $0.50 a button. If a total of 250 buttons were sold, how much did they charge for each button the first year?

28 A campus organization chartered a bus for $60 to go on a field trip, the cost to be divided equally among those attending. At the last minute 10 more members decided to attend, and the cost was reduced $0.50 per person. How many club members went on the trip?

29 A man purchased some shares of stock for $1,560. Later, when the price had gone up $24/share, he sold all but 10 of them for $1,520. How many shares had he bought?

30 A group of men bought some property for $24,000. Later one of the men sold his share to the others for $5,000, and when the additional cost was divided among the others, it was $3,000 less than each had paid originally. How many men were involved in the first investment?

31 A swimming pool holds 1,800 cu ft of water. It can be drained at a rate of 15 cu ft/min faster than it can be filled. If it takes 20 min longer to fill it than to drain it, find the drainage rate.

32 A family planned to spend $600 on their summer vacation. By camping part of the time, they found they could reduce their average daily expense by $10 and vacation 3 days longer. How long were they gone?

EXERCISE 8.8   REVIEW

Solve the equations in Probs. 1 to 3 by factoring, those in Probs. 4 to 6 by completing the square, and those in Probs. 7 to 10 by use of the quadratic formula.

1 $8x^2 - 2x - 15 = 0$ 2 $3x^2 + 22x + 24 = 0$ 3 $10x^2 - 19x + 6 = 0$
4 $12x^2 - 7x - 12 = 0$ 5 $15x^2 + 19x - 10 = 0$ 6 $20x^2 + 33x + 10 = 0$
7 $8x^2 - 2x - 15 = 0$ 8 $x^2 - 4x - 7 = 0$ 9 $4x^2 - 4x + 17 = 0$
10 $9x^2 - 24x + 25 = 0$

Solve the equations in Probs. 11 and 12 for y in terms of x.

11 $y^2 + (5 - x)y - 2x^2 - x + 6 = 0$ 12 $y^2 + (x - 7)y - 12x^2 + 7x + 10 = 0$

Solve the equations in Probs. 13 to 28 for real values of x.

13 $x^4 - 5x^2 + 4 = 0$ 14 $x^6 + 26x^3 - 27 = 0$
15 $2x^{-4} - 33x^{-2} + 70 = 0$ 16 $x^{4/5} - 10x^{2/5} + 9 = 0$
17 $3x^{-2/3} - 7x^{-1/3} + 2 = 0$ 18 $(x^2 - 4x)^2 + 7(x^2 - 4x) + 12 = 0$
19 $\sqrt{3x - 2} = 2$ 20 $\sqrt{2x + 10} = 4$
21 $\sqrt{3x + 1} = 2\sqrt{x - 1}$ 22 $\sqrt{3x - 2}/\sqrt{3x - 5} = 2$
23 $\sqrt{x^2 + 3x + 7} = x + 2$ 24 $\sqrt{3x + 4} = x - 2$
25 $\sqrt{3x + 15} - \sqrt{2 - x} = 1$ 26 $\dfrac{\sqrt{3x + 1} + 4}{\sqrt{2x + 2} + 1} = 2$
27 $\sqrt{x + 4} + \sqrt{x + 7} = \sqrt{3 - 2x}$ 28 $\sqrt{1 + \sqrt{x}} = \sqrt{x} - 1$

Find the sum and the product of the roots in each of Probs. 29 and 30.

29 $2x^2 - 3x - 9 = 0$ 30 $5x^2 + 8x + 3 = 0$

Determine the nature of the roots of the equation in each of Probs. 31 to 34.

31 $2x^2 - x - 6 = 0$ 32 $9x^2 - 6x + 1 = 0$
33 $x^2 - 4x + 1 = 0$ 34 $4x^2 - 12x + 25 = 0$

35 Determine k so that the product of the roots of $x^2 + (k - 1)x + 2k + 1 = 0$ is twice their sum.

36 Determine k so that one root of $x^2 + (3k - 1)x - 7 = 0$ is the negative of the other.

37 Find a quadratic equation whose roots are $3 + \sqrt{5}$ and $3 - \sqrt{5}$.

38 Show that the roots of $ax^2 + bx + c = 0$ are the reciprocals of those of $cx^2 + bx + a = 0$.

39 Show that the roots of $ax^2 + bx + a = 0$ are reciprocals of each other.

40 Show that the sum of the reciprocals of the roots of a quadratic equation is the sum of the roots divided by their product.

41 Show that the average of the roots of $3x^2 + 5x + 1 = 0$ is equal to the root of $6x + 5 = 0$.

CHAPTER NINE

Systems of Equations and of Inequalities

Most problems that confront scientists, especially mathematicians, physicists, and engineers, involve several variables that are related in various ways. If the relations between the variables can be expressed as equations, then under suitable conditions, replacements for the variables can be found that satisfy the requirements of the problem. In this chapter we shall be chiefly concerned with methods for determining sets of ordered pairs of numbers whose components satisfy two equations in two variables.

9.1 / EQUATIONS IN TWO VARIABLES

If $f(x, y)$ and $g(x, y)$ stand for expressions in x and y, then

$$f(x, y) = g(x, y) \tag{1}$$

Solution
Solution pair
Equivalent
is an equation in two variables. An ordered pair of numbers that satisfy $f(x, y) = g(x, y)$ is called a *solution* or *solution pair* of the equation.

Two equations in two variables are *equivalent* if every solution pair of each is a solution pair of the other.

208

In Chap. 5 we discussed the operations which, if performed on a given equation, yield an equation that is equivalent to the one given. The same operations can be employed in dealing with an equation in two variables and can be used to solve the equation for one variable in terms of the other.

We now consider the equation

$$x^2 - 3x = y - x + 2 \tag{2}$$

We solve this equation for y in terms of x by adding $-y - x^2 + 3x$ to each member and dividing by -1. Thus we get

$$y = x^2 - 2x - 2 \tag{3}$$

This equation is the same as (1) of Sec. 6.6, which defined the function whose graph is shown in Fig. 6.3 and is shown again in Fig. 9.1. Consequently, the coordinates of each point on the graph is a solution pair of (3), and also of (2), since (2) and (3) are equivalent. This illustrates the fact that the set of all solution pairs, or the solution set, of an equation in two variables is the function, or the relation, defined by the equation. Thus, using the notation employed in Chap. 6, we can designate the set of all solution pairs of (2) as follows:

$$\{(x, y) \mid x^2 - 3x = y - x + 2\} \tag{4}$$

We next consider the equation

$$x - y = -2 \tag{5}$$

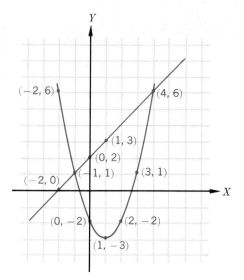

Figure 9.1

The set of solution pairs of (5) is

$$\{(x, y)\,|\,x - y = -2\} \tag{6}$$

Since any ordered pair of numbers that is a solution pair of (2) and also of (5) belongs to the set (4) and to the set (6), it follows that the set of all such ordered pairs is the intersection of sets (4) and (6). We call this

Simultaneous solution set   set of pairs the *simultaneous solution set* of (2) and (5), and in set notation it is

$$\{(x, y)\,|\,x^2 - 3x = y - x + 2\} \cap \{(x, y)\,|\,x - y = -2\} \tag{7}$$

Since the numbers in each ordered pair of a function are the coordinates of a point on the graph of the function, the ordered pairs that belong to the interesection of sets (4) and (6) are the coordinates of the points of intersection of the graphs of the two functions. As stated previously, the graph of (4) is the curve in Fig. 9.1. In order to obtain the graph of (6), we solve $x - y = -2$ for y and get

$$y = x + 2$$

We know from the discussion in Sec. 6.6 that the graph of this equation is a straight line, and we determine the line by using the following table of corresponding values:

x	0	1	-2
y	2	3	0

After the points determined by the above pairs are plotted and a straight line is drawn through them, we obtain the line in Fig. 9.1. By referring to the figure, we see that the graphs intersect at the points $(4, 6)$ and $(-1, 1)$; hence the simultaneous solution set of (2) and (5) is $\{(-1, 1), (4, 6)\}$.

9.2 / GRAPH OF AN EQUATION IN TWO VARIABLES

In Chap. 6 we explained the method for associating an ordered pair of numbers with a point in the cartesian plane and discussed the method for obtaining the graph of a function. In most of the functions discussed in Chap. 6, the rule that established the correspondence between the elements of each ordered pair was an equation in two variables, and we obtained the set of ordered pairs in the function by finding the set of solution pairs of the equation. The methods employed in this section will be similar to those used in Chap. 6.

Graph of an equation in two variables

The graph of an equation in two variables is the totality of points in the cartesian plane such that the coordinates of each form a solution pair of the equation.

In this chapter we shall consider only linear and quadratic equations in two variables, using information from analytic geometry that enables us to determine the general nature of the graph of a particular equation by considering only the degree of the equation and the coefficients in it.

LINEAR EQUATIONS IN TWO VARIABLES
An equation of the type

Linear equation

$$ax + by = c \tag{1}$$

is a linear equation in two variables. It is proved in analytic geometry that the graph of Eq. (1) is a straight line. Since a straight line is fully determined by the position of two points on it, we need only find two solution pairs of Eq. (1) in order to construct the graph. If $c \neq 0$, these two solution pairs are easily obtained by assigning zero to x and solving for y and then by assigning zero to y and solving for x. As a check, it is advisable to get a third solution pair in which neither x nor y is zero. If $c = 0$, then a solution pair is $(0, 0)$, and the graph passes through the origin. In this case it is advisable to assign a positive and a negative number to x and compute each corresponding value of y, and then we have the coordinates of three points on the line. In the illustrative examples in the following sections we shall discuss the graphs of several linear equations and explain the methods for constructing them.

QUADRATIC EQUATIONS IN TWO VARIABLES
The general form of a quadratic equation in two variables is

$$Ax^2 + Bxy + Cy^2 + Dx + Ey = F \tag{2}$$

It is proved in analytic geometry that the graph of Eq. (2) is either a circle, an ellipse, a hyperbola, or a parabola. In certain degenerate cases, the graph of (2), if it exists, is a straight line, two straight lines, or a single point. For example, the graph of $x^2 + y^2 = 0$ is the single point $(0, 0)$, since no other pair of real numbers satisfies the equation. Furthermore the graph of $x^2 + y^2 = -1$ does not exist since x^2 and y^2 are nonnegative numbers. In this chapter we shall deal with the special cases of Eq. (2) listed below:

Equation		Graph
$ax^2 + by^2 = c$	where $\begin{cases} a > 0 \\ b > 0 \\ c > 0 \end{cases}$	An ellipse if $a \neq b$ and a circle if $a = b$
$ax^2 - by^2 = c$	where $\begin{cases} a > 0 \\ b > 0 \\ c \neq 0 \end{cases}$	A hyperbola
$axy = b$	where $b \neq 0$	A hyperbola
$y = ax^2 + bx + c$	where $a \neq 0$	A parabola
$x = ay^2 + by + c$	where $a \neq 0$	A parabola

The graphs of these equations are shown in Fig. 9.2.

    If the nature of the graph is known, it is necessary to obtain a relatively few solution pairs of the equation in order to construct the graph. The method for constructing the graph of a parabola was explained in Chap. 6.

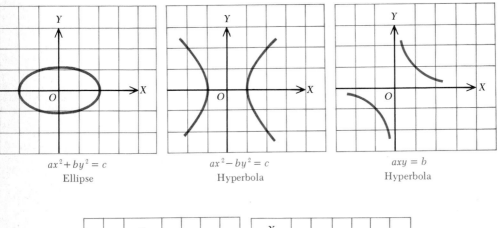

$ax^2 + by^2 = c$
Ellipse

$ax^2 - by^2 = c$
Hyperbola

$axy = b$
Hyperbola

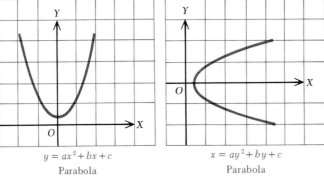

$y = ax^2 + bx + c$
Parabola

$x = ay^2 + by + c$
Parabola

Figure 9.2

In the following examples we illustrate the procedure for obtaining the graph of an ellipse and a hyperbola.

EXAMPLE 1 Construct the graph of $4x^2 + 9y^2 = 36$.

Solution Since the equation is of the type $ax^2 + by^2 = c$, the graph is an ellipse. We solve the equation for y and get

$$y = \pm\tfrac{2}{3}\sqrt{9 - x^2}$$

From this equation we calculate the following table of corresponding values of x and y:

x	-3	-2	-1	0	1	2	3
y	0	±1.5	±1.9	±2	±1.9	±1.5	0

When the points determined by the pairs of numbers in this table are plotted and a smooth curve is drawn through them, we obtain the ellipse in Fig. 9.3a.

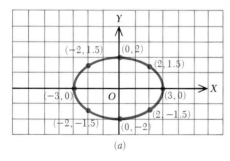

(a)

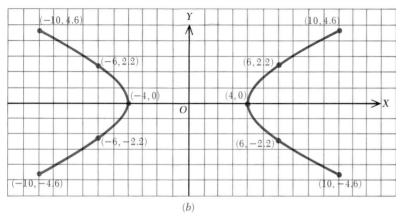

(b)

Figure 9.3

EXAMPLE 2   Construct the graph of $x^2 - 4y^2 = 16$.

Solution   The equation is of type $ax^2 - by^2 = c$, and so the graph is a hyperbola.   We construct the graph as follows.   Solve the equation for y and get

$$y = \pm\tfrac{1}{2}\sqrt{x^2 - 16}$$

Assign the set of numbers $\{-10, -8, -6, -5, -4, 4, 5, 6, 8, 10\}$ to x, calculate each corresponding value of y, and tabulate these values.   We obtain

x	-10	-8	-6	-5	-4	4	5	6	8	10
y	±4.6	±3.5	±2.2	±1.5	0	0	±1.5	±2.2	±3.5	±4.6

We plot the points and draw a smooth curve through them to get the hyperbola in Fig. 9.3b.

EXERCISE 9.1 **Graphs of Equations in Two Variables**

Construct the graph of the function defined by the equation in each of Probs. 1 to 40.

1	$x - y = 3$	**2**	$2x - y = 5$	**3**	$3x + y = 6$
4	$4x + y = 9$	**5**	$x + y = 4$	**6**	$x + 2y = 7$
7	$x - 4y = -4$	**8**	$x - 3y = 8$	**9**	$3x - 2y = 6$
10	$2x + 5y = 10$	**11**	$2x + 3y = 7$	**12**	$3x - 4y = -2$
13	$4x + y = 0$	**14**	$x + 3y = 0$	**15**	$3x - 2y = 0$
16	$5x - 2y = 0$	**17**	$y^2 = 8x$	**18**	$y^2 = -4x$
19	$y^2 - 6y = 4x - 9$	**20**	$y^2 - 2y = 4x$	**21**	$x^2 = 16y$
22	$x^2 - 4x = 4y - 8$	**23**	$x^2 - 6x = 4y - 1$	**24**	$x^2 = 8y$
25	$x^2 + y^2 = 4$	**26**	$x^2 + y^2 = 9$	**27**	$x^2 + y^2 = 25$
28	$x^2 + y^2 = 16$	**29**	$x^2 + 4y^2 = 4$	**30**	$9x^2 + y^2 = 9$
31	$9x^2 + 4y^2 = 36$	**32**	$4x^2 + 25y^2 = 100$	**33**	$4x^2 - y^2 = 4$
34	$x^2 - 9y^2 = 9$	**35**	$4x^2 - 9y^2 = 36$	**36**	$25x^2 - 4y^2 = 100$
37	$xy = 5$	**38**	$xy = -3$	**39**	$x + xy = 6$
40	$y - xy = -4$				

9.3 / SIMULTANEOUS SOLUTION BY GRAPHICAL METHODS

Simultaneous solution set   The *simultaneous solution set* of two equations in x and y is the set of ordered pairs of numbers such that each equation is a true statement if x is replaced by the first number in each pair of the set and y is replaced by the second.

Point of
intersection

Since the pair of coordinates of each point on the graph of an equation in two variables is a solution pair of the equation, the pair of coordinates of each *point of intersection* of the graphs of two equations is a solution pair of each equation. Consequently, if S is the simultaneous solution set of two equations in two variables, if S_1 is the solution set of the first equation, and if S_2 is the solution set of the second, then

$$S = S_1 \cap S_2$$

Of course, if the graphs do not intersect, the simultaneous solution set is the empty set $\varnothing$.

We can therefore obtain the simultaneous solution set of two equations in two unknowns by constructing the graphs of the equations and then estimating the coordinates of their points of intersection. We illustrate the method with three examples.

EXAMPLE 1 Find graphically the simultaneous solution set of the equations

$$3x - 4y = 8 \tag{1}$$
$$2x + 5y = 10 \tag{2}$$

Solution Since each of the given equations is linear, their graphs are straight lines. We obtain the coordinates of three points on each graph by assigning 0 and 4 to x and 0 to y in each equation, and solving for the other variable. We get the following tables of corresponding values:

x	0	$\frac{8}{3}$	4
y	-2	0	1

for Eq. (1) and

x	0	5	4
y	2	0	$\frac{2}{5}$

for Eq. (2). We next plot the points whose coordinates are the pairs of corresponding numbers in each table, draw a straight line through each set, and get the lines in Fig. 9.4. By inspection, we see that these lines intersect at a point whose coordinates to the nearest tenth are $(3.5, 0.6)$. Therefore we say that the simultaneous solution set of Eqs. (1) and (2) appears to be $\{(3.5, 0.6)\}$. If we replace x and y in the given equations by 3.5 and 0.6, respectively, we get

$$3(3.5) - 4(0.6) = 10.5 - 2.4 = 8.1 \qquad \text{from (1)}$$
$$2(3.5) + 5(0.6) = 7 + 3 = 10 \qquad \text{from (2)}$$

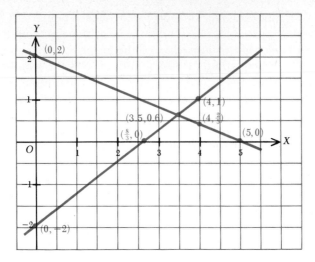

Figure 9.4

Since the right members of (1) and (2) are 8 and 10, respectively, we are fairly safe in assuming that the simultaneous solution set $\{(3.5,\ 0.6)\}$ is correct to one decimal place.

EXAMPLE 2   Find graphically the simultaneous solution set of

$$5x^2 + 8y^2 = 220 \tag{3}$$
$$3x + 4y = 14 \tag{4}$$

Solution   Equation (3) is of the type $ax^2 + by^2 = c$, and its graph is therefore an ellipse. We obtain the intercepts and additional points by assigning $-6, -2, 0, 2,$ and 6 to x, and 0 to y. The corresponding pairs of numbers thus obtained are:

x	-6.6	-6	-2	0	2	6	6.6
y	0	± 2.2	± 5	± 5.2	± 5	± 2.2	0

We plot the points whose coordinates are the pairs of corresponding numbers in this table, draw a smooth curve through them, and thus obtain the ellipse in Fig. 9.5.

The graph of Eq. (4) is the straight line in Fig. 9.5 that is determined by the points whose coordinates are the corresponding numbers in the table below.

x	0	2	$4\frac{2}{3}$
y	$3\frac{1}{2}$	2	0

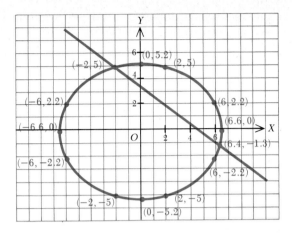

Figure 9.5

From the figure, the coordinates of the upper intersection of the two graphs appear to be $(-2, 5)$; we estimate that the coordinates of the lower intersection to one decimal place are $(6.4, -1.3)$. Therefore, insofar as we are able to estimate from the graph, the simultaneous solution set of Eqs. (3) and (4) to one decimal place is $\{(-2, 5), (6.4, -1.3)\}$.

We test the accuracy of this solution set by replacing x and y in each given equation by the appropriate number in each solution pair. Using $(-2, 5)$, we have

$$5(-2)^2 + 8(5)^2 = 20 + 200 = 220 \qquad \text{from (3)}$$
$$3(-2) + 4(5) = -6 + 20 = 14 \qquad \text{from (4)}$$

Consequently, the solution pair $(-2, 5)$ is exact. Similarly, using $(6.4, -1.3)$ and calculating each result to one decimal place, we get

$$5(6.4)^2 + 8(-1.3)^2 = 204.8 + 13.5 = 218.3 \qquad \text{from (3)}$$
$$3(6.4) + 4(-1.3) = 19.2 - 5.2 = 14 \qquad \text{from (4)}$$

Hence, since the right members of (3) and (4) are 220 and 14, we have reason to believe that our estimate $(6.4, -1.3)$ is correct to one decimal place. By algebraic methods that will be explained later, we can obtain the solution set $\{(-2, 5), (\frac{122}{19}, -\frac{25}{19})\}$, and to one decimal place the latter pair is equal to $(6.4, -1.3)$.

If the graphs of the equations in a given system do not intersect, the graphical method will not reveal the solution set, since only real-number pairs are obtained graphically. By algebraic methods to be discussed later, we can show that either the simultaneous solution set is the empty set $\varnothing$ or the elements of the set are imaginary numbers. Furthermore, except in certain cases,† if a straight line intersects one of the curves in Fig. 9.2 at all,

---

† The exceptional cases occur when the line is tangent to the curve or when it is parallel to the axis of symmetry of a parabola. The axis of symmetry of a parabola is the line L that bisects every chord of the parabola that is perpendicular to L.

it will intersect it at two points. Thus, in general, a system consisting of a linear and a quadratic equation will have two solution pairs in the solution set.

EXAMPLE 3   Obtain the simultaneous solution set of the equations

$$y = x^2 - 4 \tag{5}$$
$$3x^2 + 8y^2 = 75 \tag{6}$$

by the graphical method.

Solution   Equation (5) is of the type $y = ax^2 + bx + c$, with $a = 1 > 0$, $b = 0$, and $c = -4$. Hence the graph is a parabola opening upward (see Fig. 9.2). We construct the parabola by means of the following table of corresponding values and show the graph in Fig. 9.6.

x	-3	-2	-1	0	1	2	3
y	5	0	-3	-4	-3	0	5

Equation (6) is of the type $ax^2 + by^2 = c$, with $a = 3$, $b = 8$, and $c = 75$. The graph is therefore an ellipse (see Fig. 9.2). We construct the graph by means of the corresponding numbers tabulated below, and thus obtain the ellipse in Fig. 9.6.

x	-5	-4	-2	-1	0	1	2	4	5
y	0	± 1.8	± 2.8	± 3	± 3.1	± 3	± 2.8	± 1.8	0

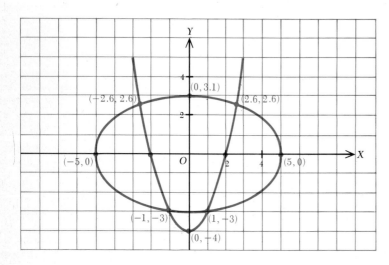

Figure 9.6

The graphs intersect at points whose coordinates appear to be $(1, -3)$, $(-1, -3)$, $(-2.6, 2.6)$, and $(2.6, 2.6)$. Consequently the solution set of the two equations is $\{(1, 3), (-1, -3), (-2.6, 2.6), (2.6, 2.6)\}$.

9.4 / INDEPENDENT, INCONSISTENT, AND DEPENDENT EQUATIONS

The graphs of two linear equations in two unknowns may be two intersecting straight lines, two parallel lines, or two coincident lines. If the two lines intersect in only one point, the simultaneous solution set contains only one ordered pair of numbers and the equations are said to be *independent*. If the two lines are parallel and distinct, there is no point of intersection; therefore no solution pair of one equation is a solution pair of the other, and so the simultaneous solution set is the empty set $\varnothing$. In this case the equations are called *inconsistent*. If the two lines coincide, then every solution pair of one equation is a solution pair of the other and the equations are said to be *dependent*. Figure 9.7 illustrates the three possibilities.

We now derive a simple criterion that enables us to decide whether two linear equations in two unknowns are independent, inconsistent, or dependent. We consider the equations

$$ax + by = c \tag{1}$$
$$Ax + By = C \tag{2}$$

The graphs of (1) and (2) intersect the X axis at $R(c/a, 0)$ and $S(C/A, 0)$, respectively, and the Y axis at the points $T(0, c/b)$ and $U(0, C/B)$, respectively, as illustrated in Fig. 9.7*b*. We can see that the graphs of Eqs. (1) and (2) are parallel if the segments RT and SU are parallel, and these two

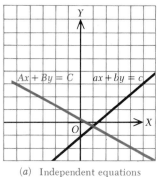

(*a*) Independent equations

$$\frac{A}{a} \neq \frac{B}{b}$$

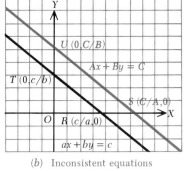

(*b*) Inconsistent equations

$$\frac{A}{a} = \frac{B}{b} \neq \frac{C}{c}$$

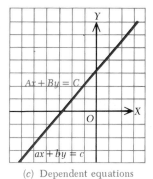

(*c*) Dependent equations

$$\frac{A}{a} = \frac{B}{b} = \frac{C}{c}$$

Figure 9.7

segments are parallel if and only if the triangles ORT and OSU are similar. The triangles are similar if and only if $OR/OS = OT/OU$.

Now $OR/OS = c/a \div C/A = Ac/aC$, and $OT/OU = c/b \div C/B = Bc/bC$. Consequently, $OR/OS = OT/OU$ if and only if $Ac/aC = Bc/bC$. If we multiply each member of the last equation by C/c, we obtain $A/a = B/b$. Therefore, the graphs of (1) and (2) are parallel if and only if $A/a = B/b$. If in addition to $A/a = B/b$ we have $A/a = B/b = C/c = k$, then $A = ak$ and $C = ck$. Hence, $C/A = ck/ak = c/a$, and the points R and S coincide. Therefore, since the graphs are parallel and have at least one point in common, they coincide. Hence we have the following theorem:

Independent, inconsistent, and dependent systems   *Two linear equations $ax + by = c$ and $Ax + By = C$ are independent if and only if $A/a \neq B/b$, inconsistent if and only if $A/a = B/b \neq C/c$, and dependent if and only if $A/a = B/b = C/c$.*

The following three examples illustrate the application of the above theorem.

The equations

$$2x - 3y = 4$$
$$5x + 2y = 8$$

are independent, since $\frac{2}{5} \neq -\frac{3}{2}$. There is only one solution.

The equations

$$3x - 9y = 1$$
$$2x - 6y = 2$$

are inconsistent, since $\frac{3}{2} = -9/(-6) \neq \frac{1}{2}$. There are no solutions.

The equations

$$2x - 4y = 12$$
$$3x - 6y = 18$$

are dependent, since $\frac{2}{3} = -4/(-6) = \frac{12}{18}$. The number of solutions is infinite.

EXERCISE 9.2 Simultaneous Solution by Graphing — Types of Linear Systems

Find the solution set of the pair of equations in each of Probs. 1 to 20 to one decimal place by graphical methods.

1 $y^2 = 2x$	**2** $y^2 = -8x$	**3** $y^2 = 9x$
$y = x - 2$	$y = 2x$	$y = x + 1$
4 $y^2 = -10x$	**5** $x^2 + 4y^2 = 4$	**6** $x^2 + 9y^2 = 9$
$y = 2x - 1$	$y = -x$	$y = -2x$
7 $9x^2 + 4y^2 = 36$	**8** $4x^2 + 25y^2 = 100$	**9** $x^2 - 4y^2 = -4$
$y = 2x + 1$	$y = 3x + 2$	$2x = y$

10 $4x^2 - 9y^2 = 36$ 11 $xy = 8$ 12 $xy = 8$
   $2x - 3 = y$ $y = -3x$ $y = 3x$

13 $x^2 + y^2 = 9$ 14 $x^2 + y^2 = 4$ 15 $x^2 + y^2 = 1$
   $x^2 + 4y^2 = 14$ $x^2 + 9y^2 = 9$ $9x^2 - 4y^2 = 36$

16 $x^2 + y^2 = 4$ 17 $y^2 = 6x$ 18 $y^2 = 2x$
   $9x^2 - 16y^2 = 144$ $x^2 + 4y^2 = 4$ $9x^2 + y^2 = 9$

19 $x^2 = -y$ 20 $x^2 = -3y$
   $y^2 - 4x^2 = 4$ $9y^2 - x^2 = 9$

Determine whether the system in each of Probs. 21 to 36 is dependent, independent, or inconsistent. Solve each independent pair graphically.

21 $2x + 3y = 5$ 22 $3x - 4y = 8$ 23 $5x + 7y = 11$
   $3x + 2y = 7$ $3x + 5y = -5$ $3x + 4y = 8$

24 $4x - 5y = 1$ 25 $3x + 5y = 4$ 26 $5x - y = 2$
   $3x - 4y = -1$ $6x + 10y = 8$ $15x - 3y = 6$

27 $2x + 3y = 5$ 28 $7x + 2y = 1$ 29 $5x - 3y = 8$
   $4x + 6y = 10$ $35x + 10y = 5$ $3x + 2y = 4$

30 $4x - 5y = 11$ 31 $4x - 3y = 9$ 32 $2x + 5y = -2$
   $3x + y = 10$ $3x + 4y = 3$ $3x - y = 8$

33 $2x + 3y = 4$ 34 $5x - y = 3$ 35 $8x + 10y = 14$
   $4x + 6y = 7$ $10x - 2y = 5$ $4x + 5y = 6$

36 $20x - 15y = 45$
   $4x - 3y = 8$

9.5 / ALGEBRAIC METHODS

The first step in the algebraic procedure for solving two equations in two variables is to combine the two equations in such a way as to obtain one equation in one variable whose roots are each one of the numbers in an ordered pair of the solution set. This process is called *eliminating a variable*. After one number in each solution pair is determined, the other can be obtained by substitution. The methods that we shall discuss in the subsequent sections of this chapter are elimination by addition or subtraction, elimination by substitution, and elimination by a combination of addition or subtraction and substitution.

Eliminating a variable

9.6 / ELIMINATION BY ADDITION OR SUBTRACTION

We shall illustrate the principle involved in the elimination of a variable by addition or subtraction by considering the system of two linear equations

$$ax + by = c \tag{1}$$
$$dx + ey = f \tag{2}$$

If (x', y') is a solution pair of each of these equations, then

$$\begin{aligned} ax' + by' &= c \\ dx' + ey' &= f \end{aligned} \qquad \text{by the definition of a solution pair}$$

Then by the axiom shown by Eq. (2.6),

$$\begin{aligned} max' + mby' &= mc \\ ndx' + ney' &= nf \end{aligned}$$

By the axiom shown by Eq. (2.8), the sum of the left members of the above equations is equal to the sum of the right members. Hence we have

$$(ma + nd)x' + (mb + ne)y' = mc + nf$$

If we determine m and n so that $mb + ne = 0$, we have

$$(ma + nd)x' = mc + nf$$

Then

$$x' = \frac{mc + nf}{ma + nd}$$

Thus we have the value of x', and we can obtain y' by replacing x' by the above fraction in either (1) or (2) and solving for y'. We shall illustrate the application of this method with several examples.

TWO LINEAR EQUATIONS

The solution set of any given system of two linear equations in two variables can be obtained by the above method, provided, of course, that the system is consistent. We illustrate the method in Example 1 below.

EXAMPLE 1 Solve the system

$$3x + 4y = -6 \tag{1}$$
$$5x + 6y = -8 \tag{2}$$

by the method of addition or subtraction.

Solution We arbitrarily select y as the variable to be eliminated, and we notice first that the lcm of the coefficients of y, 4 and 6, is 12. Furthermore, $12 \div 4 = 3$ and $12 \div 6 = 2$. Then we proceed as follows:

$9x + 12y = -18$	multiplying (1) by 3	(3)
$\dfrac{10x + 12y = -16}{-x \qquad\quad = -2}$	multiplying (2) by 2	(4)
	subtracting each member of (4) from the corresponding member of (3)	
$x = 2$	multiplying each member by -1	

Hence the first number in the solution pair is 2. To obtain the second number in the solution pair, we replace x by 2 in either (1) or (2) and solve for y. We arbitrarily select (1) for this purpose and get

$$6 + 4y = -6 \qquad \text{replacing } x \text{ by 2 in (1)}$$
$$4y = -12 \qquad \text{adding } -6 \text{ to each member}$$
$$y = -3 \qquad \text{dividing each member by 4}$$

Consequently, the solution set is $\{(x, y)\} = \{(2, -3)\}$.

Check We replace x and y by 2 and -3, respectively, and get

$$6 + (-12) = -6 \qquad \text{from (1)}$$
$$10 + (-18) = -8 \qquad \text{from (2)}$$

This computation verifies that $\{(2, -3)\}$ is the solution set.

Steps in solving two linear equations in two variables   The steps in the process of solving two linear equations in two variables by the method of addition or subtraction can be summarized as follows:

1   Select the variable that will be easier to eliminate.

2   Find the lcm of the two coefficients of this variable.

3   Multiply both members of each equation by the quotient of the lcm and the coefficient of the selected variable (step 1) in that equation.

4   Add or subtract the corresponding members of each equation obtained in step 3, depending upon whether the terms involving the selected variable have unlike or like signs.

5   Solve the resulting equation for the variable that remains.

6   Substitute the number obtained in step 5 into one of the given equations, and solve for the other variable.

7   Write the solution set in the form $\{(x, y)\} = \{(\_\_, \_\_)\}$, filling the blank before the comma with the value of x obtained above and the blank following the comma with the value of y.

8   Check by replacing each variable in each equation by the appropriate number in the solution set.

    The method of elimination by addition or subtraction can be used as the first step in the process of obtaining the solution sets for the two types of systems of quadratic equations discussed in the remainder of this section.

TWO EQUATIONS OF THE TYPE $Ax^2 + Cy^2 = F$
The procedure for obtaining the solution set of a system of two equations of the type $Ax^2 + Cy^2 = F$ is the same as that used in Example 1, except that, in general, we should obtain four solution pairs instead of just one. We illustrate the process in Example 2.

EXAMPLE 2 Obtain the solution set of the system of equations

$$2x^2 + 3y^2 = 21 \tag{1}$$
$$3x^2 - 4y^2 = 23 \tag{2}$$

Solution We arbitrarily select y as the variable to be eliminated and proceed as follows:

$$\begin{array}{ll}
8x^2 + 12y^2 = 84 & \text{multiplying (1) by 4} \qquad\qquad (3) \\
\underline{9x^2 - 12y^2 = 69} & \text{multiplying (2) by 3} \qquad\qquad (4) \\
17x^2 \qquad\;\; = 153 & \text{equating the sums of the corresponding} \\
& \text{members of (3) and (4)} \\
x^2 = 9 & \text{solving for } x^2 \\
\\
x = \pm 3 & \text{equating the square roots of } x^2 \text{ and } 9
\end{array}$$

We obtain the corresponding values of y by replacing x by 3 and -3 in (1) and solving for y. In either case, we get

$$\begin{array}{ll}
18 + 3y^2 = 21 & \\
3y^2 = 3 & \text{adding } -18 \text{ to each member} \\
y^2 = 1 & \\
y = \pm 1 &
\end{array}$$

Hence the solution set is $\{(3, 1), (3, -1), (-3, 1), (-3, -1)\}$.

Check If we replace x and y in the two given equations by 3 and 1, respectively, we get

$$\begin{array}{ll}
18 + 3 = 21 & \text{from (1)} \\
27 - 4 = 23 & \text{from (2)}
\end{array}$$

The other solution pairs can be checked in a similar manner. Figure 9.8 shows the graphs of the two equations and the coordinates of their points of intersection.

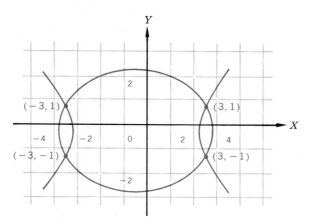

Figure 9.8

TWO EQUATIONS OF THE TYPE $Ax^2 + Cy^2 + Dx = F$ **or** $Ax^2 + Bxy + Dx = F$

The first step in the process of solving two equations of the type $Ax^2 + Cy^2 + Dx = F$ is to eliminate y^2 by addition or subtraction. Next, we solve the resulting equation for x. Finally, we replace x in one of the given equations by the roots obtained in the second step, solve the resulting equation for y, and then write each value of x and the corresponding value of y in the solution set. The procedure is illustrated in Example 3.

EXAMPLE 3 Solve the system of equations

$$3x^2 - 2y^2 - 6x = -23 \qquad (1)$$
$$x^2 + y^2 - 4x = 13 \qquad (2)$$

Solution Since each of the given equations contains one term in y^2 and no other term involving y, we eliminate y^2 and then complete the process of solving as follows:

$3x^2 - 2y^2 - 6x = -23$	(1) recopied	(1)
$2x^2 + 2y^2 - 8x = 26$	multiplying (2) by 2	(3)
$5x^2 \quad\quad -14x = 3$	equating the sums of the left and the right members of (1) and (3)	(4)
$5x^2 - 14x - 3 = 0$	adding -3 to each member of (4)	(5)

$$x = \frac{14 \pm \sqrt{196 + 60}}{10} \qquad \text{solving (5) by the quadratic formula}$$

$$= \frac{14 \pm \sqrt{256}}{10}$$

$$= \frac{14 \pm 16}{10}$$

$$= 3 \text{ and } -\tfrac{1}{5}$$

$$9 + y^2 - 12 = 13 \qquad \text{replacing } x \text{ by 3 in (2)}$$
$$y^2 = 16 \qquad \text{adding 3 to each member}$$
$$y = \pm 4$$

Hence, when $x = 3$, $y = \pm 4$.

$$(-\tfrac{1}{5})^2 + y^2 - [4 \times (-\tfrac{1}{5})] = 13 \qquad \text{replacing } x \text{ by } -\tfrac{1}{5} \text{ in (2)}$$
$$\tfrac{1}{25} + y^2 + \tfrac{4}{5} = 13 \qquad \text{performing the indicated operations}$$
$$1 + 25y^2 + 20 = 325 \qquad \text{multiplying each member by 25}$$
$$25y^2 = 304 \qquad \text{adding } -21 \text{ to each member}$$

$$y^2 = \frac{304}{25}$$

$$y = \pm\sqrt{\frac{304}{25}} = \pm\frac{4\sqrt{19}}{5}$$

Consequently, if $x = -\frac{1}{5}$, then $y = \pm(4\sqrt{19})/5$. Therefore, the solution set is $\{(3, 4), (3, -4), [-\frac{1}{5}, (4\sqrt{19})/5], [-\frac{1}{5}, -(4\sqrt{19})/5]\}$. Each solution pair can be checked by replacing x and y in the given equations by the appropriate number from the solution pair. The graphs of (1) and (2) are shown in Fig. 9.9, together with the coordinates of their points of intersection.

We solve two equations of the type $Ax^2 + Bxy + Dx = F$ by first eliminating xy and then solving the resulting equation for x. The corresponding value of y can then be found by substitution. The method is illustrated in Example 4.

EXAMPLE 4   Obtain the solution set of the system of equations

$$x^2 + 4xy - 7x = 12 \tag{6}$$
$$3x^2 - 4xy + 4x = 15 \tag{7}$$

Solution

$x^2 + 4xy - 7x = 12$	(6) recopied	(6)
$3x^2 - 4xy + 4x = 15$	(7) recopied	(7)
$\overline{4x^2 \qquad -3x = 27}$	equating the sums of the corresponding members of (6) and (7)	(8)
$4x^2 - 3x - 27 = 0$	adding -27 to each member of (8)	(9)

$$x = \frac{3 \pm \sqrt{9 + 432}}{8} \qquad \text{solving (9) by the quadratic formula}$$

$$= \frac{3 \pm \sqrt{441}}{8}$$

$$= \frac{3 \pm 21}{8}$$

$$= 3 \text{ and } -\tfrac{9}{4}$$

Finally, we replace x in (6) successively by 3 and $-\frac{9}{4}$, solve for y, and get $y = 2$ and $\frac{47}{48}$. Consequently, the solution set is $\{(3, 2), (-\frac{9}{4}, \frac{47}{48})\}$. The graphs of (6) and (7), together with the coordinates of their points of intersection, are shown in Fig. 9.10.

The procedure for solving two equations of the type $Ax^2 + Cy^2 + Ey = F$ is the same as that used in Example 3, except that we eliminate x^2 as the first step and solve the resulting equation for y. To solve two equations of the type $Bxy + Cy^2 + Ey = F$, we proceed as in Example 4. In this case, however, the equation obtained after eliminating xy involves the variable y instead of x.

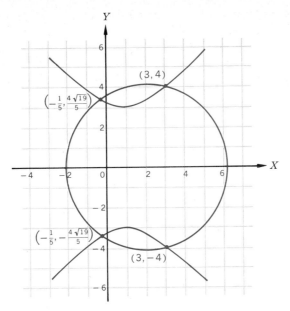

Figure 9.9

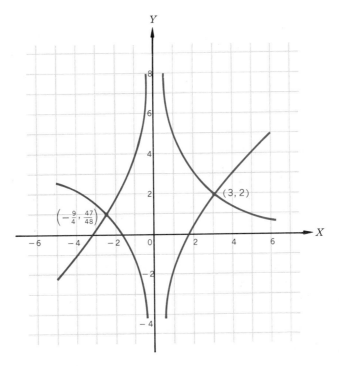

Figure 9.10

EXERCISE 9.3 Elimination by Addition and Subtraction

Find the points of intersection of the pair of equations in each of the following problems by use of addition and subtraction. In Probs. 13 to 16 begin by solving for $1/x$ and $1/y$.

1 $2x + y = 5$
  $x + 3y = 5$

2 $x + 4y = 7$
  $2x - 3y = 3$

3 $3x + 2y = 7$
  $2x + y = 4$

4 $5x - y = 4$
  $3x + 5y = 8$

5 $3x + 2y = 1$
  $4x - 3y = 7$

6 $4x + 5y = 2$
  $3x - 2y = 13$

7 $2x + 3y = 1$
  $3x + 4y = 3$

8 $7x - 9y = -5$
  $3x - 8y = 2$

9 $4x + 3y = 3$
  $6x + 6y = 5$

10 $9x + 4y = 1$
  $12x + 6y = 1$

11 $5x - 2y = 1$
  $6x + 5y = -21$

12 $7x + 8y = 9$
  $4x + 5y = 6$

13 $1/x + 1/y = 7$
  $2/x + 3/y = 19$

14 $1/x + 2/y = 8$
  $3/x - 4/y = -1$

15 $2/x + 5/y = -7$
  $3/x + 8/y = -12$

16 $3/x + 2/y = -3$
  $5/x + 6/y = -1$

17 $x^2 + y^2 = 1$
  $2x^2 + 3y^2 = 2$

18 $x^2 + y^2 = 2$
  $3x^2 + 4y^2 = 7$

19 $x^2 + y^2 = 5$
  $3x^2 - 2y^2 = -5$

20 $x^2 + y^2 = 13$
  $5x^2 - 3y^2 = -7$

21 $x^2 + 4y^2 = 5$
  $9x^2 - y^2 = 8$

22 $9x^2 + 4y^2 = 72$
  $x^2 - 9y^2 = -77$

23 $x^2 + 9y^2 = 9$
  $4x^2 - 25y^2 = 36$

24 $x^2 + 25y^2 = 100$
  $25x^2 - 4y^2 = -16$

25 $x^2 + y^2 + 2x = 9$
  $x^2 + 4y^2 + 3x = 14$

26 $x^2 + y^2 - 4x = -2$
  $9x^2 + y^2 + 18x = 136$

27 $x^2 + y^2 - 6x = -4$
  $x^2 - y^2 + 5x = 5$

28 $x^2 + y^2 - 8x = -8$
  $x^2 - 4y^2 + 6x = 0$

29 $4x^2 + y^2 - 9y = -4$
  $4x^2 - 9y^2 - 2y = -219$

30 $4x^2 - 9y^2 + 7y = -18$
  $x^2 + 4y^2 - 5y = 7$

31 $9x^2 + 4y^2 - 17y = 21$
  $4x^2 - y^2 - 2y = 1$

32 $16x^2 - 9y^2 - 10y = 0$
  $9x^2 + 4y^2 + 12y = 1$

33 $x^2 + xy + 4x = -4$
  $3x^2 - 2xy + x = 4$

34 $x^2 - xy + 5x = 4$
  $2x^2 - 3xy + 10x = -2$

35 $2x^2 + 3xy + x = -2$
  $3x^2 - 2xy + 4x = 1$

36 $x^2 + 4xy - 7x = 10$
  $x^2 + 3xy - 6x = 7$

37 $3y^2 - 5xy + y = 4$
  $2y^2 - 4xy + y = 2$

38 $y^2 + 3xy + 3y = -8$
  $3y^2 + xy + y = 8$

39 $5y^2 - 5xy + 4y = 18$
  $-20y^2 + 25xy + 13y = -4$

40 $2y^2 - xy + 2y = 1$
  $y^2 + xy + 2y = 6$

9.7 / ELIMINATION BY SUBSTITUTION

If, in a system of two equations in two variables, one equation can be solved for one variable in terms of the other, this variable can be eliminated by substitution. We shall assume that the variables are x and y, that one equation can be solved for y in terms of

x, and that the solution is in the form $y = f(x)$. We then replace y in the other equation by $f(x)$ and obtain an equation involving only x. We shall designate this equation by $F(x) = 0$. Next we solve $F(x) = 0$ for x, and the roots thus obtained will be the first number in each pair in the solution set. We complete the process by substituting each root of $F(x) = 0$ in $f(x)$ to obtain each corresponding value of y. Finally, we arrange the corresponding values of x and y in pairs, with the first number in each pair as the value of x, and thus obtain the solution set.

We shall illustrate the method with three examples. In the first example, the method is applied to two linear equations, and in the second example, it is applied to a system containing a linear and a quadratic equation. The third example illustrates the method for solving a system of two quadratics in two variables in which one equation is easily solvable for one variable in terms of the other.

EXAMPLE 1   Solve the equations

$$5x + 3y = 5 \tag{1}$$
$$4x + y = 11 \tag{2}$$

simultaneously by the method of elimination by substitution.

Solution   We first notice that (2) is readily solvable for y in terms of x, and the solution is

$$y = 11 - 4x \qquad \text{obtained by adding } -4x \text{ to each member} \tag{3}$$

We now replace y by $11 - 4x$ in (1) and complete the process of solving as indicated below.

$5x + 3(11 - 4x) = 5$	replacing y by $11 - 4x$ in (1)
$5x + 33 - 12x = 5$	by the distributive axiom
$5x - 12x = 5 - 33$	adding -33 to each member
$-7x = -28$	combining terms
$x = 4$	dividing by -7

We now replace x by 4 in (3) and get

$$y = 11 - 16$$
$$y = -5$$

Therefore, the solution set is $\{(4, -5)\}$.

Check   By (1), $(5 \times 4) + [3 \times (-5)] = 20 - 15 = 5$. By (2), $(4 \times 4) + (-5) = 16 - 5 = 11$.

EXAMPLE 2 Obtain the solution set of the system of equations

$$x^2 + 2y^2 = 54 \qquad (4)$$
$$2x - y = -9 \qquad (5)$$

by the method of substitution.

Solution We first solve (5) for y and get

$$y = 2x + 9 \qquad (6)$$

Next, we replace y by $2x + 9$ in (4) and obtain

$$x^2 + 2(2x + 9)^2 = 54$$

which we solve as follows:

$$
\begin{aligned}
x^2 + 2(4x^2 + 36x + 81) &= 54 && \text{squaring } 2x + 9 \\
x^2 + 8x^2 + 72x + 162 &= 54 && \text{by the left-hand distributive theorem} \\
x^2 + 8x^2 + 72x + 162 - 54 &= 0 && \text{adding } -54 \text{ to each member} \\
9x^2 + 72x + 108 &= 0 && \text{combining similar terms} \\
x^2 + 8x + 12 &= 0 && \text{dividing by 9} \\
(x + 6)(x + 2) &= 0 && \text{factoring} \\
x &= -6 && \text{setting } x + 6 = 0 \text{ and solving} \\
x &= -2 && \text{setting } x + 2 = 0 \text{ and solving}
\end{aligned}
$$

We now replace x in (6) by -6 and then by -2, solve for y, and get

$$y = [2 \times (-6)] + 9 = -12 + 9 = -3$$
$$y = [2 \times (-2)] + 9 = -4 + 9 = 5$$

Therefore, the solution set of the system is $\{(-6, -3), (-2, 5)\}$.

Check Replacing x and y in each equation by -6 and 3, respectively, we have from (4)

$$(-6)^2 + [2 \times (-3)^2] = 36 + 18 = 54$$

and from (5)

$$[2 \times (-6)] - (-3) = -12 + 3 = -9$$

Similarly, for $(-2, 5)$, we have from (4)

$$(-2)^2 + (2 \times 5^2) = 4 + 50 = 54$$

and from (5)

$$[2 \times (-2)] - 5 = -4 - 5 = -9$$

EXAMPLE 3   Solve the system of equations

$$4x^2 - 2xy - y^2 = -5 \tag{7}$$
$$y + 1 = -x^2 - x \tag{8}$$

by the method of substitution.

Solution   We first solve (8) for y in terms of x and get

$$y = -x^2 - x - 1 \tag{9}$$

Next we replace y in (7) by the right member of (9) and have

$$4x^2 - 2x(-x^2 - x - 1) - (-x^2 - x - 1)^2 = -5 \tag{10}$$

Now, if we perform the indicated operations in (10), combine similar terms, and then add 5 to each member, we get

$$-x^4 + 3x^2 + 4 = 0 \tag{11}$$

This is an equation in quadratic form, and we solve it as follows:

$x^4 - 3x^2 - 4 = 0$	dividing each member of (11) by -1 (12)
$(x^2 - 4)(x^2 + 1) = 0$	factoring the left member of (12)
$x^2 = 4$	setting $x^2 - 4 = 0$ and solving for x^2
$x = \pm 2$	equating the square roots
$x^2 = -1$	setting $x^2 + 1 = 0$ and solving for x^2
$x = \pm\sqrt{-1}$	equating the square roots
$x = \pm i$	since $\sqrt{-1} = i$

Now we replace x in (9) successively by $2, -2, i,$ and $-i$ and obtain the corresponding value of y. This procedure yields the following results:

$y = -2^2 - 2 - 1$	replacing x by 2
$\quad = -7$	
$y = -(-2)^2 - (-2) - 1$	replacing x by -2
$\quad = -3$	
$y = -i^2 - i - 1$	replacing x by i
$\quad = 1 - i - 1$	since $i^2 = -1$
$\quad = -i$	
$y = -(-i)^2 - (-i) - 1$	replacing x by $-i$
$\quad = 1 + i - 1$	
$\quad = i$	

Consequently, the solution set is $\{(2, -7), (-2, -3), (i, -i), (-i, i)\}$.

Since the coordinates of any point in a cartesian plane are real numbers, $(i, -i)$ and $(-i, i)$ do not represent points on the graph of either equation, although each is a solution pair of both equations. The graphs

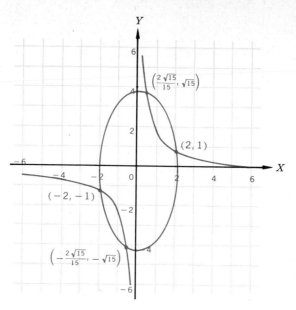

Figure 9.11

of (7) and (8) are shown in Fig. 9.11, and in the figure we see that $(2, -7)$ and $(-2, -3)$ are the points of intersection.

EXERCISE 9.4 **Elimination by Substitution**

Find the solution set of each of the following pairs of equations by substitution.

1	$2x - y = 3$		**2**	$3x + y = 5$		**3**	$3x + 4y = 15$
	$3x + 2y = 8$			$2x + 5y = 12$			$2x - y = -1$
4	$4x + 5y = -1$		**5**	$x - 3y = 7$		**6**	$4x + 7y = 1$
	$3x + y = 2$			$2x + 5y = -8$			$x - 3y = 5$
7	$5x - 2y = 6$		**8**	$x - 2y = 3$		**9**	$3x + 5y = 6$
	$x + 3y = 8$			$4x - 3y = 2$			$2x - 7y = 4$
10	$7x + 5y = 11$		**11**	$5x - 2y = 26$		**12**	$6x + 5y = 37$
	$5x + 7y = 1$			$2x + 5y = -7$			$5x - 3y = -5$
13	$2x + y = 5$		**14**	$5x - y = 2$		**15**	$4x - y = 3$
	$y^2 - 2y = -3x + 5$			$y^2 - 2y = x + 2$			$y^2 - 2y = x - 2$
16	$3x + y = 1$		**17**	$x + 3y = -2$		**18**	$x - y = 4$
	$y^2 - 3y = 7x + 3$			$4x^2 + 3y^2 = 7$			$5x^2 + y^2 = 24$
19	$x - 2y = 7$		**20**	$x + 3y = 5$		**21**	$2x + y = 8$
	$2x^2 + 3y^2 = 29$			$3x^2 + 2y^2 = 11$			$x^2 - y^2 = 5$
22	$3x + y = -8$		**23**	$x - 3y = 1$		**24**	$x - 2y = -5$
	$2x^2 - 5y^2 = 13$			$4x^2 - 9y^2 = 4$			$9x^2 - 2y^2 = -9$

25　$x^2 - 6xy + 20 = 0$　　　26　$x^2 + 4xy + 2y^2 = 46$　　27　$x^2 - 12xy + 2y^2 = -72$
　　$x - 2y^2 - 3y + 4 = 0$　　　　$y^2 - 2y - x = 1$　　　　　　$x = 2y^2 + 6y$

28　$x^2 + 20xy + y^2 = 100$　　　　　29　$4y^2 - 16xy - 6x + 3y = 720$
　　$y^2 - 10y - x = 0$　　　　　　　　$y + x^2 - 2x = 0$

30　$x^2 - 2xy + y^2 + 6x - 6y = 0$　　31　$3x^2 - 12xy = 0$
　　$x^2 + x = y$　　　　　　　　　　　$x + 1 = 3y^2 + 2y$

32　$x^2 - 2xy + y^2 = 144$　　　　　33　$x^2 - 2xy + y^2 = 9$
　　$y = 3x^2 + x$　　　　　　　　　　$y^2 + y = 4 + x$

34　$x^2 - 2xy - y + x = 68$　　　　35　$x^2 - 8xy - 4y + x = -14$
　　$x = 2y^2 + y$　　　　　　　　　　$x = y^2 + 4y$

36　$x^2 - 2xy + y^2 = 1$　　　　　　37　$x^2 + 2y^2 = 9$
　　$x + 2 = y^2 + y$　　　　　　　　　$xy = 2$

38　$2x^2 + 3y^2 = 21$　　　　　　　39　$x^2 + xy - y^2 = -1$
　　$xy = 3$　　　　　　　　　　　　　$xy = 2$

40　$4x^2 + xy - 3y^2 = -6$　　　　　41　$2x^2 + 3y^2 = 29$
　　$xy = 2$　　　　　　　　　　　　　$xy = 3$

42　$4x^2 - 3y^2 = -32$　　　　　　43　$2x^2 + 3xy - y^2 = -16$
　　$xy = -8$　　　　　　　　　　　　$xy = -3$

44　$x^2 + 2xy + 2y^2 = 17$
　　$xy = -5$

9.8 / ELIMINATION BY A COMBINATION OF METHODS

Frequently, the computation involved in solving the system of equations

$$F(x, y) = K \tag{1}$$
$$f(x, y) = k \tag{2}$$

by substitution is very tedious. In such cases, it may be more efficient to use a method that depends upon a theorem in analytic geometry. It states that the graph of

$$mF(x, y) + nf(x, y) = mK + nk \tag{3}$$

passes through the intersections of the graphs of Eqs. (1) and (2). Hence the solution set of the system (1) and (2) is the same as the solution set of the system composed of either (1) or (2) and (3) for all nonzero replacements of m and n.

Now if we can so determine m and n in (3) that the resulting equation is easily solvable for either unknown in terms of the other, we can obtain the solution set of (1) and (2) by solving either (1) or (2) with (3) by substitution. We shall discuss two classes of systems of quadratic equations in two unknowns that can be solved by this method.

TWO EQUATIONS OF THE TYPE $Ax^2 + Bxy + Cy^2 = D$

The first step in solving two equations of the type $Ax^2 + Bxy + Cy^2 = D$ is to eliminate the constant terms by addition or subtraction. That is, we combine the two equations in such a way as to obtain an equation in which the constant term is zero. We then solve this equation for one unknown in terms of the other and complete the process by substitution. For example, if we solve the equation for y in terms of x, we usually obtain two equations of the type $y = rx$ and $y = tx$, where r and t are constants. We then replace y in one of the given equations successively by rx and tx and solve each resulting equation for x. We then find the value of y that corresponds to each root thus obtained by replacing x in $y = rx$ and $y = tx$ by the root. We shall illustrate the process with an example.

EXAMPLE 1   Find the simultaneous solution set of the equations

$$3x^2 - 4xy + 2y^2 = 3 \tag{4}$$
$$2x^2 - 6xy + y^2 = -6 \tag{5}$$

Solution   We eliminate the constant terms:

$6x^2 - 8xy + 4y^2 =$	6	multiplying (4) by 2 $\qquad$ (6)
$2x^2 - 6xy + y^2 = -6$		Eq. (5) recopied
$8x^2 - 14xy + 5y^2 =$	0	equating the corresponding sums $\quad$ (7)

We now solve Eq. (7) for y in terms of x by the quadratic formula with $a = 5$, $b = -14x$, and $c = 8x^2$, and obtain

$$y = \frac{14x \pm \sqrt{196x^2 - 160x^2}}{10}$$

$$= \frac{14x \pm 6x}{10} = \frac{20x}{10} \text{ and } \frac{8x}{10}$$

Consequently

$$y = 2x \tag{8}$$
$$y = \frac{4x}{5} \tag{9}$$

We continue the process by replacing y in either (4) or (5) separately by $2x$ and $4x/5$ and solving the resulting equation for x. Substituting $y = 2x$ in (4), we get

$$3x^2 - 4x(2x) + 2(2x)^2 = 3$$
$$3x^2 - 8x^2 + 8x^2 = 3 \qquad \text{performing the indicated operations}$$
$$3x^2 = 3 \qquad \text{combining terms}$$
$$x^2 = 1$$
$$x = \pm 1$$

We now replace x in (8) by ± 1 and get $y = 2(\pm 1) = \pm 2$. Hence two pairs in the solution set are $(1, 2)$ and $(-1, -2)$.

Next we replace y by $4x/5$ in (4) and solve the resulting equation for x. Thus we obtain

$$3x^2 - 4x\,\frac{4x}{5} + 2\left(\frac{4x}{5}\right)^2 = 3$$

$$3x^2 - \frac{16x^2}{5} + \frac{32x^2}{25} = 3$$

$$75x^2 - 80x^2 + 32x^2 = 75 \qquad \text{multiplying each member by 25}$$

$$27x^2 = 75$$

$$x^2 = \tfrac{75}{27} = \tfrac{25}{9}$$

$$x = \pm\tfrac{5}{3}$$

Finally, we replace x by $\pm\tfrac{5}{3}$ in (9) and get $y = \tfrac{4}{5}(\pm\tfrac{5}{3}) = \pm\tfrac{4}{3}$. Therefore, two additional solution pairs are $(\tfrac{5}{3}, \tfrac{4}{3})$ and $(-\tfrac{5}{3}, -\tfrac{4}{3})$.

Consequently the complete simultaneous solution set is $\{(1,2),$ $(-1,-2),\ (\tfrac{5}{3}, \tfrac{4}{3}),\ (-\tfrac{5}{3}, -\tfrac{4}{3})\}$, which can be checked by the usual methods.

If the constant term in either of the given equations is zero, as in the system

$$2x^2 - 3xy - 2y^2 = 0$$
$$x^2 + 2xy + 5y^2 = 17$$

it is not necessary to perform the first step in Example 1. We immediately solve the first equation for y in terms of x, or x in terms of y, and then complete the solution by the method of substitution.

TWO EQUATIONS OF THE TYPE $Ax^2 + Ay^2 + Dx + Ey = F$

If each of the given equations is of the type $Ax^2 + Ay^2 + Dx + Ey = F$, we can eliminate the second-degree terms by addition or subtraction and obtain a linear equation in x and y. We can then solve this equation with one of the given equations by the method illustrated in Example 2 of Sec. 9.8.

EXAMPLE 2   Obtain the simultaneous solution set of

$$3x^2 + 3y^2 + \ \ x - 2y = 20 \qquad (10)$$
$$2x^2 + 2y^2 + 5x + 3y = 9 \qquad (11)$$

Solution

$$
\begin{array}{ll}
6x^2 + 6y^2 + \ \ 2x - \ \ 4y = 40 & \text{multiplying (10) by 2} \qquad (12) \\
6x^2 + 6y^2 + 15x + \ \ 9y = 27 & \text{multiplying (11) by 3} \qquad (13) \\
\hline
 -13x - 13y = 13 & \text{equating the differences of} \qquad (14)
\end{array}
$$

the corresponding members
of (12) and (13)

Now we solve Eq. (14) simultaneously with Eq. (11) and complete the process of solving as indicated below.

$$y = -x - 1 \qquad \text{solving (14) for } y \qquad (15)$$
$$2x^2 + 2(-x-1)^2 + 5x + 3(-x-1) = 9 \qquad \text{replacing } y \text{ by } -x-1$$
$$\text{in (11)}$$

$$2x^2 + 2x^2 + 4x + 2 + 5x - 3x - 3 - 9 = 0 \qquad \text{performing the indicated}$$
operations and adding
-9 to each member

$$4x^2 + 6x - 10 = 0 \qquad \text{combining similar terms}$$
$$2x^2 + 3x - 5 = 0 \qquad \text{dividing by 2}$$
$$x = 1 \text{ and } -\tfrac{5}{2} \qquad \text{by the quadratic formula}$$

We find the corresponding values of y as follows:

$$y = -1 - 1 = -2 \qquad \text{replacing } x \text{ by 1 in (15)}$$
$$y = \tfrac{5}{2} - 1 = \tfrac{3}{2} \qquad \text{replacing } x \text{ by } -\tfrac{5}{2} \text{ in (15)}$$

Consequently the solution set is $\{(1, -2), (-\tfrac{5}{2}, \tfrac{3}{2})\}$, which can be checked in the usual manner.

9.9 / SYMMETRIC EQUATIONS

Symmetric equation

An equation in two variables is *symmetric* if the equation is not altered when the variables are interchanged. For example,

$$Ax^2 + Bxy + Ay^2 + Dx + Dy = F \qquad (1)$$

is symmetric since interchanging x and y does not change the equation. If we transform Eq. (1) by means of the equations

$$x = u + v \qquad (2)$$
$$y = u - v$$

we get
$$(2A + B)u^2 + (2A - B)v^2 + 2Du = F \qquad (3)$$

Therefore, the transformation (2) will convert two symmetric equations to two equations of the type (3), which can be solved by the method of Example 3 of Sec. 9.6. We illustrate the procedure with the following example.

EXAMPLE 1   Obtain the simultaneous solution set of the equations

$$7x^2 + 2xy + 7y^2 - 8x - 8y = 108 \qquad (4)$$
$$3x^2 - 2xy + 3y^2 + 4x + 4y = 68 \qquad (5)$$

Solution   We first replace x by $u + v$ and y by $u - v$ in each of (1) and (2), divide by 4, and get

$$4u^2 + 3v^2 - 4u = 27 \qquad \text{from (4)} \qquad (6)$$
$$u^2 + 2v^2 + 2u = 17 \qquad \text{from (5)} \qquad (7)$$

Since each of (6) and (7) contains one term involving v^2 and no other term involving v, we can eliminate v^2 by addition or subtraction and complete the procedure of solving as follows:

$$8u^2 + 6v^2 - 8u = 54 \quad\quad \text{multiplying (6) by 2} \quad\quad (8)$$
$$3u^2 + 6v^2 + 6u = 51 \quad\quad \text{multiplying (7) by 3} \quad\quad (9)$$
$$\overline{5u^2 \qquad\quad -14u = 3} \quad\quad \begin{array}{l}\text{equating the differences} \quad (10)\\ \text{of the corresponding}\\ \text{members of (8) and (9)}\end{array}$$

$$5u^2 - 14u - 3 = 0 \quad\quad \begin{array}{l}\text{adding } -3 \text{ to each}\\ \text{member of (10)}\end{array}$$

$$u = \frac{14 \pm \sqrt{196 + 60}}{10} \quad\quad \text{by the quadratic formula}$$

$$u = \frac{14 \pm 16}{10} = 3 \text{ and } -\tfrac{1}{5}$$

We now obtain the corresponding values of v by replacing u in Eq. (7) by each of the above values, as indicated below.

$$9 + 2v^2 + 6 = 17 \quad\quad \text{replacing } u \text{ by 3 in (7)}$$
$$2v^2 = 2$$
$$v^2 = 1$$
$$v = \pm 1$$

Therefore, if $u = 3$, $v = \pm 1$. Now we replace (u, v) by $(3, 1)$ and $(3, -1)$ in $x = u + v$ and $y = u - v$ and get

$$x = 3 + 1 = 4 \quad\quad y = 3 - 1 = 2$$
and
$$x = 3 - 1 = 2 \quad\quad y = 3 - (-1) = 4$$

Hence, two pairs in the solution set are $(4, 2)$ and $(2, 4)$. If we replace u by $-\tfrac{1}{5}$ in (7), we have

$$\tfrac{1}{25} + 2v^2 - \tfrac{2}{5} = 17$$
$$1 + 50v^2 - 10 = 425$$
$$50v^2 = 434$$
$$v^2 = \tfrac{434}{50} = \tfrac{217}{25}$$
$$v = \pm \frac{\sqrt{217}}{5}$$

Finally,
$$x = -\frac{1}{5} + \frac{\sqrt{217}}{5} \quad\quad y = -\frac{1}{5} - \frac{\sqrt{217}}{5}$$

and
$$x = -\frac{1}{5} - \frac{\sqrt{217}}{5} \quad\quad y = -\frac{1}{5} - \left(\frac{-\sqrt{217}}{5}\right)$$

Therefore the simultaneous solution set is

$$\left\{ (4, 2),\ (2, 4),\ \left(\frac{-1 + \sqrt{217}}{5}, \frac{-1 - \sqrt{217}}{5}\right),\ \left(\frac{-1 - \sqrt{217}}{5}, \frac{-1 + \sqrt{217}}{5}\right) \right\}$$

EXERCISE 9.5 Elimination by a Combination of Methods, Symmetric Equations

Find the solution set of each of the following pairs of equations.

1 $x^2 + xy - 2y^2 = 0$
   $11x^2 + 3xy - 11y^2 = 27$

2 $4x^2 + 3xy + 2y^2 = 3$
   $2x^2 + 6xy + y^2 = -3$

3 $2x^2 + 4xy - 3y^2 = 3$
   $x^2 + xy - 2y^2 = 4$

4 $5x^2 - 3xy - 9y^2 = 3$
   $3x^2 + xy - 3y^2 = 1$

5 $x^2 + 4xy - 17y^2 = -20$
   $x^2 + xy - 6y^2 = 0$

6 $5x^2 - 2xy - 2y^2 = 1$
   $7x^2 - 3xy - 3y^2 = 1$

7 $x^2 - 2xy - 2y^2 = 2$
   $2x^2 - 4xy - 3y^2 = 3$

8 $x^2 - 6xy + 4y^2 = 1$
   $x^2 - 10xy + 8y^2 = -4$

9 $2x^2 - xy - 3y^2 = 0$
   $x^2 - 2xy + 2y^2 = 5$

10 $x^2 + 2xy - 5y^2 = -5$
    $2x^2 - xy - 6y^2 = 4$

11 $3x^2 - xy - 3y^2 = 4$
    $5x^2 + xy - 2y^2 = 8$

12 $6x^2 - 5xy + 2y^2 = 72$
    $5x^2 + 4xy - 3y^2 = -45$

13 $x^2 + 2xy + 2y^2 = 10$
    $2x^2 + xy + 22y^2 = 50$

14 $x^2 + 15xy + 9y^2 = -5$
    $7x^2 + 9xy + 27y^2 = 25$

15 $x^2 + xy + 2y^2 = 8$
    $5x^2 + 8xy - 4y^2 = 32$

16 $2x^2 - 4xy + 3y^2 = 3$
    $x^2 + 7xy - 3y^2 = 15$

17 $x^2 + y^2 - 2x = 1$
    $x^2 + y^2 - 6y = -1$

18 $x^2 + y^2 - 3x - 2y = 4$
    $x^2 + y^2 - 2x - y = 5$

19 $x^2 + y^2 + 2x - y = 2$
    $x^2 + y^2 - x + 2y = 5$

20 $x^2 + y^2 - 3x = 1$
    $x^2 + y^2 - y = 9$

21 $6x^2 + 6y^2 - 5x + 3y = -1$
    $9x^2 + 9y^2 - 7x + 4y = -1$

22 $x^2 + y^2 - x + 3y = 0$
    $2x^2 + 2y^2 + 3x + 5y = -3$

23 $x^2 + y^2 - 3x = 1$
    $x^2 + y^2 + 3y = 7$

24 $x^2 + y^2 - 2x - 3y = 1$
    $3x^2 + 3y^2 - 6x - 4y = 13$

25 $x^2 + y^2 - 5x + y = -4$
    $x^2 + y^2 - 3x + 2y = 1$

26 $4x^2 + 4y^2 - 11x = 19$
    $3x^2 + 3y^2 - 11y = 17$

27 $3x^2 + 3y^2 - 13x - y = -2$
    $5x^2 + 5y^2 - 16x + 4y = 25$

28 $x^2 + y^2 + 2x + 2y = 6$
    $2x^2 + 2y^2 + 3x + 3y = 10$

29 $x^2 - 3xy + y^2 + 3x + 3y = 20$
    $3x^2 - 4xy + 3y^2 - 6x - 6y = 5$

30 $2x^2 + xy + 2y^2 - 5x - 5y = 57$
    $x^2 - xy + y^2 + 7x + 7y = 84$

31 $2x^2 + 2xy + 2y^2 - x - y = 22$
    $3x^2 + 2xy + 3y^2 + 4x + 4y = 52$

32 $4x^2 - 5xy + 4y^2 - 6x - 6y = -8$
    $2x^2 - 3xy + 2y^2 - 2x - 2y = -2$

33 $x^2 + xy + y^2 - 6x - 6y = -11$
    $5x^2 - 4xy + 5y^2 - 3x - 3y = 26$

34 $7x^2 - 6xy + 7y^2 + 8x + 8y = 144$
    $2x^2 - xy + 2y^2 + 3x + 3y = 44$

35 $5x^2 + 4xy + 5y^2 + 3x + 3y = 74$
    $x^2 + 2xy + y^2 - 6x - 6y = -8$

36 $5x^2 + 4xy + 5y^2 - 13x - 13y = 144$
    $2x^2 + 2xy + 2y^2 - 9x - 9y = 32$

37 $xy + 2x + 2y = 21$
    $2xy - x - y = 12$

38 $3xy + 4x + 4y = -2$
    $xy + x + y = -1$

39 $xy + x + y = 11$
    $2xy - 3x - 3y = -3$

40 $3xy - 2x - 2y = 4$
    $xy - x - y = 0$

9.10 / GRAPHICAL SOLUTION OF A SYSTEM OF INEQUALITIES

In this section we shall explain the use of the graphical method for finding the solution set of a linear inequality in two variables and of a system of linear inequalities in two variables. We shall consider only $x > 0$, $x < 0$, $y > ax + b$, and $y < ax + b$, since any linear inequality is equivalent to one of these types.

In most of our discussion we shall be concerned with finding the solution set of a statement such as $y \geq ax + b$. This is a combination of an inequality and an equation, but we shall refer to it as an inequality. As in our previous sections, we shall use the notation $P(x, y)$ to stand for the point whose coordinates are (x, y).

In order to find the solution set of $y \geq ax + b$, we begin by constructing the graph of $y = ax + b$. Now if $P(x, y)$ is on the graph of $y = ax + b$ and if $y' > y$, then $P(x, y')$ is above the graph. Furthermore, since $y' > y$, it follows that $y' > ax + b$. Therefore, the solution set of $y \geq ax + b$ is $\{(x, y) \mid P(x, y)$ is on or above the graph of $y = ax + b\}$. Furthermore, for similar reasons, the solution set of $y \leq ax + b$ is $\{(x, y) \mid P(x, y)$ is on or below the graph of $y = ax + b\}$.

In keeping with the above discussion, the solution set of $y \leq 3x - 2$ is $\{x, y) \mid P(x, y)$ is on or below the graph of $y = 3x - 2\}$. Hence P is in the region indicated by the arrows in Fig. 9.12.

In the remainder of this section we shall be concerned with determining the region in which $P(x, y)$ lies if (x, y) belongs to the simultaneous

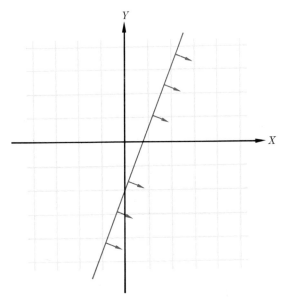

Figure 9.12

solution set of two or more inequalities. The method consists of the following steps: first, draw the graph of each related equation; second, for each line, indicate by arrows the region in which $P(x, y)$ lies if (x, y) is a solution pair of the related inequality; third, determine the intersection of the regions obtained in the second step. We shall refer to this region as the *region determined by the inequalities* or the *feasible region*.

EXAMPLE   Find the region determined by $x \geq 0$, $y \geq 0$, $y \geq 3x - 3$, and $y \leq 0.5x + 2$.

Solution   We begin by sketching the graphs of $x = 0$, $y = 0$, $y = 3x - 3$, and $y = 0.5x + 2$. The first one is the Y axis, the second is the X axis, the third is through $(0, -3)$ and $(1, 0)$, and the last is through $(0, 2)$ and $(4, 4)$. (The graphs are shown in Fig. 9.13.) We now indicate the half plane determined by each inequality by use of arrows. We thus find that the region determined by the given set of inequalities is the boundary and interior of the quadrilateral with vertices at the origin, $Q(1, 0)$, $I(2, 3)$, and $R(0, 2)$.

Convex region   If all points of the line segment PQ are in a set S of the xy plane whenever P and Q are in it, we say that the set S is a *convex set of points* or a *convex region*.

According to this definition, a circle and the quadrilateral $OQIR$ of Fig. 9.13 are convex sets, as is a half plane.

Polygonal set   The intersection of two or more closed half planes is called a *polygonal set of points*.

Finite polygonal region   If a polygonal set of points has a finite area, then it is called a *finite*
Convex polygon   *polygonal region* and its boundary is called a *convex polygon*.

The quadrilateral $OQIR$ in Fig. 9.13 is a convex polygon.

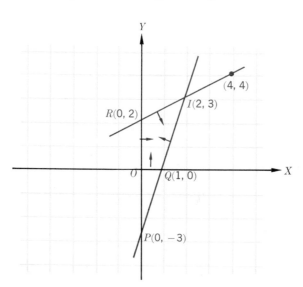

Figure 9.13

EXERCISE 9.6 Convex Polygons

Indicate the half-plane region determined by the inequality in each of Probs. 1 to 4.

1 $x \geq -1$ 2 $y \leq 3$ 3 $-2x + y \leq 4$ 4 $3x - y \leq 2$

Show the convex region determined by the set of inequalities in each of Probs. 5 to 12.

5 $x \geq 0, y \geq 0, x + 2y \leq 4$ 6 $x \leq 0, y \leq 3, 2x + y \geq -6$
7 $x \geq 1, y \leq 2, x - y \leq 2$ 8 $x \leq 3, y \geq 1, x - y \geq 0$
9 $x + y \leq 1, x - y \geq -1, x + y \geq -1, x - y \leq 1$
10 $2x - y + 2 \geq 0, x + y - 3 \leq 0, 3x - 2y \leq 6$
11 $x + y \leq 2, x - y \leq 2, 2x - y \leq 2, 3x - y \geq -1$
12 $5x - 2y \leq 6, 5x - 2y \geq 4, x + y \geq 0, 5x + 2y \leq 10$

Show that the graphs of the equations in each of Probs. 13 to 16 may bound a non-convex region by picking two points P and Q in the chosen region and showing that not all points on the segment PQ are in the region.

13 $x = 0, y = 0, y = 3x + 2, y = x - 3$
14 $x = 1, y = -1, x - 3y + 7 = 0, 4x + y + 6 = 0$
15 $x = 2, y = 1, x + 3y + 3 = 0, 3x + 2y = 6$
16 $2x - y + 4 = 0, x + y + 2 = 0, x = -4, x = 1, y = -3$

Show that the sets of inequalities in each of Probs. 17 to 20 do not determine a finite polygonal region.

17 $x \geq 0, y \leq -1, 3x + y - 5 \leq 5$
18 $4x - y - 7 \leq 0, x + 5y - 7 \leq 0, 5x + 4y + 7 \leq 0$
19 $x \geq -1, y \geq 1, x - y - 1 \geq 0$
20 $x + 5y - 12 \geq 0, x - 2y + 1 \leq 0, 3x + y - 4 \geq 0$

Show that the quadrilateral with the given points as vertices in each of Probs. 21 to 24 is not a convex set.

21 $(-3, 1), (0, 2), (3, 2), (-1, 4)$
22 $(-2, 1), (0, -2), (-1, 0), (3, 2)$
23 $(-1, 2), (1, -1), (0, 1), (3, 3)$
24 $(1, 0), (4, 5), (-1, 4), (1, 3)$

Find the vertices of the polygonal region determined by the inequalities in each of Probs. 25 to 32.

25 $3x - 4y - 12 \leq 0, x + y - 4 \leq 0, 5x - 2y \geq 6$
26 $4x + y \leq 7, 2x + 5y + 1 \geq 0, x - 2y + 5 \geq 0$
27 $x + y - 3 \leq 0, x - 2y - 3 \leq 0, 5x + 2y - 3 \geq 0$
28 $7x - 2y + 6 \leq 0, 5x + y + 14 \geq 0, 2x - 3y + 9 \geq 0$
29 $x + y - 3 \leq 0, 3x - y - 9 \leq 0, y + 3 \geq 0, 7x - 4y + 23 \geq 0$
30 $x + 5y - 11 \leq 0, 5x + y - 7 \leq 0, x - 5y - 17 \leq 0, 7x + y + 25 \geq 0$

31   $x + 3y - 7 \leq 0,\ 4x - y - 28 \leq 0,\ x + 9y + 30 \geq 0,\ 4x - 7y - 9 \geq 0$

32   $3x + 4y - 23 \leq 0,\ 2x - y - 8 \geq 0,\ 2x + 3y \geq 0,\ x - y - 10 \leq 0$

9.11 / LINEAR PROGRAMMING

Maximum
Minimum
Extrema

If there is a largest and a smallest value of a function, they are called the *maximum* and the *minimum*, respectively, and are often referred to as the *extrema*. For example, the maximum value of $f = \{x \mid y = x^2,\ -1 \leq x \leq 3\}$ occurs for $x = 3$ and is $3^2 = 9$; the minimum is zero, for $x = 0$.

If a problem involves two variables x and y, if the conditions restrict the domain to a region S determined by a set of linear inequalities, and if a given linear combination of x and y is to be an extremum subject to the restricting inequalities, then the determination of (x, y) and/or the extremum is called *linear programming*.

Linear programming

The inequalities that determine the region S are called *restraints*, the region S is called the set of *feasible solutions*, and the linear function f is called the *objective function*.

Feasible solutions
Objective
function

We shall now state without proof a theorem on linear programming, then give a strictly mathematical application, and follow that by one from business.

THEOREM   **If S is a convex polygon, if $f(x, y) = ax + by + c$ with domain S, and if B is the polygonal boundary of S, then f has a maximum and a minimum in S and they occur at vertices of B.**

EXAMPLE 1   Find the maximum and minimum values of $f(x, y) = 2x - 3y + 4$ in the region S with vertices at $(0, 5),\ (6, 4),\ (7, 1)$ and $(-3, -2)$.

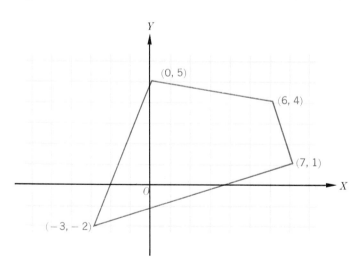

Figure 9.14

Solution   Since the region S, shown in Fig. 9.14 with the four given points as vertices, is convex and f is a linear function in two variables, we know by use of the theorem of this section that the extrema of f are attained at vertices of the boundary. Therefore, we shall evaluate f at each vertex. The values are $f(0, 5) = (2 \times 0) - (3 \times 5) + 4 = -11$; $f(6, 4) = (2 \times 6) - (3 \times 4) + 4 = 4$; $f(7, 1) = (2 \times 7) - (3 \times 1) + 4 = 15$; and $f(-3, -2) = [2 \times (-3)] - [3 \times (-2)] + 4 = 4$. Consequently, the maximum value of f in S is $f(7, 1) = 15$ and the minimum value is $f(0, 5) = -11$.

EXAMPLE 2   A maker of animal shoes specializes in horseshoes, mule shoes, and oxen shoes and can produce 200 sets of shoes per unit of time. He has standing orders for 60 sets of horseshoes and 20 sets of oxen shoes and can sell at most 150 sets of horseshoes and 50 sets of mule shoes. How many sets of each type should he produce to make a maximum profit provided his profit on a set of shoes is \$0.40 for horseshoes, \$0.50 for mule shoes, and \$0.30 for oxen shoes?

Solution   If we represent the number of sets of horseshoes produced by x and of mule shoes by y, then $200 - x - y$ is the number of sets of oxen shoes. Consequently, his profit in terms of dollars is

$$f(x, y) = 0.4x + 0.5y + 0.3(200 - x - y)$$
$$= 0.1x + 0.2y + 60$$

This objective function is subject to the restraints imposed by the problem. They are:

$x \geq 60$	since he has a standing order for 60 sets of horseshoes
$200 - x - y \geq 20$	since he has a standing order for 20 sets of oxen shoes
$x \leq 150$	since he cannot sell more than 150 sets of horseshoes
$y \leq 50$	since he cannot sell more than 50 sets of mule shoes

$x \geq 0$ $y \geq 0$ $200 - x - y \geq 0$
since all three types are for sale

Figure 9.15 shows these restraints graphically, along with the vertices of the polygonal boundary of the set of feasible solutions as obtained by solving

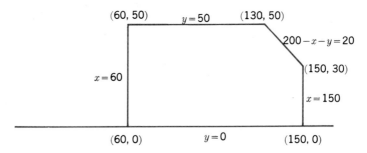

Figure 9.15

the equations of the pairs of bounding lines that meet in a vertex. We now evaluate $f(x, y) = 0.1x + 0.2y + 60$ at each vertex and find that $f(60, 0) = 66$, $f(150, 0) = 75$, $f(150, 30) = 81$, $f(130, 50) = 83$, and $f(60, 50) = 76$. Consequently, the shoemaker has the greatest profit if he makes 130 sets of horseshoes, 50 sets of mule shoes, and 20 sets of oxen shoes.

EXERCISE 9.7 Extrema and Linear Programming

For each system of restraints or linear inequalities in Probs. 1 to 16, (1) find the set of feasible solutions graphically; (2) select a point of S and show that each inequality is satisfied by its coordinates; (3) select a point not in S and show that it does not satisfy at least one of the inequalities; and (4) find the maximum and minimum values of the objective function, and the coordinates of the vertices at which they are attained.

1 $x \geq 0,\ y \geq 0,\ 3x + 7y - 21 \leq 0,\ f(x, y) = 2x + 3y + 1$

2 $x \leq 1,\ y \leq 2,\ 2x + y - 2 \geq 0,\ f(x, y) = 3x - 5y + 6$

3 $x \geq -2,\ y \leq 1,\ x - 2y + 2 \leq 0,\ f(x, y) = -5x + 2y + 7$

4 $x \leq 2,\ y \geq -2,\ 2x - y - 2 \geq 0,\ f(x, y) = -4x + 3y$

5 $x \geq 0,\ x + 2y - 4 \leq 0,\ x - 2y \leq 0,\ f(x, y) = -x + 4y - 3$

6 $x + 3y - 6 \leq 0,\ y \geq 1,\ x - y + 2 \geq 0,\ f(x, y) = -3x - 2y + 11$

7 $3x - 2y - 5 \leq 0,\ 4x - y + 2 \geq 0,\ 3x + y - 2 \leq 0,\ f(x, y) = 4x - 2y - 1$

8 $3x - y - 7 \leq 0,\ x - 4y + 5 \geq 0,\ 2x + 3y - 1 \geq 0,\ f(x, y) = 7x + 6y - 13$

9 $x \geq 0,\ y \leq 0,\ x + y + 1 \geq 0,\ x - y - 3 \leq 0,\ f(x, y) = 3x - 4y + 8$

10 $x \leq 0,\ y \geq 0,\ 2x + y + 2 \geq 0,\ x - 2y + 6 \geq 0,\ f(x, y) = 5x - y - 1$

11 $x \leq 2,\ y \leq 1,\ 2x - y + 1 \geq 0,\ x + 3y + 4 \geq 0,\ f(x, y) = 3x - 2y + 7$

12 $x \leq 2,\ y \geq -1,\ 2x - y + 1 \geq 0,\ x - 2y + 2 \geq 0,\ f(x, y) = x + 4y + 9$

13 $x - 4y + 10 \geq 0,\ 3x + y - 8 \geq 0,\ x + 5y + 6 \geq 0,\ 5x + 2y - 16 \leq 0,\ f(x, y) = 2x + 5y - 3$

14 $x \leq 3,\ 2x - 3y - 9 \leq 0,\ 4x + y + 3 \geq 0,\ x - 4y + 5 \geq 0,\ f(x, y) = x - y - 2$

15 $2x - 3y + 2 \geq 0,\ 4x + y - 10 \leq 0,\ x - 3y - 9 \leq 0,\ 3x + y + 3 \geq 0,\ f(x, y) = x + 3y - 4$

16 $x + 3y - 3 \leq 0,\ 3x - y - 9 \leq 0,\ x + 4y + 10 \geq 0,\ 3x - 2y + 2 \geq 0,\ f(x, y) = 2x - 7y - 5$

17 A builder has 42 units of material and 32 units of labor available for use during a given period and can build houses of type A or of type B or some of both. How many of each should he build so as to make a maximum profit if he makes $1,000 on each house of type A and $400 on each house of type B? Assume that a house of type A requires 7 units of material and 1 of labor, whereas a house of type B requires 1 unit of material and 2 units of labor. What is the maximum profit?

18 Repeat Prob. 17 with only 19 units of labor available.

19 Repeat Prob. 17 with 99 units of material and 55 of labor available.

20 Repeat Prob. 17 with the profit on type A as $1,700 and on type B as $800.

21 A farmer raises rice and soybeans. The rice requires 7 units of insecticide and 5 units of fertilizer per acre, whereas the soybeans need 6 units of each. In order to maximize his profit, how many acres of each should he plant if the profit per acre on rice is $25 and on soybeans is $47? Assume he has 660 units of insecticide and 600 units of fertilizer available.

22 Repeat Prob. 21 with the rice requiring 8 units of insecticide and 4 units of fertilizer per acre.

23 Repeat Prob. 21 with the soybeans requiring 7 units of fertilizer and 5 units of insecticide per acre.

24 Repeat Prob. 21 with the profit per acre as $34 for rice and $40 for soybeans.

25 During the time that two machines are not otherwise needed, they are used to make fire pokers and tongs. It is anticipated that machines A and B will be free for 10 and 7.5 hours, respectively, during a certain period. Pokers require 1 hour of time from machine A and 1 hour from machine B and sell for a profit of $3, whereas tongs require 2 hours of time from machine A and 1.5 hours from machine B and sell for a profit of $5.50. How many of each should be made so as to make the profit a maximum?

26 A truck farmer has 20 acres available for planting with peppers, rhubarb, and tomatoes. He has reason to think that he can make a profit of $300/acre on peppers, $225/acre on rhubarb, and $250/acre on tomatoes. He cannot take care of more than 10 acres of peppers, more than 12 acres of rhubarb, or more than 7 acres of tomatoes. In order to make a maximum profit, how many acres of each should he grow?

27 An animal food is to be a mixture of products A and B. The content and cost of 1 lb of each is:

Product	Protein, g	Fat, g	Carbohydrates, g	Cost
A	180	2	240	$0.50
B	36	8	200	$0.24

How much of each product should be used to minimize the cost if each bag must contain at least 612 g of protein and 22 g of fat and at most 1,880 g of carbohydrates?

28   A manufacturer produces three types of products in 500 hours. Product A requires 4 hour/unit, product B requires 6 hour/unit, and product C requires 2 hour/unit. The profits per unit of output are $50 on A, $80 on B, and $30 on C. He has an order for 25 A units, 30 B units, 70 C units. If he can sell all he can produce, what combination should he produce to maximize profits, assuming that he cannot produce more than 120 A units and 40 B units?

9.12 / PROBLEMS LEADING TO SYSTEMS OF EQUATIONS

A word problem can frequently be best solved if more than one unknown is used and more than one equation is employed. The method of setting up the equations involves letting the unknowns represent quantities which are called for in the problem. The general rule is that *the number of equations formed must be equal to the number of unknowns introduced.*

EXAMPLE 1   A real-estate dealer received $1,200 in rents on two dwellings in 1976, and one of the dwellings brought $10/month more than the other. How much did he receive per month for each if the more expensive house was vacant for 2 months?

Solution   On inspecting the problem for equations that we might form, we find two basic relations: the connection between the separate rentals and the connection between the monthly rentals and the income per year. Since one house rented for $10 more than the other, we let

x = the monthly rental on the more expensive house, in dollars
y = the monthly rental on the other house, in dollars

and then
$$x - y = 10 \tag{1}$$

Furthermore, since the first house was rented for 10 months and the other was rented for 12 months, we know that $10x + 12y$ is the total amount received in rentals. Hence,

$$10x + 12y = 1,200 \tag{2}$$

We now have the two equations (1) and (2) in the variables x and y, and we shall solve them simultaneously by eliminating y. The solution follows.

$$
\begin{array}{lll}
12x - 12y = 120 & (1) \times 12 & (3) \\
10x + 12y = 1,200 & & (2) \\
\hline
22x \qquad\;\; = 1,320 & (3) + (2) & \\
x \qquad\quad = 60 & \text{dividing both members by 22} &
\end{array}
$$

By substituting 60 for x in (1), we get

$$60 - y = 10$$
$$-y = 10 - 60 = -50 \qquad \text{adding } -60 \text{ to each member}$$
$$y = 50 \qquad \text{dividing both members by } -1$$

Therefore, the monthly rentals were $60 and $50, respectively.

EXAMPLE 2   A tobacco dealer mixed 12 lb of one grade of tobacco with 10 lb of another grade to obtain a blend worth $54. He then made a second blend worth $61 by mixing 8 lb of the first grade with 15 lb of the second grade. Find the price per pound of each grade.

In this problem we find two basic relations that we can use to form two equations. Therefore we let

$$x = \text{price per pound, in dollars, of the first grade}$$
$$y = \text{price per pound, in dollars, of the second grade}$$

and then

$$12x + 10y = 54 \qquad \text{by using the numbers of pounds as coefficients and} \qquad (4)$$
$$8x + 15y = 61 \qquad \text{the values of the blends as the constant terms} \qquad (5)$$

We shall solve these two equations simultaneously by eliminating y and solving for x. The steps in the solution follow.

$$36x + 30y = 162 \qquad (4) \times 3 \qquad (6)$$
$$\underline{16x + 30y = 122} \qquad (5) \times 2 \qquad (7)$$
$$20x \qquad = 40 \qquad (6) - (7) \qquad (8)$$
$$x \qquad = 2$$

By substituting 2 for x in (5), we get

$$16 + 15y = 61$$
$$15y = 45$$
$$y = 3$$

Therefore, the prices of the two grades are $2/lb and $3/lb. The solution can be checked by substitution in (4) or (5).

EXAMPLE 3   Two airfields A and B are 720 mi apart, and B is due east of A. A plane flew from A to B in 1.8 hours and then returned to A in 2 hours. If the wind blew with a constant velocity from the west during the entire trip, find the speed of the plane in still air and the speed of the wind.

Solution   The essential point in solving such a problem is that the wind helps the plane in one direction and hinders it in the other. We therefore have the basis for two equations that involve the speed of the plane, the speed of the wind, and the time for the trip. We let

$x =$ speed of the plane in still air, in miles per hour
$y =$ speed of the wind, in miles per hour

Then, since the wind blew constantly from the west,

$x + y =$ speed of the plane eastward from A to B (wind helping)
$x - y =$ speed of the plane westward from B to A (wind hindering)

The distance traveled each way was 720 mi, and we set up equations in time:

$$\frac{720}{x + y} = 1.8, \text{ time required for first half of trip} \qquad (9)$$

$$\frac{720}{x - y} = 2, \text{ time required for second half of trip} \qquad (10)$$

Now we multiply both members of (9) by $5(x + y)$ and of (10) by $x - y$ and get

$$3,600 = 9x + 9y \qquad (11)$$
$$720 = 2x - 2y \qquad (12)$$

We solve (11) and (12) simultaneously by first eliminating y.

$$\begin{array}{lll} 7,200 = 18x + 18y & (11) \times 2 & (13) \\ 6,480 = 18x - 18y & (12) \times 9 & (14) \\ \hline 13,680 = 36x & (13) + (14) & \\ x = 380 & & \end{array}$$

On substituting 380 for x in (11), we have

$$3,600 = (9 \times 380) + 9y$$
$$3,600 = 3,480 + 9y$$
$$y = 20$$

Hence, the speed of the plane in still air was 380 mi/hour, and the speed of the wind was 20 mi/hour. The solution can be checked by substitution in (9) or (10).

EXAMPLE 4 The cost for building a rectangular vat with a square base was $128. The base cost $0.30/sq ft, and the sides cost $0.20/sq ft. Find the dimensions of the vat if the combined area of the base and sides was 512 sq ft.

Solution We let $x =$ length of one side of the base

and $y =$ depth

Then $x^2 =$ area of the base

and $$4xy = \text{area of the sides}$$

Hence, $$x^2 + 4xy = 512 \tag{15}$$

Furthermore, since the costs of the base and sides were \$0.30/sq ft and \$0.20/sq ft, respectively, then

$$0.30x^2 = 0.3x^2 = \text{cost of base, in dollars}$$

and $$0.80xy = 0.8xy = \text{cost of the sides, in dollars}$$

Therefore, $$0.3x^2 + 0.8xy = 128 \tag{16}$$

Hence we have the equations

$$x^2 + \quad 4xy = 512 \tag{15}$$
$$0.3x^2 + 0.8xy = 128 \tag{16}$$

Since each equation contains a term in xy and no other term involving y, we eliminate xy and complete the solution as follows:

$0.2x^2 + 0.8xy = 102.4$	multiplying (15) by 0.2	(17)
$0.3x^2 + 0.8xy = 128$	(16) recopied	(16)
$-0.1x^2 = -25.6$	equating the differences of the members of (16) and (17)	
$x^2 = 256$	multiplying by -10	
$x = 16$		

Note that if $x^2 = 256$, then $x = \pm 16$, but we discard -16 since the dimensions are positive numbers.

Now we replace x by 16 in (15), solve for y, and get

$$256 + 64y = 512$$
$$64y = 256$$
$$y = 4$$

Hence the vat is 16 ft wide by 16 ft long by 4 ft deep.

EXERCISE 9.8 Problems Solvable by Means of Simultaneous Equations

1 A high school club earned a net profit of \$45.80 selling candy apples and suckers, which cost them \$0.08 apiece, at a basketball game. If they sold 480 candy apples and 610 suckers and a candy apple and a sucker together sold for \$0.25, what was the selling price of each?

2 Mr. Stewart spent \$2.99 for $2\frac{1}{2}$ qt of cream to make ice cream. If whipping cream cost \$0.33/half pt and half-and-half cost \$0.35/pt, how much of each type was used?

3   One term Fred received 22 grade points for making A or B in each of the six subjects he studied. If each A was worth four grade points and each B was worth three grade points, in how many subjects did he make A and in how many did he make B?

4   A music teacher charged $5 for each $\frac{1}{2}$-hour organ lesson and $3.50 for each $\frac{1}{2}$-hour piano lesson. If in 4 hours of teaching she earned $32.50, how many lessons did she teach on each instrument?

5   Tickets for a banquet were $4 for a single ticket or $7.50 for a couple. If 144 people attended the banquet and $549 was collected from ticket sales, how many couples and how many singles attended?

6   A ranger inspecting a forest trail walked at a rate of $3\frac{1}{2}$ mi/hour. A second ranger inspecting another portion of the trail walked at a rate of 3 mi/hour. If the trail was 42 mi long and the complete inspection required 13 hours, how far did each ranger walk?

7   An apartment building contained 20 units, consisting of one-bedroom apartments which rented for $110/month and two-bedroom apartments which rented for $135/month. If the rental from 17 apartments one month was $2,045 and three apartments were vacant, how many of each type were rented?

8   Frank found that he could drive from the campus to his home in 5 hours by averaging 55 mi/hour. However, on one trip, after he had averaged 55 mi/hour for awhile, he encountered bad weather and was forced to reduce his speed to 40 mi/hour. If that trip required $5\frac{3}{4}$ hours, how many miles did he travel at each speed?

9   On a television quiz program, each contestant was given $100 at the start of the program. For each question he answered correctly, he was given a bonus of $100; for each one he missed, he was penalized $25. If a contestant attempted 14 questions and ended the game with $875, how many questions were answered correctly and how many were missed?

10 A man had 6 gal of paint to cover 2,380 sq ft of fencing. One gallon will cover 470 sq ft with one coat and 250 sq ft with two coats. If he used all his paint, how many square feet received one coat and how many received two coats?

11 Two different routes between two cities differ by 20 mi. Two men made the trip between the cities in exactly the same time. One traveled the shorter route at 50 mi/hour, and the other traveled the longer route at 55 mi/hour. Find the length of each route.

12 Mr. Conner estimated that it cost him $0.55/day to drive to work when he took three passengers, all of whom paid the same daily fee. When two more passengers joined the group, Mr. Conner cut the fee each paid by $0.10 and found that he earned $0.05 on each trip.

Find the total cost of each trip and the fee that the first three passengers paid.

13 On the first day of homecoming weekend, a campus organization earned $650 by selling 450 college pennants and 100 corsages. On the second day they sold the 150 pennants they had left, but the remaining 50 corsages had wilted, so that they lost as much per corsage as they had earned on the previous day. If on the second day they earned $50, how much did they earn on each pennant and on each corsage sold the first day?

14 A biology class of 35 students took a field trip that included a hike of 8 mi. Part of the class also investigated a side trail, which added 3 mi to their hike. If the class walked a total of 331 human-miles, how many students took each hike? (If a group of 20 people walk 10 miles, the group has walked $10 \times 20 = 200$ human-miles.)

15 One year Bill worked at a part-time job some of the 9 months he was in school. The months he worked he was able to save $65/month, but the months he was idle he had to dip into his savings at the rate of $150/month. If at the end of the school year he had $60 less in his savings account than he had at the beginning, how many months did he work and how many months was he idle?

16 Jack and Dick signed a lease on an apartment for 9 months. At the end of 6 months, Dick got married and moved out. He paid the landlord an amount equal to the difference between double-occupancy rental and single-occupancy rental for the remaining 3 months, and Jack paid the single-occupancy rate for the 3 months. If the 9-month rental cost Jack $435 and Dick $285, what were the single and double monthly rates?

17 The sum of the seven digits in a telephone number is 30. Counting from the left, the first three digits and the last digit are the same. The fourth digit is twice the first, the sum of the fifth digit and the first is 7, and the sixth digit is twice the fifth. Find the number.

18 One summer Joe earned $12/day and Tom earned $14/day. Together the boys earned $1,142. How long did each boy work if Tom worked 11 days more than Joe?

19 Three volunteers assembled 741 newsletters for bulk mailing. The first could assemble 124/hour, the second 118/hour, and the third 132/hour. They worked a total of 6 hours. If the first worked 2/hours, how long did each of the others work?

20 A class of 32 students was made up of people who were 18, 19, and 20 years of age. The average of their ages was 18.5. How many of each age were in the class if the number of 18-year-olds was six more than the combined number of 19- and 20-year-olds?

21 A merchant sold a square carpet and a rectangular carpet whose length was $\frac{3}{2}$ the width, and the combined area of the two was 375 sq ft. The price of the first carpet was \$10/sq yd, and of the second \$12/sq yd. If he received \$50 more for the square piece than for the other, find the dimensions of each.

22 A tract of land is in the form of a trapezoid with two angles equal to 90°. Two sides which meet at the vertex of one right angle are equal, and the oblique side is 50 rods in length. The area of the tract is 2,200 sq rods, and the perimeter is 200 rods. Find the lengths of the unknown sides.

23 The cost of the material for a rectangular bin with a square base and an open top was \$0.24/sq ft for the base and \$0.16/sq ft for the sides. The total material cost was \$27.84. Find the dimensions of the bin if the combined area of the base and sides was 156 sq ft.

24 A rectangle of area 48 sq ft is inscribed in a circle of area $78\frac{4}{7}$ sq ft. Find the dimensions of the rectangle. (Use $\frac{22}{7}$ for pi.)

25 A rectangular pasture with an area of 6,400 sq rods is divided into three smaller pastures by two fences parallel to the shorter sides. The widths of two of the smaller pastures are the same, and the width of the third is twice that of the others. Find the dimensions of the original pasture if the perimeter of the larger of the subdivisions is 240 rods.

26 A rectangular field has an irrigation well at the midpoint of a longer side. Ditches 5×10^2 rods long run from the well to each opposite corner. If the area of the field is 24×10^4 sq rods, find the dimensions.

27 A civic club adopted a project that would cost \$960. Before the project was completed, 16 new members joined the club and agreed to pay their share of the cost of the project. The cost per member was thereby reduced by \$2. Find the original number of members and the original cost per member.

28 A block of stock was bought for \$11,000. After 1 year the buyer received a dividend of \$2.20/share and a stock dividend of 20 shares. He then sold the stock for \$2 more per share than it cost and made a profit of \$1,980 on the transaction. Find the number of shares bought and the price of each share.

29 A circle is drawn inside a rectangle and is tangent to each of the longer sides at their midpoints. The difference between the areas of the rectangle and circle is 126 sq ft, and the sum of their perimeters is 112 ft. Find the dimensions of the rectangle. (Use $\frac{22}{7}$ for pi.)

30 A student worked for 90 days in the summer. During the first 60 days he worked 570 hours and earned \$960 in the daytime and \$270 at night. During the last 30 days he worked 8 hours each day and

3 hours each night and earned $750. Find the hourly wage in the daytime and the hourly wage at night, provided that each is an integral number of dollars.

31   A piece of wire 152 in. long is cut into two pieces. One piece is bent into a square, and the other into a circle. If the combined area of the square and the circle is 872 sq in., find the side of the square and the radius of the circle.

32   A swimming pool is in the shape of a rectangle with a semicircle at each end. The area of the pool is $329\frac{1}{7}$ sq yd, and the perimeter is $66\frac{2}{7}$ yd. Find the width and overall length of the pool.

EXERCISE 9.9 **REVIEW**

Construct the graph of the function defined by the equation in each of Probs. 1 to 8.

1	$3x - 4y = 12$	**2**	$2x + 3y = 6$	**3**	$y^2 = 4x$
4	$x^2 = -8y$	**5**	$x^2 + y^2 = 16$	**6**	$x^2 + 4y^2 = 16$
7	$x^2 - 4y^2 = 16$	**8**	$xy = 3$		

Graphically solve the pair of equations in each of Probs. 9 to 12.

9 $y^2 = 4x$ **10** $x^2 + y^2 = 20$ **11** $x^2 + y^2 = 7$
     $y = 2x$ $y = 2x$ $x^2 + 4y^2 = 16$

12 $4x^2 + y^2 = 4$
     $x^2 = 6y$

Determine whether the system in each of Probs. 13 to 18 is dependent, independent, or inconsistent, and solve each independent pair.

13	$3x + 4y = 6$	**14**	$2x - y = 5$	**15**	$2x + y = 5$
	$6x + 8y = 12$		$6x - 3y = 13$		$2x - y = -1$
16	$4x + 5y = -1$	**17**	$x + 5y = 7$	**18**	$2x + 3y = 0$
	$5x - 3y = 8$		$3x - 2y = 4$		$3x - 4y = 17$

Find the solution set of the pair of equations in each of Probs. 19 to 25.

19	$1/x + 1/y = 5$	**20**	$x^2 + y^2 = 5$
	$2/x + 3/y = 13$		$2x^2 + 3y^2 = 14$
21	$2y^2 + 3xy + y = -2$	**22**	$3x^2 + 2x = y + 1$
	$3y^2 - 2xy + 4y = 1$		$y^2 - 4xy = 0$
23	$x^2 - 3xy - 2y^2 = 6$	**24**	$x^2 + xy + y^2 - 6x - 6y = -11$
	$xy = -3$		$5x^2 - 4xy + 5y^2 - 3x - 3y = 26$
25	$xy + x + y = 11$		
	$2xy - 3x - 3y = -3$		

26   Show that the following set of inequalities do not bound a finite polygonal region: $x \geq 0$, $y \geq 0$, $x \leq 1$, $x + y \geq 2$.

Find the extrema of f in the region determined by the inequalities in Probs. 27 and 28.

27   $y \geq 0$, $-2x + y + 6 \geq 0$, $x + 3y \leq 10$, $3x - y \geq 0$, $f(x, y) = 2x + 3y - 4$

28   $x \geq 0$, $y \geq 0$, $3x - 2y - 9 \leq 0$, $x - 5y + 10 \geq 0$, $f(x, y) = 7x - y + 9$

29   A manufacturer of small engines makes a 3-hp and a 5-hp model. They make a profit of \$12 on each of the smaller engines, and \$17 on each of the larger. They can make 500 or less of the smaller per month and 300 or less of the larger per month, but not more than a total of 700 engines. How many of each type should the company produce in order to make a maximum profit, and what is that profit?

30   Repeat Prob. 29 with a profit of \$15 for each 3-hp engine and \$14 for each 5-hp engine.

CHAPTER TEN *Matrices and Determinants*

The concepts of a matrix and of a determinant are important ones in mathematics. Determinants were invented or devised independently by Kiowa, a Japanese, in 1683 and Leibnitz, a German, in 1693 and were rediscovered in 1750 by Cramer, a Swiss, who used them for solving systems of linear equations. Matrices were invented by Cayley, an Englishman, during the nineteenth century and are used in solving systems of linear equations and in connection with computers.

10.1 / MATRICES AND THEIR BASIC PROPERTIES

A rectangular array of numbers of the type

$$\begin{bmatrix} a_{11} & a_{12} & \cdots & a_{1g} & \cdots & a_{1n} \\ a_{21} & a_{22} & \cdots & a_{2g} & \cdots & a_{2n} \\ \cdots\cdots\cdots\cdots\cdots\cdots\cdots\cdots\cdots\cdots\cdots \\ a_{i1} & a_{i2} & \cdots & a_{ij} & \cdots & a_{in} \\ \cdots\cdots\cdots\cdots\cdots\cdots\cdots\cdots\cdots\cdots\cdots \\ a_{m1} & a_{m2} & \cdots & a_{mj} & \cdots & a_{mn} \end{bmatrix}$$

$m \times n$ matrix   is called an $m \times n$ matrix. If $m = n$, it is called a *square matrix*. As seen from the array, an $m \times n$ matrix consists of m rows (horizontal) and n columns

Element   (vertical). Each number in the matrix is called an *element* of the matrix. In the above array, a_{ij} is the element in the ith row and jth column. Sometimes $A = (a_{ij})$ is used to designate the matrix. The number i belongs to $\{1, 2, 3, \ldots, m\}$, and j is an element of $\{1, 2, 3, \ldots, n\}$. The symbol $A_{m \times n}$ is often used to indicate that A is an $m \times n$ matrix.

Equality of matrices   Two m by n matrices are equal if and only if each element of one is equal to the corresponding element of the other. For example,

$$\begin{bmatrix} 3 & 2 & 1 \\ 2 & \sqrt{9} & 7 \end{bmatrix} = \begin{bmatrix} 3 & \sqrt{4} & 1 \\ \sqrt[3]{8} & 3 & 7 \end{bmatrix}$$

but

$$\begin{bmatrix} 5 & -3 \\ 4 & -2 \end{bmatrix} = \begin{bmatrix} 2x+1 & y \\ z+2 & w-1 \end{bmatrix}$$

if and only if $5 = 2x + 1, -3 = y, 4 = z + 2$, and $-2 = w - 1$; hence, if and only if $x = 2, y = -3, z = 2$, and $w = -1$.

Sum of two matrices   The sum of two $m \times n$ matrices A and B is the $m \times n$ matrix obtained by adding corresponding elements of A and B. The sum is defined if and only if both matrices are $m \times n$. Consequently the sum of

$$A = \begin{bmatrix} 1 & 3 & 5 \\ 2 & 4 & 6 \end{bmatrix} \quad \text{and} \quad B = \begin{bmatrix} -2 & 7 & 8 \\ 9 & -5 & -6 \end{bmatrix}$$

is $$A + B = \begin{bmatrix} 1+(-2) & 3+7 & 5+8 \\ 2+9 & 4+(-5) & 6+(-6) \end{bmatrix} = \begin{bmatrix} -1 & 10 & 13 \\ 11 & -1 & 0 \end{bmatrix}$$

The product $A_{m \times p} B_{p \times n}$ The product AB of two matrices is defined if and only if the number of columns in A is equal to the numbers of rows in B. The product $(AB)_{m \times n}$ of the matrices $A_{m \times p}$ and $B_{p \times n}$ is the matrix with the element $c_{ij} = \Sigma_{k=1}^{p} a_{ik} b_{kj}$ in the ith row and jth column. The element c_{ij} is found by adding the product of the first element a_{i1} in the ith row of A and the first element b_{1j} in the jth column of B, the product of the second element a_{i2} in the ith row of A and the second element b_{2j} in the jth column of $B, \ldots$, and the product of the pth element a_{ip} in the ith row of A and the pth element a_{pj} in the jth column of B. If AB and BA are both defined, they may or may not be equal.

EXAMPLE 1   Find the product $(AB)_{2 \times 2}$ if

$$A_{2 \times 3} = \begin{bmatrix} a_{11} & a_{12} & a_{13} \\ a_{21} & a_{22} & a_{23} \end{bmatrix} \quad \text{and} \quad B_{3 \times 2} = \begin{bmatrix} b_{11} & b_{12} \\ b_{21} & b_{22} \\ b_{31} & b_{32} \end{bmatrix}$$

Solution If we follow the procedure outlined above, we get

$$(AB)_{2\times2} = \begin{bmatrix} a_{11}\,b_{11} + a_{12}\,b_{21} + a_{13}\,b_{31} & a_{11}\,b_{12} + a_{12}\,b_{22} + a_{13}\,b_{32} \\ a_{21}\,b_{11} + a_{22}\,b_{21} + a_{23}\,b_{31} & a_{21}\,b_{12} + a_{22}\,b_{22} + a_{23}\,b_{32} \end{bmatrix}$$

EXAMPLE 2 Find the products AB and BA if

Solution
$$A = \begin{bmatrix} 3 & -2 \\ 1 & 4 \end{bmatrix} \quad \text{and} \quad B = \begin{bmatrix} 5 & 1 \\ -6 & 3 \end{bmatrix}$$

$$AB = \begin{bmatrix} (3)(5) + (-2)(-6) & (3)(1) + (-2)(3) \\ (1)(5) + (4)(-6) & (1)(1) + (4)(3) \end{bmatrix}$$

$$= \begin{bmatrix} 27 & -3 \\ -19 & 13 \end{bmatrix}$$

Similarly, we find that

$$BA = \begin{bmatrix} 16 & -6 \\ -15 & 24 \end{bmatrix}$$

EXAMPLE 3 Find AB if

$$A = \begin{bmatrix} a_1 & a_2 & a_3 \\ & & \\ b_1 & b_2 & b_3 \end{bmatrix} \quad \text{and} \quad B = \begin{bmatrix} x \\ y \\ z \end{bmatrix}$$

Solution By the usual procedure, we have

$$AB = \begin{bmatrix} a_1x + a_2y + a_3z \\ b_1x + b_2y + b_3z \end{bmatrix}$$

The product kA   The product of a scalar k and the matrix A is denoted by kA and is the matrix obtained by multiplying each element of A by the constant k. Symbolically $kA = k(a_{ij}) = (ka_{ij})$. For example, if

$$A = \begin{bmatrix} 1 & 3 & 6 \\ 0 & -2 & 5 \end{bmatrix} \quad \text{then} \quad 4A = \begin{bmatrix} 4 & 12 & 24 \\ 0 & -8 & 20 \end{bmatrix}$$

Identity matrix   If A is an $n \times n$ matrix and if each entry in the main diagonal (the diagonal from upper left to lower right) is one and all other elements are zero, the matrix is called an identity *matrix* since it acts as an identity for matrix multiplication. It is designated by I. Thus

$$I_2 = \begin{bmatrix} 1 & 0 \\ 0 & 1 \end{bmatrix} \quad \text{and} \quad I_3 = \begin{bmatrix} 1 & 0 & 0 \\ 0 & 1 & 0 \\ 0 & 0 & 1 \end{bmatrix}$$

are identity matrices.

Zero matrix   If each element of a matrix is zero, the matrix is called a *zero matrix* and is designated by O.

Transpose of a matrix   The matrix obtained by interchanging the rows and columns of a matrix A is called the *transpose* of A and is designated by A^T. Consequently, the transpose of

$$A = \begin{bmatrix} 3 & 1 & 2 \\ 4 & 0 & -5 \end{bmatrix} \quad \text{is} \quad A^T = \begin{bmatrix} 3 & 4 \\ 1 & 0 \\ 2 & -5 \end{bmatrix}$$

EXERCISE 10.1 Properties of Matrices

Find the values of a, b, c, and d so that the statement in each of Probs. 1 to 4 is true.

1   $\begin{bmatrix} 2 & a & b & -3 \end{bmatrix} = \begin{bmatrix} c & 4 & 3 & d \end{bmatrix}$

2   $\begin{bmatrix} a & -2 \\ 5 & d \end{bmatrix} = \begin{bmatrix} 4 & c \\ b & -3 \end{bmatrix}$ 3 $\begin{bmatrix} 7 & 6 \\ -3 & 5 \end{bmatrix} = \begin{bmatrix} a+3 & c+5 \\ b-1 & d+4 \end{bmatrix}$

4   $\begin{bmatrix} a+b & 4 & 3 \\ c+d & -2 & 2b+d \end{bmatrix} = \begin{bmatrix} 4 & 2a-c & b \\ -2 & c-d & 6 \end{bmatrix}$

Use $A = \begin{bmatrix} 3 & -1 & 2 \\ 4 & 0 & 5 \end{bmatrix}$ *and* $B = \begin{bmatrix} -2 & 3 & 0 \\ 1 & 5 & -4 \end{bmatrix}$ *to evaluate the matrices in Probs. 5 to 16.*

5 $3A$	6 $0B$	7 $-B$	8 $2A$
9 $A+B$	10 $B+A$	11 $2A+3B$	12 $3A-4B$
13 A^T	14 B^T	15 A^T-B	16 $A+B^T$

If $A_i = \begin{bmatrix} a_i & b_i & c_i \\ d_i & e_i & f_i \end{bmatrix}$ *for* $i = 1, 2, 3$ *are three* 2×3 *matrices, and h and k are real, nonzero numbers, prove the statement in each of Probs. 17 to 24.*

17   $A_1 + A_3 = A_3 + A_1$ 18 $h(A_1 + A_2) = hA_1 + hA_2$
19   $(h+k)A_1 = hA_1 + kA_1$ 20 $A_2 + (-A_2) = 0$
21   $A_1 + (A_2 + A_3) = (A_1 + A_2) + A_3$ 22 $A_1 + 0 = A_1$
23   If $hA_1 + kA_2 = hA_1 + kA_3$, then $A_2 = A_3$.
24   If $hA_1 - kA_2 = hA_1 - kA_3$, then $A_2 = A_3$.

Find the matrix X in each of Probs. 25 to 28.

25   $X - \begin{bmatrix} 3 & 1 \\ 0 & -2 \end{bmatrix} = \begin{bmatrix} 4 & 1 \\ 0 & 3 \end{bmatrix}$ 26 $X - 2\begin{bmatrix} 2 & -1 \\ 3 & 5 \end{bmatrix} = 3\begin{bmatrix} 1 & 2 \\ -2 & -3 \end{bmatrix}$

27   $2X - 3\begin{bmatrix} 2 & 0 \\ 4 & -8 \end{bmatrix} = 2\begin{bmatrix} -3 & 5 \\ -7 & 12 \end{bmatrix}$ 28 $3X + \begin{bmatrix} 3 & -8 \\ 11 & 0 \end{bmatrix} = 3\begin{bmatrix} 1 & -3 \\ 4 & -2 \end{bmatrix}$

Find the indicated products in Probs. 29 to 32, and prove the statements in Probs. 33 to 36, if

$$A = \begin{bmatrix} 3 & 0 & 1 \\ 2 & -1 & 4 \\ -2 & 3 & 5 \end{bmatrix} \quad and \quad B = \begin{bmatrix} -1 & 4 & -3 \\ 3 & -2 & 5 \\ 0 & 1 & 2 \end{bmatrix}$$

29 AB

30 $(AB)^T$

31 $A^T B^T$

32 $(A^T B)^T$

33 $A^T B = (B^T A)^T$

34 $B^T A^T = (AB)^T$

35 $A(B^T) \neq A^T B$

36 $(AB)^T \neq A^T B^T$

Find the indicated products in each of Probs. 37 to 40.

37 $\begin{bmatrix} 2 & 1 & -3 \\ 3 & 0 & 1 \\ -2 & 2 & 5 \end{bmatrix} \begin{bmatrix} 2 \\ 0 \\ 1 \end{bmatrix}$

38 $\begin{bmatrix} 1 & -2 \\ 4 & 3 \\ 3 & 5 \end{bmatrix} \begin{bmatrix} 3 \\ 2 \end{bmatrix}$

39 $\begin{bmatrix} 1 & 2 & 4 \\ 3 & -1 & 0 \end{bmatrix} \begin{bmatrix} 1 & 3 \\ 2 & -1 \\ 4 & 0 \end{bmatrix}$

40 $\begin{bmatrix} 4 & -2 \\ 1 & 3 \\ 0 & 5 \end{bmatrix} \begin{bmatrix} 2 & -1 \\ 3 & 0 \end{bmatrix}$

Prove the statements in Probs. 41 to 44 for A, B, and C any 2 × 2 matrices.

41 $(AB)^T = B^T A^T$

42 $(A+B)^T = A^T + B^T$

43 $A(B+C) = AB + AC$

44 $A(BC) = (AB)C$

10.2 / MATRICES AND SYSTEMS OF LINEAR EQUATIONS

If we have the system of linear equations

$$\begin{aligned}
a_{11}x_1 + a_{12}x_2 + \cdots + a_{1n}x_n &= k_1 \\
a_{21}x_2 + a_{22}x_2 + \cdots + a_{2n}x_n &= k_2 \\
&\cdots \cdots \cdots \cdots \\
a_{m1}x_n + a_{m2}x_2 + \cdots + a_{mn}x_n &= k_m
\end{aligned} \tag{10.1}$$

we can (1) interchange any two equations; (2) multiply any equation by a nonzero constant; (3) add a nonzero multiple of any equation to any other equation and have a system that is equivalent to the given system; i.e., we can have a system of equations with the same solution as the given system.

The matrix whose elements are the coefficients in the system of equations (10.1) and whose elements occur in the relative position in the system

Coefficient matrix is called the *coefficient matrix*. If the constant terms are adjoined on the right of the coefficient matrix, as an $(n + 1)$st column, the new matrix is

Augmented matrix   called the *augmented matrix.* Therefore, if the system of equations is

$$x + y + z = 2$$
$$2x + 5y + 3z = 1 \qquad\qquad (1)$$
$$3x - y - 2z = -1$$

the matrix of the coefficients is

$$A = \begin{bmatrix} 1 & 1 & 1 \\ 2 & 5 & 3 \\ 3 & -1 & -2 \end{bmatrix}$$

and
$$B = \begin{bmatrix} 1 & 1 & 1 & 2 \\ 2 & 5 & 3 & 1 \\ 3 & -1 & -2 & -1 \end{bmatrix}$$

is the augmented matrix.

If we know the augmented matrix, we have each equation of the system just as clearly as if the variables and equality signs were written in.

In keeping with the relation between the system of equations and the augmented matrix and statements (1), (2), and (3) just after equations (10.1), we can save time and space by performing the following *row operations* on the augmented matrix of the system of equations:

1   *Interchange any two rows of the augmented matrix.*
2   *Multiply any row of the augmented matrix by a nonzero constant.*
3   *Add a nonzero multiple of any row to any other row.*

We shall make use of these row operations to change the augmented matrix into one that represents a set of equations equivalent to the given set and which becomes the identity matrix after deleting the right-hand column. This is desirable since such a matrix is immediately translated into a system of equations with only one variable in each equation.

EXAMPLE   We shall illustrate the procedure by working with the system (1) above.

Solution   The matrix B is the augmented matrix for the system (1). We want to perform row operations on it so as to replace each element of the main diagonal by a 1 and the elements below and above the main diagonal by zeros.

We normally begin by getting a 1 as the first element in the main diagonal, but it is already 1; hence we write

$$B_1 = \begin{bmatrix} 1 & 1 & 1 & 2 \\ 2 & 5 & 3 & 1 \\ 3 & -1 & -2 & -1 \end{bmatrix}$$

We now get zero as the second element in the first column by subtracting twice each element of row one from the corresponding element of row two. This operation is indicated by $R_2' = R_2 - 2R_1$. We get a zero in the third row of column one by performing $R_3' = R_3 - 3R_1$. Thus,

$$B_2 = \begin{bmatrix} 1 & 1 & 1 & 2 \\ 0 & 3 & 1 & -3 \\ 0 & -4 & -5 & -7 \end{bmatrix} \qquad \begin{matrix} R_2' = R_2 - 2R_1 \\ R_3' = R_3 - 3R_1 \end{matrix}$$

$$C_1 = \begin{bmatrix} 1 & 1 & 1 & 2 \\ 0 & 1 & \frac{1}{3} & -1 \\ 0 & -4 & -5 & -7 \end{bmatrix} \qquad R_2' = \tfrac{1}{3}R_2$$

$$C_2 = \begin{bmatrix} 1 & 0 & \frac{2}{3} & 3 \\ 0 & 1 & \frac{1}{3} & -1 \\ 0 & 0 & -\frac{11}{3} & -11 \end{bmatrix} \qquad \begin{matrix} R_1' = R_1 - R_2 \\ \\ R_3' = R_3 + 4R_2 \end{matrix}$$

$$D_1 = \begin{bmatrix} 1 & 0 & \frac{2}{3} & 3 \\ 0 & 1 & \frac{1}{3} & -1 \\ 0 & 0 & 1 & 3 \end{bmatrix} \qquad R_3' = -\tfrac{3}{11}R$$

$$D_2 = \begin{bmatrix} 1 & 0 & 0 & 1 \\ 0 & 1 & 0 & -2 \\ 0 & 0 & 1 & 3 \end{bmatrix} \qquad \begin{matrix} R_1' = R_1 - \tfrac{2}{3}R_3 \\ R_2' = R_2 - \tfrac{1}{3}R_3 \end{matrix}$$

The matrix D_2 represents the system of equations $x = 1$, $y = -2$, $z = 3$. The solution set $\{1, -2, 3\}$ for the system of equations can be checked by substituting in the given system.

   If we had cared to do so, we could have stopped with matrix D_1 since the system of equations represented by it is readily solved. The last row in D_1 represents the equation $z = 3$. The other two equations from D_1 are $y + \tfrac{1}{3}z = -1$ and $x + \tfrac{2}{3}z = 3$. From the first of these, we get $y = -1 - 1 = -2$, and from the second equation we find that $x = 3 - 2 = 1$. Consequently, as found earlier, the solution set of the given system is $\{1, -2, 3\}$. In most situations the equation obtained from the first row of the matrix corresponding to D_1 will contain more than two variables, but the value of all except one of them can be found from earlier equations.

10.3 / THE INVERSE OF A MATRIX

Inverse of a matrix   The inverse of a matrix M is represented by M^{-1} and is the matrix such that $MM^{-1} = I$. We shall not do so, but it can be shown that if the matrix

$$\begin{bmatrix} a_{11} & a_{12} & \cdots & a_{1n} & 1 & 0 & \cdots & 0 \\ a_{21} & a_{22} & \cdots & a_{2n} & 0 & 1 & \cdots & 0 \\ \cdot & \cdot & \cdot & \cdot & \cdot & \cdot & \cdot & \cdot \\ a_{n1} & a_{n2} & \cdots & a_{nn} & 0 & 0 & \cdots & 1 \end{bmatrix}$$

is transformed, by the use of row operations, into

$$\begin{bmatrix} 1 & 0 & \cdots & 0 & A_{11} & A_{12} & \cdots & A_{1n} \\ 0 & 1 & \cdots & 0 & A_{21} & A_{22} & \cdots & A_{2n} \\ \cdot & \cdot & \cdot & \cdot & \cdot & \cdot & \cdot & \cdot \\ 0 & 0 & \cdots & 1 & A_{n1} & A_{n2} & \cdots & A_{nn} \end{bmatrix}$$

then, the inverse of

$$M = \begin{bmatrix} a_{11} & a_{12} & \cdots & a_{1n} \\ a_{21} & a_{22} & \cdots & a_{2n} \\ \cdot & \cdot & \cdot & \cdot \\ a_{n1} & a_{n2} & \cdots & a_{nn} \end{bmatrix}$$

is

$$M^{-1} = \begin{bmatrix} A_{11} & A_{12} & \cdots & A_{1n} \\ A_{21} & A_{22} & \cdots & A_{2n} \\ \cdot & \cdot & \cdot & \cdot \\ A_{n1} & A_{n2} & \cdots & A_{nn} \end{bmatrix}$$

EXAMPLE   Find the inverse of

$$M = \begin{bmatrix} 1 & 2 & 5 \\ 2 & 3 & 8 \\ -1 & 1 & 2 \end{bmatrix}$$

Solution   We begin by augmenting M on the right with a 3×3 identity matrix. Thus we have

$$\begin{bmatrix} 1 & 2 & 5 & 1 & 0 & 0 \\ 2 & 3 & 8 & 0 & 1 & 0 \\ -1 & 1 & 2 & 0 & 0 & 1 \end{bmatrix}$$

We now must use row operations in order to get a 3×3 identity matrix on the left. For this, we perform the operations indicated below.

$$B_1 = A$$

$$B_2 = \begin{bmatrix} 1 & 2 & 5 & 1 & 0 & 0 \\ 0 & -1 & -2 & -2 & 1 & 0 \\ 0 & 3 & 7 & 1 & 0 & 1 \end{bmatrix} \qquad \begin{array}{l} R_2' = R_2 - 2R_1 \\ R_3' = R_3 + R_1 \end{array}$$

$$C_1 = \begin{bmatrix} 1 & 2 & 5 & 1 & 0 & 0 \\ 0 & 1 & 2 & 2 & -1 & 0 \\ 0 & 3 & 7 & 1 & 0 & 1 \end{bmatrix} \qquad R_2' = -R_2$$

$$C_2 = \begin{bmatrix} 1 & 0 & 1 & -3 & 2 & 0 \\ 0 & 1 & 2 & 2 & -1 & 0 \\ 0 & 0 & 1 & -5 & 3 & 1 \end{bmatrix} \qquad \begin{array}{l} R'_1 = R_1 - 2R_2 \\ \\ R'_3 = R_3 - 3R_2 \end{array}$$

$$D_1 = C_2$$

$$D_2 = \begin{bmatrix} 1 & 0 & 0 & 2 & -1 & -1 \\ 0 & 1 & 0 & 12 & -7 & -2 \\ 0 & 0 & 1 & -5 & 3 & 1 \end{bmatrix} \qquad \begin{array}{l} R'_1 = R_1 - R_3 \\ R'_2 = R_2 - 2R_3 \end{array}$$

Consequently, the inverse of the given matrix M is

$$M^{-1} = \begin{bmatrix} 2 & -1 & -1 \\ 12 & -7 & -2 \\ -5 & 3 & 1 \end{bmatrix}$$

and it can be verified by showing that $MM^{-1} = I$.

EXERCISE 10.2 Solution Sets, Inverses

Use row operations on matrices to find the solution set of the system of equations in each of Probs. 1 to 16.

1 $x - 2y = 0$
  $2x + y = 5$

2 $2x + 3y = 0$
  $3x - y = 11$

3 $2x + y = -9$
  $x - 3y = 13$

4 $5x - 4y = 7$
  $4x - 3y = 5$

5 $3x + 2y + z = 8$
  $2x + y + 3z = 7$
  $x + 3y + 2z = 9$

6 $x + 3y + z = 4$
  $x - 5y + 2z = 7$
  $3x + y - 4z = -9$

7 $2x + 3y + z = -1$
  $5x + 7y - z = 5$
  $4x + 3y = 5$

8 $3x - 2y + 3z = 4$
  $5x + 4z = 3$
  $2x + 7y = -8$

9 $x - 2y + 3z = -5$
  $3x + y = 1$
  $2x - 3y + z = -4$

10 $3x - 4y + 2z = -6$
  $4x + 3y - 2z = 18$
  $2x - 3y + 4z = -10$

11 $2x + 3y - 6z = -3$
  $4x - 3y = 1$
  $4x + 3y + 12z = 13$

12 $x + 2z = 0$
  $3x + 5y - 8z = 3$
  $5x - 7y - 4z = 0$

13 $x + z - w = 0$
  $x - y - z = -2$
  $x - y - w = -3$
  $y + z - w = 1$

14 $x + 2z - w = 6$
  $y + z - w = 2$
  $z + 2w = -1$
  $y + z + w = 0$

15 $x - y - w - z = -1$
  $x + y - w + z = -5$
  $x + y + w + z = 1$
  $x + y + w - z = 9$

16 $x + y + z = 1$
  $x + y + w = 2$
  $x + z + w = 0$
  $y + z + 3w = 0$

17-32 Find the inverse of the matrix of the coefficients of the system of equations in each of Probs. 1 to 16.

10.4 / INVERSIONS

In any permutation of integers, an *inversion* occurs if an integer precedes a smaller integer. For example, in the permutation 5 4 3 1 2 7 6 note that 7 precedes 6; 5 precedes 4, 3, 1, and 2; 4 precedes 3, 1, and 2; 3 precedes 1 and 2. Hence, the permutation contains $1 + 4 + 3 + 2 = 10$ inversions.

The following theorem is necessary for the development of the theory of determinants of order greater than 2.

Theorem on inversions   *If two adjacent integers in any permutation are interchanged, the number of inversions in the permutation is increased or decreased by 1.*

Proof   The integers after the two that are interchanged are not affected; consequently if s and t are the two integers of the permutation that are interchanged, an inversion is introduced if $s < t$, and one is removed if $s > t$.

The following theorem will be employed in deriving the properties of determinants:

THEOREM   *If*

$$e_1, e_2, e_3, \ldots, e_n \tag{1}$$

is a permutation of the integers

$$1, 2, 3, \ldots, n \tag{2}$$

and if k inversions occur in (1), *then* (1) *can be transformed into* (2) *by k interchanges of consecutive terms.*

The meaning of this theorem and the method for proving it are illustrated below.

In the permutation 4 1 5 3 2; we note that 5 precedes 3 and 2; 4 precedes 1, 3, 2; and 3 precedes 2. Hence, we have $2 + 3 + 1 = 6$ inversions. To arrange these integers in numerical order, we proceed as follows:

1   Start with the largest integer, 5, that is out of place. Move it over 3 and 2. Thus, by *two interchanges*, get 4 1 3 2 5.

2   Move the next largest integer, 4, over 1, 3, and 2 and get 1 3 2 4 5. This involves *three interchanges*.

3   Finally, interchange 3 and 2 and obtain 1 2 3 4 5.

Consequently, the desired arrangement is accomplished by $3 + 2 + 1 = 6$ *interchanges.*

10.5 / DETERMINANTS OF ORDER n

We represent an element of a determinant of order n by a_{ij}. In the double subscript, i denotes the row in which the element occurs, and j denotes the column. Thus a_{13} is the element in the first row and third column. In terms of this notation

Determinant
of order n

$$D_n = \begin{vmatrix} a_{11} & a_{12} & \cdots & a_{1j} & \cdots & a_{1n} \\ a_{21} & a_{22} & \cdots & a_{2j} & \cdots & a_{2n} \\ \cdots & \cdots & \cdots & \cdots & \cdots & \cdots \\ a_{i1} & a_{i2} & \cdots & a_{ij} & \cdots & a_{in} \\ \cdots & \cdots & \cdots & \cdots & \cdots & \cdots \\ a_{n1} & a_{n2} & \cdots & a_{nj} & \cdots & a_{nn} \end{vmatrix} \qquad (10.2)$$

is a number called a *determinant of order n*. The number is represented by a homogeneous polynomial obtained by:

Procedure for
expanding a
determinant

1 Forming every possible product by taking one and only one factor from each row and each column of D_n

2 Arranging the factors in each product so that the first, or row, subscripts are in order of magnitude

3 Prefixing each product by a plus or minus sign according as the number of inversions in the second subscript is even or odd

4 Taking the algebraic sum of the products obtained by performing steps 1 to 3

The polynomial obtained by the above procedure is called the *expansion* of the determinant D_n.

The following theorem is a direct consequence of the above definition.

THEOREM   *The expansion of a determinant of order n has* $n! = n(n-1)(n-2) \cdots (2)(1)$ *terms.*

Proof   Each term in the expansion of D_n will be of the form $a_{1e_1} a_{2e_2} a_{3e_3} \cdots a_{ne_n}$, where $e_1 e_2 e_3 \cdots e_n$ is a permutation of the integers $1, 2, 3, \cdots, n$, and there will be one term in the expansion for each permutation of these integers. Since $n!$ permutations can be formed from n different elements taken n at a time, there will be $n!$ terms in the expansion of D_n.

We shall illustrate the application of the above four steps by using them to obtain the expansion of

$$D_3 = \begin{vmatrix} a_{11} & a_{12} & a_{13} \\ a_{21} & a_{22} & a_{23} \\ a_{31} & a_{32} & a_{33} \end{vmatrix}$$

We first form 3! permutations of the integers 1, 2, 3. Thus, we get 123, 231, and 312 and also 321, 132, and 213. Each of the first three has an even number of inversions, and each of the last three has an odd number. Now we form the products of three factors using 1, 2, and 3, respectively, as the first subscripts in each product and one of the above permutations as the second subscripts. Thus we obtain

$$D = a_{11}a_{22}a_{33} + a_{12}a_{23}a_{31} + a_{13}a_{21}a_{32} - a_{13}a_{22}a_{31} - a_{11}a_{23}a_{32} - a_{12}a_{21}a_{33}$$

It should be noticed that the number of inversions in the second subscripts of the first three terms is even, and in the second three terms odd. Hence, the first three terms are preceded by a positive sign, and the second three terms by a negative sign.

The above method for obtaining the expansion of a determinant is rather tedious and would be much more so for determinants of order greater than 3. Fortunately, there is another method which can be applied to determinants of order greater than 3, and we discuss this method in the next section.

10.6 / EXPANSION OF DETERMINANTS OF ORDER n

We shall give some definitions and notation for use in connection with determinants.

Minor   The *minor* of the element a_{ij} of a determinant D_n is the determinant that remains after deleting the ith row and the jth column of D_n and is designated by $m(a_{ij})$.

Cofactor   The *cofactor* of a_{ij} is the minor of a_{ij} if $i + j$ is an even number, and the negative of the minor if $i + j$ is an odd number. We designate the cofactor of aij by A_{ij}.

According to this definition,

$$\mathbf{A}_{ij} = (-1)^{i+j}m(a_{ij}) \tag{10.3}$$

Expansion of a   The expansion of a determinant D_n of order n can be expressed in determinant by   terms of the minors of the elements of the ith row in the following way:
minors

$$\boldsymbol{D}_n = (-1)^{i+1}\boldsymbol{a}_{i1}\boldsymbol{m}(\boldsymbol{a}_{i1}) + (-1)^{i+2}\boldsymbol{a}_{i2}\boldsymbol{m}(\boldsymbol{a}_{i2}) + (-1)^{i+3}\boldsymbol{a}_{i3}\boldsymbol{m}(\boldsymbol{a}_{i3})$$
$$+ \cdots + (-1)^{i+j}\boldsymbol{a}_{ij}\boldsymbol{m}(\boldsymbol{a}_{ij}) + \cdots + (-1)^{i+n}\boldsymbol{a}_{in}\boldsymbol{m}(\boldsymbol{a}_{in}) \tag{10.4}$$

By use of (10.3) we can express (10.4) in terms of cofactors as follows:

$$\boldsymbol{D}_n = \boldsymbol{a}_{i1}\boldsymbol{A}_{i1} + \boldsymbol{a}_{i2}\boldsymbol{A}_{i2} + \boldsymbol{a}_{i3}\boldsymbol{A}_{i3} + \cdots + \boldsymbol{a}_{ij}\boldsymbol{A}_{ij} + \cdots + \boldsymbol{a}_{in}\boldsymbol{A}_{in} \tag{10.5}$$

Similarly, the expansion of D_n in terms of the minors and cofactors of the elements of the jth column are, respectively.

$$D_n = (-1)^{1+j}a_{1j}m(a_{1j}) + (-1)^{2+j}a_{2j}m(a_{2j}) + (-1)^{3+j}a_{3j}m(a_{3j})$$
$$+ \cdots + (-1)^{i+j}a_{ij}m(a_{ij}) + \cdots + (-1)^{n+j}a_{nj}m(a_{nj}) \qquad (10.6)$$

and

$$D_n = a_{1j}A_{1j} + a_{2j}A_{2j} + a_{3j}A_{3j} + \cdots + a_{ij}A_{ij} + \cdots + a_{nj}A_{nj} \qquad (10.7)$$

The proof for a determinant of order n is very abstract and will not be given here.

We shall now expand

$$D_3 = \begin{vmatrix} a_{11} & a_{12} & a_{13} \\ a_{21} & a_{22} & a_{23} \\ a_{31} & a_{32} & a_{33} \end{vmatrix}$$

in terms of the elements of the first row by use of (10.5) and see that we get the same expansion as was obtained in Sec. 10.5. We shall write out each minor instead of merely indicating it. Thus, we have

$$D_3 = (-1)^{1+1}a_{11}\begin{vmatrix} a_{22} & a_{23} \\ a_{32} & a_{33} \end{vmatrix} + (-1)^{1+2}a_{12}\begin{vmatrix} a_{21} & a_{23} \\ a_{31} & a_{33} \end{vmatrix} + (-1)^{1+3}a_{13}\begin{vmatrix} a_{21} & a_{22} \\ a_{31} & a_{32} \end{vmatrix}$$

$$= a_{11}(a_{22}a_{33} - a_{23}a_{32}) - a_{12}(a_{21}a_{33} - a_{23}a_{31}) + a_{13}(a_{21}a_{32} - a_{22}a_{31})$$
$$= a_{11}A_{11} + a_{12}A_{12} + a_{13}A_{13}$$

EXAMPLE   Obtain the value of

$$D = \begin{vmatrix} 2 & 5 & 4 & 3 \\ 3 & 2 & 5 & 1 \\ 4 & 0 & 2 & 1 \\ 3 & 0 & 3 & 2 \end{vmatrix}$$

Solution   We expand in terms of the minors of the second column and obtain

$$D = (-1)^{1+2}(5)\begin{vmatrix} 3 & 5 & 1 \\ 4 & 2 & 1 \\ 3 & 3 & 2 \end{vmatrix} + (-1)^{2+2}(2)\begin{vmatrix} 2 & 4 & 3 \\ 4 & 2 & 1 \\ 3 & 3 & 2 \end{vmatrix} + 0 + 0$$

We next expand each of the third-order determinants in terms of the minors of the elements of the first row and obtain

$$D = -5[3(4-3) - 5(8-3) + 1(12-6)]$$
$$+ 2[2(4-3) - 4(8-3) + 3(12-6)]$$
$$= -5(3 - 25 + 6) + 2(2 - 20 + 18)$$
$$= -5(-16) + 2(0) = 80$$

EXERCISE 10.3 Expansion of Second- and Third-Order Determinants

Evaluate the determinant in each of Probs. 1 to 40.

1 $\begin{vmatrix} 2 & 3 \\ 1 & 5 \end{vmatrix}$ 2 $\begin{vmatrix} 3 & 2 \\ 4 & 1 \end{vmatrix}$ 3 $\begin{vmatrix} 5 & -1 \\ 0 & 2 \end{vmatrix}$ 4 $\begin{vmatrix} 0 & 2 \\ 1 & 0 \end{vmatrix}$

5 $\begin{vmatrix} 7 & 3 \\ 2 & 1 \end{vmatrix}$ 6 $\begin{vmatrix} 3 & 4 \\ 2 & 6 \end{vmatrix}$ 7 $\begin{vmatrix} 2 & 3 \\ 5 & 8 \end{vmatrix}$ 8 $\begin{vmatrix} 9 & 8 \\ 7 & 6 \end{vmatrix}$

9 $\begin{vmatrix} 3 & 0 \\ 2 & -1 \end{vmatrix}$ 10 $\begin{vmatrix} -4 & 0 \\ 2 & 3 \end{vmatrix}$ 11 $\begin{vmatrix} 2 & -3 \\ 4 & 0 \end{vmatrix}$ 12 $\begin{vmatrix} -3 & -1 \\ 5 & -2 \end{vmatrix}$

13 $\begin{vmatrix} 5 & -7 \\ a & -2 \end{vmatrix}$ 14 $\begin{vmatrix} 7 & b \\ 6 & a \end{vmatrix}$ 15 $\begin{vmatrix} 3 & -2 \\ b & a \end{vmatrix}$ 16 $\begin{vmatrix} a & 2 \\ -3 & b \end{vmatrix}$

17 $\begin{vmatrix} a & a \\ b & c \end{vmatrix}$ 18 $\begin{vmatrix} a & b \\ c & d \end{vmatrix}$ 19 $\begin{vmatrix} a & -b \\ c & a \end{vmatrix}$ 20 $\begin{vmatrix} a & c \\ c & b \end{vmatrix}$

21 $\begin{vmatrix} 1 & 6 & 0 \\ 2 & 5 & 3 \\ 3 & 4 & 0 \end{vmatrix}$ 22 $\begin{vmatrix} 2 & 0 & 4 \\ 0 & 5 & 7 \\ 3 & 1 & 0 \end{vmatrix}$ 23 $\begin{vmatrix} 5 & 4 & 1 \\ 2 & 0 & 3 \\ 0 & 3 & 6 \end{vmatrix}$ 24 $\begin{vmatrix} 4 & 0 & 2 \\ 0 & 2 & 3 \\ 5 & 0 & 1 \end{vmatrix}$

25 $\begin{vmatrix} 3 & 4 & 1 \\ 1 & 4 & 3 \\ 4 & 3 & 1 \end{vmatrix}$ 26 $\begin{vmatrix} 2 & 1 & 5 \\ 1 & 1 & 5 \\ 5 & 5 & 1 \end{vmatrix}$ 27 $\begin{vmatrix} 3 & 1 & -2 \\ 4 & 0 & -4 \\ 2 & -1 & -3 \end{vmatrix}$ 28 $\begin{vmatrix} 1 & 6 & 5 \\ 2 & 5 & 4 \\ 3 & 4 & 3 \end{vmatrix}$

29 $\begin{vmatrix} 1 & 6 & 0 \\ 4 & 2 & 7 \\ 0 & 5 & 3 \end{vmatrix}$ 30 $\begin{vmatrix} 1 & 4 & 1 \\ 2 & 2 & 2 \\ 1 & 4 & 3 \end{vmatrix}$ 31 $\begin{vmatrix} 2 & 4 & 3 \\ 5 & 0 & 5 \\ 3 & 4 & 2 \end{vmatrix}$ 32 $\begin{vmatrix} 1 & 3 & 4 \\ -5 & 2 & 5 \\ -3 & -4 & 3 \end{vmatrix}$

33 $\begin{vmatrix} 0 & 1 & 2 \\ 1 & 0 & 1 \\ 2 & 1 & 0 \end{vmatrix}$ 34 $\begin{vmatrix} 1 & 2 & -3 \\ 3 & 1 & -2 \\ 2 & -3 & 1 \end{vmatrix}$ 35 $\begin{vmatrix} 0 & 1 & 6 \\ 1 & 0 & 4 \\ 8 & 3 & 1 \end{vmatrix}$ 36 $\begin{vmatrix} 0 & 0 & 3 \\ 0 & -2 & 4 \\ 1 & -5 & 6 \end{vmatrix}$

37 $\begin{vmatrix} m & t & m \\ a & a & a \\ t & s & t \end{vmatrix}$ 38 $\begin{vmatrix} s & k & t \\ b & k & s \\ s & k & t \end{vmatrix}$ 39 $\begin{vmatrix} t & t & t \\ 1 & w & 0 \\ 0 & 0 & 0 \end{vmatrix}$ 40 $\begin{vmatrix} m & -1 & b \\ m & a & a \\ m & a & m \end{vmatrix}$

10.7 / PROPERTIES OF DETERMINANTS

Although the expansion of a determinant by minors enables us to express the determinant in terms of determinants of lower order, the computation in calculating the value of a determinant of order 4 or more by use of this method would be very tedious. The computation can be greatly simplified if we use the following seven properties:

1 *If the rows of one determinant are the same as the columns of another, the two determinants are equal.*

Proof We shall prove the theorem for determinants of the second and third order. If we expand the two determinants

$$D_2 = \begin{vmatrix} a_{11} & a_{12} \\ a_{21} & a_{22} \end{vmatrix} \quad \text{and} \quad D_2' = \begin{vmatrix} a_{11} & a_{21} \\ a_{12} & a_{22} \end{vmatrix}$$

we get $a_{11}a_{22} - a_{12}a_{21}$ in both cases since multiplication is commutative. Hence, the theorem is true for $n = 2$. We shall now prove it for $n = 3$ by considering

$$D_3 = \begin{vmatrix} a_{11} & a_{12} & a_{13} \\ a_{21} & a_{22} & a_{23} \\ a_{31} & a_{32} & a_{33} \end{vmatrix} \quad \text{and} \quad D_3' = \begin{vmatrix} a_{11} & a_{21} & a_{31} \\ a_{12} & a_{22} & a_{32} \\ a_{13} & a_{23} & a_{33} \end{vmatrix}$$

If we expand D_3 in terms of the elements of the first row and D_3' in terms of the elements of the first column, we obtain $a_{11}A_{11} + a_{12}A_{12} + a_{13}A_{13}$ in each case. Hence the theorem is true for $n = 3$.

We shall prove that the properties stated in the remainder of this section hold for determinants of order 4. The arguments used, however, are general, and can be applied to a determinant of any given order.

2 *If two columns (or rows) of a determinant are interchanged, the value of the determinant is equal to the negative of the value of the given determinant.*

Proof The proof of this statement is as follows. By (10.6), with $n = 4$, the expansion of the determinant D in terms of the minors of the elements of the jth column is

$$D = (-1)^{1+j}a_{1j}m(a_{1j}) + (-1)^{2+j}a_{2j}m(a_{2j}) + (-1)^{3+j}a_{3j}m(a_{3j}) + (-1)^{4+j}a_{4j}m(a_{4j})$$

where the exponent of -1 is the sum of the numbers of the row and column in which the element a_{ij} appears.

Now if we interchange the jth column with the column immediately to the left, we obtain a new determinant D'. This operation changes neither the elements of the jth column nor the minors of the elements. It does, however, decrease the number of the column by 1; hence, the jth column of D becomes the $(j-1)$st column of D'. Therefore, the expansion of D' will be the same as the expansion of D except that the exponent of -1 in each term will be decreased by 1. Hence, $D = -D'$.

If the two columns interchanged are not adjacent, we can prove that this interchange can be accomplished by an odd number of interchanges of adjacent columns. For example, to interchange the first and fourth columns of D, we interchange the fourth column successively with the third, the second, and the first, placing it immediately at the left of the first column. Next we interchange the first column successively with the second and third, placing it in the position vacated by the fourth. Hence, we have made $3+2=5$ interchanges, and therefore have five changes in sign. Consequently, if two nonadjacent columns of a determinant are interchanged, the value of the determinant obtained will be the negative of the value of the original determinant.

3 *If the corresponding elements of two columns (or rows) of a determinant are identical, the value of the determinant is zero.*

Proof The proof of this statement is as follows. If any two columns of the determinant D are identical, and if we obtain the determinant D' by interchanging these two columns, then $D=-D'$. On the other hand, since the two columns interchanged are identical, $D=D'$. Therefore $D=-D$, and it follows that $D=0$.

4 *If the elements of a column (or row) of a determinant are multiplied by k, the value of the determinant is multiplied by k.*

Proof To prove this statement, we multiply the elements of the jth column of the determinant D by k and obtain the determinant D''. If we expand D'' in terms of the minors of the elements of the jth columns, we obtain

$$D'' = \text{sum of products } (-1)^{i+j}ka_{ij}m(a_{ij}), \quad i=1, 2, 3, 4, \ldots, n$$
$$= k\,[\text{sum of products } (-1)^{i+j}a_{ij}m(a_{ij}), i=1, 2, 3, 4, \ldots, n]$$
$$= kD$$

5 *If the elements of the jth column of a determinant D are of the form $a_{ij}+b_{ij}$, then D is the sum of the determinants D' and D'' in which all the columns of D, D', and D'' are the same except the jth; furthermore, the jth column of D' is $a_{ij}, i=1,2,3,4,\ldots,n$, and the jth column of D'' is $b_{ij}, i=1,2,3,4,\ldots,n$.*

Proof We shall prove that this property is valid for a determinant of the fourth order in which the third column is of the form $a_{ij}+b_{ij}$. If

$$D = \begin{vmatrix} a_{11} & a_{12} & a_{13}+b_{13} & a_{14} \\ a_{21} & a_{22} & a_{23}+b_{23} & a_{24} \\ a_{31} & a_{32} & a_{33}+b_{33} & a_{34} \\ a_{41} & a_{42} & a_{43}+b_{43} & a_{44} \end{vmatrix}$$

and we expand D in terms of the minors of the elements of the third column, we obtain

$$D = (a_{13}+b_{13})m(a_{13}) - (a_{23}+b_{23})m(a_{23}) + (a_{33}+b_{33})m(a_{33})$$
$$- (a_{43}+b_{43})m(a_{43})$$
$$= [a_{13}m(a_{13}) - a_{23}m(a_{23}) + a_{33}m(a_{33}) - a_{43}m(a_{43})]$$
$$+ [b_{13}m(a_{13}) - b_{23}m(a_{23}) + b_{33}m(a_{33}) - b_{43}m(a_{43})]$$

By (10.6) the first and second bracketed expressions are respectively the expansions of

$$D' = \begin{vmatrix} a_{11} & a_{12} & a_{13} & a_{14} \\ a_{21} & a_{22} & a_{23} & a_{24} \\ a_{31} & a_{32} & a_{33} & a_{34} \\ a_{41} & a_{42} & a_{43} & a_{44} \end{vmatrix} \quad \text{and} \quad D'' = \begin{vmatrix} a_{11} & a_{12} & b_{13} & a_{14} \\ a_{21} & a_{22} & b_{23} & a_{24} \\ a_{31} & a_{32} & b_{33} & a_{34} \\ a_{41} & a_{42} & b_{43} & a_{44} \end{vmatrix}$$

Therefore, $D = D' + D''$.

6 *If in a given determinant D the elements of the kth column a_{ik}, $i = 1, 2, 3, \ldots$, n, are replaced by $a_{ik} + ta_{ij}$, where a_{ij}, $i = 1, 2, 3, \ldots$, n, are the elements of the jth column, the determinant obtained is equal to D.*

Proof We shall prove that this property is true for a determinant of order 4 in which we multiply each element of the fourth column by t and add the product to the corresponding element of the second column. The same method can be used to prove that the property holds for any two determinants of any order. If

$$D = \begin{vmatrix} a_{11} & a_{12} & a_{13} & a_{14} \\ a_{21} & a_{22} & a_{23} & a_{24} \\ a_{31} & a_{32} & a_{33} & a_{34} \\ a_{41} & a_{42} & a_{43} & a_{44} \end{vmatrix} \quad \text{and} \quad D' = \begin{vmatrix} a_{11} & a_{12} + ta_{14} & a_{13} & a_{14} \\ a_{21} & a_{22} + ta_{24} & a_{23} & a_{24} \\ a_{31} & a_{32} + ta_{34} & a_{33} & a_{34} \\ a_{41} & a_{42} + ta_{44} & a_{43} & a_{44} \end{vmatrix}$$

then, by properties 4 and 5.

$$D' = \begin{vmatrix} a_{11} & a_{12} & a_{13} & a_{14} \\ a_{21} & a_{22} & a_{23} & a_{24} \\ a_{31} & a_{32} & a_{33} & a_{34} \\ a_{41} & a_{42} & a_{43} & a_{44} \end{vmatrix} + t \begin{vmatrix} a_{11} & a_{14} & a_{13} & a_{14} \\ a_{21} & a_{24} & a_{23} & a_{24} \\ a_{31} & a_{34} & a_{33} & a_{34} \\ a_{41} & a_{44} & a_{43} & a_{44} \end{vmatrix}$$

Now the first determinant in the above pair is equal to D, and the second is equal to zero, since two columns are identical. Therefore, $D' = D$.

By property 1, the above property is true if the word "column" is replaced by "row."

By a repeated application of property 6 to a determinant D of order n, we can obtain a determinant in which all of the elements except one of some row or column are zeros, and the determinant thus obtained will be equal to D. We may then expand the determinant in terms of the minors of the row or column that contains the zeros and thus have the given determinant equal to the product of a constant and a determinant of order $n - 1$. We shall illustrate the method with two examples. We shall use the notation $C_k + tC_j$ to indicate that each element in the jth column of D is multiplied by t and the product is added to the corresponding element of the kth column, and we shall use a similar notation for rows.

EXAMPLE 1   Find the value of

$$D = \begin{vmatrix} 2 & 1 & 4 & 1 \\ 4 & 2 & 6 & 2 \\ 3 & 5 & 2 & 3 \\ 7 & 3 & 1 & 3 \end{vmatrix}$$

Solution   We first notice that the second and fourth columns have three elements in common; hence, we perform $C_2 - C_4$ and get

$$D = \begin{vmatrix} 2 & 0 & 4 & 1 \\ 4 & 0 & 6 & 2 \\ 3 & 2 & 2 & 3 \\ 7 & 0 & 1 & 3 \end{vmatrix}$$

Now we expand in terms of the minors of the elements of the second column and get

$$D = -2 \begin{vmatrix} 2 & 4 & 1 \\ 4 & 6 & 2 \\ 7 & 1 & 3 \end{vmatrix}$$

Finally, we perform the operation $R_2 - 2R_1$, expand the determinant thus obtained in terms of the minors of the elements of the second row, and get

$$D = -2 \begin{vmatrix} 2 & 4 & 1 \\ 0 & -2 & 0 \\ 7 & 1 & 3 \end{vmatrix} = (-2)(-2) \begin{vmatrix} 2 & 1 \\ 7 & 3 \end{vmatrix} = 4(6-7) = -4$$

EXAMPLE 2   Obtain the value of the determinant

$$D = \begin{vmatrix} 2 & 3 & 5 & 1 \\ 4 & 2 & 3 & 5 \\ 3 & 1 & 4 & 2 \\ 5 & 4 & 2 & 3 \end{vmatrix}$$

Solution   If we examine the above determinant, we see that we cannot obtain a determinant with three zeros in either one row or one column by adding or subtracting corresponding terms in either rows or columns. We can, however, obtain a determinant in which the elements in the first row are 0, 0, 0, 1 by performing successively the operations $C_1 - 2C_4$, $C_2 - 3C_4$, and $C_3 - 5C_4$ and then writing column 4 unchanged. Thus we get

$$D = \begin{vmatrix} 0 & 0 & 0 & 1 \\ -6 & -13 & -22 & 5 \\ -1 & -5 & -6 & 2 \\ -1 & -5 & -13 & 3 \end{vmatrix}$$

If we now expand in terms of the minors of the elements of the first row, we get

$$D = -1 \begin{vmatrix} -6 & -13 & -22 \\ -1 & -5 & -6 \\ -1 & -5 & -13 \end{vmatrix}$$

Now we notice that the first two terms in the second and third rows are the same, so we perform the operation $R_2 - R_3$ and get

$$D = -1 \begin{vmatrix} -6 & -13 & -22 \\ 0 & 0 & 7 \\ -1 & -5 & -13 \end{vmatrix} = (-1)(-7) \begin{vmatrix} -6 & -13 \\ -1 & -5 \end{vmatrix} = 7(30 - 13) = 119$$

7 *If each element of a row is multiplied by the cofactor of the corresponding element of another row and the products are added, the sum is zero.*

Proof We shall prove this theorem for a third-order determinant, but the procedure can be applied to one of any order. We shall consider the elements of the first row of

$$D = \begin{vmatrix} a_{11} & a_{12} & a_{13} \\ a_{21} & a_{22} & a_{23} \\ a_{31} & a_{32} & a_{33} \end{vmatrix}$$

and the cofactors of the elements of the third row and prove that

$$a_{11}A_{31} + a_{12}A_{32} + a_{13}A_{33} = 0 \tag{1}$$

Since

$$A_{31} = \begin{vmatrix} a_{12} & a_{13} \\ a_{22} & a_{23} \end{vmatrix} \qquad A_{32} = -\begin{vmatrix} a_{11} & a_{13} \\ a_{21} & a_{23} \end{vmatrix} \qquad A_{33} = \begin{vmatrix} a_{11} & a_{12} \\ a_{21} & a_{22} \end{vmatrix}$$

the left member of (1) is the expansion of

$$\begin{vmatrix} a_{11} & a_{12} & a_{13} \\ a_{11} & a_{12} & a_{13} \\ a_{21} & a_{22} & a_{23} \end{vmatrix}$$

in terms of the elements of the first row. The value of the determinant is zero since two rows are identical.

EXERCISE 10.4 Properties of Determinants

By use of the properties of determinants, prove the statement in each of Probs. 1 to 20 without expanding.

1 $\begin{vmatrix} t & i & m \\ w & a & s \\ b & a & d \end{vmatrix} = \begin{vmatrix} t & w & b \\ i & a & a \\ m & s & d \end{vmatrix}$

2 $\begin{vmatrix} 3 & p & m \\ n & a & p \\ i & k & e \end{vmatrix} = \begin{vmatrix} 3 & n & i \\ p & a & k \\ m & p & e \end{vmatrix}$

3 $\quad 2\begin{vmatrix} t & o & d \\ d & l & a \\ b & a & t \end{vmatrix} = \begin{vmatrix} t & o & d \\ 2d & 2l & 2a \\ b & a & t \end{vmatrix}$

4 $\quad 3\begin{vmatrix} a & i & m \\ t & o & o \\ b & a & d \end{vmatrix} = \begin{vmatrix} a & 3t & b \\ i & 3o & a \\ m & 3o & d \end{vmatrix}$

5 $\quad \begin{vmatrix} 1 & 2 & 3 \\ 3 & 0 & 9 \\ 5 & 6 & 15 \end{vmatrix} = 0$

6 $\quad \begin{vmatrix} 5 & 6 & 4 \\ 4 & 2 & 3 \\ 6 & 3 & 7 \end{vmatrix} = \begin{vmatrix} 5 & 4 & 6 \\ 6 & 2 & 3 \\ 4 & 3 & 7 \end{vmatrix}$

7 $\quad \begin{vmatrix} 1 & 3 & 2 \\ 2 & 2 & 4 \\ 3 & 5 & 6 \end{vmatrix} = \begin{vmatrix} 1 & 2 & 3 \\ 0 & 1 & 7 \\ 3 & 6 & 9 \end{vmatrix}$

8 $\quad \begin{vmatrix} 1 & 2 & 3 \\ 2 & 5 & 6 \\ 5 & 8 & 15 \end{vmatrix} = \begin{vmatrix} 2 & 1 & 3 \\ 2 & 5 & 8 \\ 4 & 2 & 6 \end{vmatrix}$

9 $\quad \begin{vmatrix} 3 & 4 & 1 \\ 5 & 0 & 2 \\ 2 & 3 & 5 \end{vmatrix} = -1\begin{vmatrix} 4 & 3 & 1 \\ 0 & 5 & 2 \\ 3 & 2 & 5 \end{vmatrix}$

10 $\quad \begin{vmatrix} 1 & 2 & 4 \\ 3 & 1 & 3 \\ 7 & 4 & 0 \end{vmatrix} = 3\begin{vmatrix} 2 & 1 & 4 \\ 1 & 3 & 7 \\ -4 & -3 & 0 \end{vmatrix}$

11 $\quad \begin{vmatrix} 1 & 5 & 2 \\ 3 & 4 & 3 \\ 4 & 7 & 1 \end{vmatrix} = \begin{vmatrix} 1 & 5 & 2 \\ 3 & 3 & 3 \\ 4 & 5 & 1 \end{vmatrix} + \begin{vmatrix} 1 & 0 & 2 \\ 3 & 1 & 3 \\ 4 & 2 & 1 \end{vmatrix}$

12 $\quad \begin{vmatrix} 3 & 3 & 3 \\ 5 & 0 & 1 \\ 2 & 1 & 4 \end{vmatrix} = \begin{vmatrix} 1 & 0 & 2 \\ 5 & 0 & 1 \\ 2 & 1 & 4 \end{vmatrix} + \begin{vmatrix} 2 & 3 & 1 \\ 5 & 0 & 1 \\ 2 & 1 & 4 \end{vmatrix}$

13 $\quad \begin{vmatrix} 2 & 1 & 5 \\ 1 & 5 & 2 \\ 5 & 2 & 1 \end{vmatrix} = \begin{vmatrix} 3 & 1 & 5 \\ 6 & 5 & 2 \\ 7 & 2 & 1 \end{vmatrix}$

14 $\quad \begin{vmatrix} 1 & 2 & 3 \\ 4 & 5 & 6 \\ 7 & 8 & 9 \end{vmatrix} = \begin{vmatrix} 1 & 2 & 2 \\ 4 & 5 & 2 \\ 7 & 8 & 2 \end{vmatrix}$

15 $\quad 1\begin{vmatrix} 2 & 8 \\ 3 & 9 \end{vmatrix} + 2(-1)^3\begin{vmatrix} 1 & 7 \\ 3 & 9 \end{vmatrix} + 3\begin{vmatrix} 1 & 7 \\ 2 & 8 \end{vmatrix} = 0$

16 $\quad 4\begin{vmatrix} 2 & 5 \\ 3 & 6 \end{vmatrix} + 3(-1)\begin{vmatrix} 3 & 5 \\ 4 & 6 \end{vmatrix} + 6\begin{vmatrix} 3 & 2 \\ 4 & 3 \end{vmatrix} = 0$

17 $\quad \begin{vmatrix} a & b & c \\ d & e & f \\ g & h & i \end{vmatrix} = -\begin{vmatrix} a & c & b \\ d & f & e \\ g & i & h \end{vmatrix}$

18 $\quad \begin{vmatrix} a & b & c \\ d & e & f \\ g & h & i \end{vmatrix} = \begin{vmatrix} a+b-c & b & c \\ d+e-f & e & f \\ g+h-i & h & i \end{vmatrix}$

19 $\quad \begin{vmatrix} x_1 & y_1 & z_1 \\ x_2 & y_2 & z_2 \\ x_3 & y_3 & z_3 \end{vmatrix} = \begin{vmatrix} x_1+ax_2 & y_1+ay_2 & z_1+az_2 \\ x_2 & y_2 & z_2 \\ x_3 & y_3 & z_3 \end{vmatrix}$

20 $\quad \begin{vmatrix} a & b \\ c & d \end{vmatrix} = \begin{vmatrix} a+2b & b \\ c+2d & d \end{vmatrix} = \begin{vmatrix} a & b-7a \\ c & d-7c \end{vmatrix}$

Use Property 6 (properties of determinants) to find the value of the determinant in each of Probs. 21 to 24.

21 $\quad \begin{vmatrix} 1 & 8 & 1 \\ 1 & 3 & 3 \\ 8 & 3 & 24 \end{vmatrix}$

22 $\quad \begin{vmatrix} 1 & -1 & 1 \\ 4 & 4 & -7 \\ -2 & -3 & 4 \end{vmatrix}$

23 $\begin{vmatrix} 2 & 9 & 4 \\ 8 & 2 & -3 \\ -2 & 1 & 2 \end{vmatrix}$
  24 $\begin{vmatrix} 6 & 2 & 10 \\ 2 & 0 & 5 \\ 2 & 3 & -4 \end{vmatrix}$

Find the value or values of x so that the statement in each of Probs. 25 to 36 is true.

25 $\begin{vmatrix} 3 & 2 \\ 6 & x \end{vmatrix} = 0$
  26 $\begin{vmatrix} 3 & 2 \\ x & 12 \end{vmatrix} = 0$
  27 $\begin{vmatrix} x & -2 \\ 1 & x-1 \end{vmatrix} = 0$

28 $\begin{vmatrix} 2x+7 & 4 \\ 1 & x \end{vmatrix} = 0$
  29 $\begin{vmatrix} x & 2 & 3 \\ 1 & 6 & 1 \\ 3 & x & 0 \end{vmatrix} = -46$
  30 $\begin{vmatrix} 3 & 5 & -1 \\ 2 & 1 & 1 \\ 4 & x & 2 \end{vmatrix} = 0$

31 $\begin{vmatrix} 5 & 2 & x \\ 1 & 3 & -1 \\ x & 2 & 1 \end{vmatrix} = -4$
  32 $\begin{vmatrix} 1 & x & 1 \\ x & 1 & 1 \\ 3 & 2 & 2 \end{vmatrix} = 0$
  33 $\begin{vmatrix} 1 & x & 2 \\ 2 & 1 & x \\ 1 & 0 & 3 \end{vmatrix} = -4$

34 $\begin{vmatrix} 2 & 0 & x \\ 3 & 1 & x \\ x & 2 & 1 \end{vmatrix} = -13$
  35 $\begin{vmatrix} x & 1 & x \\ 1 & x & 2 \\ 3 & x & 1 \end{vmatrix} = -7$
  36 $\begin{vmatrix} x & 2 & 3 \\ x & x & 1 \\ 2 & 0 & 1 \end{vmatrix} = -8$

10.8 / CRAMER'S RULE

In the eighteenth century, the Swiss mathematician Cramer devised a rule for obtaining the simultaneous solution set of a system of linear equations. We shall explain this rule and its use in this section. His rule states that if the determinant of the coefficients is not zero, the solution set of the system of equations

$$a_{11}x_1 + a_{12}x_2 + a_{13}x_3 + \cdots + a_{1j}x_j + \cdots + a_{1n}x_n = e_1$$
$$a_{21}x_1 + a_{22}x_2 + a_{23}x_3 + \cdots + a_{2j}x_j + \cdots + a_{2n}x_n = e_2$$
$$a_{31}x_1 + a_{32}x_2 + a_{33}x_3 + \cdots + a_{3j}x_j + \cdots + a_{3n}x_n = e_3$$
$$\cdots\cdots\cdots\cdots\cdots\cdots\cdots\cdots\cdots\cdots\cdots\cdots$$
$$a_{i1}x_1 + a_{i2}x_2 + a_{i3}x_3 + \cdots + a_{ij}x_j + \cdots + a_{in}x_n = e_i$$
$$\cdots\cdots\cdots\cdots\cdots\cdots\cdots\cdots\cdots\cdots\cdots\cdots\cdots\cdots$$
$$a_{n1}x_1 + a_{n2}x_2 + a_{n3}x_3 + \cdots + a_{nj}x_j + \cdots + a_{nn}x_n = e_n$$

is
$$\left\{ \frac{N(x_1)}{D}, \frac{N(x_2)}{D}, \frac{N(x_3)}{D}, \ldots, \frac{N(x_j)}{D}, \ldots, \frac{N(x_n)}{D} \right\}$$

where
$$D = \begin{vmatrix} a_{11} & a_{12} & a_{13} & \cdots & a_{1j} & \cdots & a_{1n} \\ a_{21} & a_{22} & a_{23} & \cdots & a_{2j} & \cdots & a_{2n} \\ a_{31} & a_{32} & a_{33} & \cdots & a_{3j} & \cdots & a_{3n} \\ \cdots & \cdots & \cdots & & \cdots & & \cdots \\ a_{i1} & a_{i2} & a_{i3} & \cdots & a_{ij} & \cdots & a_{in} \\ \cdots & \cdots & \cdots & & \cdots & & \cdots \\ a_{n1} & a_{n2} & a_{n3} & \cdots & a_{nj} & \cdots & a_{nn} \end{vmatrix}$$

and the determinant $N(x_j)$, for $j = 1, 2, 3, \ldots, n$, is obtained by replacing the jth column of D by the column of constant terms e_i, for $i = 1, 2, 3, \ldots, n$.

We shall prove Cramer's rule for a system of three linear equations. The method used is applicable regardless of the number of equations. We shall consider the system

$$a_1x + b_1y + c_1z = d_1 \tag{1}$$
$$a_2x + b_2y + c_2z = d_2 \tag{2}$$
$$a_3x + b_3y + c_3z = d_3 \tag{3}$$

The determinant of the coefficients is

$$D = \begin{vmatrix} a_1 & b_1 & c_1 \\ a_2 & b_2 & c_2 \\ a_3 & b_3 & c_3 \end{vmatrix}$$

We define the determinants formed when the constant terms are substituted for the coefficients of x, y, and z in D as

$$D_x = \begin{vmatrix} d_1 & b_1 & c_1 \\ d_2 & b_2 & c_2 \\ d_3 & b_3 & c_3 \end{vmatrix} \qquad D_y = \begin{vmatrix} a_1 & d_1 & c_1 \\ a_2 & d_2 & c_2 \\ a_3 & d_3 & c_3 \end{vmatrix} \qquad D_z = \begin{vmatrix} a_1 & b_1 & d_1 \\ a_2 & b_2 & d_2 \\ a_3 & b_3 & d_3 \end{vmatrix}$$

We shall show that, if $D \neq 0$, then the solution set $\{(x, y, z)\}$ of the system is $\{(D_x/D, D_y/D, D_z/D)\}$. For this purpose, we multiply the first column of D by x, and by Property 4, we have

$$xD = \begin{vmatrix} a_1x & b_1 & c_1 \\ a_2x & b_2 & c_2 \\ a_3x & b_3 & c_3 \end{vmatrix} \tag{4}$$

Next, we multiply the elements of the second column of the above determinant by y, and the elements of the third column by z and add the products to the elements of the first column. Thus by Property 6 we get

$$xD = \begin{vmatrix} a_1x + b_1y + c_1z & b_1 & c_1 \\ a_2x + b_2y + c_2z & b_2 & c_2 \\ a_3x + b_3y + c_3z & b_3 & c_3 \end{vmatrix}$$

$$= \begin{vmatrix} d_1 & b_1 & c_1 \\ d_2 & b_2 & c_2 \\ d_3 & b_3 & c_3 \end{vmatrix} \qquad \text{since } a_jx + b_iy + c_iz = d_j,$$
$$\text{for } i = 1, 2, 3$$

$$xD = D_x$$

Consequently, $x = D_x/D$.

By a similar argument, we can show that $y = D_y/D$ and $z = D_z/D$.

We shall now verify that these values of x, y, and z satisfy Eq. (1). If we replace x, y, and z respectively in the left member of (1) by these values, we have

$$\frac{1}{D}\Big(a_1 D_x + b_1 D_y + c_1 D_z\Big) \tag{5}$$

and we must show that this expression is equal to d_1. If we expand D_x in terms of the elements of the first column, we obtain

$$D_x = d_1 \begin{vmatrix} b_2 & c_2 \\ b_3 & c_3 \end{vmatrix} - d_2 \begin{vmatrix} b_1 & c_1 \\ b_3 & c_3 \end{vmatrix} + d_3 \begin{vmatrix} b_1 & c_1 \\ b_2 & c_2 \end{vmatrix} = d_1 A_1 + d_2 A_2 + d_3 A_3$$

since the second-order determinants together with the sign that precedes each are respectively the cofactors of a_1, a_2, and a_3 in D. Therefore,

$$a_1 D_x = a_1 d_1 A_1 + a_1 d_2 A_2 + a_1 d_3 A_3 \tag{6}$$

Similarly,
$$b_1 D_y = b_1 d_1 B_1 + b_1 d_2 B_2 + b_1 d_3 B_3 \tag{7}$$

and
$$c_1 D_z = c_1 d_1 C_1 + c_1 d_2 C_2 + c_1 d_3 C_3 \tag{8}$$

We now equate the sums of the left and right members of (6), (7), and (8) and get

$$\begin{aligned} a_1 D_x + b_1 D_y + c_1 D_z &= d_1(a_1 A_1 + b_1 B_1 + c_1 C_1) \\ &+ d_2(a_1 A_2 + b_1 B_2 + c_1 C_2) \\ &+ d_3(a_1 A_3 + b_1 B_3 + c_1 C_3) \\ &= d_1 D + d_2(0) + d_3(0) = d_1 D \end{aligned}$$

since the expression in the first parentheses is the expansion of D in terms of the elements of the first column, and the expression in each of the second and third parentheses is the sum of the products of the elements of the first column of D and the cofactors of the elements of the second and third columns, respectively, of D. Therefore by Property 4 the two latter sums are zero. Consequently, expression (5) is equal to $d_1 D/D = d_1$. Therefore, since (5) is the left member of Eq. (1) for $x = D_x/D, y = D_y/D$, $z = D_z/D$, these values satisfy the equation. By a similar method we can verify that these values also satisfy Eqs. (2) and (3). Hence the simultaneous solution set of Eqs. (1), (2), and (3) is $\{(D_x/D, D_y/D, D_z/D)\}$.

By use of a similar argument, we can show that the solution set of the equations

$$\begin{aligned} a_1 x + b_1 y &= d_1 \\ a_2 x + b_2 y &= d_2 \end{aligned}$$

is $\{(D_x/D, D_y/D)\}$, provided $D \neq 0$ and

$$D = \begin{vmatrix} a_1 & b_1 \\ a_2 & b_2 \end{vmatrix} \qquad D_x = \begin{vmatrix} d_1 & b_1 \\ d_2 & b_2 \end{vmatrix} \qquad \text{and} \qquad D_y = \begin{vmatrix} a_1 & d_1 \\ a_2 & d_2 \end{vmatrix}$$

If the determinant of the coefficients D is zero, the system of equations is not independent, and the solution set may be the empty set $\varnothing$, or it may contain an infinitude of elements.

We illustrate the use of Cramer's rule with three examples.

EXAMPLE 1   Find the simultaneous solution set of the equations

$$3x - 6y - 2 = 0 \qquad\qquad (9)$$
$$4x + 7y + 3 = 0 \qquad\qquad (10)$$

Solution   by use of Cramer's rule.

We first add 2 to each member of (9) and -3 to each member of (10) and get

$$3x - 6y = 2$$
$$4x + 7y = -3$$

We now obtain the solution set by the following steps:

Step 1   Form the determinant D whose elements are the coefficients of the unknowns in the order in which they appear, and get

$$D = \begin{vmatrix} 3 & -6 \\ 4 & 7 \end{vmatrix} = 21 + 24 = 45$$

Step 2   Replace the column of coefficients of x in D by the constant terms and get

$$D_x = \begin{vmatrix} 2 & -6 \\ -3 & 7 \end{vmatrix} = 14 - 18 = -4$$

Step 3   Replace the column of coefficients of y in D by the constant terms and get

$$D_y = \begin{vmatrix} 3 & 2 \\ 4 & -3 \end{vmatrix} = -9 - 8 = -17$$

Step 4   By Cramer's rule,

$$x = \frac{D_x}{D} = \frac{-4}{45} = -\frac{4}{45}$$

$$y = \frac{D_y}{D} = \frac{-17}{45} = -\frac{17}{45}$$

Hence the simultaneous solution set is $\{(-\frac{4}{45}, -\frac{17}{45})\}$.

Check   Replacing x and y in the given equations by the appropriate elements of the solution set, we have

$$3(-\tfrac{4}{45}) - 6(-\tfrac{17}{45}) - 2 = \frac{-12 + 102 - 90}{45} = 0 \qquad \text{from (9)}$$

$$4(-\tfrac{4}{45}) + 7(-\tfrac{17}{45}) + 3 = \frac{-16 - 119 + 135}{45} = 0 \qquad \text{from (10)}$$

EXAMPLE 2   Use Cramer's rule to solve the system of equations

$$3x + y - 2z = -3$$
$$2x + 7y + 3z = 9$$
$$4x - 3y - z = 7$$

Solution   The terms in the left members are arranged in the proper order, and only the constant terms appear in the right members. Hence, we proceed as follows:

Step 1
$$D = \begin{vmatrix} 3 & 1 & -2 \\ 2 & 7 & 3 \\ 4 & -3 & -1 \end{vmatrix}$$

$$= 3 \begin{vmatrix} 7 & 3 \\ -3 & -1 \end{vmatrix} - 1 \begin{vmatrix} 2 & 3 \\ 4 & -1 \end{vmatrix} - 2 \begin{vmatrix} 2 & 7 \\ 4 & -3 \end{vmatrix}$$

$$= 3(-7 + 9) - 1(-2 - 12) - 2(-6 - 28)$$
$$D = 6 + 14 + 68 = 88$$

Step 2
$$D_x = \begin{vmatrix} -3 & 1 & -2 \\ 9 & 7 & 3 \\ 7 & -3 & -1 \end{vmatrix}$$

$$= -3 \begin{vmatrix} 7 & 3 \\ -3 & -1 \end{vmatrix} - 1 \begin{vmatrix} 9 & 3 \\ 7 & -1 \end{vmatrix} - 2 \begin{vmatrix} 9 & 7 \\ 7 & -3 \end{vmatrix}$$

$$= -3(-7 + 9) - 1(-9 - 21) - 2(-27 - 49)$$
$$= -6 + 30 + 152 = 176$$

Step 3
$$D_y = \begin{vmatrix} 3 & -3 & -2 \\ 2 & 9 & 3 \\ 4 & 7 & -1 \end{vmatrix}$$

$$= 3 \begin{vmatrix} 9 & 3 \\ 7 & -1 \end{vmatrix} + 3 \begin{vmatrix} 2 & 3 \\ 4 & -1 \end{vmatrix} - 2 \begin{vmatrix} 2 & 9 \\ 4 & 7 \end{vmatrix}$$

$$= 3(-9 - 21) + 3(-2 - 12) - 2(14 - 36)$$
$$= -90 - 42 + 44 = -88$$

Step 4
$$D_z = \begin{vmatrix} 3 & 1 & -3 \\ 2 & 7 & 9 \\ 4 & -3 & 7 \end{vmatrix}$$

$$= 3 \begin{vmatrix} 7 & 9 \\ -3 & 7 \end{vmatrix} - 1 \begin{vmatrix} 2 & 9 \\ 4 & 7 \end{vmatrix} - 3 \begin{vmatrix} 2 & 7 \\ 4 & -3 \end{vmatrix}$$

$$= 3(49 + 27) - 1(14 - 36) - 3(-6 - 28)$$
$$= 228 + 22 + 102 = 352$$

Step 5

$$x = \frac{D_x}{D} = \frac{176}{88} = 2$$

$$y = \frac{D_y}{D} = \frac{-88}{88} = -1$$

$$z = \frac{D_z}{D} = \frac{352}{88} = 4$$

by Cramer's rule. Hence the solution set is $\{(2, -1, 4)\}$; it can be checked by the usual method.

EXAMPLE 3   Show that the following equations are not independent:

$$\begin{array}{rrr} 5x + 4y + 11z = 3 \\ 6x - 4y + 2z = 1 \\ x + 3y + 5z = 2 \end{array}$$

Solution
$$D = \begin{vmatrix} 5 & 4 & 11 \\ 6 & -4 & 2 \\ 1 & 3 & 5 \end{vmatrix} = 5\begin{vmatrix} -4 & 2 \\ 3 & 5 \end{vmatrix} - 4\begin{vmatrix} 6 & 2 \\ 1 & 5 \end{vmatrix} + 11\begin{vmatrix} 6 & -4 \\ 1 & 3 \end{vmatrix}$$

$$= 5(-20 - 6) - 4(30 - 2) + 11(18 + 4)$$
$$D = -130 - 112 + 242 = 0$$

Hence, since $D = 0$, the equations are not independent, and no unique solution set exists.

EXERCISE 10.5 Cramer's Rule

Use Cramer's rule to find the solution set of each of the following systems of equations.

1   $\begin{array}{r} x + y = 3 \\ 2x - 3y = 1 \end{array}$

2   $\begin{array}{r} 2x + 3y = 12 \\ 3x - y = 7 \end{array}$

3   $\begin{array}{r} 5x + y = 8 \\ 3x - 2y = -3 \end{array}$

4   $\begin{array}{r} 3x + 2y = 16 \\ 2x + 5y = 29 \end{array}$

5   $\begin{array}{r} 2x + 9y = -3 \\ 3x + 2y = 7 \end{array}$

6   $\begin{array}{r} 3x - 4y = -5 \\ 2x + 3y = 8 \end{array}$

7   $\begin{array}{r} 2x + 9y = 4 \\ 3x - 2y = 6 \end{array}$

8   $\begin{array}{r} 5x + 3y = 1 \\ 2x - 5y = -12 \end{array}$

9   $\begin{array}{r} 2x - 7y = 19 \\ 3x + 2y = -9 \end{array}$

10   $\begin{array}{r} 5x - 4y = 2 \\ 3x - 2y = 0 \end{array}$

11   $\begin{array}{r} 3x - y = 1 \\ 2x - 3y = 10 \end{array}$

12   $\begin{array}{r} 5x + 3y = 0 \\ 3x + 5y = -16 \end{array}$

13   $\begin{array}{r} 2x - 3y = 4 \\ 4x - 6y = 8 \end{array}$

14   $\begin{array}{r} 3x + y = 2 \\ 6x + 2y = 4 \end{array}$

15   $5x - 8y = 7$
   $15x - 24y = 20$

16   $4x + 3y = 4$
   $12x + 9y = 15$

17   $x - y = a - b$
   $2x + 3y = 2a + 3b$

18   $2x + 3y = 5a - 3b$
   $3x - 2y = a + 2b$

19   $ax + by = a^2 + b^2$
   $bx - ay = -a^2 - b^2$

20   $ax + by = a^2 + b^2$
   $2ax - by = 2a^2 - b^2 + 6ab$

21   $3x + 2y + 4z = 13$
   $2x - 3y + z = 1$
   $5x + 4y - 2z = 5$

22   $5x + 4y - 2z = 8$
   $4x - 2y + 5z = 21$
   $2x + 5y + 4z = 21$

23   $7x + 3y + 4z = 15$
   $2x + 6y + 3z = 1$
   $7x + 3y + 2z = 13$

24   $2x - 5y - 8z = 12$
   $5x - y + 3z = 7$
   $3x + 4y + 5z = 7$

25   $6x + 5y + 4z = 8$
   $7x - 5y + 3z = 26$
   $5x - 2y - 6z = -9$

26   $3x + y + 4z = 13$
   $6x + 2y - 3z = -29$
   $2x + 3y + 2z = 3$

27   $3x + 5y + 4z = 2$
   $5x + 4y + 3z = 7$
   $4x + 2y + 5z = 3$

28   $7x + 5y + 9z = 10$
   $6x + 4y + 7z = 7$
   $5x + 3y + 4z = 1$

29   $2x + 3y - z = 0$
   $x + y + 2z = 9$
   $3x + 4y + z = 8$

30   $x - y + 2z = 6$
   $x - 2y - z = 1$
   $x + 5y = -4$

31   $x + 2y + 3z = 19$
   $3x + 6y + 4z = 32$
   $x + 2y - 2z = -3$

32   $3x + 2y + z = 9$
   $2x + y + 2z = 12$
   $x + y - z = -3$

33   $3x + y = 9$
   $2x + 3z = -2$
   $3y - z = 11$

34   $3x + z = 7$
   $2x - 3y = -1$
   $2y + 3z = -9$

35   $5x + 3z = 1$
   $3x + 2y = 4$
   $4y - 5z = 11$

36   $y + 4z = -14$
   $3x + 2y = 11$
   $2x + 3z = 1$

37   $2x - 3y + 4z = 3a - 3b$
   $x - y + z = a - b$
   $3x - 2y + 2z = 3a - 2b$

38   $x + y + z = 0$
   $x + 2y + 3z = a + b$
   $2x - 2y + 5z = -4a + 7b$

39   $2x + 2y - 3z = a$
   $3x - 3y + z = a + 6b$
   $x - 2y + 2z = a + 3b$

40   $x + 2y + 3z = a + 3b$
   $2x - y + 2z = a - 4b$
   $4x + 2y + 9z = a$

EXERCISE 10.6 **REVIEW**

1   Find x and y so that $\begin{bmatrix} 2x + y & -4 \\ 14 & 5 \end{bmatrix} = \begin{bmatrix} 7 & x - 2y \\ 14 & 5 \end{bmatrix}$.

2   Find x, y, and z so that $\begin{bmatrix} x + y & 7 & 0 \\ 2 & y - z & 6 \\ -1 & 5 & 8 \end{bmatrix} = \begin{bmatrix} 3 & 7 & 0 \\ 2 & 3 & 6 \\ x + 2z & 5 & 8 \end{bmatrix}$.

Perform the operations indicated in Probs. 3 to 6.

3 $\begin{bmatrix} 2 & x-y & 4 \\ 3 & y & -6 \end{bmatrix} + \begin{bmatrix} -2 & y+x & x-4 \\ 0 & 1-y & x+6 \end{bmatrix}$

4 $\begin{bmatrix} 2 & 4 \\ 0 & 3 \\ 1 & -2 \end{bmatrix} \begin{bmatrix} 0 & 1 & 4 & 2 \\ -3 & 2 & 7 & 5 \end{bmatrix}$

5 $\begin{bmatrix} a & 5 & b \\ b & -2 & 0 \\ c & 1 & 3 \end{bmatrix} + \begin{bmatrix} 1-a & -5 & 1-b \\ 2-b & 2 & a \\ 3-c & -2 & -4 \end{bmatrix}$

6 $\begin{bmatrix} 1 & 0 & 2 \\ 3 & -1 & 4 \end{bmatrix} \begin{bmatrix} 2 & 3 & -1 & 5 \\ 1 & 0 & 2 & -3 \\ -3 & 4 & 1 & 0 \end{bmatrix}$

Use $A = \begin{bmatrix} 3 & 2 & -2 \\ 0 & -1 & 3 \\ 1 & 4 & 5 \end{bmatrix}$ *and* $B = \begin{bmatrix} 2 & -1 & -1 \\ 12 & -7 & -2 \\ -5 & 3 & 1 \end{bmatrix}$ *to find the quantities called for in Probs. 7 to 9.*

7 $(AB)^T$ **8** $A^T B^T$ **9** B^{-1} by row operations.

10 Prove that the value of B^{-1} found in Prob. 9 is correct by showing that $BB^{-1} = I$.

11 Show that I is its own inverse.

Find the value of the determinant in each of Probs. 12 to 17.

12 $\begin{vmatrix} 3 & 2 \\ 1 & 4 \end{vmatrix}$ **13** $\begin{vmatrix} 5 & 0 \\ -1 & 3 \end{vmatrix}$ **14** $\begin{vmatrix} -2 & 3 \\ 4 & 6 \end{vmatrix}$

15 $\begin{vmatrix} 1 & 2 & 3 \\ 1 & 2 & 3 \\ 3 & 2 & 1 \end{vmatrix}$ **16** $\begin{vmatrix} 1 & 3 & 5 \\ 3 & 1 & 3 \\ 5 & 3 & 1 \end{vmatrix}$ **17** $\begin{vmatrix} 0 & 2 & 1 \\ 0 & 1 & 0 \\ 1 & 2 & 2 \end{vmatrix}$

Prove the statements in Probs. 18 to 23 without expanding, by use of the properties of determinants.

18 $\begin{vmatrix} 2 & 5 & 0 \\ 1 & 4 & -2 \\ 3 & 0 & 6 \end{vmatrix} = \begin{vmatrix} 2 & 1 & 3 \\ 5 & 4 & 0 \\ 0 & -2 & 6 \end{vmatrix}$ **19** $\begin{vmatrix} 3 & 4 & -2 \\ 1 & 0 & 7 \\ 6 & 5 & 8 \end{vmatrix} = -1 \begin{vmatrix} 1 & 3 & 6 \\ 0 & 4 & 5 \\ 7 & -2 & 8 \end{vmatrix}$

20 $\begin{vmatrix} 3 & 0 & 6 \\ -1 & 4 & -2 \\ 5 & 0 & 10 \end{vmatrix} = \begin{vmatrix} 1 & -2 & 4 \\ 3 & -6 & 5 \\ 4 & -8 & 16 \end{vmatrix}$ **21** $6 \begin{vmatrix} 3 & 5 & -1 \\ 2 & 4 & 3 \\ 1 & 0 & 2 \end{vmatrix} = \begin{vmatrix} 9 & 15 & -3 \\ 2 & 4 & 3 \\ 2 & 0 & 4 \end{vmatrix}$

22 $\begin{vmatrix} 3 & 5 & 2 \\ 1 & 3 & -4 \\ 0 & 2 & 6 \end{vmatrix} = \begin{vmatrix} 3 & 4 & 2 \\ 1 & 1 & -4 \\ 0 & -1 & 6 \end{vmatrix} + \begin{vmatrix} 3 & 1 & 2 \\ 1 & 2 & -4 \\ 0 & 3 & 6 \end{vmatrix}$

23 $\begin{vmatrix} 2 & 3 & 1 \\ 1 & -1 & 2 \\ -3 & 2 & 3 \end{vmatrix} = \begin{vmatrix} 2 & 9 & 1 \\ 1 & 2 & 2 \\ -3 & -7 & 3 \end{vmatrix}$

Find replacements for x, y, and z by use of determinants, so that the statements in Probs. 24 to 30 are true.

24 $\begin{vmatrix} x & 5 \\ 4 & 3 \end{vmatrix} = 1$

25 $\begin{vmatrix} x & 1 & 3 \\ 2 & -1 & 5 \\ 2 & -3 & 4 \end{vmatrix} = 1$

26 $\begin{vmatrix} x & 0 & 3 \\ -1 & 2 & x \\ 4 & -2 & 1 \end{vmatrix} = 6$

27 $\begin{aligned} 3x + y &= 3 \\ 2x - 3y &= 13 \end{aligned}$

28 $\begin{aligned} 2x - 3y &= -9 \\ 3x + y &= 14 \end{aligned}$

29 $\begin{aligned} x + y + z &= 4 \\ x \quad\ + 2z &= 11 \\ 2x - y \quad &= 4 \end{aligned}$

30 $\begin{aligned} 2x + 3y - z &= 10 \\ 3x - 2y + 4z &= -7 \\ x + 5y - 2z &= 11 \end{aligned}$

31 Solve the system of Prob. 29 by use of row operations on matrices.

32 Solve the system of Prob. 30 by use of row operations on matrices.

CHAPTER ELEVEN

Logarithms

Logarithms were invented in the seventeenth century, and they are very useful and efficient for arithmetical computation. They are important in the application of mathematics to chemistry, physics, and engineering and are indispensable for some parts of advanced mathematics. In this chapter we shall develop some of the properties of logarithms and show how they are used in numerical computation. We shall also use logarithms to obtain the solution sets of certain types of equations. The theory of logarithms is based on the laws of exponents, and the reader is advised to review these laws at this time.

11.1 / APPROXIMATIONS

In Sec. 1.3 we stated that $\sqrt{18}$ correct to five decimal places is 4.24264 and that $\sqrt{18}$ cannot be expressed *exactly* as a decimal fraction. This situation illustrates the difference between an exact number and an approximation. The number $\sqrt{18}$ is exact, and 4.24264 is an approximation to $\sqrt{18}$. Frequently, in arithmetic computation an approximation to a fraction is used. For example, 0.667 might be used for $\frac{2}{3}$. The exact ratio of the circumference of a circle to the diameter is designated by the Greek letter π, but in computation, the approximation 3.1416 is usually employed.

A measurement depends upon several variable conditions, such as the accuracy of the instrument employed, the temperature and atmospheric conditions, and the skill and care of the person making the measurement. Consequently, all numbers obtained by measurement are approximations. If the thickness of a sheet of metal is measured with a micrometer calibrated in thousandths of an inch and found to be 0.053, it is understood that $0.0525 \leq$ exact thickness < 0.0535. Similarly, if the length and width of a rectangle are measured with a ruler calibrated in tenths of a centimeter and found to be 10.2 cm and 5.1 cm, respectively, then

$$10.15 \leq \text{exact length} < 10.25 \quad \text{and} \quad 5.05 \leq \text{exact width} < 5.15 \quad (1)$$

Now, in connection with the statement that the area is $10.2 \times 5.1 = 52.02$ sq cm, the question arises, "How many digits in 52.02 are accurate?" In order to answer this question, we refer to (1) and see that $10.15 \times 5.05 \leq$ exact area $< 10.25 \times 5.15$, or

$$51.2575 \leq \text{exact area} < 52.7875 \quad (2)$$

Since the first number in (2) is slightly more than 51 and the last is nearly 53, we cannot be sure of the exact area even to two digits. We shall say, however, that the exact area is probably nearer to 52 sq cm than to any other two-digit number. This discussion illustrates the following definition:

Significant digits   In a number that represents an approximation, the digits known to be correct are called *significant*.

The digits 1, 2, 3, 4, 5, 6, 7, 8, and 9 in a number obtained by measurement are always significant, as are any zeros between any two of them. Zeros whose only function is to place the decimal point are never significant. Such zeros occur between the decimal point and the first nonzero digit and only in numbers between 1 and -1.

In computation involving approximate numbers, we do not employ a digit if its accuracy is doubtful. The process of eliminating these digits is

Rounding off   called *rounding off a number* to n significant digits and consists of the following steps:

1   If the decimal point is to the right of the nth digit, replace all digits between the nth digit and the decimal point by zeros and discard all digits to the right of the decimal point.

2   If the decimal point is to the left of the nth digit, discard all digits to the right of the nth digit.

3   If the first discarded digit is 5, 6, 7, 8, or 9, increase the nth digit by 1.

4   If the first discarded digit is 0, 1, 2, 3, or 4, the nth digit is not changed.

EXAMPLE 1 In rounding off 48.157 to three digits, we first notice that the decimal point is to the left of the third digit. Hence, we discard all digits to the right of the third. We increase the last digit retained by 1, since the first digit discarded is 5. Then, we get 48.2 as the value of 48.157 rounded off to three digits.

EXAMPLE 2 In rounding off 3.842 to two digits, we observe that the decimal point is to the left of the second digit. Hence, we discard all digits to the right of the second. We do not change the second digit, since the first digit discarded is 4. Therefore, we get 3.8 by rounding off 3.842 to two digits.

EXAMPLE 3 In rounding off 538,762 to four digits, we see that the decimal point is to the right of the fourth digit. Hence, we replace all digits to the right of the fourth by zeros; furthermore, we increase the fourth digit by 1 since the first discarded digit is 6. Thus, 538,800 is the number obtained by rounding off 538,762 to four digits.

Computation with approximate numbers

As illustrated above, the result of a computation depends upon the number of significant digits in the numbers involved. We shall now present three rules that are employed in such cases. In the statement of these rules, we employ the word "accurate." The degree of accuracy of a number is determined by the number of significant digits in it.

Rule 1 *If the numbers of significant digits in M and in N are equal or differ by 1, round off the product MN and the quotient M/N to the number of significant digits in the less accurate one of M and N.*

Rule 2 *If the numbers of significant digits in M and in N differ by 2 or more, round off the one with the largest number of significant digits so that it contains only one more significant digit than the other and then proceed as in Rule 1.*

EXAMPLE 4 Find the product of 27.3 and 145.8, and round it off to the correct number of significant digits.

Solution We notice that 27.3 contains three significant digits and 145.8 contains four, so we obtain the product and round it off to three significant digits; thus:

$$27.3 \times 145.8 = 3,980.34 = 3,980 \qquad \text{rounded off to three significant digits}$$

In the product 3,980, the final zero is not significant.

EXAMPLE 5 Obtain the quotient of 583.27 and 34.7 and round it off to the correct number of significant digits.

Solution In this case, the number of significant digits in the dividend 583.27 is two more than the number in the divisor 34.7. Consequently, we round 583.27 off to 583.3 since the latter number contains one more significant digit than 34.7. Finally, we proceed with the division and round the quotient off to three digits. We thus obtain

$$\frac{583.27}{34.7} = \frac{583.3}{34.7} = 16.81 = 16.8 \qquad \text{rounded off to three digits}$$

In addition, we are not primarily interested in the number of significant digits in the addends but rather in the precision of each addend. The precision of an approximate number depends upon the number of decimal places in it. For example, we say that 34.2 in. is precise to tenths of an inch and 34.23 in. is precise to hundredths of an inch. Of course, 34 in. is precise to inches. Consequently, in addition, we are concerned with the number of decimal places in each addend since we add digits having the same place value. For this reason, the following rule is usually employed in adding approximate numbers.

Rule 3 *Round off each addend to one more decimal place than there is in the addend which has the least number of decimal places. Add the resulting numbers, and round off the last digit in the sum. A similar procedure is used in subtraction.*

EXAMPLE 6 Find the sum of 28.72, 3.683, 7.2, and 23.7864.

Solution The third number contains only one decimal place; hence the others should be rounded off to two decimal places before adding. If this is done, we get the addends and sum shown below, and after rounding off to one decimal place, we obtain the sum 63.4.

$$\begin{array}{r} 28.72 \\ 3.68 \\ 7.2 \\ \underline{23.79} \\ 63.39 \end{array}$$

11.2 / SCIENTIFIC NOTATION

In the previous section we stated the conditions under which some zeros in an approximate number are significant and others are not. We did not discuss the significance of the final zero or zeros in a number. The fact is that in some cases the final zero or zeros in a number are significant and in other cases they are not. For example, if the approximate number 1,269.8 is rounded off to four digits, we have 1,270 and the zero is significant. If, however, 1,269.8 is rounded off to two digits, we have 1,300 and neither zero is significant. This ambiguity with respect to the final zeros in a number is removed if the number is expressed in *scientific notation*. A number is in scientific notation if it is expressed as the product of two factors, with the first factor a number equal to or greater than 1 but less than 10 and with all its digits significant, and with the other factor an integral power of 10. For example, the numbers 1,270 and 1,300 mentioned above expressed in scientific notation are 1.270×10^3 and 1.3×10^3, respectively. Note that in the first number, all four digits are significant; in the second, only two digits are significant. Before explaining the procedure for expressing a number in scientific notation, we shall define the reference position for the decimal point.

Reference position for the decimal point

The *reference position* for the decimal point in a number N is the position between the first significant digit and the second digit in N.

For example, the reference position for the decimal point in each of the following numbers is indicated by a caret:

$$2 _\wedge 16.34 \qquad 3 _\wedge 021 \qquad 0.004 _\wedge 623$$

The scientific notation for the number N is $N'(10^c)$, with all significant digits in N appearing in N' and with the decimal point in N' in reference position. The exponent c is equal to the number of digits between the reference position and the decimal point in N, and it is positive or negative according as the decimal point is to the right or to the left of the reference position.

The definition is illustrated below.

Number	Scientific Notation
312	3.12×10^2
1,235	1.235×10^3
0.0621	6.21×10^{-2}
0.00004326	4.326×10^{-5}
1,250 (zero significant)	1.250×10^3
3,650 (zero not significant)	3.65×10^3

EXAMPLE Convert $M = 3,270$ (zero not significant) and $N = 43$ to scientific notation, and then find the product MN. Leave the result in scientific notation.

Solution $M = 3.27 \times 10^3$, $N = 4.3 \times 10$. Hence

$$
\begin{aligned}
MN &= 3.27 \times 10^3 \times 4.3 \times 10 \\
&= 3.27 \times 4.3 \times 10^4 \\
&= 14.061 \times 10^4 \\
&= 14 \times 10^4 \\
&= 1.4 \times 10 \times 10^4 \\
&= 1.4 \times 10^5
\end{aligned}
$$

since $3.27 \times 4.3 = 14.061$
rounding off to two digits
since $14 = 1.4 \times 10$

EXERCISE 11.1 Approximations

Express the number in each of Probs. 1 to 20 in scientific notation. Assume that all final zeros are significant unless otherwise stated.

1	2,856	2	59,437
3	648	4	17
5	21.3	6	402.1
7	3,030	8	826.3
9	0.475	10	0.09160
11	0.00704	12	0.7429
13	8,430	14	6,700
15	3,060	16	5,000

17 8,430 (zero not significant)
18 6,700 (last zero not significant)
19 3,060 (last zero not significant)
20 5,000 (last zero not significant)

Round off the number in each of Probs. 21 to 36 to four significant digits. Use scientific notation if needed.

21	596.348	22	67.1253	23	7.8435	24	0.58449
25	380.271	26	29.9248	27	0.012345	28	0.0012354
29	37,628	30	4,973.7	31	23,456.7	32	975,864
33	5,555.5	34	55,555	35	48,357.1	36	483.571

Assume that all the numbers in Probs. 37 to 68 are approximations, and perform the indicated operation after appropriate rounding off. Round off each result to the proper number of digits, and express it in scientific notation.

37	$(3.4)(5.7)$	38	$(4.9)(6.2)$	39	$(7.1)(1.6)$
40	$(8.3)(3.8)$	41	$(45.6)(6.53)$	42	$(3.72)(2.56)$
43	$(80.3)(57.9)$	44	$(613)(7.24)$	45	$(7.6)(8.14)$
46	$(38.4)(4.8)$	47	$(58.3)(3.852)$	48	$(71.86)(6.81)$
49	$(4.7)(3.845)$	50	$(97)(19.64)$	51	$(80.07)(6.3)$
52	$(54.31)(7.9)$	53	$9.6 \div 3.7$	54	$43 \div 3.4$
55	$74 \div 4.18$	56	$64.1 \div 5.9$	57	$87 \div 59.67$
58	$73 \div 3.204$	59	$918.3 \div 29$	60	$37 \div 0.4076$

61	$3.7 + 8.14 + 5.43$	62	$78.3 + 3.27 + 1.86$
63	$96.81 - 43.52 + 19.1$	64	$49.7 + 32.84 - 17.63$
65	$5.803 + 2.71 - 6.3$	66	$8.6435 - 1.72 + 2.3752$
67	$8.732 - 4.1 + 3.72$	68	$99.546 + 3.74 - 4.5$

11.3 / DEFINITIONS

In the statement

$$2^3 = 8 \tag{11.1}$$

we use the term "exponent" to indicate the relationship that exists between 3 and 2. However, 3 is also related to 8 in Eq. (11.1), in that it is the exponent of the power to which 2 must be raised to produce 8; we indicate this relationship by the term "logarithm." In other words, in Eq. (11.1), 3 is the exponent of 2 and also the logarithm of 8, or more precisely, the logarithm to the base 2 of 8. In this chapter, we shall show how exponents, or logarithms, can be used to simplify numerical computation and to solve certain types of equations that are not solvable by more elementary methods.

As a basis for our subsequent discussion, we present the following definition:

Logarithm

The *logarithm* of a positive number for a given base is the exponent that indicates the power to which the base must be raised in order to obtain the number.

Symbolic definition of logarithm

The abbreviated form of the statement, "The logarithm to the base b of N is L" is $\log_b N = L$. If we use this notation, we can express the definition in symbolic form as

$$\log_b N = L \text{ if and only if } b^L = N \tag{11.2}$$

Thus, we have, by definition, $b^{\log_b N} = N$

Note that the abbreviation "log" appears without a period and that the symbol for the base appears as a subscript. The following are examples of logarithms:

$$\log_8 64 = 2 \qquad \text{since } 8^2 = 64$$
$$\log_4 64 = 3 \qquad \text{since } 4^3 = 64$$
$$\log_{81} 9 = \tfrac{1}{2} \qquad \text{since } 81^{1/2} = 9$$
$$\log_a 1 = 0 \qquad \text{since } a^0 = 1$$

From these examples it can be seen that a logarithm may be integral or fractional. Furthermore, in many cases, if two of the three letters in Eq. (11.2) are known, the third can be found by inspection.

EXAMPLE 1   Find the value of N if $\log_7 N = 2$.

Solution   If we convert the above logarithmic statement to the exponential form, we have $7^2 = N$, and then it is obvious that $N = 49$.

EXAMPLE 2 If $\log_b 125 = 3$, find the value of b.

Solution By use of Eq. (11.2), we have

$$b^3 = 125$$
$$b = \sqrt[3]{125} = 5 \qquad \text{solving for } b$$

EXAMPLE 3 Find a if $\log_{27} 3 = a$.

Solution Again, by using Eq. (11.2), we have

$$3 = 27^a = (3^3)^a$$

Hence, it follows that $3a = 1$ and $a = \frac{1}{3}$.

11.4 / PROPERTIES OF LOGARITHMS

In this section we shall employ the laws of exponents and the definition of a logarithm to derive three important properties of logarithms. In Sec. 11.9 we shall show how to use these properties in numerical computation.

We first show how to find the logarithm of a product of two numbers in terms of the logarithms of the two numbers.

If we are given

$$\log_b M = m \qquad \text{and} \qquad \log_b N = n \qquad\qquad (11.3)$$

then by Eq. (11.2) we have

$$M = b^m \qquad \text{and} \qquad N = b^n \qquad\qquad (11.4)$$

Hence, $\quad MN = (b^m)(b^n)$
$$= b^{m+n} \qquad \text{by Eq. (2.38)}$$

and $\quad \log_b MN = m + n \qquad \text{by Eq. (11.2)}$
$$= \log_b M + \log_b N \qquad \text{by Eq. (11.3)}$$

Consequently, we have the following property:

Logarithm of a product **1** *The logarithm of the product of two numbers is equal to the sum of the logarithms of the numbers.*

EXAMPLE 1 If $\log_{10} 3 = 0.4771$ and $\log_{10} 4 = 0.6021$, find $\log_{10} 12$.

Solution
$$\log_{10} 12 = \log_{10}(3 \times 4) = \log_{10} 3 + \log_{10} 4$$
$$= 0.4771 + 0.6021$$
$$= 1.0792$$

This means that $12 = 10^{1.0792}$.

Property 1 can be extended to three or more numbers by the following process:

$$\log_b MNP = \log_b (MN)(P)$$
$$= \log_b M + \log_b N + \log_b P$$

Again, by using Eqs. (11.4), we have

$$\frac{M}{N} = \frac{b^m}{b^n}$$
$$= b^{m-n} \qquad \text{by Eq. (2.47)}$$

Hence, $\qquad \log_b \dfrac{M}{N} = m - n \qquad$ by Eq. (11.2)

$$= \log_b M - \log_b N \qquad \text{by Eq. (11.3)}$$

Therefore, we have the following property:

Logarithm of a quotient **2** *The logarithm of the quotient of two numbers is equal to the logarithm of the numerator minus the logarithm of the denominator.*

EXAMPLE 2 If $\log_{10} 3 = 0.4771$ and $\log_{10} 2 = 0.3010$, find $\log_{10} (3 \div 2)$.

Solution
$$\log_{10} (3 \div 2) = \log_{10} \tfrac{3}{2}$$
$$= \log_{10} 3 - \log_{10} 2$$
$$= 0.4771 - 0.3010$$
$$= 0.1761$$

Finally, if we raise both members of $M = b^m$ to the kth power, we have

$$M^k = (b^m)^k$$
$$= b^{km} \qquad \text{by Eq. (2.40)}$$

Therefore, $\qquad \log_b M^k = km \qquad$ by Eq. (11.2)

$$= k \log_b M \qquad \text{by Eq. (11.3)}$$

Thus, we have the following property:

Logarithm of a power **3** *The logarithm of a power of a number is equal to the product of the exponent of the power and the logarithm of the number.*

EXAMPLE 3 Find the logarithm of the square of 3.

Solution
$$\log_{10} 3^2 = 2 \log_{10} 3$$
$$= 2 \times 0.4771$$
$$= 0.9542$$

NOTE: Since a root of a number can be expressed as a fractional power, Property 3 can be used to find the logarithm of a root of a number. Thus,

$$\log_b \sqrt[r]{M} = \log_b (M)^{1/r} = \frac{1}{r} \log_b M$$

EXAMPLE 4   Find the logarithm of the cube root of 2.

Solution

$$\begin{aligned}
\log_{10} \sqrt[3]{2} &= \log_{10} 2^{1/3} \\
&= \tfrac{1}{3} \log_{10} 2 \\
&= \tfrac{1}{3} \times 0.3010 \\
&= 0.1003
\end{aligned}$$

For the convenience of the reader, we shall restate the above three properties in symbolic form, and we shall hereafter refer to them by the numbers of the equations.

$$\log_b MN = \log_b M + \log_b N \qquad \text{Property 1} \qquad \textbf{(11.5)}$$

$$\log_b \frac{M}{N} = \log_b M - \log_b N \qquad \text{Property 2} \qquad \textbf{(11.6)}$$

$$\log_b M^k = k \log_b M \qquad \text{Property 3} \qquad \textbf{(11.7)}$$

EXERCISE 11.2 Conversion of Exponential and Logarithmic Forms

Change the statement in each of Probs. 1 to 16 to logarithmic form, and the statement in each of Probs. 17 to 28 to exponential form, by making use of the definition of the logarithm of a number to a given base as expressed in Eq. (11.2).

1	$2^3 = 8$	2	$3^4 = 81$	3	$4^5 = 1{,}024$
4	$7^2 = 49$	5	$5^{-1} = \frac{1}{5}$	6	$6^{-2} = \frac{1}{36}$
7	$3^{-5} = \frac{1}{243}$	8	$2^{-4} = \frac{1}{16}$	9	$(\frac{1}{3})^{-3} = 27$
10	$(\frac{1}{2})^{-5} = 32$	11	$(\frac{3}{5})^{-3} = \frac{125}{27}$	12	$(\frac{4}{7})^{-2} = \frac{49}{16}$
13	$16^{3/4} = 8$	14	$27^{2/3} = 9$	15	$64^{5/6} = 32$
16	$25^{3/2} = 125$	17	$\log_3 9 = 2$	18	$\log_5 125 = 3$
19	$\log_2 64 = 6$	20	$\log_7 343 = 3$	21	$\log_5 \frac{1}{5} = -1$
22	$\log_3 \frac{1}{27} = -3$	23	$\log_5 \frac{1}{25} = -2$	24	$\log_{10} 0.001 = -3$
25	$\log_{27} 9 = \frac{2}{3}$	26	$\log_{36} 216 = \frac{3}{2}$	27	$\log_8 32 = \frac{5}{3}$
28	$\log_{32} 64 = \frac{6}{5}$				

In each of Probs. 29 to 52, find the replacement for L, N, or b so that the statement is true.

29	$\log_4 16 = L$	30	$\log_5 125 = L$	31	$\log_7 49 = L$

32 $\log_2 64 = L$ **33** $\log_9 27 = L$ **34** $\log_8 16 = L$
35 $\log_{81} 729 = L$ **36** $\log_{81} 27 = L$ **37** $\log_5 N = 3$
38 $\log_3 N = 4$ **39** $\log_7 N = 2$ **40** $\log_6 N = 1$
41 $\log_8 N = \frac{2}{3}$ **42** $\log_{16} N = \frac{3}{4}$ **43** $\log_{49} N = \frac{3}{2}$
44 $\log_4 N = \frac{5}{2}$ **45** $\log_b 125 = 3$ **46** $\log_b 64 = 6$
47 $\log_b 81 = 4$ **48** $\log_b 49 = 2$ **49** $\log_b \frac{8}{125} = 3$
50 $\log_b \frac{4}{9} = 2$ **51** $\log_b \frac{216}{343} = 3$ **52** $\log_b \frac{243}{32} = 5$

Given that $\log_2 4 = 2$, $\log_2 8 = 3$, $\log_2 64 = 6$, $\log_2 256 = 8$, and $\log_2 8,192 = 13$, obtain the logarithm of the combination of numbers in each of Probs. 53 to 76.

53 $\log_2 (4 \times 256)$ **54** $\log_2 (4 \times 8 \times 64)$
55 $\log_2 (8 \times 64 \times 256)$ **56** $\log_2 (4 \times 256 \times 8,192)$
57 $\log_2 (4 \times 64 \times 8,192)$ **58** $\log_2 (8 \times 64 \times 8,192)$

59 $\log_2 (32 \times 64 \times 256)$ **60** $\log_2 (2 \times 16 \times 128)$ **61** $\log_2 \dfrac{8,192}{8}$

62 $\log_2 \dfrac{256}{4}$ **63** $\log_2 \dfrac{128}{2}$ **64** $\log_2 \dfrac{1,024}{32}$

65 $\log_2 8^3$ **66** $\log_2 256^2$ **67** $\log_2 8,192^2$

68 $\log_2 512^4$ **69** $\log_2 \sqrt{256}$ **70** $\log_2 \sqrt[3]{8,192}$

71 $\log_2 \sqrt[5]{64}$ **72** $\log_2 \sqrt[4]{8,192}$ **73** $\log_2 \dfrac{4^2 \times 8}{256}$

74 $\log_2 \dfrac{4^3 \times 256}{8,192}$ **75** $\log_2 \dfrac{8\sqrt{256}}{64}$ **76** $\log_2 \dfrac{4\sqrt{8,192}}{256}$

11.5 / COMMON, OR BRIGGS, LOGARITHMS

Common, or Briggs, logarithms

We stated previously that logarithms can be employed efficiently in numerical computation, and we shall use logarithms to the base 10 for this purpose. If the base is 10, the logarithm of a number is called the *common*, or *Briggs*, logarithm.

Most common logarithms of numbers that are not integral powers of 10 cannot be computed by elementary methods. Tables have been prepared that enable us to obtain a decimal approximation to the common logarithm of any positive number. These tables can also be used to find a number if its common logarithm is known. The method for using these tables will be explained in the next three sections.

If c is a real number, then $10^c > 0$. Hence the common logarithm of zero or of a negative number does not exist as a real number.

It is customary to omit the subscript indicating the base in the notation for a common logarithm. Consequently, in the statement $\log N = L$, it will be understood hereafter that the base is 10.

11.6 / CHARACTERISTIC AND MANTISSA

If we express a positive number $N \neq 1$ in scientific notation, we have

$$N = N'(10^c) \qquad \text{where } 1 \leq N' < 10 \text{ and } c \text{ is an integer} \qquad (1)$$

We shall now consider the following three situations:

1   If $N \geq 10$, then in (1), $c \geq 1$.   For example, $231 = 2.31 \times 10^2$.

2   If $N < 1$, then $c < 0$.   For example, $0.0231 = 2.31 \times 10^{-2}$.

3   If $1 < N < 10$, then in (1), $N = N'$ and $c = 0$.   For example, $2.31 = 2.31 \times 10^0$, since $10^0 = 1$.

Now, if we equate the common logarithms of the members of (1), we have

$$
\begin{aligned}
\log N &= \log N' + \log 10^c & \text{by Eq. (11.5)} \\
&= \log N' + c \log 10 & \text{by Eq. (11.7)} \\
&= \log N' + c & \text{since } \log 10 = 1
\end{aligned}
$$

Thus, by the commutative property of addition, we have

$$\log N = c + \log N' \qquad (2)$$

Since $1 \leq N' < 10$, it follows that $10^0 \leq N' < 10^1$, and hence $0 \leq \log N' < 1$. By use of Table A.1 in the Appendix, we can obtain a decimal approximation to the common logarithm of any number between 0 and 1 (correct to four decimal places).   Consequently, by referring to (2), we see that the common logarithm of any positive number not equal to 1 can be expressed approximately as an integer plus a positive decimal fraction.   Since $1 = 10^0$, $\log 1 = 0$.   Thus, for $\log 1$, the integer is zero and the decimal fraction is zero.   We are now in position to state the following definition:

Characteristic and mantissa
If the common logarithm of a positive number is expressed as an integer plus a nonnegative decimal fraction, the integer is called the *characteristic* of the logarithm and the decimal fraction is called the *mantissa*.

In the expression for $\log N$ in (2), the characteristic is the integer c and the mantissa is $\log N'$, where c is the exponent of 10 in the scientific notation for N.   In Sec. 11.2 we demonstrated that c is numerically equal to the number of digits between the reference position and the decimal point in N and that it is positive or negative according as the decimal point is to the right or to the left of the reference position.   Therefore, we have the following rule for determining the characteristic of the common logarithm of a positive number.

Rule for determining the characteristic
The characteristic of the common logarithm of a positive number N is numerically equal to the number of digits between the reference position and the decimal point in N and is positive or negative according as the decimal point is to the right or to the left of the reference position.

The following examples illustrate the method for finding the characteristic.

EXAMPLE 1   The reference position in 436.78 is between 4 and 3.   Hence there are two digits, 3 and 6, between the reference position and the decimal point. Furthermore, the decimal point is to the right of the reference position. Therefore, the characteristic of the common logarithm of 436.78 is 2.

EXAMPLE 2   The characteristic of the common logarithm of 4.3678 is zero since the decimal point is in reference position.

EXAMPLE 3   The decimal point in 0.0043678 is three places to the left of the reference position.   Therefore, the characteristic of log 0.0043678 is −3.

Since the position of the decimal point in the number N affects only the integer c in the scientific notation for N and the mantissa of $\log N$ is $\log N'$, the mantissa of N depends only upon the sequence of digits in N.

If $N \geq 1$, the characteristic of $\log N$ is zero or a positive integer.   Therefore, $\log N$ can be written as a single number.   For example, if the mantissa of $\log 23678$ is 0.3743, then $\log 236.78 = 2 + 0.3743 = 2.3743$, and $\log 2.3678 = 0 + 0.3743 = 0.3743$.

With a positive number less than 1, however, we have a different situation.   For example, $\log 0.0023678 = -3 + 0.3743$, and we can express this logarithm as a single number only in this way: $-3 + 0.3743 = -(3 - 0.3743) = -2.6257$.   However, since $-2.6257 = -2 - 0.6257$, the decimal fraction is negative and, consequently, is not a mantissa.   For this reason, we use the following device for dealing with such situations: If the characteristic of $\log N$ is $-c$, $c > 0$, we express $-c$ in the form $n - 10$, $n > 0$; then we write the mantissa at the right of n.   Consequently, we express $\log 0.0023678$ in the form $7.3743 - 10$ since $7 - 10 = -3$.   Similarly, $\log 0.23678 = 9.3743 - 10$, $\log 0.023678 = 8.3743 - 10$, and $\log 0.0000023678 = 4.3743 - 10$.

11.7 / USE OF TABLES TO OBTAIN THE MANTISSA

In this section we shall explain the method of finding the mantissa of a logarithm by use of Table A.1 in the Appendix.   We shall first discuss numbers of only three digits† and as a specific example shall consider 3.27.   As we stated in Sec. 11.6, the mantissa

---

† In this section, when we count the number of digits in a decimal fraction, we start with the first nonzero digit.

of the logarithm of a number is not affected by the position of the decimal point in the number. Hence, for the present, we shall disregard the decimal point. We now turn to Table A.1 and look in the column headed by N on the left side of the page for the first two digits, 32, of the number 327. Then in line with this and across the page in the column headed by the third digit, 7, we find the entry 5145. Except for the decimal point† before it, this is the desired mantissa. Since the decimal point in 3.27 is in the reference position, the characteristic of the logarithm is zero. Hence, $\log 3.27 = 0.5145$.

As a second example, consider 0.00634. Again, we temporarily disregard the decimal point and find 63 in the column headed by N. In line with this and in the column headed by 4, we find the entry 8021. By the rule in Sec. 11.6, the characteristic of the logarithm of 0.00634 is -3, which we write as $7 - 10$. Hence, $\log 0.00634 = 7.8021 - 10$.

If a number is composed of fewer than three digits, we mentally annex one or two zeros at the right and proceed as before. For example, to get the mantissa of the logarithm of 72, we look up 720; to get the mantissa of the logarithm of 3, we look up 300.

If all the digits in a number after the third are zeros, we disregard them in the process of getting the mantissa.

In the discussion that follows, it will be necessary to use the expression "the mantissa of the logarithm of N" frequently. For the sake of brevity, we shall abbreviate the expression to ml N.

Linear interpolation

If a number is composed of four digits, we obtain the mantissa by a method known as *linear interpolation*. We shall explain and illustrate the method with two examples.

EXAMPLE 1 Find ml 412.8.

Solution Since the number 412.8 has four digits, we must use the interpolation method to find ml 412.8. Since 412.8 is between 412 and 413, ml 412.8 is between ml 412 and ml 413. Furthermore, 413 differs from 412 by 1, and 412.8 differs from 412 by 0.8. Hence, 412.8 differs from 412 by 0.8 of the difference between 412 and 413. We assume that ml 412.8 differs from ml 412 by 0.8 of the difference between ml 412 and ml 413. We now turn to the table and find that ml 413 = 0.6160 and that ml 412 = 0.6149.

Hence, since $0.6160 - 0.6149 = 0.0011$, ml 412.8 differs from ml 412 by

$$0.8 \times 0.0011 = 0.00088 = 0.0009 \qquad \text{to four decimal places}$$

Therefore, ml 412.8 = 0.6149 + 0.0009 = 0.6158.

The above procedure can be condensed into the following form, in which the calculations can be performed easily:

---

† No decimal points are printed in the table of mantissas. Hence, when a mantissa is obtained from the table, a decimal point must be placed to the left of it.

$$1 \begin{bmatrix} \text{ml } 413 & = 0.6160 \\ 0.8 \begin{bmatrix} \text{ml } 412.8 = \\ \text{ml } 412 & = 0.6149 \end{bmatrix} 0.0011 \end{bmatrix}$$

$$0.8 \times 0.0011 = 0.00088 = 0.0009 \qquad\qquad \text{to four decimal places}$$
$$\text{ml } 412.8 = 0.6149 + 0.0009 = 0.6158$$

Hence, since the decimal point in 412.8 is two places to the right of the reference position, we have log 412.8 = 2.6158.

EXAMPLE 2 Find ml 0.006324.

Solution Since the position of the decimal point has no effect on the mantissa, it follows that

$$\text{ml } 0.006324 = \text{ml } 632.4$$

Then the process of obtaining the latter mantissa is the same as that in Example 1. We note that 632.4 is between 632 and 633. Hence, we have

$$1 \begin{bmatrix} \text{ml } 633 & = 0.8014 \\ 0.4 \begin{bmatrix} \text{ml } 632.4 = \\ \text{ml } 632 & = 0.8007 \end{bmatrix} 0.0007 \end{bmatrix}$$

$$0.4 \times 0.0007 = 0.00028 = 0.0003 \qquad \text{to four decimal places}$$

Consequently,   ml 632.4 = 0.8007 + 0.0003 = 0.8010

Since the decimal point in 0.006324 is three places to the left of the reference position, we have log 0.006324 = 7.8010 − 10.

The interpolation process consists of simple operations which, after some practice, can be performed mentally, thus saving considerable time. We suggest the following steps for finding the mantissa of a four-digit number:

Steps in interpolation 1 Temporarily place the decimal point between the third and fourth digits of the given number.

2 Find the difference between the mantissas of the logarithms of the two 3-digit numbers between which the given number lies.

3 Multiply the difference between the two mantissas by the fourth digit of the given number considered as a decimal fraction.

4 Add the product obtained in step 3 to the smaller of the mantissas in step 2.

If a number contains more than four digits, we round it off to four places and proceed as before. For example, to get ml 17.6352, we find ml 17.64.†

EXERCISE 11.3 **Characteristic and Mantissa**

The number given in each of Probs. 1 to 8 is log N; find the mantissa and the characteristic.

1	1.3746	2	8.4137	3	2.5709	4	0.5028
5	$7.3814 - 10$	6	-3.3814	7	-2.7667	8	$8.7667 - 10$

Find the characteristic of the common logarithm of the number in each of Probs. 9 to 24.

9	48.2	10	4.82	11	7,843	12	603.4
13	5.973	14	30.57	15	76,596.3	16	76.5963
17	0.905	18	0.095	19	0.0001	20	0.023
21	0.1004	22	0.8177	23	0.003003	24	0.0087

Find the common logarithm of the number in each of Probs. 25 to 68.

25	764	26	8.03	27	57.2	28	63.9
29	47.2	30	30.8	31	107	32	2.53
33	29	34	96	35	4.5	36	6
37	4,682	38	159.3	39	78.47	40	5,974
41	253.7	42	8,628	43	38.75	44	444.5
45	1.23759	46	978.6742	47	5346.73	48	6545.68
49	0.03	50	0.076	51	0.008	52	0.0072
53	0.875	54	0.0964	55	0.483	56	0.00597
57	0.1238	58	0.04796	59	0.08084	60	0.8763
61	0.87654	62	0.87645	63	0.04683	64	0.001358
65	0.0234567	66	0.0234576	67	0.5798642	68	0.00542329

11.8 / USE OF TABLES TO FIND *N* WHEN log *N* IS GIVEN

The next problem in the use of the tables is the process of finding a number when its logarithm is given. We shall illustrate the process in several examples and shall explain each step as we proceed.

† Since the result of a computation obtained by use of a four-place table of logarithms is accurate to only four significant digits, we round off all numbers involved in the computation to four digits before applying logarithms.

EXAMPLE 1   Find N if $\log N = 1.6191$.

Solution   The first step in finding N when $\log N = 1.6191$ is to find the mantissa, 0.6191, in the body of the tables. Hence, we look through the tables until we find the mantissas starting with 61, and then we look through these until we locate 6191. We see that it is in line with 41 (in the column headed by N) and is in the column headed by 6. Thus, N is made up of the digits 416, and the next step is to place the decimal point. Since the characteristic of $\log N$ is 1, the decimal point is one place to the right of the reference position and hence is between 1 and 6. Therefore, $N = 41.6$.

If ml N is not listed in the tables, we must resort to interpolation. By the use of a four-place table, we cannot obtain accurately more than the first four digits in N. We shall show in the following discussion that the first three digits are obtained from the table and the fourth is determined by interpolation. If the characteristic of $\log N$ indicates that N contains more than four digits, then zeros are added after the fourth, or scientific notation is used.

EXAMPLE 2   Find N if $\log N = 5.4978$.

Solution   To find N when $\log N = 5.4978$, we shall let T represent the number composed of the first four digits in N and shall determine T. Finally, we shall place the decimal point by considering the characteristic and thus get N. The mantissa 0.4978 is not listed in the table, but the two mantissas nearest to it are 0.4969 and 0.4983. These two mantissas are ml 3,140 and ml 3,150, respectively. (We add the zero in each case in order to obtain four places for use in interpolation.) Furthermore, since $4,983 - 4,969 = 14$ and $4,978 - 4,969 = 9$, it follows that 4,978 is $\frac{9}{14}$ of the way from 4,969 to 4,983. Hence, T is approximately $\frac{9}{14}$ of the way from 3,140 to 3,150. Furthermore,

$$\tfrac{9}{14} \times 10 = \tfrac{90}{14} = 6.4 = 6 \qquad \text{to one digit because the fourth place}$$
$$\text{is the limit of our accuracy}$$

and $3{,}140 + 6 = 3{,}146$. Hence, $T = 3{,}146$. Since the characteristic of $\log N$ is 5, the decimal point in N is five places to the right of the reference position. Hence, $N = 314{,}600$.

The steps in the above process are shown in the condensed form given below.

$$\log N = 5.4978$$
$$14\begin{bmatrix} 0.4983 = \text{ml } 3{,}150 \\ 9\begin{bmatrix} 0.4978 = \text{ml } T \\ 0.4969 = \text{ml } 3{,}140 \end{bmatrix}10 \end{bmatrix}$$

$$\tfrac{9}{14} \times 10 = 6.4 = 6 \qquad \text{to one digit}$$
$$3{,}140 + 6 = 3{,}146 = T$$

Hence $$N = 314{,}600 = 3.146 \times 10^5$$

since the characteristic of $\log N$ is 5.

EXAMPLE 3   Find N if $\log N = 8.6736 - 10$.

Solution   To determine N if $\log N = 8.6736 - 10$, we again let T represent the number composed of the first four digits in N. The two entries in the table that are nearest to 0.6736 are $0.6730 = \text{ml } 4{,}710$ and $0.6739 = \text{ml } 4{,}720$. By using these in the interpolation process, we have

$$9\left[\begin{array}{c} 0.6739 = \text{ml } 4{,}720 \\ 6\left[\begin{array}{c} 0.6736 = \text{ml } T \\ 0.6730 = \text{ml } 4{,}710 \end{array} \right. \end{array} \right]10$$

$$\tfrac{6}{9} \times 10 = 6.7 = 7 \qquad \text{to the nearest integer}$$

Hence, $$4{,}710 + 7 = 4{,}717 = T$$

Therefore, $$N = 0.04717 = 4.717 \times 10^{-2}$$

since the characteristic of $\log N$ is $8 - 10 = -2$.

EXERCISE 11.4 Given log N, to Find N

If log N is as given in each of Probs. 1 to 20, find N to three significant digits. Express N in scientific notation in Probs. 9 to 20.

1	0.5843	2	1.4409	3	2.6053	4	2.4330
5	1.7574	6	2.8439	7	2.9754	8	0.1367
9	3.5119	10	4.9253	11	3.8842	12	3.6345
13	9.9042 − 10	14	7.5515 − 10	15	8.8162 − 10	16	6.4425 − 10
17	8.2695 − 10	18	6.8082 − 10	19	7.5999 − 10	20	9.6749 − 10

In each of Probs. 21 to 32, log N is given. Find the value of N to three significant digits by using the entry in the table that is at least as near the given value of log N as any other entry in the table. Express N in scientific notation in Probs. 25 to 32.

21	1.3029	22	2.5764	23	0.7708	24	2.1843
25	4.6238	26	3.5136	27	3.2772	28	4.3255
29	8.7203 − 10	30	7.5154 − 10	31	9.8023 − 10	32	6.4004 − 10

By use of interpolation, find the value of N to four significant digits if log N is the number given in each of Probs. 33 to 48. Express N in scientific notation in Probs. 37 to 48.

33	2.4358	**34**	1.5995	**35**	0.2764	**36**	1.1776
37	4.8138	**38**	3.1863	**39**	5.2323	**40**	5.7009
41	8.3286 − 10	**42**	7.3322 − 10	**43**	9.4123 − 10	**44**	6.9195 − 10
45	8.6202 − 10	**46**	7.0257 − 10	**47**	8.4396 − 10	**48**	6.2345 − 10

11.9 / LOGARITHMIC COMPUTATION

As we stated previously, one of the most useful applications of logarithms is in the field of numerical computation. We shall presently explain the methods involved, by means of several examples. However, before considering special problems, we wish to call attention again to the fact that results obtained by the use of four-place tables are correct at most to four places. If the numbers in any computation problem contain only three places, the result is dependable to only three places. If a problem contains a mixture of three-place and four-place numbers, we cannot expect more than three places of the result to be correct, so we round the result off to three places. Hence, in the problems that follow, we shall not obtain any answer to more than four nonzero places, and sometimes not that far. There exist tables from which logarithms can be obtained to five, six, seven, and even more places. If results that are correct to more than four places are desired, such tables should be used. The methods which we have presented may be applied to a table with any number of places.

We shall now present examples with explanations that illustrate the methods for using logarithms for (1) products and quotients, (2) powers and roots, (3) miscellaneous computation problems.

In all computation problems, we use the properties of logarithms in Sec. 11.4 to find the logarithm of the result. Then the value of the result can be obtained from the table.

PRODUCTS AND QUOTIENTS

We shall show the process of using logarithms to find products and quotients by three examples.

EXAMPLE 1   Use logarithms to find the value of R if $R = 8.56 \times 3.47 \times 198$.

Solution   Since R is equal to the product of three numbers, by Eq. (11.5) $\log R$ is equal to the sum of the logarithms of the three factors. Hence, we shall obtain the logarithm of each of the factors, add them together, and thus have $\log R$. Then we can use the table to get R. Before turning to the table, it is advisable to make an outline, leaving blanks in which to enter the logarithms as they are found. It is also advisable to arrange the outline so that the logarithms to be added are in a column. We suggest the following plan:

$$\begin{array}{l} \log 8.56 = \\ \log 3.47 = \\ \underline{\log\ 198 =} \\ \quad \log R = \qquad\qquad \text{enter sum here} \\ \quad\quad R = \end{array}$$

Next we enter the characteristics in each of the blanks and have

$$\begin{array}{l} \log 8.56 = 0. \\ \log 3.47 = 0. \\ \underline{\log\ 198 = 2.} \\ \quad \log R = \\ \quad\quad R = \end{array}$$

Now we turn to the tables, get the mantissas, and, as each is found, enter it in the proper place in the outline. Then we perform the addition, and finally we determine R by the method of Sec. 11.8. The completed solution then appears as

$$\begin{array}{l} \log 8.56 = 0.9325 \\ \log 3.47 = 0.5403 \\ \underline{\log\ 198 = 2.2967} \\ \quad \log R = 3.7695 \\ R\ /\ 5{,}880 = 5.88(10^3) \qquad \text{to three digits} \end{array}$$

NOTE: Each of the numbers in the problem contains only three digits; hence, we can determine only three digits in R. Since the mantissa 7695 is between the two entries 7694 and 7701 and nearer the former than the latter,† the first three digits in R are 588, the number corresponding to the mantissa 0.7694. The characteristic of $\log R$ is 3. Hence, the decimal point is three places to the right of the reference position. Therefore, we use scientific notation and place the decimal point by multiplying by the proper power of 10.

We have written the outline of the solution three times in order to show how the outline appears at the conclusion of each step. In practice, it is necessary to write the outline only once, since each operation requires the filling of separate blanks.

_____

† That is, the fourth digit would be less than 5 and is therefore discarded (Sec. 11.3, step 4).

EXAMPLE 2   Use logarithms to find R, where $R = (337 \times 2.68) \div (521 \times 0.763)$.

Solution   In this problem, R is a quotient in which the dividend and divisor are each the product of two numbers. Hence, we shall add the logarithms of the two numbers in the dividend, then add the logarithms of the two in the divisor, subtract the latter sum from the former, and thus obtain $\log R$. We suggest the following outline for the solution:

$$
\begin{array}{ll}
\log 337 = & \\
\underline{\log 2.68 =} & \\
\log \text{dividend} = & \text{enter sum here} \\
\log 521 = & \\
\underline{\log 0.763 =} & \\
\log \text{divisor} = & \text{enter sum here} \\
\log R = & \text{enter the difference of two sums here} \\
R = &
\end{array}
$$

After the characteristics are entered and the mantissas are found and listed in the proper places, the problem can be completed as below.

$$
\begin{array}{rl}
\log 337 = & 2.5276 \\
\underline{\log 2.68 =} & \underline{0.4281} \\
\log \text{dividend} = & 2.9557 \\
\log 521 = & 2.7168 \\
\underline{\log 0.763 =} & \underline{9.8825 - 10} \\
\log \text{divisor} = & 12.5993 - 10 = 2.5993 \\
\log R = & 0.3564 \\
R = & 2.27
\end{array}
$$

EXAMPLE 3   Use logarithms to evaluate $R = 2.68/33.2$.

Solution
$$
\begin{array}{rl}
\log 2.68 = & 0.4281 \\
\underline{\log 33.2 =} & \underline{1.5211} \\
\log R = & -1.0930
\end{array}
$$

where $\log R$ is obtained by subtracting the second logarithm from the first. This is a correct value of $\log R$, but since the fractional part 0930 is negative, it is not a mantissa, and the value of R cannot be obtained from the table. We can avoid this type of difficulty by adding $10 - 10$ to $\log 2.68$ before performing the subtraction. Then we have

$$
\begin{array}{rl}
\log 2.68 = & 10.4281 - 10 \\
\underline{\log 33.2 =} & \underline{1.5211} \\
\log R = & 8.9070 - 10 \\
R = & 0.0807
\end{array}
$$

We use the device shown in Example 3 whenever a necessary subtraction of one logarithm from another leads to a negative remainder. Thus, in order to subtract $9.2368 - 10$ from 2.6841, we add $10 - 10$ to the latter and have

$$
\begin{array}{r}
12.6841 - 10 \\
9.2368 - 10 \\
\hline
3.4473
\end{array}
$$

Similarly, in performing the indicated subtraction in $(7.3264 - 10) - (9.4631 - 10)$, we would again obtain a negative number. Hence, we add $10 - 10$ to $7.3264 - 10$ and proceed with the subtraction as below.

$$
\begin{array}{r}
17.3264 - 20 \\
9.4631 - 10 \\
\hline
7.8633 - 10
\end{array}
$$

POWERS AND ROOTS
We shall show the process of using logarithms to find powers and roots by three examples.

EXAMPLE 4 Obtain the value of $R = (3.74)^5$ by using logarithms.

Solution We use Eq. (11.7) and have

$$
\begin{aligned}
\log R &= \log (3.74)^5 \\
&= 5(\log 3.74) \\
&= 5(0.5729) \\
&= 2.8645 \\
R &= 732 \qquad \text{from Table A.1}
\end{aligned}
$$

We can also use Eq. (11.7) to obtain the root of a number by means of logarithms and by making use of the fact that $\sqrt[n]{N} = N^{1/n}$. The method is illustrated in the following example:

EXAMPLE 5 Find the cube root of 62.3 by using logarithms.

Solution If $R = \sqrt[3]{62.3}$, we rewrite the problem in the exponential form and get

$$
\begin{aligned}
\log R &= \log (62.3)^{1/3} \\
&= \tfrac{1}{3} \log 62.3 \qquad \text{by Eq. (11.7)} \\
&= \tfrac{1}{3}(1.7945) \\
&= 0.5982 \\
R &= 3.96 \qquad \text{from Table A.1}
\end{aligned}
$$

In applying Eq. (11.7) to the problem of extracting a root of a decimal fraction, we employ a device similar to that described in Example 3 in order to avoid a troublesome situation.

___

EXAMPLE 6   Find the sixth root of 0.0628 by using logarithms.

Solution   If $R = \sqrt[6]{0.0628}$, we have

$$\log R = \tfrac{1}{6} \log 0.0628$$
$$= \frac{8.7980 - 10}{6}$$

If we perform the division indicated above, we get

$$\log R = 1.4663 - 1.6667$$
$$= -0.2004$$

Thus, we have a negative fraction and cannot obtain R from the table. We can avoid a situation of this sort by adding $50 - 50$ to $\log 0.0628$, obtaining $\log 0.0628 = 58.7980 - 60$.   Now we have

$$\log R = \frac{58.7980 - 60}{6}$$

$$= 9.7997 - 10$$

The last logarithm is in the customary form, and by referring to the table we find that

$$R = 0.631 \qquad \text{from Table A.1}$$

MISCELLANEOUS PROBLEMS

Many computation problems require a combination of the processes of multiplication, division, raising to powers, and the extraction of roots. We shall illustrate the general procedure for solving such problems by an example.

___

EXAMPLE 7   Use logarithms to find R if $R = \sqrt[5]{\dfrac{\sqrt{2.689} \times 3.478}{(52.18)^2 \times 51.67}}$

Solution   Since all the numbers in this problem contain four digits, we must obtain the value of R to four places.   Furthermore, we must use interpolation to obtain the mantissas.   The steps in the solution are indicated in the following suggested outline:

$$\log \sqrt{2.689} = \tfrac{1}{2} \log 2.689 = \tfrac{1}{2}(\quad) =$$
$$\log 3.478 \qquad\qquad\qquad\qquad =$$

$$\log \text{dividend} = \qquad \text{sum here}$$
$$\log (52.18)^2 = 2 \log 52.18 = 2(\quad) =$$
$$\log 51.67 \qquad\qquad\qquad\qquad =$$

$$\log \text{divisor} = \qquad \text{sum here}$$

5 ⌐

└──── difference

$$\log R =$$
$$R =$$

We now enter the characteristics in the proper places, then turn to Table A.1, get the mantissas and enter each in the space left for it, and complete the solution. Then the outline appears as below.

$$\log \sqrt{2.689} \;= \tfrac{1}{2} \log 2.689 = \tfrac{1}{2} \times 0.4296 = 0.2148$$
$$\log 3.478 \qquad\qquad\qquad\qquad\qquad = 0.5413$$

$$\log \text{dividend} = 10.7561\dagger - 10$$
$$\log (52.18)^2 = 2 \log 52.08 = 2 \times 1.7175 = 3.4350$$
$$\log 51.67 \qquad\qquad\qquad\qquad\quad = 1.7132$$

$$\log \text{divisor} = \underline{5.1482}$$
$$5\,\overline{|45.6079} - 50\ddagger$$
$$\log R = 9.1216 - 10$$
$$R = 0.1323$$

EXERCISE 11.5 Computation

Use the properties of logarithms to express the logarithm in each of Probs. 1 to 8 as the sum or difference of logarithms of first powers of numbers.

1 $\log \dfrac{a}{b+x}$ 2 $\log \dfrac{a-x}{b}$ 3 $\log ab(c-x)$ 4 $\log (b-x)(c-y)$

5 $\log ax^4$ 6 $\log a^3\sqrt{b+x}$ 7 $\log \dfrac{b\sqrt{x-a}}{c\sqrt{x+a}}$ 8 $\log \dfrac{a^2 x^5}{b\sqrt{ax-b}}$

† Note that we add $10 - 10$ here so that we can subtract 5.1482.
‡ We add $40 - 40$ here so that we can divide by 5.

By use of the computation theorems, write the expression in each of Probs. 9 to 16 as the logarithm of a single number.

9 $\log a + \log b$ 10 $\log b - \log c$
11 $\log a + \log b - \log c$ 12 $\log a - \log b - \log c$
13 $\log a + 2 \log b - \log c$ 14 $\log a - 3 \log b + 2 \log c$
15 $2 \log a + 3 \log b - \frac{1}{2} \log (2a + 3b)$
16 $3 \log a - \frac{1}{2} \log b + \frac{1}{3} \log (a - b)$

By use of the computation theorems, perform the operations indicated in Probs. 17 to 84. Obtain each result to the justified degree of accuracy, and use scientific notation as needed.

17 $(3.47)(23.6)$ 18 $(58.6)(13.8)$ 19 $(80.3)(76.2)$
20 $(47.1)(17.4)$ 21 $976 \div 342$ 22 $243 \div 679$

23 $607 \div 874$ 24 $582 \div 258$ 25 $\dfrac{(326)(503)}{797}$

26 $\dfrac{(905)(3.28)}{47.6}$ 27 $\dfrac{(85.3)(0.714)}{32.7}$ 28 $\dfrac{(7.96)(0.843)}{0.579}$

29 $\dfrac{487}{(52.6)(89.3)}$ 30 $\dfrac{0.375}{(0.0238)(5.27)}$ 31 $\dfrac{7.84}{(23.6)(3.47)}$

32 $\dfrac{985}{(286)(3.41)}$ 33 $(2.79)(7.92)(9.27)$

34 $(87.5)(3.46)(5.61)$ 35 $(34.2)(0.342)(68.4)$
36 $(9.84)(78.3)(349)$ 37 5.98^3
38 47.1^2 39 0.319^2 40 0.987^3
41 $(276)(3.85)^4$ 42 $(329)(4.73)^2$ 43 $(97.3)(5.81)^3$
44 $(0.132)(6.47)^3$ 45 $\sqrt{76.2}$ 46 $\sqrt[3]{497}$
47 $\sqrt[5]{2.38}$ 48 $\sqrt[7]{985}$ 49 $\sqrt[3]{0.737}$
50 $\sqrt[4]{0.0126}$ 51 $\sqrt[7]{0.843}$ 52 $\sqrt[6]{0.714}$
53 $\sqrt{475}\,\sqrt[3]{0.346}$ 54 $\sqrt[3]{27.3}\,\sqrt[7]{0.0863}$ 55 $\sqrt[5]{938}\,\sqrt[3]{0.839}$

56 $\sqrt{8.14}\,\sqrt[3]{0.349}$ 57 $\dfrac{\sqrt{239}}{\sqrt[3]{0.404}}$ 58 $\dfrac{\sqrt[3]{763}}{\sqrt[7]{0.0567}}$

59 $\dfrac{\sqrt{0.802}}{\sqrt[5]{644}}$ 60 $\dfrac{\sqrt[3]{0.0439}}{\sqrt{0.713}}$ 61 $(34.72)(5.039)$

62 $(0.8077)(1,153)$ 63 $(0.3427)(0.01863)$ 64 $(71.18)(60.33)$

65 $\dfrac{2,843}{1,596}$ 66 $\dfrac{9,437}{7,562}$

67 $\dfrac{3,491}{8,743}$ 68 $\dfrac{2,399}{8,276}$

69 $(4,713)(0.23)(0.0483)$ 70 $(9.13)(807.2)(0.28)$
71 $(0.19)(0.938)(4574)$ 72 $(8.47)(0.85)(7.342)$

73 $\dfrac{38.61}{297}$ 74 $\dfrac{973.2}{37.6}$

75 $\dfrac{283}{57.14}$ 76 $\dfrac{8.03}{87.41}$

77 $\dfrac{27.32^{2/3}}{34.1}$ 78 $\dfrac{4.13}{2.874^{4/3}}$

79 $\dfrac{0.236^{2/7}}{5.473}$ 80 $\dfrac{\sqrt{3.142}}{0.481^{3/4}}$

81 $\sqrt[3]{\dfrac{57.23(0.4256)^2}{82.17}}$ 82 $\sqrt[3]{\dfrac{(31.84)^2\sqrt[3]{0.9173}}{276.3}}$

83 $\sqrt[7]{\dfrac{0.7133(7.234)^2}{67.32}}$ 84 $\sqrt{\dfrac{\sqrt[3]{0.7131}\ \sqrt{0.4962}}{0.05005}}$

11.10 / LOGARITHMS TO BASES OTHER THAN 10

As stated earlier in this chapter, any positive number except 1 can be used as a base for a system of logarithms. However, there are only two systems of logarithms in general use: the common, or Briggs, system and the natural, or napierian, system. The former is the system more convenient for numerical computation, and the latter is very important in more advanced mathematics and its applications. The base for the napierian system is the irrational number $e = 2.718 \ldots$ to three decimal places. Tables of napierian logarithms can be found in most technical handbooks and manuals of mathematical tables. However, the logarithm of a number to any base can be expressed in terms of the logarithm of the number to another base. In particular, the logarithm of a number to any base can be obtained by use of a table of common logarithms, and it is the purpose of this section to explain how this can be done. The theorem on which the process rests is as follows:

Theorem on logs to any two bases

If a and b are any two bases, then

$$\log_a N = \frac{\log_b N}{\log_b a} \qquad (11.8)$$

To prove Eq. (11.8), we let

$$N = a^y$$

Then $\log_a N = \log_a a^y$

$= y$ by Eq. (11.2)

Furthermore, $\log_b N = \log_b a^y$

$= y \log_b a$ by Eq. (11.7)

$= \log_a N \log_b a$ since $y = \log_a N$

Now, by solving this equation for $\log_a N$, we have

$$\log_a N = \frac{\log_b N}{\log_b a}$$

Relation between natural and Briggs logarithms

As a corollary to the above theorem, we have the following:
The relation between $\log_{10} N$ and $\log_e N$ is given by

$$\mathbf{\log_e N = \frac{\log_{10} N}{\log_{10} e}} \qquad\qquad \textbf{(11.9)}$$

To prove the corollary, we let $a = e$ and $b = 10$ in Eq. (11.8). We then have the equation as stated in the corollary.

Since $\log_{10} e = 0.4343$ and $1/0.4343 = 2.3026$, Eq. (11.9) can be expressed in the form

$$\log_e N = 2.3026 \log_{10} N$$

EXAMPLE   By means of Eq. (11.8), find the value of $\log_7 236$.

Solution   If we substitute in Eq. (11.8), we get

$$\log_7 236 = \frac{\log_{10} 236}{\log_{10} 7}$$

$$= \frac{2.3729}{0.8451} = 2.808$$

EXERCISE 11.6 Logarithms to Bases Other Than 10

By use of a table of common logarithms, evaluate each of the following to three digits.

1	$\log_2 1.12$	2	$\log_2 91.6$	3	$\log_2 5.43$	4	$\log_2 13.7$
5	$\log_e 23.4$	6	$\log_e 9.04$	7	$\log_e 8.17$	8	$\log_e 1{,}350$
9	$\log_3 13.1$	10	$\log_3 7.03$	11	$\log_3 8.63$	12	$\log_3 26.5$
13	$\log_5 0.217$	14	$\log_5 1{,}066$	15	$\log_5 18.3$	16	$\log_5 139$
17	$\log_{11} 119$	18	$\log_{11} 1{,}891$	19	$\log_{11} 7.13$	20	$\log_{11} 276$
21	$\log_{13} 7.03$	22	$\log_{23} 987$	23	$\log_{12} 352$	24	$\log_{17} 90.27$

11.11 / EXPONENTIAL AND LOGARITHMIC EQUATIONS

An *exponential equation* is an equation in which a variable occurs in one or more exponents. A *logarithmic equation* is an equation which involves the logarithm of a function of the variable. For example, $3^x = 7$ and $3^{x+1} = 5^{x-2}$ are exponential equations, whereas $\log x + \log (x - 1) = 2$ is a logarithmic equation.

In general, exponential and logarithmic equations cannot be solved by the methods heretofore discussed, but many such equations can be solved by use of the properties of logarithms. The following examples illustrate the procedure.

EXAMPLE 1 Solve the equation $3^{x+4} = 5^{x+2}$.

Solution If we take the common logarithm of each member of the above equation, we have by Eq. (11.7), $(x + 4) \log 3 = (x + 2) \log 5$, and we complete the solution as follows:

$$x \log 3 + 4 \log 3 = x \log 5 + 2 \log 5$$
$$x(\log 3 - \log 5) = 2 \log 5 - 4 \log 3 \qquad \text{adding } -x \log 5 - 4 \log 3$$
$$\text{to each member}$$
$$= \log 25 - \log 81 \qquad \text{by Eq. (11.7)}$$
$$x = \frac{\log 25 - \log 81}{\log 3 - \log 5} \qquad \text{solving for } x$$
$$= \frac{1.3979 - 1.9085}{0.4771 - 0.6990}$$
$$= \frac{-0.5106}{-0.2219}$$
$$= 2.301$$

EXAMPLE 2 Solve $\log_6 (x + 3) + \log_6 (x - 2) = 1$.

Solution By applying Eq. (11.5) to the given equation, we have

$$\log_6 (x + 3)(x - 2) = 1$$
$$(x + 3)(x - 2) = 6^1 \qquad \text{by Eq. (11.2)}$$
$$x^2 + x - 6 = 6 \qquad \text{performing the indicated operations}$$
$$x^2 + x - 12 = 0 \qquad \text{adding } -6 \text{ to each member}$$
$$(x + 4)(x - 3) = 0$$
$$x = -4 \qquad \text{setting the factors separately equal to zero}$$
$$x = 3$$

However, if $x = -4$ is substituted in the given equation, we get $\log_6 (-1) + \log_6 (-6) = 1$. Since we have not defined the logarithm of a negative number, we discard $x = -4$. Hence the only solution consistent with our definition of logarithms is $x = 3$.

EXAMPLE 3 Solve $y = \log_e (x + \sqrt{x^2 + 1})$ for x.

Solution By use of Eq. (11.2), we can change the given equation to the exponential form and obtain

$$e^y = x + \sqrt{x^2 + 1}$$

We next rationalize the above equation and solve for x as follows:

$e^y - x = \sqrt{x^2 + 1}$	isolating the radical
$e^{2y} - 2e^y x + x^2 = x^2 + 1$	squaring each member
$-2e^y x + x^2 - x^2 = -e^{2y} + 1$	adding $-e^{2y} - x^2$ to each member

$$x = \frac{-e^{2y} + 1}{-2e^y}$$

$$= \frac{e^y - e^{-y}}{2} \qquad \text{dividing each member of the fraction by } e^y$$

EXAMPLE 4   Solve the system

$$5^{x-2y} = 100 \tag{1}$$
$$3^{2x-y} = 10 \tag{2}$$

for x and y.

Solution   If we equate the logarithms of the members of (1) and of (2), we get

$$(x - 2y) \log 5 = 2 \tag{3}$$
$$(2x - y) \log 3 = 1 \tag{4}$$

Therefore,

$$x - 2y = \frac{2}{\log 5} = \frac{2}{0.6990} \tag{3'}$$

$$2x - y = \frac{1}{\log 3} = \frac{1}{0.4771} \tag{4'}$$

If we perform the computation that is indicated in the right member, we have

$$x - 2y = 2.86 \tag{3''}$$
$$2x - y = 2.10 \tag{4''}$$

By multiplying each member of (4″) by 2 and subtracting from the corresponding member of (3″), we obtain

$$-3x = -1.34$$
$$x = 0.447$$

By substituting this value for x in (3″) and solving for y, we get

$$-2y = 2.86 - 0.447$$
$$= 2.413$$
$$y = -1.206$$

Therefore, the solution set of the system is $\{0.447, -1.206\}$.

EXERCISE 11.7 Exponential and Logarithmic Equations

Solve the equation in each of Probs. 1 to 16 for x or n in terms of y.

1 $y = e^x$ **2** $y = e^{-x}$ **3** $y = 2e^{3x}$ **4** $y = 3e^{-x}$

5 $y = \dfrac{r^n}{a}$ **6** $y = ar^{n+1}$ **7** $y = 2(c+r)^n$ **8** $y = 3(c+r)^{-n}$

9 $y = \log_e (x + \sqrt{x^2 - 1}), \; x \geq 1$

10 $y = \log_e \sqrt{(x-2)(x+1)}, \; x > 2$

11 $y = \log_e (x + \sqrt{x^2 + 1})$ **12** $y = \log_e [x + \sqrt{(x-1)(x+2)}], \; x > 1$

13 $y = \dfrac{e^x + e^{-x}}{2}$ **14** $y = \dfrac{e^x - e^{-x}}{2}$

15 $y = \log_e \dfrac{e^x - e^{-x}}{2}$ **16** $y = \log_e \dfrac{e^x + e^{-x}}{2}$

Find the solution of the equation in each of Probs. 17 to 40.

17 $5^x = 625$ **18** $3^x = 27$ **19** $2^x = 64$ **20** $7^x = 343$

21 $6^{x+1} = 36$ **22** $7^{2x-1} = 2{,}401$ **23** $5^{3x-2} = 625$ **24** $3^{2x+1} = 243$

25 $3^{x^2-1} = 27$ **26** $4^{x^2+3x} = 256$ **27** $2^{x^2-x-2} = 1$ **28** $5^{x^2-4x+5} = 25$

29 $107^x = 3.16$ **30** $3.16^x = 107$

31 $4.71^x = 0.0283$ **32** $219^{x-1} = 24.7$

33 $\log_2 (x+2) + \log_2 (x-1) = 2$ **34** $\log_2 (x+4) + \log_2 (x+8) = 5$

35 $\log_3 (x+4) + \log_3 (x+2) = 1$ **36** $\log_5 (2x+1) + \log_5 (3x-1) = 2$

37 $\log_3 (x+11) - \log_3 (x+3) = 2$ **38** $\log_2 (3x+1) - \log_2 (x-3) = 3$

39 $\log_7 (x^2 - x - 6) - \log_7 (x+2) = 1$

40 $\log_6 (x^2 + 2x + 15) - \log_6 (x+2) = 1$

Find the solution pair of the system of equations in each of Probs. 41 to 48.

41 $2^{x-2y} = 8$
$x + 2y = 5$

42 $3^{4x-y} = 27$
$2x + 3y = 5$

43 $5^{x+2y} = 28$
$4x + 2y = 3$

44 $2^{2x+3y} = 48$
$x + y = 2$

45 $10^{x+2y} = 2$
$\log 3x - \log 2y = 1$

46 $10^{3x+y} = 40$
$\log x - \log 3y = 1$

47 $\log x + \log y = 4$
$\log 2x - \log 5y = 1$

48 $\log x + \log y = 2$
$\log 2x - \log y = 1$

11.12 / THE GRAPHS OF log $_a x$ AND a^x

The graphs of $y = \log_a x$ and $y = a^x$ reveal several important properties of these equations. Furthermore, the graphical method of solution must be applied to logarithmic and exponential equations that are not solvable by algebraic methods. Hence it is im-

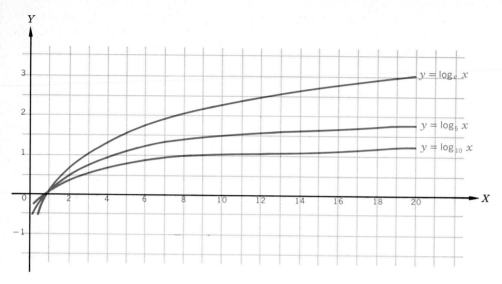

Figure 11.1

portant to be able to sketch the graph of the above equations rapidly. In Fig. 11.1 we show three graphs of the equation $y = \log_a x$, with $a = 10$, $a = e$, and $a = 5$. The following table of values computed to one decimal place was used for constructing the graphs. The values were obtained by use of a table of common logarithms and Eqs. (11.9) and (11.8).

x	0.2	0.4	0.6	0.8	1	2	4	6	8	10	20
$\log_{10} x$	−0.7	−0.4	−0.2	−0.1	0	0.3	0.6	0.8	0.9	1	1.3
$\log_e x$	−1.6	−0.9	−0.5	−0.2	0	0.7	1.4	1.8	2.1	2.3	3.0
$\log_5 x$	−1	−0.6	−0.3	−0.14	0	0.4	0.9	1.1	1.3	1.4	1.8

Properties of $\log_a x$ The graphs reveal the following properties of $\log_a x$:

1   $\mathrm{Log}_a x$ is not defined for negative values of x.

2   The value of $\log_a x$ increases as x increases.

3   $\mathrm{Log}_a x$ is equal to zero if x is equal to 1.

4   $\mathrm{Log}_a x$ is negative for values of x less than 1 and positive for values of x greater than 1.

5   The larger the value of a, the smaller the value of y for a given value of x.

     In order to illustrate the method for obtaining the graph of $y = a^x$, we shall show how to compute a table of corresponding values of x and y, using a table of common logarithms, and we shall construct the graph of this equation. For example, if $x = 1.5$ and $a = e$, then

$$y = e^{1.5}$$

and
$$\begin{aligned} \log y &= 1.5(\log e) \\ &= 1.5(\log 2.718) \\ &= 1.5 \times 0.4343 \\ &= 0.6514 \end{aligned}$$

Hence,
$$y = 4.5$$

Furthermore, if $x = -2$,

$$\begin{aligned} \log y &= -2(\log e) \\ &= -2 \times 0.4343 \\ &= -0.8686 \\ &= 9.1314 - 10 \end{aligned}$$

Therefore,
$$y = 0.14$$

or 0.1 for plotting purposes.

In a similar manner, we obtain the following table, which was used in constructing the graph in Fig. 11.2. This figure also shows the graphs of $y = 2^x$ and $y = \log_2 x$.

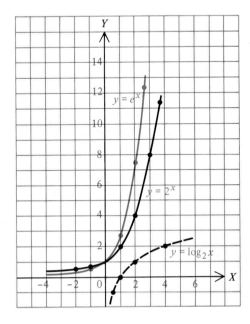

Figure 11.2

x	-4	-2	-1	0	1	1.5	2	2.5	3
e^x	0.02	0.1	0.4	1	2.7	4.5	7.4	12.2	20

The graph of $y = 2^x$ is closely related to the graph of $y = \log_2 x$. If we let $f(x) = 2^x$ and $g(x) = \log_2 x$, then

$$f[g(x)] = f(\log_2 x) = 2^{\log_2 x} = x$$

and $$g[f(x)] = \log_2 f(x) = \log_2 2^x = x \log_2 2 = x$$

These two equations show that $g = f^{-1}$; hence, $y = 2^x$ and $y = \log_2 x$ define inverse functions; as mentioned at the end of Sec. 6.9, each graph is the reflection of the other in the line $y = x$. Similar comments apply to $y = a^x$ and $y = \log_a x$.

The following example illustrates the application of graphical methods to systems of equations that involve logarithms.

EXAMPLE   Solve the equations

$$y = \log x$$
$$3x + 2y = 6$$

Solution   The equations cannot be solved simultaneously by algebraic methods. However, if we construct the graphs of the two equations (Fig. 11.3), we can estimate their point of intersection and thus obtain an approximate solution. In this figure the abscissa of the point of intersection is between 1.8 and 1.9. In the first equation, when $x = 1.8$, $y = 0.26$; and when $x = 1.9$, $y = 0.28$. Furthermore, between these two points, the graph is approximately a straight line. In the second equation, if $x = 1.8$, $y = 0.3$; and if $x = 1.9$, $y = 0.15$. We now enlarge the scale, plot the above points, and join the two corresponding pairs with straight lines (see Fig. 11.4). These two lines intersect at a point whose coordinates are approximately $x = 1.83$, $y = 0.26$. By substituting in the original equations, it can be verified that this pair of values is an approximate solution.

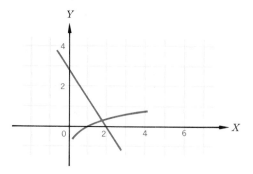

Figure 11.3

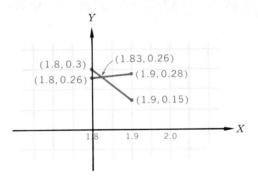

Figure 11.4

EXERCISE 11.8 Graphs of Logarithmic and Exponential Equations

Construct the graph of the equation in each of Probs. 1 to 24.

1 $y = \log_5 x$	2 $y = \log_6 x$	3 $y = \log_7 x$
4 $y = \log_{13} x$	5 $y = \log_{10} 3x$	6 $y = \log_{10} 5x$
7 $y = \log_{10} x^3$	8 $y = \log_{10} x^2$	9 $y = 2^{2x}$
10 $y = 7^x$	11 $y = 5^{3x}$	12 $y = 3^{2x}$
13 $y = \log_{10}(x + 1)$		14 $y = \log_{10}(x - 3)$
15 $y + 2 = \log_{10} x$		16 $y - 4 = \log_{10}(x + 6)$
17 $y = 3^{x+4}$		18 $y = 3^{x-2}$
19 $y - 3 = 3^x$		20 $y + 6 = 3^x$
21 $4 - y = \log_{10}(2x + 5)$		22 $3 - 4y = \log_{10}(3x - 1)$
23 $2y/3 = 3^{2x+1}$		24 $6 - 2y = 3^{4-3x}$

Find, by use of the graphical method, the solution to one decimal place of each system of equations given in Probs. 25 to 32.

25 $y = 2^x$	26 $y = 3^x$	27 $y = 10^x$
$x + y = 2$	$2x + y = 2$	$y = -1 + 4x - 4x^2$
28 $y = 5^x$	29 $y = \log_{10} x$	30 $y = \log_{10} x$
$y = 1 - x^2$	$x - 7y = 5$	$x - y = 1$
31 $y = \log_{10} x$	32 $y = \log_{10} x$	
$x + 2y = 0$	$y = 4x^2 - 4x$	

11.13 / A WORD ABOUT CALCULATORS

There are calculators that vary in size from those that will fit into a pocket to those that will not fit into a room. The simplest of them perform the four fundamental operations of addition, subtraction, multiplication, and division, whereas the more complex ones perform almost unbelievable operations. One of the authors owns a machine that measures just over 3 by 5 in. in area and is an inch thick. In addition to performing the four fundamental operations, it can be used to find reciprocals, square roots, powers of *e*, natural and common logarithms, the trigonometric functions, and the inverse trigonometric functions, and it has a memory.

EXAMPLE 1　In order to find the product of two numbers, we put one of them on the display dial, push the button for multiplication, put the other number on the display dial, push the = button, and the product appears on the display dial. Thus, the exact product of the numbers 42.3 and 3,756 is 158,878.8.

EXAMPLE 2　In order to find the logarithm of a number, we put the number on the display dial, push the F button, and then push the log button. The logarithm of the number appears on the display dial. Thus, log 456.8 is 2.6597, and $\log_e 456.8 = 6.1242$, which is 2.3026 times the common logarithm.

EXAMPLE 3　If we know the natural logarithm of a number and want the number, we put the logarithm on the display dial, push the F button, then push the e^x button, and read the number on the display dial. Thus if ln x is 2.7142, we put this number on the display dial, push the F button, then the e^x button, and see that the number is 15.09.

EXAMPLE 4　If we know the common logarithm of a number and want the number, we multiply the common logarithm by 2.3026, push the F button, then the e^x button, and read the number on the display dial.

EXAMPLE 5　In order to find the quotient of two numbers by use of a calculator, we get the logarithm of the denominator, put it in the memory, get the logarithm of the numerator, push the − button, then the = button and read the logarithm of the quotient on the display dial. We then find the number as in Example 4.

EXERCISE 11.9 Use of Calculators

Perform the operations indicated in Probs. 1 to 32 with a calculator if one is available. Also try some of the problems in Exercise 11.5 with a calculator.

1　$56.3 + 4.72$	2　$3.59 + 81.72$	3　$9.83 + 7.21$
4　$98.423 + 62.596$	5　$80.21 - 63.85$	6　$83.49 - 28.73$
7　$54.38 - 71.89$	8　$84.1 - 95.7$	9　$(5.23)(34.7)$

10 $(89.4)(3.27)$	11 $(26.1)(37.3)$	12 $(9.81)(28.3)$
13 $\dfrac{76.4}{23.8}$	14 $\dfrac{9.83}{28.1}$	15 $\dfrac{78.3}{107}$
16 $\dfrac{4724}{86.19}$	17 $\sqrt{8.13}$	18 $\sqrt{987}$
19 $\sqrt{63.7}$	20 $\sqrt{342}$	21 $\log 597$
22 $\log 73.1$	23 $\log 0.63$	24 $\log 9{,}845$
25 $\ln 2.81$†	26 $\ln 7.36$	27 $\ln 21.84$
28 $\ln 0.314$	29 $e^{2.3}$	30 $e^{4.71}$
31 $e^{0.72}$	32 $e^{1.862}$	

Find N in each of Probs. 33 to 40.

33 $\ln N = 1.81$	34 $\ln N = 3.02$	35 $\ln N = 2.76$
36 $\ln N = 0.189$	37 $\log N = 2.75$	38 $\log N = 1.95$
39 $\log N = 0.763$	40 $\log N = 0.14$	

EXERCISE 11.10   REVIEW

Express the number in each of Probs. 1 to 3 in scientific notation.

1 $7{,}863$	2 0.0591	3 $4{,}086.4$

Round off the number in each of Probs. 4 to 6 to three significant digits.   Use scientific notation as needed.

4 584.6	5 $5{,}846$	6 $5{,}723.49$

Assume that all numbers in Probs. 7 to 12 are approximations.   Perform the indicated operations by use of logarithms and with a calculator if one is available.

7 $(34.2)(5.86)$	8 $(9.03)(7.6)$	9 $\dfrac{84.7}{12}$
10 $\dfrac{58.92}{348.7}$	11 $7.63 + 5.1$	12 $8.47 - 9.13$

Change the equation in each of Probs. 13 to 16 from exponential to logarithmic form or from logarithmic to exponential.

13 $7^3 = 343$	14 $5^{-2} = \frac{1}{25}$	15 $\log_3 81 = 4$	16 $\log_{64} 4 = \frac{1}{3}$

In each of Probs. 17 to 22, find L, b, or N.

17 $\log_5 125 = L$	18 $\log_8 4 = L$	19 $\log_3 N = 4$
20 $\log_7 N = 3$	21 $\log_b 216 = 3$	22 $\log_b \frac{1}{81} = -2$

† NOTE: $\ln = \log_e$

Find the common logarithm of the number in each of Probs. 23 to 25.

23   84.73   **24**   0.7136   **25**   2,357

Find the four-digit number whose common logarithm is given in each of Probs. 26 to 28.

26   1.3426   **27**   2.5947   **28**   8.1433 − 10

Perform the operations indicated in Probs. 29 to 34 by use of logarithms and, if available, by use of a calculator.

29 $(34.7)(29.63)$ **30** $(0.8439)(764)$ **31** $68,475 \div 327$

32 $5.976 \div 1.703$ **33** $\sqrt{849}$ **34** $\sqrt{0.714}$

35   Evaluate $\log_3 57.2$.

36   Evaluate $\log_{11} 829$.

37   Solve $y = 5e^{2x}$ for x in terms of y.

Solve the equation in each of Probs. 38 to 41 for x.

38 $2^{x^2-x} = 64$ **39** $7.13^x = 36.4$ **40** $\log_3 x + \log_3 (x + 2) = 1$

41 $\log_7 (x^2 - x - 6) - \log_7 (x + 2) = -1$

Solve the pair of equations in each of Probs. 42 and 43 for x and y graphically.

42 $y = \log_{10} x$ **43** $y = 10^x$

$\quad\;\;$ $x + y = 6$ $\quad\;\;$ $x + y = 3$

CHAPTER TWELVE

Complex Numbers

In Sec. 7.4, we stated that a complex number is a number of the form $a + bi$, where a and b are real and i represents $\sqrt{-1}$. We also discussed quadratic equations whose roots are complex numbers. The work of Gauss and Argand in the eighteenth century aroused an interest in complex numbers, which have since become of considerable importance in mathematics, physics, electronics, and electrical engineering. In this chapter we shall give another definition of a complex number and make some use of a third form for such numbers.

12.1 / COMPLEX NUMBERS AS ORDERED PAIRS

We found in Chap. 2 that the set of real numbers is closed under several operations, including addition. The set of real numbers is not closed if the operation is taking square roots. This is clear since the square of zero is zero and the square of a positive or a negative number is positive. Consequently, the square root of a negative number cannot be a real number. Therefore, we shall introduce a new type of number that permits taking the square root of a negative number.

Complex number   An ordered pair (a, b) of real numbers is called a *complex number*. It can be associated with the point (a, b) of the coordinate plane.

In terms of this notation, the complex number that would have been written as $2 + 3i$ in Sec. 7.4 is written now as $(2, 3)$. It may be thought of as having a real component 2 and imaginary component 3.

We state as a matter of definition that

Equal complex
numbers

$$(a, b) = (c, d) \text{ if and only if } a = c \text{ and } b = d \qquad (12.1)$$

Thus two complex numbers are equal if and only if they are associated with the same point. Therefore, $(x, y) = (3, 2)$ if and only if $x = 3$ and $y = 2$.

We shall now define the sum and product of two complex numbers by stating that

Sum

$$(a, b) + (c, d) = (a + c, b + d) \qquad (12.2)$$

Product   and

$$(a, b) \cdot (c, d) = (ac - bd, ad + bc) \qquad (12.3)$$

The definitions of addition and multiplication above are motivated by the $a + bi$ notation and will be justified after Eq. (12.5).

EXAMPLE 1   By use of Eq. (12.2), we have $(3, 2) + (4, -1) = (3 + 4, 2 - 1) = (7, 1)$.

EXAMPLE 2   By use of Eq. (12.3), we see that

$$(3, 2) \cdot (4, -1) = [3(4) - 2(-1), 3(-1) + 2(4)]$$
$$= (12 + 2, -3 + 8) = (14, 5)$$

We shall now consider a special case of Eqs. (12.2) and (12.3). If $b = d = 0$, then Eq. (12.2) becomes $(a, 0) + (c, 0) = (a + c, 0)$ and Eq. (12.3) becomes $(a, 0) \cdot (c, 0) = (ac, 0)$. Hence, complex numbers with the second of the ordered pairs equal to zero behave as real numbers do relative to addition and multiplication. Consequently, we define $(a, 0)$ as follows:

The complex number $(a, 0)$ is equal to the real number a and
$$\text{at times will be written as } a \qquad (12.4)$$

We shall designate the number $(0, 1)$ by i and apply Eq. (12.3). Thus, we have

$$i^2 = (0, 1) \cdot (0, 1) = (0 \cdot 0 - 1 \cdot 1, 0 \cdot 1 + 1 \cdot 0) = (-1, 0) = -1$$

Hence, i is a square root of -1. The other one is $-i$.

We now obtain a relation between complex numbers in the number-pair form (a, b) and in the binomial form $a + bi$ of Sec. 7.4. If a and b are real numbers and $i = (0, 1)$, then

$$a + bi = (a, 0) + (b, 0)(0, 1)$$
$$= (a, 0) + (b \cdot 0 - 0 \cdot 1, b \cdot 1 + 0 \cdot 0)$$
$$= (a, 0) + (0, b) = (a, b)$$

We now know that

(a, b) and $a + bi$ are two forms of the same number (12.5)

Real part

Imaginary part

The number a is called the *real part*, and b the *imaginary part*, of the number $(a, b) = a + bi$.

Assuming each step is valid, we have $(a + bi) + (c + di) = (a + c) + (b + d)i$ and $(a + bi)(c + di) = a(c + di) + (bi)(c + di) = ac + adi + bci + bdi^2 = ac + (ad + bc)i + bd(-1) = (ac - bd) + (ad + bc)i$.

The justification of each step above is that the set of complex numbers with addition and multiplication defined by Eqs. (12.2) and (12.3) is a field (see Sec. 2.17). We shall not prove this.

Calculations with complex numbers are generally easier if they are written $a + bi$ instead of (a, b). The latter notation has the following advantage, though: It emphasizes the fact that complex or imaginary numbers are just as "real" as real numbers, since they are simply two real numbers written in a definite order.

Subtraction and division of complex numbers are defined as the inverse of addition and multiplication, respectively. We say $(a, b) - (c, d)$ is (x, y) is to mean that $(a, b) = (c, d) + (x, y) = (c + x, d + y)$. By Eq. (12.1) this means that $a = c + x$ and $b = d + y$, and so $x = a - c$ and $y = b - d$.

Difference

Thus, $$(a, b) - (c, d) = (a - c, b - d) \tag{12.6}$$

EXAMPLE 3

By use of Eq. (12.6), we get

$$(2, 3) - (5, -2) = (2 - 5, 3 - [-2]) = (-3, 5)$$

We could similarly define the quotient $(a, b)/(c, d) = (x, y)$ to mean $(a, b) = (c, d)(x, y)$ and use Eqs. (12.1) and (12.3) to show that

Quotient

$$\frac{(a, b)}{(c, d)} = \left(\frac{ac + bd}{c^2 + d^2}, \frac{bc - ad}{c^2 + d^2}\right) \tag{12.7}$$

It is simpler, however, to multiply numerator and denominator of the left member of (12.7) by $(c, -d)$. Now

$$\frac{(a, b)}{(c, d)} \cdot \frac{(c, -d)}{(c, -d)} = \frac{(ac - b[-d], a[-d] + bc)}{(c[c] - d[-d], c[-d] + d[c])} \qquad \text{by Eq. (12.3)}$$

$$= \frac{(ac + bd, bc - ad)}{(c^2 + d^2, 0)} \qquad \text{simplifying}$$

$$= \left(\frac{ac + bd}{c^2 + d^2}, \frac{bc - ad}{c^2 + d^2}\right) \qquad \text{by Eq. (12.4)}$$

We have defined the four fundamental operations in the set of complex numbers $\{(a, b)\}$ and have shown that (a, b) and $a + bi$ are the same number. Consequently, applying Eqs. (12.2), (12.3), (12.6), and (12.7) to $a + bi$ and $c + di$, we have

$$(a + bi) + (c + di) = (a + c) + (b + d)i \tag{12.2}$$
$$(a + bi)(c + di) = (ac - bd) + (ad + bc)i \tag{12.3}$$
$$(a + bi) - (c + di) = (a - c) + (b - d)i \tag{12.6}$$
$$\frac{(a + bi)}{(c + di)} = \left(\frac{ac + bd}{c^2 + d^2}, \frac{bc - ad}{c^2 + d^2}\right) \tag{12.7}$$

EXAMPLE 4

$$\frac{3 - 2i}{3 - 4i} = \frac{(3 - 2i)(3 + 4i)}{(3 - 4i)(3 + 4i)} = \frac{9 + 12i - 6i - 8i^2}{9 + 16} = \frac{17}{25} + \frac{6i}{25}$$

EXAMPLE 5

The product $[(2 + 3i) + (3 - i)](-2 + i)$ may be calculated in two ways. Adding first gives

$$(5 + 2i)(-2 + i) = -10 + 5i - 4i + 2i^2 = -12 + i$$

while using the distributive law gives

$$(2 + 3i)(-2 + i) + (3 - i)(-2 + i) =$$
$$(-4 + 2i - 6i - 3) + (-6 + 3i + 2i + 1) = -12 + i$$

Conjugate

The complex number $(a, -b) = a - bi$ is called the *conjugate* of (a, b) or $a + bi$. It was used in obtaining the quotient in Eq. (12.7). It has other nice properties also. If $z = a + bi$, then we write the conjugate of z as $\bar{z} = \overline{a + bi} = a - bi$. Thus $z + \bar{z} = (a + bi) + (a - bi) = 2a$, and so the real part a of a complex number is given by

$$a = \frac{z + \bar{z}}{2} \tag{1}$$

Similarly $z - \bar{z} = (a + ib) - (a - ib) = 2bi$, so the imaginary part of a complex number is

$$b = \frac{z - \bar{z}}{2i} \tag{2}$$

Conjugate
of a sum
Conjugate
of a product

If $z = a + bi$ and $w = c + di$, then $\overline{z + w} = \overline{(a + bi) + (c + di)} = \overline{a + c + (b + d)i} = a + c - (b + d)i = a + c - bi - di = (a - bi) + (c - di) = \bar{z} + \bar{w}$. It follows that the *conjugate of a sum* is the sum of the conjugates.
In a similar manner it may be shown that the *conjugate of a product* is the product of the conjugates. These two facts are used in Sec. 13.7.

EXAMPLE 6

$$\overline{(3 - i)(2 + 5i)} = \overline{6 + 5 - 2i + 15i} = \overline{11 + 13i} = 11 - 13i$$

and

$$\overline{(3 - i)}\,\overline{(2 + 5i)} = (3 + i)(2 - 5i) = 6 + 5 + 2i - 15i = 11 - 13i$$

EXERCISE 12.1 Fundamental Operations on Complex Numbers

For which values of x and y are the two complex numbers in each of Probs. 1 to 8 equal?

1 $(x, 5), (2, y)$ **2** $(x-2, y), (1, -4-y)$

3 $(3-x, 1+y), (x-5, 3)$ **4** $(x-y, -1), (4, 2x+y)$

5 $x+5i, 3-yi$ **6** $2x-4i, y+2xi$

7 $x+y+yi, 3y-5+(5+x)i$ **8** $2x-y+3i, 11+3xi+3yi$

Perform the indicated operations in each of Probs. 9 to 48. Leave each answer in the same form as the problem.

9 $(2, 4) + (-1, 3)$ **10** $(3, -1) + (2, 5)$

11 $(-\frac{5}{2}, 3) + (-\frac{1}{2}, -1)$ **12** $(-3, -4) + (2, 4)$

13 $(2 + 4i) + (-3 + i)$ **14** $(-\frac{2}{3} + 5i) + (\frac{8}{3} + 2i)$

15 $(3 + 2i) + (4 - 2i)$ **16** $(2 - 4i) + (-2 + 5i)$

17 $(3, -1) - (2, 4)$ **18** $(3, 2) - (1, 0)$

19 $(4, -2) - (4, -3)$ **20** $(-2, -4) - (4, -3)$

21 $(2 + 3i) - (-1 + 5i)$ **22** $5i - (2 + 3i)$

23 $(-4 - i) - (-4 + i)$ **24** $(3 + i) - (3 - i)$

25 $(3, 4)(1, 4)$ **26** $(2, -1)(-2, 5)$

27 $(-2, -1)(3, -2)$ **28** $(-1, 4)(3, 3)$

29 $(2 - i)(1 + 3i)$ **30** $(3 + 2i)(3 - 2i)$

31 $(4i)(6 - i)$ **32** $(-3 + 4i)(2 + 5i)$

33 $\dfrac{(4, 5)}{(2, 1)}$ **34** $\dfrac{(2, -3)}{(5, 1)}$

35 $\dfrac{(3, 5)}{(5, 3)}$ **36** $\dfrac{(-2, 1)}{(3, -4)}$

37 $\dfrac{3 + i}{2 + 5i}$ **38** $\dfrac{2 + 3i}{5 - 2i}$

39 $\dfrac{1 + 3i}{1 - 3i}$ **40** $\dfrac{5 - 3i}{2 - 7i}$

41 $(2 + 3i)\overline{(-2 + i)}$ **42** $\overline{(1 - i)}(2 + 5i)$

43 $(3 + 2i) \div \overline{(5 - 3i)}$ **44** $\overline{(2 + i)} \div (3 + i)$ **45** $(2 + i)^3$

46 $(3 - i)^3$ **47** $(1 + i)^4$ **48** $(2 + 3i)^4$

49 Show that the addition of complex numbers defined by Eq. (12.2) is associative.

50 Show that the multiplication of complex numbers defined by (12.3) is commutative.

51 Show that $(1, 0)$ is the identity for multiplication.

52 Show that the distributive law $z_1(z_2 + z_3) = z_1z_2 + z_1z_3$ holds for complex numbers $z_1 = (a, b)$, $z_2 = (c, d)$, $z_3 = (e, f)$.

53 Show that $i^5 = i$ and $i^6 = i^2 = -1$. What are i^7 and i^8?

54 Show that for any complex number z, we have $\bar{\bar{z}} = z$.

55 Prove that $\overline{z_1 z_2} = \bar{z}_1 \bar{z}_2$.

56 Prove that $(\overline{z^n}) = (\bar{z})^n$.

12.2 / COMPLEX NUMBERS AS VECTORS

The rest of this chapter requires trigonometry, although many of the problems in Exercise 12.2 may be done without it. For those students not familiar with trigonometry, a resumé of relevant facts in the Appendix will allow efficient computation. The material in the chapter involves a clever geometrical way of finding products, quotients, powers, and roots of complex numbers.

Vectors may be treated in two ways. The first way is abstract and is embodied in the (a, b) notation, which we have already treated. The second way is more concrete and geometrical. Any quantity that has magnitude and direction is called a *vector* or a *vector quantity*. It can be represented by a directed line segment, as shown in Fig. 12.1.

Vector quantity
Vector

We shall identify a vector by a boldface letter such as **v**. Since such a letter cannot be made in writing, it is customary to use a letter with an arrow above it. Thus, if **v** were used in print, $\vec{v}$ would be used in writing. The vector **v** in Fig. 12.1 is determined by the ordered number pair (x, y), where x and y are real numbers. In fact, x and y are called the *components of* **v**, and **v** is their *resultant*.

Components
Resultant

We have seen in this section that an ordered number pair determines a vector, and we found in Sec. 12.1 that an ordered number pair is a complex number. Consequently, we conclude that *a complex number is a vector.*

Absolute value

If **v** $= (x, y)$ *is a vector or complex number, then its absolute value or modulus is designated by* $|\mathbf{v}|$ *or* r *and is defined by*

$$|\mathbf{v}| = r = \sqrt{x^2 + y^2} \tag{12.8}$$

Amplitude
Argument

The angle from the positive real axis to the vector is called the *amplitude* or *argument* of the complex number. It is designated by θ in Fig. 12.1 and is determined by

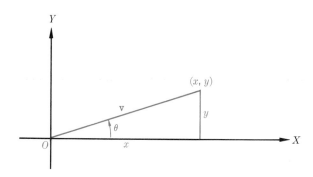

Figure 12.1

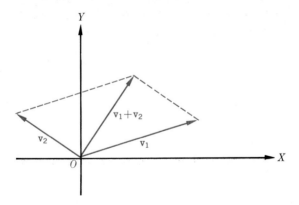

Figure 12.2

$$\theta = \arctan \frac{y}{x} \quad \text{or} \quad \tan \theta = \frac{y}{x} \qquad (12.9)$$

where the signs of x and y determine the quadrant that θ is in.

We defined the sum of two complex numbers or vectors in (12.2) in algebraic terms, and we shall now define the sum or resultant in geometric terms.

Sum The *sum* or *resultant* of two vectors v_1 and v_2 is the vector $v_1 + v_2$, which begins at the origin as do v_1 and v_2 and terminates at the opposite vertex of a parallelogram that has v_1 and v_2 as two of its adjacent sides.

This definition is illustrated in Fig. 12.2. Notice that v_2 may be represented also by the line segment w since they have the same magnitude and direction, even though they begin at different points. Thus, $w = v_2$.

To multiply a complex number by a real number, we write $c(a, b) = (ca, cb)$. We shall now see how to show this product as a vector. Each component of (ca, cb) is c times the corresponding component of (a, b); this is shown in Fig. 12.3.

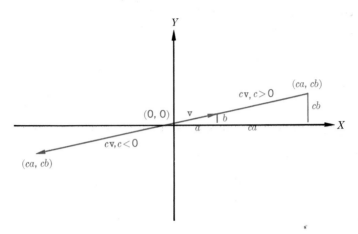

Figure 12.3

EXAMPLE Find the absolute value and argument of $(3, -4)$.

Solution The absolute value of $\mathbf{v} = (3, -4)$, by use of Eq. (12.8), is

$$|\mathbf{v}| = r = \sqrt{3^2 + (-4)^2} = 5$$

and by Eq. (12.9) the argument is

$$\theta = \arctan\left(\frac{-4}{3}\right) = 306°50'$$

12.3 / POLAR REPRESENTATION

By use of Fig. 12.1 and the definitions in Sec. 1 of the Appendix, we see that $\cos\theta = x/r$ and $\sin\theta = y/r$. Hence, $x = r\cos\theta$ and $y = r\sin\theta$. Therefore, the complex number

$$z = x + iy$$
$$= r\cos\theta + ir\sin\theta$$

Consequently,

$$z = r(\cos\theta + i\sin\theta) \qquad (12.10)$$

Trigonometric, or polar, form of z

We shall frequently use the notation "cis θ" to stand for $\cos\theta + i\sin\theta$.

The right member of Eq. (12.10) is called the *trigonometric*, or *polar*, form of z, and it is very useful in the process of multiplication and division.

EXAMPLE 1 Express the complex number $1 - i\sqrt{3}$ in polar form.

Solution We shall first obtain the values of r and θ:

$$r = \sqrt{1^2 + (\sqrt{3})^2} = \sqrt{1 + 3} = 2 \qquad \text{by Eq. (12.8)}$$

$$\theta = \arctan\frac{\sqrt{3}}{1} \qquad \text{by Eq. (12.9)}$$

$$= 60° \qquad \text{by Eq. (a.8), Appendix}$$

Consequently, $1 + i\sqrt{3} = 2(\cos 60° + i\sin 60°) = 2 \text{ cis } 60°$.

EXAMPLE 2 Express $4 - 5i$ in polar form.

Solution The absolute value of $4 - 5i$ is

$$r = \sqrt{4^2 + (-5)^2} = \sqrt{16 + 25} = \sqrt{41} \qquad \text{by Eq. (12.8)}$$

The argument is

$$\theta = \arctan\left(\frac{-5}{4}\right) \qquad \text{by Eq. (12.9)}$$

Consequently, $\quad 4 - 5i = \sqrt{41}\left[\cos\left(\arctan\dfrac{-5}{4}\right) + i \sin\left(\arctan\dfrac{-5}{4}\right)\right]$

$$= \sqrt{41} \, \text{cis} \left(\arctan\dfrac{-5}{4}\right)$$

Using a table of trigonometric functions, we find that $\arctan\frac{5}{4} = \arctan 1.25 = 51°20'$. Then, since the point representing $4 - 5i$ is in the fourth quadrant, we have

$$\theta = 360° - 51°20' = 308°40'$$

Hence, $4 - 5i = \sqrt{41} \, \text{cis} \, 308°40'$.

EXERCISE 12.2 Vectors, Polar Representation

Plot the points that represent the complex number and its conjugate in each of Probs. 1 to 20, and draw the vectors.

1	$(2, 4)$	2	$(-1, 2)$	3	$(4, -3)$	4	$(-1, -5)$
5	$-3 + i$	6	$-2i$	7	$8 + 3i$	8	$2 - 5i$
9	$2(3 + i)$	10	$-4(6 - 5i)$	11	$5(-2 + 5i)$	12	$-3(-1 + 7i)$
13	$(3 + 2i) + (-4 + i)$			14	$(-3 + 5i) + (-2 - i)$		
15	$(4, -1) - (2, 3)$			16	$(-1, 1) - (-3, 2)$		
17	$2(3 + i) + 3(2 - i)$			18	$5(-2 + 3i) - 3(2 + 5i)$		
19	$4(1, 0) - 2(5, 2)$			20	$3(3, 2) + 2(5, -3)$		

Find the absolute value and argument of the complex number in each of Probs. 21 to 32.

21	$3 + 4i$	22	$-5 + 12i$	23	$(-6, -8)$
24	$(15, -8)$	25	$(2, 2)$	26	$(1, -\sqrt{3})$
27	$-3 + i\sqrt{3}$	28	$-2\sqrt{3} - 2i$	29	$(-2 + i) + (3 - 2i)$
30	$(1 - i) - (3 + i)$		31	$2(2 - 3i) - 3(1 + i)$	
32	$5(3 - i) + 2(-6 + 5i)$				

Express the complex number in each of Probs. 33 to 52 in polar form. Use the exact angle if possible; otherwise use the nearest degree.

33	$3 + 3i$	34	$-2 + 2i$	35	$-5 - 5i$	36	$3 - 3i$
37	$\sqrt{3} + i$	38	$\sqrt{3} - i$	39	$-1 + i\sqrt{3}$	40	$-1 - i\sqrt{3}$
41	$5i$	42	6	43	$-2i$	44	-5
45	$8 - 15i$	46	$-4 + 3i$	47	$7 + 24i$	48	$-11 - 60i$
49	$5 + 2i$	50	$-2 - 5i$	51	$-4 + i$	52	$3 - 7i$

53 Let $z = a + bi$ and $w = c + di$. Show that

$$|zw| = |z| \cdot |w|$$

HINT: Why is it equivalent to show $|zw|^2 = |z|^2|w|^2$?

54 Using the same notation as in Prob. 53, show that

$$\left|\frac{z}{w}\right| = \frac{|z|}{|w|}$$

55 Verify the equations in Probs. 53 and 54 for $z = 2 + 3i$ and $w = -1 + 5i$.

56 Repeat Prob. 55 for $z = -4 + i$ and $w = 3 - 2i$.

57 Show that for any two complex numbers z and w

$$|z + w| \le |z| + |w|$$

Refer to Fig. 12.2 and decide why this is called the *triangle inequality*.
HINT: Justify the following steps and complete the argument [$\text{Re}(z\bar{w})$
means the real part of $z\bar{w}$].

$$
\begin{aligned}
|z + w|^2 &= (z + w)(\bar{z} + \bar{w}) \\
&= z\bar{z} + z\bar{w} + w\bar{z} + w\bar{w} \\
&= |z|^2 + z\bar{w} + \overline{z\bar{w}} + |w|^2 \\
&= |z|^2 + 2\text{Re}(z\bar{w}) + |w|^2 \\
&\le |z|^2 + 2|z\bar{w}| + |w|^2 \\
&= |z|^2 + 2|z| \cdot |w| + |w|^2 \\
&= (|z| + |w|)^2
\end{aligned}
$$

58 Show that for any two complex numbers

$$||z| - |w|| \le |z - w|$$

HINT: $|z| = |(z - w) + w| \le |z - w| + |w|$.

59 Verify the inequalities in Probs. 57 and 58 for $z = 3 + 5i$ and $w = -2 + i$.

60 Repeat Prob. 59 for $z = -2 - 3i$ and $w = 4 - i$.

12.4 / THE PRODUCT OF TWO COMPLEX NUMBERS IN POLAR FORM

In this section and in Sec. 12.5, we
shall show that the product and the quotient of two complex numbers are
obtained very readily if the numbers are expressed first in polar form.
We shall consider the two complex numbers

$$z = (x, y) = x + iy = r(\cos \theta + i \sin \theta) = r \text{ cis } \theta$$

and $$w = (u, v) = u + iv = R(\cos \phi + i \sin \phi) = R \text{ cis } \phi$$

By Eq. (12.3), we have

$$zw = (x, y)(u, v) = (xu - yv, xv + yu) = (xu - yu) + i(xv + yu)$$

Now, if we replace x, y, u, and v by $r \cos \theta$, $r \sin \theta$, $R \cos \phi$, and $R \sin \phi$, respectively, and simplify, we have

$$zw = (r \cos \theta)(R \cos \phi) - (r \sin \theta)(R \sin \phi) + i[(r \cos \theta)(R \sin \phi) + \\ (r \sin \theta)(R \cos \phi)]$$
$$= rR[(\cos \theta \cos \phi - \sin \theta \sin \phi) + i(\sin \theta \cos \phi + \cos \theta \sin \phi)]$$

Furthermore, $\qquad \cos \theta \cos \phi - \sin \theta \sin \phi = \cos (\theta + \phi) \qquad$ by Eq. (a.29)

and $\qquad \cos \theta \sin \phi + \sin \theta \cos \phi = \sin \theta \cos \phi + \cos \theta \sin \phi$
$$= \sin (\theta + \phi) \qquad \text{by Eq. (a.28)}$$

Therefore, since $zw = (r \text{ cis } \theta)(R \text{ cis } \phi)$, we have

$$\textbf{(r cis } \theta \textbf{)(R cis } \phi \textbf{)} = \textbf{rR cis } (\theta + \phi) \qquad \textbf{(12.11)}$$

Consequently, we have the following theorem:

THEOREM Product of two complex numbers	*The absolute value of the product of two complex numbers is the product of their absolute values. An argument of the product of two complex numbers is the sum of their arguments.*

Since the product of two complex numbers is itself a complex number, we can obtain the product of three or more complex numbers by a repeated application of this theorem.

EXAMPLE Find the product of $1 + i$, $1 + \sqrt{3}i$, and $\sqrt{3} - i$.

Solution We first express each of these numbers in polar form and get

$$1 + i = \sqrt{2}(\cos \arctan 1 + i \sin \arctan 1)$$
$$= \sqrt{2}(\cos 45° + i \sin 45°) \qquad \text{by Eq. (a.6)}$$
$$= \sqrt{2} \text{ cis } 45°$$
$$1 + \sqrt{3}i = 2(\cos \arctan \sqrt{3} + i \sin \arctan \sqrt{3})$$
$$= 2 \text{ cis } 60° \qquad \text{by Eq. (a.8)}$$

Since the point representing $\sqrt{3} - i$ is the fourth quadrant, the argument of $\sqrt{3} - i$ is $360°$ minus the acute angle whose tangent is $1/\sqrt{3}$. By Eq. (a.4), this angle is $30°$. Hence,

$$\sqrt{3} - i = 2 \text{ cis } (360° - 30°) = 2 \text{ cis } 330°$$

Therefore,

$$(1 + i)(1 + \sqrt{3}i)(\sqrt{3} - i) = (\sqrt{2} \text{ cis } 45°)(2 \text{ cis } 60°)(2 \text{ cis } 330°)$$
$$= 4\sqrt{2} \text{ cis } (45° + 60° + 330°)$$
$$= 4\sqrt{2} \text{ cis } 435°$$
$$= 4\sqrt{2} \text{ cis } (360° + 75°)$$
$$= 4\sqrt{2} \text{ cis } 75° \qquad \text{by Eqs. (a.22) and (a.21)}$$

12.5 / THE QUOTIENT OF TWO COMPLEX NUMBERS IN POLAR FORM

The quotient $z/w = q$ means that $z = wq$. If $z = r \text{ cis } \theta$, $w = R \text{ cis } \phi$, and $q = A \text{ cis } \alpha$, then by Eq. (12.11), $z = wq$ becomes

$$r \text{ cis } \theta = (R \text{ cis } \phi)(A \text{ cis } \alpha) = RA \text{ cis } (\phi + \alpha)$$

Consequently $r = RA$ and $\theta = \phi + \alpha$ provides a solution, any other solution being different only by a multiple of 360° in the angle. Solving for A and α gives

$$A = \frac{r}{R} \qquad \alpha = \theta - \phi$$

and so we have

$$\frac{r \text{ cis } \theta}{R \text{ cis } \phi} = \frac{r}{R} \text{ cis } (\theta - \phi) \qquad \qquad \textbf{(12.12)}$$

Therefore, we have the following theorem:

THEOREM

Quotient of two complex numbers

The absolute value of the quotient of two complex numbers is equal to the absolute value of the dividend divided by the absolute value of the divisor. An argument of the quotient is equal to the argument of the dividend minus the argument of the divisor.

The formulas in Eqs. (12.11) and (12.12) have the following geometrical interpretations: To multiply two complex numbers, add their angles and measure off a distance rR on the terminal side of the resulting angle. To divide, subtract the angles and measure off r/R on the terminal side.

EXAMPLE   Express $1 + \sqrt{3}i$ and $1 + i$ in polar form and obtain their quotient.

Solution   If we express each of the given complex numbers in polar form, we have

$$\frac{1 + \sqrt{3}i}{1 + i} = \frac{2 \text{ cis arctan } \sqrt{3}}{\sqrt{2} \text{ cis arctan } 1}$$

$$= \frac{2 \text{ cis } 60°}{\sqrt{2} \text{ cis } 45°} \qquad \text{since by Eqs. (a.8) and (a.6)}$$
$$\qquad \text{arctan } \sqrt{3} = 60° \text{ and arctan } 1 = 45°$$

$$= \frac{2}{\sqrt{2}} \text{ cis } (60° - 45°) \qquad \text{by Eq. (11.12)}$$

$$= \sqrt{2} \text{ cis } 15°$$

By using tables we may express this quotient as $1.4142 \ (0.9659 + 0.2588i) = 1.3660 + 0.3660i$.

EXERCISE 12.3 Products and Quotients

Perform the indicated operations in Probs. 1 to 20 and leave the answer in polar form.

1 $[2(\cos 26° + i \sin 26°)][4(\cos 191° + i \sin 191°)]$
2 $[\sqrt{2}(\cos 321° + i \sin 321°)][\sqrt{12}(\cos 6° + i \sin 6°)]$
3 $[\frac{3}{2}(\cos 140° + i \sin 140°)][6(\cos 108° + i \sin 108°)]$
4 $[1.52(\cos 225° + i \sin 225°)][3.79(\cos 45° + i \sin 45°)]$
5 $(3 \text{ cis } 28°)(2 \text{ cis } 44°)(2 \text{ cis } 82°)$
6 $(4 \text{ cis } 47°)(\text{cis } 132°)(4 \text{ cis } 78°)$
7 $(1.2 \text{ cis } 36°)(2.6 \text{ cis } 108°)(3 \text{ cis } 126°)$
8 $(6 \text{ cis } 212°)(2 \text{ cis } 97°)(10 \text{ cis } 88°)$
9 $5(\cos 128° + i \sin 128°) \div 2.5(\cos 38° + i \sin 38°)$
10 $8(\cos 247° + i \sin 247°) \div 2(\cos 112° + i \sin 112°)$
11 $12 \text{ cis } 308° \div \text{cis } 158°$
12 $4.8 \text{ cis } 219° \div 3.2 \text{ cis } 84°$
13 $(6 \text{ cis } 84°)(8 \text{ cis } 37°) \div 12 \text{ cis } 92°$
14 $(14 \text{ cis } 138°)(15 \text{ cis } 322°) \div 35 \text{ cis } 205°$
15 $(\sqrt{3} \text{ cis } 248°)(\sqrt{6} \text{ cis } 238°) \div \sqrt{2} \text{ cis } 188°$
16 $(1.4 \text{ cis } 310°)(0.625 \text{ cis } 147°) \div \text{cis } 403°$
17 $8 \text{ cis } 48° \div [(2 \text{ cis } 30°)(3 \text{ cis } 7°)]$
18 $24 \text{ cis } 408° \div [(3 \text{ cis } 126°)(\sqrt{2} \text{ cis } 200°)]$
19 $\sqrt{6} \text{ cis } 210° \div [(2 \text{ cis } 45°)(3 \text{ cis } 75°)]$
20 $5.1 \text{ cis } 282° \div [(0.17 \text{ cis } 137°)(0.3 \text{ cis } 45°)]$

Perform the indicated operations in Probs. 21 to 32 graphically.

21 $(1 + i)(1 + i\sqrt{3})$
22 $(1 - i)(\sqrt{3} + i)$
23 $(3 + i\sqrt{3})(-2 - 2i)$
24 $(-4 + 4i)(-1 - i\sqrt{3})$
25 $(\sqrt{3} + i\sqrt{3}) \div (\sqrt{3} + i)$
26 $(-3 - i\sqrt{3}) \div (-1 + i)$

27　$(2-2i) \div (-1+i\sqrt{3})$　　　　　　28　$(-\sqrt{3}-i) \div (2+2i)$

29　$(\text{cis } 37°)(\text{cis } 74°)$　　　　　　　　30　$\text{cis } 144° \div \text{cis } 16°$

31　$(3 \text{ cis } 55°)(2 \text{ cis } 81°) \div 12 \text{ cis } 73°$

32　$(4 \text{ cis } 122°) \div [(\text{cis } 38°)(1.5 \text{ cis } 46°)]$

Perform the following operations in polar form. Give the answer in polar and a + bi form. Use two decimal places in all intermediate and final approximations; use angles to the nearest degree.

33　$(1+i\sqrt{3})(\sqrt{3}+i)$　　　　　　34　$(\sqrt{2}+i\sqrt{2})(-\sqrt{2}+i\sqrt{2})$

35　$(-1+i\sqrt{3})(\sqrt{2}+i\sqrt{2})$　　　36　$(-1+i\sqrt{3})(-1-i\sqrt{3})$

37　$(2+3i)(5+2i)$　　　　　　　38　$(2+7i)(5+8i)$

39　$(1.4+1.1i)(3.8+2.7i)$　　　40　$(3.6+1.7i)(2.2+3.4i)$

41　$(1+i)(-2+2i)(1+i\sqrt{3})$　　42　$(\sqrt{2}+i\sqrt{2})(2+i\sqrt{12})(\sqrt{2}-i\sqrt{6})$

43　$(4+3i)(12+5i)(15+8i)$　　44　$(3+4i)(24+7i)(5+12i)(0.01)$

45　i^3　　46　i^7　　　　　　47　i^6　　48　i^{10}

49　$(-1+i\sqrt{3})^3$　　　　　　50　$\left(\dfrac{3}{2}+\dfrac{3i\sqrt{3}}{2}\right)^3$

51　$(\sqrt{5}-i\sqrt{5})^4$　　　　　　52　$(1+i)^4$

53　$(3+4i) \div (12+5i)$　　　　54　$(8+15i) \div (24+7i)$

55　$(2+3.1i) \div (3.6+3.9i)$　　56　$(4.23+7.81i) \div (6.23+2.89i)$

57　$(1+i)(1-i\sqrt{3}) \div (-\sqrt{3}+i)$　　58　$(1+i\sqrt{3})(\sqrt{3}+i) \div (1+i)$

59　$(-1-i) \div [(1+i\sqrt{3})(1-i\sqrt{3}]$

60　$(1+i)(1+i\sqrt{3}) \div [i(-1-i)]$

61　How does it follow from Eq. (12.11) that the absolute value of the product of two complex numbers is the product of their absolute values? Compare this proof to the one in Prob. 53 of Exercise 12.2.

62　Prove Eq. (12.11) by using ordered pairs.

63　Prove Eq. (12.12) by multiplying $\dfrac{r \text{ cis } \theta}{R \text{ cis } \phi}$ by the conjugate of $R \text{ cis } \phi$ in both numerator and denominator.

64　Show that if

$$\cos \theta + i \sin \theta = \cos \phi + i \sin \phi$$

then $\theta = \phi + n \cdot (360°)$ for some integer n.

HINT: Justify the following steps:

$$\cos \theta = \cos \phi \quad \text{and} \quad \sin \theta = \sin \phi \qquad (1)$$
$$\sin \theta \cos \phi = \cos \theta \sin \phi$$
$$\sin \theta \cos \phi - \cos \theta \sin \phi = 0$$
$$\sin(\theta - \phi) = 0$$

$\theta - \phi = n \cdot (180°)$, so the only possibilities are $\theta - \phi = 0°$, $180°$, $360°$, $540°$, $720°$, . . . or their negatives. Now show that $180°$, $540°$, . . . do not satisfy Eq. (1), but $0°$ $360°$, $720°$, . . . do satisfy Eq. (1).

12.6 / DE MOIVRE'S THEOREM

In this section we shall discuss a theorem proved by Abraham De Moivre (1667–1754) that is very useful for finding a power or a root of a complex number.

If we square the complex number $z = r \operatorname{cis} \theta$, we get

$$z^2 = (r \operatorname{cis} \theta)(r \operatorname{cis} \theta)$$
$$= r^2 \operatorname{cis} (\theta + \theta) = r^2 \operatorname{cis} 2\theta \qquad \text{by Eq. (12.11)}$$

Furthermore,

$$z^3 = z^2(z) = (r^2 \operatorname{cis} 2\theta)(r \operatorname{cis} \theta)$$
$$= r^3 \operatorname{cis} (2\theta + \theta) = r^3 \operatorname{cis} 3\theta \qquad \text{by Eq. (12.11)}$$

A repeated application of this procedure leads to the statement that

De Moivre's theorem
$$[r(\cos \theta + i \sin \theta)]^n = r^n (\cos n\theta + i \sin n\theta) \qquad (12.13)$$

In abbreviated form, the above theorem becomes

$$(r \operatorname{cis} \theta)^n = r^n \operatorname{cis} n\theta$$

This statement is known as De Moivre's theorem.

The proof of this theorem for integral values of n requires the use of mathematical induction and will be given as Example 2 in Sec. 15.1. At this point, we shall assume that the theorem holds for integral values of n.

EXAMPLE 1 Find the fifth power of $\sqrt{3} + i$.

Solution
$$(\sqrt{3} + i)^5 = \left(2 \operatorname{cis} \arctan \frac{1}{\sqrt{3}}\right)^5$$

$$= (2 \operatorname{cis} 30°)^5 \qquad \text{since } \arctan \frac{1}{\sqrt{3}} = 30°$$

$$= 2^5 \operatorname{cis} 5(30°) \qquad \text{by Eq. (12.13)}$$
$$= 32 \operatorname{cis} 150°$$
$$= 32 \operatorname{cis} (180° - 30°)$$
$$= 32(-\cos 30° + i \sin 30°) \qquad \text{by Eqs. (a.19) and (a.18)}$$

$$= 32\left(-\frac{\sqrt{3}}{2} + i\frac{1}{2}\right) \qquad \text{by Eqs. (a.7) and (a.3)}$$

$$= -16\sqrt{3} + 16i$$

EXAMPLE 2　　Find the nth power of $\cos{(\theta/n)} + i \sin{(\theta/n)}$.

Solution　　De Moivre's theorem gives

$$\left(\cos\frac{\theta}{n} + i \sin\frac{\theta}{n}\right)^n = \cos n\left(\frac{\theta}{n}\right) + i \sin n\left(\frac{\theta}{n}\right)$$
$$= \cos\theta + i \sin\theta$$

12.7 / ROOTS OF COMPLEX NUMBERS

nth root　　We say the complex number z is an nth root of the complex number w if $z^n = w$. Example 2 in Sec. 12.6 shows that $\cos\theta/n + i \sin\theta/n$ is an nth root of $\cos\theta + i \sin\theta$. In the set of real numbers, there is no square root of -9, no fourth root of -16, or in fact, no even root of any negative number; furthermore, there is only one odd root of a negative number. If, however, we employ complex numbers, we can obtain n nth roots of any given number by use of De Moivre's theorem. We shall illustrate the procedure with two examples.

EXAMPLE 1　　Obtain the three cube roots of 64.

Solution　　We first express $64 = 64 + 0i$ in polar form and get

$$64 = 64(\cos 0° + i \sin 0°) = 64 \operatorname{cis} 0° \tag{1}$$

Now if $r \operatorname{cis} \theta$ is a cube root of 64, then

$$64 \operatorname{cis} 0° = (r \operatorname{cis} \theta)^3 \qquad \text{definition of cube root}$$
$$64 \operatorname{cis} 0° = r^3 \operatorname{cis} 3\theta \qquad \text{Eqs. (1) and (12.13)}$$

We next make use of the fact that if two complex numbers are equal, then their absolute values are equal and their arguments are either equal or differ by a multiple of 360°. This is evident geometrically; see also Prob. 64 in Exercise 12.3. Thus, from Eq. (2),

$$64 = r^3 \qquad \text{and} \qquad 3\theta = 0° + k \cdot 360°$$

where k is an integer. Since r is real, $r^3 = 64$ means that $r = \sqrt[3]{64} = 4$. Also

$$\theta = \frac{0° + k \cdot 360°}{3} = 0° + k \cdot 120° = k \cdot 120°$$

Taking $k = 0, 1, 2$ gives $\theta = 0°, 120°, 240°$, so the three cube roots of 64 are

$$4(\cos 0° + i \sin 0°) = 4$$
$$4(\cos 120° + i \sin 120°) = -2 + 2i\sqrt{3}$$
$$4(\cos 240° + i \sin 240°) = -2 - 2i\sqrt{3}$$

EXAMPLE 2 Find the four fourth roots of $1 + i\sqrt{3}$.

Solution Proceeding as in Example 1, we write

$$1 + i\sqrt{3} = 2(\cos 60° + i \sin 60°) = 2 \text{ cis } 60°$$
$$2 \text{ cis } 60° = (r \text{ cis } \theta)^4 = r^4 \text{ cis } 4\theta$$

$$r^4 = 2 \quad \text{and} \quad 4\theta = 60° + k \cdot 360°$$

$$r = \sqrt[4]{2} \quad \text{and} \quad \theta = \frac{60° + k \cdot 360°}{4} = 15° + k \cdot 90°$$

For $k = 0, 1, 2, 3$, we have $\theta = 15°, 105°, 195°$, and $285°$.
The corresponding values of the fourth roots of $1 + 3i$ are $2^{1/4}$ cis $15°$, $2^{1/4}$ cis $105°$, $2^{1/4}$ cis $195°$, $2^{1/4}$ cis $285°$.

If a root of a complex number is known in polar form, the approximate value in algebraic form can be obtained by using an approximation to each function value of each angle and calculating the real nth root of r by use of logarithms.

It is an interesting fact that the four fourth roots of $1 + \sqrt{3}i$ are equally spaced about the circumference of a circle of radius $r^{1/4} = 2^{1/4}$, the smallest argument being $15°$. This is a special case of the following statement:

The n nth roots of $a + bi$ are equally spaced about the circumference of a circle of radius $r^{1/n} = (\sqrt{a^2 + b^2})^{1/n}$, the smallest argument being θ/n, and the difference between adjacent angles being $360°/n$.

EXERCISE 12.4 Powers and Roots

Use De Moivre's theorem to find the indicated power in Probs. 1 to 20. Use angles to the nearest degree in all approximations.

1 $(2i)^5$	2 $(-3i)^4$	3 $(-5)^3$
4 7^3	5 $(1 + i\sqrt{3})^7$	6 $(\sqrt{3} + i)^5$
7 $(1 - i)^4$	8 $(\sqrt{3} - i)^4$	9 $(-1 - i\sqrt{3})^5$
10 $(-1 + i)^6$	11 $(1 + i)^7$	12 $(-\sqrt{3} - i)^8$
13 $(-5 + 12i)^4$	14 $(8 - 15i)^3$	15 $(-3 - 4i)^5$
16 $(24 + 7i)^2$	17 $(3 - 5i)^3$	18 $(-4 + 7i)^4$
19 $(5 + 7i)^6$	20 $(-3 - 8i)^3$	

Obtain the indicated roots in Probs. 21 to 40.

21 Cube roots of 1	22 Fourth roots of -1
23 Cube roots of i	24 Fifth roots of $32i$

25 Square roots of $1 + i$

26 Fifth roots of $-1 - i$

27 Ninth roots of $-1 + i$

28 Cube roots of $1 + i$

29 Cube roots of $\sqrt{3} + i$

30 Fourth roots of $\dfrac{1}{2} - \dfrac{i\sqrt{3}}{2}$

31 Sixth roots of $-32 - 32i\sqrt{3}$

32 Eighth roots of $-1 + i\sqrt{3}$

33 Fifth roots of $(-1 + i)^3$

34 Tenth roots of $(2 - 2i\sqrt{3})^2$

35 Fourth roots of $(1 - i\sqrt{3})^2$

36 Fifth roots of $(-\sqrt{3} + i)^2$

37 Square roots of $3 + 4i$

38 Eighth roots of $2 - 3i$

39 Sixth roots of $-3 + i$

40 Fourth roots of $5 + 12i$

41 Show, using exact values or decimal approximations, that the sum of the 4 fourth roots of -1 is zero (see Prob. 22).

42 Repeat Prob. 41 for the 5 fifth roots of $-1 - i$ (see Prob. 26).

43 Repeat Prob. 41 for the 6 sixth roots of $-32 - 32i\sqrt{3}$ (see Prob. 31).

44 Repeat Prob. 41 for the 8 eighth roots of $2 - 3i$ (see Prob. 38).

Solve the equations in Probs. 45 to 52.

45 $x^2 + 4i = 0$

46 $x^3 - 64i = 0$

47 $x^4 - i = 0$

48 $x^5 + 32i = 0$

49 $x^2 + x(-2 + i) + 1 - i = 0$

50 $x^2 + x(4 + i) + 4 + 2i = 0$

51 $x^2 - 2ix - 10 = 0$

52 $x^2 + 2ix - 2 = 0$

EXERCISE 12.5 **REVIEW**

1 Solve the equation $(x - y - 1,\ 2x) = (-3,\ 3x - y + 2)$.

2 Solve the equation $x^2 - y^2 + 2ixy = 8 - 6i$.

Perform the indicated operations in Probs. 3 to 12.

3 $(5, 8) - (2, -1)$

4 $4(2, -3) + 3(-1, 6)$

5 $(5 + 2i)(3 - i)$

6 $(5 + 2i) \div (3 - i)$

7 $\dfrac{(3, 4)}{(5, 6)}$

8 $(2, -3)\overline{(4, -5)}$

9 $(5 \text{ cis } 38°)(3 \text{ cis } 222°)$

10 $\dfrac{(3 \text{ cis } 48°)^3}{6 \text{ cis } 77°}$

11 $\overline{(\sqrt{3} + i)^5}$

12 $(1 + i\sqrt{3})^{1/6}$

13 Verify that $|2 + i| \cdot |4 - 3i| = |(2 + i)(4 - 3i)|$.

14 Verify that $|(4 + 7i) - (2 - 3i)| \le |2 + 10i|$.

15 Verify that $\overline{(2 - i)(3 + i)(1 + 2i)} = \overline{(2 - i)}\,\overline{(3 + i)}\,\overline{(1 + 2i)}$.

16   Verify that $|1 + 3i|^3 = |(1 + 3i)^3|$.

17   Verify that $\overline{(1 + 3i)^4} = \overline{(1 + 3i)}^4$.

18   Show that $|z| = |\bar{z}|$ for any complex number z.

19   Show that if $z = a + bi$ and $w = c + di$, then

$$|ac + bd| \leq |z| \cdot |w|$$

20   Verify the inequality in Prob. 19 for $z = 4 - 5i$ and $w = 1 + 6i$.

21   Find complex numbers z and w so that, in Prob. 19, the $\leq$ is actually $=$.

22   Show that $|\cos \theta + i \sin \theta| = 1$.

23   Express $\sin 60° + i \cos 60°$ in polar form.

24   Express $\sin 30° + i \sin 45°$ in polar form.

25   Prove De Moivre's theorem for negative integers n.   HINT: $(r \operatorname{cis} \theta)^n = 1/(r \operatorname{cis} \theta)^{-n}$, and $-n$ is positive.

26   Show that the sum of the 8 eighth roots of unity (that is, of 1) is 0.

CHAPTER THIRTEEN

Polynomial Equations

In this chapter we shall discuss equations of the type $f(x) = 0$, where $f(x)$ is a polynomial of degree greater than 2. Unfortunately, no general method exists for obtaining the solution set of such equations if the degree of the polynomial is greater than 4, and the direct methods for solving equations in which the polynomial is of degree 3 or 4 are long and tedious. We shall present several theorems, however, that will enable us to obtain pertinent information about the roots. We shall also discuss methods for obtaining an approximation of any irrational root to the desired degree of accuracy.

13.1 / POLYNOMIAL EQUATIONS

An equation of the type

$$a_0 x^n + a_1 x^{n-1} + a_2 x^{n-2} + \cdots + a_{n-1} x + a_n = 0 \qquad \text{(13.1)}$$

where n is a positive integer and the coefficients $a_0, a_1, a_2, \ldots, a_{n-1}, a_n$ are constants, is a polynomial equation. The coefficients a_i will often be rational in this chapter, but they may also be any real or complex numbers. The left member of (13.1) is called a *polynomial function*.

EXAMPLE   Some examples of polynomial equations are

$$x^2 + 2x + 5 = 0, \ 3x^4 + 2x^3 - 4x = 0, \ x^3 + \sqrt{3}\,x^2 - 21\pi x + \sqrt[3]{4} - \frac{1}{\sqrt{2}} = 0, \text{ and}$$

$$2i\,x^3 - (3i - 4)x^2 + x + 1 = 0.$$

We shall make extensive use of functional notation. For example, if $f(x) = 2x^3 + x^2 - 2x + 4$, then $f(-3) = 2(-3)^3 + (-3)^2 - 2(-3) + 4 = -54 + 9 + 6 + 4 = -35$ and $f(r) = 2r^3 + r^2 - 2r + 4$.

13.2 / THE REMAINDER THEOREM

The computation involved in finding the solution set of Eq. (13.1) is greatly simplified if the left member is factored into linear factors, or even if one or more linear factors are found. The remainder theorem stated and proved below is useful for this purpose and is also essential in finding the solution set of Eq. (13.1) when the left member is not readily factorable.

THEOREM

Remainder theorem

If a polynomial $f(x)$ is divided by $x - r$ until a remainder independent of x is obtained, then the remainder is equal to $f(r)$.

Before proving the remainder theorem, we shall illustrate its meaning in the following example.

EXAMPLE   If we divide $x^3 - 2x^2 - 4x + 5$ by $x - 3$, using the method of Sec. 2.15, we obtain $x^2 + x - 1$ as the quotient and 2 as the remainder. Note that the remainder is independent of x. In this problem, $x - r = x - 3$. Hence $r = 3$. Since

$$f(x) = x^3 - 2x^2 - 4x + 5$$

we have

$$f(3) = 3^3 - 2(3)^2 - 4(3) + 5 = 27 - 18 - 12 + 5 = 2$$

and this is equal to the remainder obtained above.

In a division process, we have the following relation between the dividend, the divisor, the quotient, and the remainder:

$$\text{Dividend} = (\text{quotient})(\text{divisor}) + \text{remainder}$$

Hence, if the quotient obtained by dividing $f(x)$ by $x - r$ is $Q(x)$ and if the remainder is R, we have

$$f(x) = Q(x)(x - r) + R \qquad\qquad (13.2)$$

Equation (13.2) is true for all values of x including $x = r$. Hence, if we substitute r for x in Eq. (13.2), we have

$$f(r) = Q(r)(r - r) + R = Q(r) \cdot (0) + R = R$$

This proves the remainder theorem.

13.3 / THE FACTOR THEOREM AND ITS CONVERSE

If r is a root of $f(x) = 0$, then $f(r) = 0$. Hence, by the remainder theorem, R in Eq. (13.2) is zero, and Eq. (13.2) then becomes

$$f(x) = Q(x)(x - r) \qquad (13.3)$$

Therefore, $x - r$ is a factor of $f(x)$, and we have proved the factor theorem:

Factor theorem   If r is a root of the polynomial equation $f(x) = 0$, then $x - r$ is a factor of $f(x)$.

Conversely, if $x - r$ is a factor of $f(x)$, then the remainder obtained by dividing $f(x)$ by $x - r$ is equal to zero. Hence, by the remainder theorem, $f(r) = 0$. Therefore, r is a root of $f(x) = 0$. Hence we have the converse of the factor theorem:

Converse of factor theorem   If $x - r$ is a factor of the polynomial $f(x)$, then r is a root of $f(x) = 0$.

The factor theorem and its converse may be combined by saying that $x - r$ is a factor of the polynomial $f(x)$ if and only if r is a root of $f(x) = 0$.

EXAMPLE   We may show that $x - 2$ is a factor of $f(x) = -x^3 + 2x^2 - 2x + 4$ by calculating $f(2) = -8 + 8 - 4 + 4 = 0$. By the factor theorem, $x - 2$ is a factor since $f(2) = 0$.

Notice that $x + 3$ is not a factor of $f(x)$, since $x + 3 = x - (-3)$ and $f(-3) = -(-27) + 2(9) - 2(-3) + 4 = 55$. The converse of the factor theorem guarantees that $x + 3$ is not a factor since $f(-3) \neq 0$.

EXERCISE 13.1 Remainder and Factor Theorems

By use of the remainder theorem, find the remainder obtained by dividing the polynomial in each of Probs. 1 to 12 by the binomial $x - r$ which follows it.

1	$x^3 + 7x^2 - 6x - 5,\ x - 1$	2	$x^3 - 9x^2 + 8x + 4,\ x - 3$
3	$x^3 - 8x^2 + 6x + 3,\ x + 2$	4	$x^3 + 3x^2 + 4,\ x + 3$
5	$2x^3 + 3x^2 - 5x - 8,\ x + 2$	6	$3x^3 - 2x^2 + 2x - 3,\ x + 3$
7	$2x^3 + 5x^2 - 6x + 4,\ x + 1$	8	$2x^3 - 4x^2 + 4x - 7,\ x - 2$

9 $2x^4 + 7x^3 + 4x^2 - 2x + 7, x + 3$ **10** $-3x^4 + 8x^3 + 2x^2 - 3x + 4, x + 1$

11 $x^5 - 2x^4 + 2x^3 - 3x^2 - x + 5, x - 2$

12 $16x^5 + 8x^4 + 4x^3 + 2x^2 + 4x + 1, x + \frac{1}{2}$

Use the factor theorem to show that the binomial $x - r$ is a factor of the polynomial in each of Probs. 13 to 24.

13 $2x^3 + 3x^2 - 6x + 1, x - 1$

14 $3x^3 - 9x^2 - 4x + 12, x - 3$

15 $5x^4 + 8x^3 + x^2 + 2x + 4, x + 1$

16 $3x^4 + 9x^3 - 4x^2 - 9x + 9, x + 3$

17 $-2x^5 + 11x^4 - 12x^3 - 5x^2 + 22x - 8, x - 4$

18 $3x^5 + 17x^4 + 17x^3 + 35x^2 - 4x - 20, x + 5$

19 $x^6 - x^5 - 7x^4 + x^3 + 8x^2 + 5x + 2, x + 2$

20 $2x^6 - 5x^5 + 4x^4 + x^3 - 7x^2 - 7x + 2, x - 2$

21 $x^3 - 4ax^2 + 2a^2x + a^3, x - a$

22 $x^3 - (2a + b)x^2 + (3a + 2ab)x - 3ab, x - b$

23 $x^{2n} - a^{2n}, x + a$

24 $x^{2n+1} + a^{2n+1}, x + a$

In Probs. 25 to 28, find the value of k for which the binomial $x - r$ is a factor of the polynomial.

25 $x^3 + 3x^2 + x + k, x + 1$ **26** $-x^3 + 4x^2 + kx - 2, x - 1$

27 $3x^4 - 16x^3 + kx^2 - 9x + 9, x - 3$ **28** $-3x^4 + kx^3 + 7x^2 - 11x + 5, x - 5$

In Probs. 29 to 32, use the converse of the factor theorem to show that $x - r$ is not a factor of the polynomial.

29 $-2x^3 + 4x^2 - 4x + 9, x - 2$ **30** $-3x^3 - 9x^2 + 5x + 12, x + 3$

31 $3x^4 - 8x^3 + 5x^2 + 7x - 3, x - 3$ **32** $4x^4 + 9x^3 + 3x^2 + x + 4, x + 2$

Show that the given value of r is a root of the equation in each of Probs. 33 to 40.

33 $r = 5, \ (x - 5)(x + 1)(x + 2) = 0$

34 $r = -\frac{3}{2}, \ (x - 5)(x + 2)(2x + 3) = 0$

35 $r = -\frac{2}{3}, \ (x + 3)(3x + 2)(3x + 4) = 0$

36 $r = -\frac{4}{5}, \ (2x + 3)(10x + 8)(x - 4) = 0$

37 $r = 1, \ x^3 + 3x^2 - 4x = 0$

38 $r = -1, \ x^3 + 5x^2 - 6x - 10 = 0$

39 $r = -3, \ 5x^4 + 17x^3 + 6x^2 + 9x + 27 = 0$

40 $r = 1, \ 7x^4 - 8x^3 - 9x^2 + 6x + 4 = 0$

In Probs. 41 to 44, show that the polynomial has no factor $x - r$ if r is real.

41 $x^2 + 1$ **42** $x^4 + 3x^2 + 2$

43 $x^4 + 5x^2 + 3$ **44** $x^6 + 3x^2 + 5$

13.4 / SYNTHETIC DIVISION

Synthetic division We can materially decrease the labor involved in the problem of finding the quotient and remainder when a polynomial in x is divided by $x - r$ by use of a process known as *synthetic division*. The procedure to be followed can be understood more readily by considering an example. If we use the ordinary long-division method for dividing $2x^3 + x^2 - 18x - 7$ by $x - 3$, we have

$$
\begin{array}{r}
2x^2 + 7x + 3 \\
2x^3 + x^2 - 18x - 7\,\lfloor\,x - 3 \\
\underline{(2x^3) - 6x^2} \\
7x^2 - [18x] \\
\underline{(7x^2) - 21x} \\
3x - [7] \\
\underline{(3x) - 9} \\
2
\end{array}
$$

Now let us examine this problem in order to see what can be omitted without interfering with the essential steps. In the first place, the division process requires that each term written in parentheses in the problem be exactly the same as the term just above it. Furthermore, the terms in brackets are the terms in the dividend written in a new position. If these two sets of terms are omitted, that is, not written twice, we have

$$
\begin{array}{r}
2x^2 + 7x + 3 \\
2x^3 + x^2 - 18x - 7\,\lfloor\,x - 3 \\
\underline{- 6x^2} \\
7x^2 \\
\underline{- 21x} \\
3x \\
\underline{-9} \\
2
\end{array}
$$

We can save space by placing $-21x$ and -9 on the same line with $-6x^2$, and $3x$ and 2 on the same line with $7x^2$. Furthermore, it is not necessary to write the variable, since the problem tells us what it is. Hence, a shorter form of the work is

$$
\begin{array}{r}
2 + 7 + 3 \\
2 + 1 - 18 - 7\,\lfloor\,1 - 3 \\
\underline{- 6 - 21 - 9} \\
7 + 3 + 2
\end{array}
$$

Since the method we are developing applies only to division problems in which the divisor is $x - r$, it is not necessary for the coefficient of x, which is always 1, to appear. Moreover, in subtraction, we change the sign of the subtrahend and add. This latter change becomes automatic if we replace the -3 in the divisor by $+3$. Upon carrying out these suggestions, we have

$$\frac{\begin{array}{r} 2+\ \ 7+3 \end{array}}{\begin{array}{r} 2+1-18-7 \rfloor 3 \\ \underline{+6+21+9} \\ +7+\ \ 3+2 \end{array}}$$

The final step in the process is to rewrite the 2 in the dividend as the first term in the third line. Then the first three terms in this line are the same as the coefficients in the quotient. Hence the latter can be omitted, and the problem becomes

$$\begin{array}{r} 2+1-18-7 \rfloor 3 \\ \underline{+6+21+9} \\ 2+7+\ \ 3+2 \end{array}$$

Consequently, the essential steps in the process can be carried out mechanically as follows: Write the first 2 in the third line, multiply by 3, place the product 6 under 1, and add, obtaining 7; then multiply 7 by 3, obtaining 21, which is placed under -18 and added to it; finally, multiply the last sum by 3 and add the product to -7, getting 2. Hence the coefficients in the quotient are 2, $+7$, and $+3$ and the remainder is 2. We therefore have the following *rule for synthetic division.*

Rule for
synthetic division In order to divide $F(x)$ by $x-r$ synthetically:

1 Arrange the coefficients of $F(x)$ in order of descending powers of x, supplying zero as the coefficient of each missing power.

2 Replace the divisor $x - r$ by $+r$.

3 Bring down the coefficient of the largest power of x, multiply it by r, place the product beneath the coefficient of the second largest power of x, and add the product to that coefficient. Multiply this sum by r and place it beneath the coefficient of the next largest power of x. Continue this procedure until there is a product added to the constant term.

4 The last number in the third row is the remainder, $f(r)$, and the other numbers, reading from left to right, are the coefficients of the quotient, which is of degree one less than $F(x)$.

EXAMPLE 1 Determine the quotient and the remainder obtained by dividing $2x^4 + x^3 - 16x^2 + 18$ by $x + 2$ synthetically.

Solution Since $x - r = x + 2$, we have $r = -2$. Upon writing the coefficients of the dividend in a line, supplying zero as the coefficient of the missing term in x, and carrying out the steps of synthetic division, we have

$$2 + 1 - 16 \quad\; 0 + 18 \,\lfloor\underline{-2}$$
$$\underline{-4 + \;\; 6 + 20 - 40}$$
$$2 - 3 - 10 + 20 - 22$$

Hence, the quotient is $2x^3 - 3x^2 - 10x + 20$, and the remainder is $f(-2) = -22$.

EXAMPLE 2 Use synthetic division to show that $x - 3$ is a factor of $x^3 - 2x^2 - x - 6$.

Solution By the factor theorem, we need to show that $f(3) = 0$. Using synthetic division gives

$$1 - 2 - 1 - 6 \,\lfloor\underline{3}$$
$$\underline{\quad\;\; 3 + 3 + 6}$$
$$1 + 1 + 2 \quad\; 0$$

so indeed $f(3) = 0$ as required.

EXERCISE 13.2 Synthetic Division

By use of synthetic division, find the quotient and remainder if the polynomial in each of Probs. 1 to 16 is divided by the binomial $x - r$.

1 $x^3 - 2x^2 + 3x - 5$, $x - 3$ 2 $x^3 - 3x^2 + 2x + 1$, $x - 2$
3 $2x^3 + 5x^2 + 7x + 3$, $x + 1$ 4 $3x^3 + 6x^2 + 4x + 7$, $x + 2$
5 $2x^3 + 4x^2 - 3x + 5$, $x + 3$ 6 $-3x^3 + 13x^2 + 7x + 11$, $x - 5$
7 $-2x^3 + 5x^2 + 7x - 8$, $x - 3$ 8 $-2x^3 - 5x^2 + 3x - 2$, $x + 3$
9 $x^4 + 3x^3 - 5x^2 + 7x - 9$, $x - 1$ 10 $-2x^4 + 5x^3 + 6x^2 - 11x + 2$, $x + 2$
11 $x^5 + x^4 - 7x^3 + 2x^2 + x - 1$, $x - 1$ 12 $2x^5 + 4x^4 - 3x^3 - 7x^2 + 3x - 2$, $x + 2$
13 $4x^4 - 5x^2 + 1$, $x - \frac{1}{2}$ 14 $6x^3 - 2x^2 - x - 1$, $x + \frac{2}{3}$
15 $2x^3 - ax^2 + 3a^2x - 4a^3$, $x - a$
16 $2x^4 + 4ax^3 - 2a^2x^2 - 3a^3x + 6a^4$, $x + 2a$

In Probs. 17 to 28, use synthetic division to find the specified function values.

17 Find $f(-1)$ if $f(x) = 2x^3 - 2x + 2$.
18 Find $f(-3)$ if $f(x) = x^4 + 10x - 20$.
19 Find $f(-2)$ if $f(x) = x^5 - 2x^3 + 2x$.
20 Find $f(2)$ if $f(x) = x^5 - 3x^3 - 8x$.
21 Find $f(\frac{1}{2})$ if $f(x) = 4x^4 - 5x^2 + 1$.
22 Find $f(-\frac{2}{3})$ if $f(x) = 6x^3 - 2x^2 - x - 1$.
23 Find $f(\frac{3}{4})$ if $f(x) = 12x^3 + 3x^2 - x + 3$.

24 Find $f(-\frac{3}{2})$ if $f(x) = 6x^4 + x^3 + 16x - 3$.
25 Find $f(1)$ if $f(x) = x^6 + 3x^2 - 3$.
26 Find $f(-1)$ if $f(x) = 2x^6 + 2x^3 + 2x$.
27 Find $f(2a)$ if $f(x) = 3x^3 - 5ax^2 - 3a^2x + 3a^3$.
28 Find $f(-a)$ if $f(x) = 2x^4 + 3ax^3 + 2a^2x^2 + 2a^3x + 2a^4$.

In Probs. 29 to 40, use synthetic division to show that $x - r$ is a factor of the polynomial.

29 $x - 2$, $x^3 + 3x^2 - 2x - 16$ 30 $x + 2$, $2x^3 - 3x^2 - 12x + 4$
31 $x - 3$, $2x^4 + x^3 - 14x^2 - 20x - 3$ 32 $x + 1$, $3x^4 + 4x^3 - 2x^2 + 3x + 6$
33 $x - \frac{1}{3}$, $3x^3 + 2x^2 - 4x + 1$ 34 $x - \frac{1}{4}$, $4x^3 + 3x^2 - 5x + 1$
35 $x - \frac{3}{4}$, $4x^5 - 7x^4 - 5x^3 + 2x^2 + 11x - 6$
36 $x + \frac{2}{3}$, $3x^4 - 4x^3 + 5x^2 - 4$
37 $x - i$, $x^3 - 3x^2 + x - 3$ 38 $x + i$, $x^4 - 2x^2 - 3$
39 $x - 2i$, $x^3 - 2x^2 + 4x - 8$ 40 $x + 3i$, $x^4 + 12x^2 + 27$

Find the value of k for which the remainder is as given if the first expression in each of Probs. 41 to 44 is divided by the second.

41 $x^3 - 5x^2 - 8x - k$, $x + 2$; -2 42 $x^3 - 3x^2 + kx + 5$, $x + 2$; -3
43 $x^3 + kx^2 + 6kx + 15$, $x - 3$; -12 44 $x^3 + kx^2 + (k-1)x + 5$, $x - 2$; 5

In Probs. 45 to 48, use synthetic division to find the remainder if the first expression is divided by the second.

45 $2x^3 + 3x^2 + 6x - 2$, $2x - 1$ HINT: $2x - 1 = 2(x - \frac{1}{2})$
46 $2x^3 - x^2 - 4x + 1$, $2x + 3$
47 $-3x^3 + 5x^2 + 7x - 3$, $3x - 2$
48 $4x^3 + 9x^2 - x + 2$, $4x - 3$

13.5 / NUMBER OF ROOTS

The fundamental theorem of algebra states that *every polynomial equation with complex coefficients has at least one complex root.*[†]

We shall consider the polynomial equation

$$f(x) = a_0x^n + a_1x^{n-1} + \cdots + a_{n-1}x + a = 0 \qquad (1)$$

and let r_1 be a root.   Then, by the factor theorem,

$$f(x) = Q_1(x)(x - r_1)$$

[†] For proof, see L. Ahlfors, "Complex Analysis," 2d ed., p. 99, McGraw-Hill Book Company, New York, 1966.

where $Q_1(x)$ is of degree $n - 1$. The equation $Q_1(x) = 0$ also has at least one root, and we shall let it be r_2. Then $x - r_2$ is a factor of $Q_1(x)$ and $Q_1(x) = Q_2(x)(x - r_2)$, where $Q_2(x)$ is of degree $n - 2$. Thus $f(x) = (x - r_1)(x - r_2)Q_2(x)$. If we continue this process, we can find $n - 2$ additional factors, $(x - r_3)$, $(x - r_4)$, ..., $(x - r_n)$, of $f(x)$ and have

$$f(x) = (x - r_1)(x - r_2)(x - r_3) \cdots (x - r_n)Q_n(x) \tag{2}$$

where $Q_n(x)$ is of degree $n - n = 0$ and is therefore a constant. Obviously, $Q_n(x)$ is the coefficient of x^n in the right member of (2). Hence it is equal to a_0. Then the factored form of $f(x)$ is

$$f(x) = a_0(x - r_1)(x - r_2)(x - r_3) \cdots (x - r_n) \tag{13.4}$$

By the converse of the factor theorem, $r_1, r_2, r_3, \ldots, r_n$ are roots of $f(x) = 0$ and so the equation $f(x) = 0$ has at least n roots. Furthermore, $f(x) = 0$ has no other roots, for no one of the factors in (13.4) is equal to zero for any value of x not equal to $r_1, r_2, r_3, \ldots,$ or r_n, since $r_{n+1} - r_i \neq 0$. Thus $f(x) = 0$ has at most n roots.

If s values of $r_1, r_2, r_3, \ldots, r_n$ are equal, that is, if $r_1 = r_2 = r_3 = \cdots = r_s$, then (13.4) becomes

$$f(x) = a_0(x - r_s)^s(x - r_{s+1}) \cdots (x - r_n)$$

Multiplicity

and r_s is called a *root of multiplicity* s of $f(x) = 0$. Hence we have the following theorem on the number of roots.

THEOREM

Theorem on the number of roots

A polynomial equation of degree n with complex coefficients has exactly n complex roots, where a root of multiplicity s is counted as s roots.

EXAMPLE 1   The roots of the equation

$$(x + 1)(x - 3)^2(x + \tfrac{3}{2})^2 = 0$$

are -1, and 3 with multiplicity 2, and $-\tfrac{3}{2}$ with multiplicity 2.

EXAMPLE 2   Find the four roots of $f(x) = x^4 - 8x^3 + 20x^2 - 32x + 64 = 0$ if 4 is a root of multiplicity 2.

Solution   Since 4 is a root of $f(x) = 0$, then $x - 4$ is a factor of $f(x)$. We find the quotient by synthetic division:

$$\begin{array}{r} 1 - 8 + 20 - 32 + 64 \underline{|4} \\ 4 - 16 + 16 - 64 \\ \hline 1 - 4 + 4 - 16 0 \end{array}$$

Thus $f(x) = (x - 4)(x^3 - 4x^2 + 4x - 16)$. Since 4 is a root of multiplicity 2, we divide synthetically again.

$$1 - 4 + 4 - 16 \lfloor 4$$
$$\underline{\quad 4 \quad 0 + 16}$$
$$1 \quad 0 + 4 \quad \quad 0$$

We now have $f(x) = (x-4)^2(x^2+4)$. The other two roots of $f(x) = 0$ are found by setting $x^2 + 4 = 0$. The roots are $2i$ and $-2i$. Hence the four roots of $f(x)$ are $4, 4, 2i,$ and $-2i$.

13.6 / IDENTICAL POLYNOMIALS

One way in which a polynomial $f(x)$ may be zero for all x is for each of the coefficients to be zero. The following theorem shows that, in fact, this is the only way. It is also useful in dealing with partial fractions, which allow a complicated fraction to be expressed as a sum of simpler fractions.

THEOREM   If two polynomials of degree n with complex coefficients are equal for more than n distinct values, then they are identical; that is, the coefficients of equal powers of the variable are equal.

Proof   Let $f(x) = a_0x^n + a_1x^{n-1} + \cdots + a_{n-1}x + a_n$ and $g(x) = b_0x^n + b_1x^{n-1} + \cdots + b_{n-1}x + b_n$. The hypothesis says that $f(x_i) = g(x_i)$ for $i = 1, 2, \ldots, n, n+1$. Consider $h(x) = f(x) - g(x) = (a_0 - b_0)x^n + (a_1 - b_1)x^{n-1} + \cdots + (a_n - b_n)$. Now $h(x_i) = f(x_i) - g(x_i) = 0$ if $i = 1, 2, \ldots, n$, so by the factor theorem

$$h(x) = (a_0 - b_0)(x - x_1)(x - x_2) \cdots (x - x_n) \qquad (1)$$

Furthermore, $h(x_{n+1}) = f(x_{n+1}) - g(x_{n+1}) = 0$, so by Eq. (1)

$$0 = h(x_{n+1}) = (a_0 - b_0)(x_{n+1} - x_1)(x_{n+1} - x_2) \cdots (x_{n+1} - x_n) \qquad (2)$$

Now $x_1, x_2, \ldots, x_n, x_{n+1}$ are all distinct, so $x_{n+1} - x_i \neq 0$ if $i = 1, 2, \ldots, n$, and thus by Eq. (2) we must have $a_0 - b_0 = 0$; that is, $a_0 = b_0$.

We thus have $h(x) = (a_1 - b_1)x^{n-1} + \cdots + (a_n - b_n)$, and may repeat the above argument with $x_1, x_2, \ldots, x_{n-1}, x_n$ to show that $a_1 = b_1$. Similarly $a_2 = b_2, \ldots, a_n = b_n$.

EXAMPLE 1   We will have $2x^3 + 5x^2 - 4x + 1 = Ax^3 + Bx^2 + Cx + D$ for all x if and only if $2 = A, 5 = B, -4 = C,$ and $1 = D$.

EXAMPLE 2   Find constants $A, B,$ and C so that

$$\frac{5x^2 - 6x + 9}{(x-1)(x^2 - x + 4)} = \frac{A}{x-1} + \frac{Bx + C}{x^2 - x + 4}$$

Solution We first multiply both members by the lowest common denominator, getting

$$5x^2 - 6x + 9 = A(x^2 - x + 4) + (Bx + C)(x - 1)$$
$$= x^2(A + B) + x(-A - B + C) + (4A - C)$$

Since we want these two polynomials to be equal for all values of x, then

$$5 = A + B \qquad \text{equating coefficients of } x^2$$
$$-6 = -A - B + C \qquad \text{equating coefficients of } x$$
$$9 = 4A - C \qquad \text{equating constants}$$

These equations may be solved simultaneously as in Chap. 8, the solution being $A = 2$, $B = 3$, $C = -1$. It follows that for all x except $x = 1$, where there would be division by zero,

$$\frac{5x^2 - 6x + 9}{(x - 1)(x^2 - x + 4)} = \frac{2}{x - 1} + \frac{3x - 1}{x^2 - x + 4}$$

13.7 / CONJUGATE ROOTS

It can be readily verified that $x = 2$ is one root of $x^3 - 4x^2 + 9x - 10 = 0$, and the other two roots are $x = 1 + 2i$ and $x = 1 - 2i$. Also, the roots of the quadratic equation $ax^2 + bx + c = 0$ are

$$x = -\frac{b}{2a} + \frac{1}{2a}\sqrt{b^2 - 4ac} \qquad \text{and} \qquad x = -\frac{b}{2a} - \frac{1}{2a}\sqrt{b^2 - 4ac}$$

which are conjugates if they are complex numbers and a, b, and c are real.

This illustrates the fact that the complex roots of a polynomial equation with real coefficients occur in pairs and that the members of each pair are complex conjugates.

THEOREM ON CONJUGATE ROOTS If the complex number $a + bi$, $b \neq 0$, is a root of the polynomial equation $f(x) = 0$ and f has real coefficients, then its conjugate $a - bi$ is also a root of $f(x) = 0$.

Proof Suppose
$$f(x) = a_0x^n + a_1x^{n-1} + \cdots + a_n = 0 \tag{1}$$

where $a_0, a_1, \ldots, a_n$ are real numbers, and $x = a + bi$ is a root. It was shown in Chap. 12 that the conjugate of a sum is the sum of the conjugates, and the conjugate of a product is the product of the conjugates. Since a_i is real, then $\bar{a}_i = a_i$ for $i = 0, 1, 2, \ldots, n$. Thus, taking the conjugate of each member of (1) gives

$$\overline{f(x)} = \overline{a_0x^n + a_1x^{n-1} + \cdots + a_n} = \bar{0}$$
or
$$f(\bar{x}) = a_0\bar{x}^n + a_1\bar{x}^{n-1} + \cdots + a_n = 0 \tag{2}$$

Equation (2) says that $f(\bar{x}) = 0$; hence the conjugate $\bar{x} = a - bi$ is also a root of $f(x) = 0$.

EXAMPLE   Find a third-degree polynomial with real coefficients which has roots $x = 3$ and $x = 3 - 2i$.

Solution   By the theorem on conjugate roots, we may take $x = 3 + 2i$ as a root along with $x = 3 - 2i$ and $x = 3$. By the factor theorem the desired polynomial is

$$(x-3)[x-(3-2i)][x-(3+2i)] = (x-3)(x-3+2i)(x-3-2i) =$$
$$(x-3)[(x-3)^2 + 4] = x^3 - 9x^2 + 31x - 39 \qquad (3)$$

Any constant (nonzero) times this polynomial would also be an answer. If real coefficients were not required, then $(x - 3)\ [x - (3 - 2i)]\ (x - r)$, where r is any complex number, would be an answer.

As a consequence of Eq. (13.4) and the theorem on conjugate roots, we can now state the following corollary.

Corollary   A polynomial with real coefficients can be expressed as the product of linear and irreducible quadratic factors, each factor having real coefficients. The proof is simply that if any factor in Eq. (13.4) has $r = a + bi$, then another factor will have $r = a - bi$; and, as in Eq. (3) above, $[x - (a + bi)][x - (a - bi)] = x^2 - 2ax + a^2 + b^2$ which has real coefficients.

EXERCISE 13.3 Identical Polynomials and Complex Roots

State the degree of the equation in each of Probs. 1 to 8; then find each root and give its multiplicity.

1   $(x - 2)^2(x + 3)^2(x - 4) = 0$
2   $(x + 3)^4(x - 2)^4(x - 7) = 0$
3   $(x + 1)(x - 2)^2(x + 4)^2(x - 5) = 0$
4   $(x - 4)^3(x + 2)^2(x - 1)^2(x + 10) = 0$
5   $(2x + 5)^2(3x - 1)^3 = 0$
6   $(4x - 3)^2(2x + 9)^2(x - 4)^3 = 0$
7   $(3x + 1)^3(x^2 + 1)(x + 1)^2 = 0$
8   $(4x - 7)^2(x^2 + 9)^2(2x - 1)^3 = 0$

In Probs. 9 to 28, find the roots of $f(x) = 0$ that are not given.

9   $f(x) = x^3 - 7x + 6$; one root is 2
10 $f(x) = x^3 - 5x^2 + 2x + 8$; one root is -1
11 $f(x) = x^3 + x^2 - 14x - 24$; one root is 4
12 $f(x) = x^3 - 4x^2 + x + 6$; one root is 3
13 $f(x) = x^4 + x^3 - 3x^2 - x + 2$; 1 is a root of multiplicity 2
14 $f(x) = x^4 + 2x^3 + 2x^2 + 2x + 1$; -1 is a root of multiplicity 2
15 $f(x) = x^4 - 3x^3 + 6x^2 - 12x + 8$; two roots are 1 and 2
16 $f(x) = x^4 - 2x^3 + 6x^2 - 18x - 27$; two roots are -1 and 3
17 $f(x) = x^3 + x^2 + x + 1$; one root is i
18 $f(x) = 2x^3 - 3x^2 + 8x - 12$; one root is $-2i$

19 $f(x) = 3x^3 + 5x^2 + 12x + 20$; one root is $2i$
20 $f(x) = 4x^3 - 3x^2 + 24x - 18$; one root is $i\sqrt{6}$
21 $f(x) = 2x^3 - x^2 - 2x + 6$; one root is $1 + i$
22 $f(x) = 4x^3 + 9x^2 + 22x + 5$; one root is $-1 + 2i$
23 $f(x) = 4x^3 - 19x^2 + 32x - 15$; one root is $2 - i$
24 $f(x) = 2x^3 + 7x^2 - 4x - 65$; one root is $-3 - 2i$
25 $f(x) = x^4 + 2x^3 + 9x^2 + 8x + 20$; $2i$ and $-1 + 2i$ are roots
26 $f(x) = x^5 + 5x^4 + 2x^3 - 38x^2 - 95x - 75$; $-2 + i$ is a root of multiplicity 2
27 $f(x) = x^6 + 4x^5 + 15x^4 + 24x^3 + 39x^2 + 20x + 25$; $-1 + 2i$ is a root of multiplicity 2
28 $f(x) = x^5 - 7x^4 + 24x^3 - 32x^2 + 64$; $2 + 2i$ is a root of multiplicity 2

In Probs. 29 to 36, find a polynomial of least possible degree and with real coefficients with the roots given.

29 $-\frac{1}{3}, 2i$ 30 $\frac{7}{2}, i$ 31 $1, 2 + i$ 32 $-\frac{3}{4}, 3 - i$
33 $-2 - i$, multiplicity 2   34 $-1 - 2i$, multiplicity 2
35 $2 - 2i$, multiplicity 2   36 $2i$, multiplicity 3

In Probs. 37 to 44, find the values of the constants A, B, and C so that the equation is an identity.

37 $\dfrac{2}{(x - 3)(x - 2)} = \dfrac{A}{x - 3} + \dfrac{B}{x - 2}$

38 $\dfrac{x + 5}{(x + 1)(x + 3)} = \dfrac{A}{x + 1} + \dfrac{B}{x + 3}$

39 $\dfrac{2x^2 + x + 1}{(x + 2)(3x + 1)(x + 3)} = \dfrac{A}{x + 2} + \dfrac{B}{3x + 1} + \dfrac{C}{x + 3}$

40 $\dfrac{5x^2 + 5x - 4}{(x - 1)(x + 1)(x + 2)} = \dfrac{A}{x - 1} + \dfrac{B}{x + 1} + \dfrac{C}{x + 2}$

41 $\dfrac{-2x + 11}{(x - 4)^2} = \dfrac{A}{x - 4} + \dfrac{B}{(x - 4)^2}$

42 $\dfrac{12x^2 - 9x + 20}{(2x + 3)(3x - 1)^2} = \dfrac{A}{2x + 3} + \dfrac{B}{3x - 1} + \dfrac{C}{(3x - 1)^2}$

43 $\dfrac{-2x + 5}{(x - 1)(x^2 + 2)} = \dfrac{A}{x - 1} + \dfrac{Bx + C}{x^2 + 2}$

44 $\dfrac{3x^2 - 10x + 16}{(x - 3)(x^2 + x + 1)} = \dfrac{A}{x - 3} + \dfrac{Bx + C}{x^2 + x + 1}$

45 Show that if

$$x^n + a_1 x^{n-1} + a_2 x^{n-2} + \cdots + a_{n-1}x + a_n = 0$$

has roots $r_1, r_2, \ldots, r_n$, then the sum of the roots satisfies

$$r_1 + r_2 + \cdots + r_n = -a_1 \qquad\qquad \text{(S)}$$

and the product of the roots satisfies

$$r_1 r_2 \cdots r_n = (-1)^n a_n \qquad \text{(P)}$$

HINT: Factor the polynomial; then multiply it out using $r_1, r_2, \ldots, r_n$, and compare coefficients with the original polynomial.

46 Find a polynomial with roots $2 + 3i$ and -5 and verify Eqs. (S) and (P) in Prob. 45 for this polynomial.

47 Repeat Prob. 46 for roots $3 - 2i$ and $2 + i$.

48 Repeat Prob. 46 for roots $4 - i$, $1 + 2i$, and $\frac{3}{2}$.

13.8 / BOUNDS OF THE REAL ROOTS

The theorem given in this section enables us to find a number that is greater than or equal to the largest real root of an equation, and another number that is smaller than or equal to the least root of the equation. Thus we can restrict the range in which the real roots are known to lie. Any number that is larger than or equal to the greatest root of an equation is called *an upper bound of the roots;* any number that is smaller than or equal to the least root of an equation is called a *lower bound of the roots.*

Upper bound

Lower bound

We shall now state and then prove a theorem that enables us to determine upper and lower bounds.

THEOREM

Theorem on bounds

If the coefficient of x^n in the polynomial equation $f(x) = 0$ is positive and if there are no negative terms in the third line of the synthetic division of $f(x)$ by $x - k$, $k > 0$, then k is an upper bound of the real roots of $f(x) = 0$. Furthermore, if the signs in the third line of the synthetic division of $f(x)$ by $x - (-k) = x + k$ are alternately plus and minus,† then $-k$ is a lower bound of the real roots.

Proof

In order to prove the first part of the theorem, we use Eq. (13.2) with $r = k$ and have

$$f(x) = Q(x)(x - k) + R \qquad \text{(1)}$$

By Sec. 13.4, the coefficients in $Q(x)$ and the value of R are the numbers in the third row of the division of $f(x)$ by $x - k$. If these numbers are positive or zero and if x is greater than k, then $Q(x)(x - k) + R > 0$ since $x > k > 0$ and $x - k > 0$. Hence there are no real roots of $f(x) = 0$ that are greater than k; that is, k is an upper bound of the real roots, as stated in the first part of the theorem. We omit the proof of the second part.

† When one or more zeros occur in the third line of the synthetic division of $f(x)$ by $x + k$, $-k$ is a lower bound of the real roots if each zero may be replaced by either a plus or a minus sign in a way so that the signs in the third line of the synthetic division are alternately plus and minus.

EXAMPLE 1   Find an upper and a lower bound of the real roots of the equation $x^3 - 2x^2 + 3x + 3 = 0$.

Solution   The synthetic division of the left member of $x^3 - 2x^2 + 3x + 3 = 0$ by $x - 3$ and by $x + 1$ is given below:

$$
\begin{array}{c}
1 - 2 + 3 + 3\,\lfloor 3 \\
\underline{3 + 3 + 18} \\
1 + 1 + 6 + 21
\end{array}
\qquad
\begin{array}{c}
1 - 2 + 3 + 3\,\lfloor -1 \\
\underline{-1 + 3 - 6} \\
1 - 3 + 6 - 3
\end{array}
$$

In the first case, the terms in the third row are all positive. Hence, 3 is an upper bound. In the second case, the terms in the third row are alternately plus and minus. Therefore, -1 is a lower bound of the roots.

EXAMPLE 2   Find upper and lower bounds for the real roots of the equation $x^4 - x^3 - 12x^2 - 2x + 3 = 0$.

Solution   We first divide synthetically by -3:

$$
\begin{array}{c}
1 - 1 - 12 - 2 + 3\,\lfloor -3 \\
\underline{-3 + 12 0 + 6} \\
1 - 4 0 - 2 + 9
\end{array}
$$

If we replace the 0 in the middle of the third line by $+$, then the third-row signs alternate. Thus -3 is a lower bound of the roots.
If we now divide by 4 and 5 we get

$$
\begin{array}{c}
1 - 1 - 12 - 2 + 3\,\lfloor 4 \\
\underline{4 + 12 0 - 8} \\
1 + 3 0 - 2 - 5
\end{array}
\qquad
\begin{array}{c}
1 - 1 - 12 - 2 + 3\,\lfloor 5 \\
\underline{5 + 20 + 40 + 190} \\
1 + 4 + 8 + 38 + 193
\end{array}
$$

In the division by 4, not every sign in the third row is positive or zero, so we cannot say that 4 is an upper bound of the roots. However, in the division by 5 all third-row entries are positive, so 5 is an upper bound of the roots.

The theorem in this section does not always give the best upper bound for positive roots, even among integers, but it does give an upper bound.

13.9 / DESCARTES' RULE OF SIGNS

In this section we shall present a criterion or rule that enables us to determine a number that is greater than or equal to the number of positive roots of a polynomial equation, and another that is equal to or greater than the number of negative roots.

Variation of signs

If the terms of a polynomial are arranged according to the ascending or descending powers of the variable, we say that a *variation of signs* occurs when the signs of two consecutive terms differ. For example, in the polynomial $2x^4 - 5x^3 - 6x^2 + 7x + 3$, the signs of the terms are $+--++$. Hence, there are two variations of sign, since the sign changes from positive to negative and back again to positive. Furthermore, there are three variations of sign in $x^4 - 2x^3 + 3x^2 + 6x - 4$.

We shall now state Descartes' rule of signs and then illustrate it. The proof is omitted.

Statement of Descartes' rule of signs

The number of positive roots of a polynomial equation $f(x) = 0$ with real coefficients is equal to the number of variations in sign of $f(x)$ or is less than this number by an even integer. The number of negative roots equals the number of variations in sign of $f(-x)$ or is less than this number by an even integer.

EXAMPLE 1 Find the maximum number of positive roots and the maximum number of negative roots that can be in the solution set of $x^4 - 3x^3 - 5x^2 + 7x - 3 = 0$.

Solution There are three variations of sign in $f(x) = x^4 - 3x^3 - 5x^2 + 7x - 3$. Hence, the number of positive roots of the equation $x^4 - 3x^3 - 5x^2 + 7x - 3 = 0$ does not exceed three. Furthermore, $f(-x) = x^4 + 3x^3 - 5x^2 - 7x - 3$, and this polynomial has one variation of sign. Therefore, the number of negative roots of the above equation is one.

The theorem in the previous section gives information about upper and lower bounds for the values of roots. The theorem in this section gives information about the number of positive and negative roots without saying how large any of them may be. In the application of either theorem a 0 may occur. A 0 in the third line of a synthetic division may be given either a plus sign or a minus sign in the process of determining the bounds of the root values. A 0 that occurs as a coefficient in the application of Descartes' rule of signs should be ignored.

One more fact, which is often useful in graphing, is that for a polynomial of degree n, the total number of maximum and minimum points is either $n - 1$ or is less than $n - 1$ by an even integer.

EXAMPLE 2 What information can be given about the positive and negative roots of $f(x) = x^3 - 4x^2 + 3x + 2 = 0$?

Solution There are two variations in sign of $f(x)$, so there are either two positive roots or none. Since $f(-x) = -x^3 - 4x^2 - 3x + 2$, there is one variation in sign in $f(-x)$ and so there is exactly one negative root. The synthetic divisions

$$\begin{array}{r} 1-4+3+2\underline{|-1} \\ -1+5-8 \\ \hline 1-5+8-6 \end{array} \qquad \begin{array}{r} 1-4+3+\ 2\underline{|4} \\ 4\ \ \ 0+12 \\ \hline 1\ \ \ 0+3+14 \end{array}$$

show that there is no negative root less than -1 and no positive root greater than 4.

13.10 / LOCATING THE ROOTS

The roots of

$$2x^3 - 3x^2 - 12x + 6 = 0$$

are approximately -2.1, 0.5, and 3.1, which may be checked by synthetic division. Furthermore, for $x = -3$, the curve is below the X axis since it passes through the point $(-3, -39)$; for $x = -2$, the curve is above the X axis since it passes through the point $(-2, 2)$. This is an illustration of a general truth: If the graph of $F(x)$ passes through two points on opposite sides of the X axis and if there are no gaps or breaks in the curve, it must cross the X axis between the two points. Curves that contain no gaps or that are not made up of separate or disjointed parts are called *continuous curves*. It can be proved that the graph of any polynomial is continuous, although it is beyond the scope of this book to do so. We have the following rule for locating the roots:

Rule for locating roots   *If $f(a)$ and $f(b)$ differ in sign, there is an odd number of roots of $f(x) = 0$ between $x = a$ and $x = b$. Therefore there is at least one such root.*

EXAMPLE   Locate the roots of $2x^3 - x^2 - 6x + 3 = 0$.

Solution   In the equation $2x^3 - x^2 - 6x + 3 = 0$, $f(x) = 2x^3 - x^2 - 6x + 3$. Furthermore, it can be verified that $f(-2) = -5$, $f(-1) = 6$, $f(0) = 3$, $f(1) = -2$, and $f(2) = 3$. Hence, there is an odd number of roots between -2 and -1, between 0 and 1, and between 1 and 2. Since the degree of the equation is 3, it has only three roots. Hence, there is exactly one root in each of the above intervals.

13.11 / GRAPH OF A POLYNOMIAL

The roots of a polynomial equation $f(x) = 0$ are the abscissas of the points where the graph of $y = f(x)$ crosses the X axis, for at these points $y = 0$. These abscissas are called the *zeros* of the polynomial. We shall make extensive use of the graph in the process of obtaining the roots of polynomial equations, and we shall discuss graphs of polynomials in this section.

The method of obtaining the graph of a polynomial is the same as that used in Secs. 6.6 and 9.2, except that we shall use synthetic division for obtaining the corresponding values of x and y.

EXAMPLE 1 Sketch the graph of $y = x^3 - 4x^2 + 3x + 2$.

Solution As seen in Example 2 of Sec. 13.9, there are one negative root and two or zero positive roots. We shall graph the equation only between -1 and 4, since in Example 2 these were found to be lower and upper bounds for the roots. A table of values obtained by synthetic division is given below.

x	-1	0	1	2	3	4
y	-6	2	2	0	2	14

The one negative root is between -1 and 0. Since 2 is a root, there must be another positive root; $f(2.1) = -0.079$ and $f(3) = 2$, so there must be a root between 2.1 and 3. The graph is shown in Fig. 13.1.

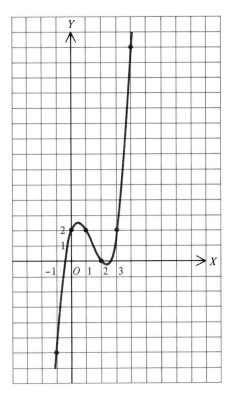

Figure 13.1

EXAMPLE 2 Construct a graph by using the table of corresponding values of x and y given below for the equation $y = x^4 - 18x^2$. Estimate the zeros of the polynomial from the graph.

x	-5	-4	-3	-2	-1	0	1	2	3	4	5
y	175	-32	-81	-56	-17	0	-17	-56	-81	-32	175

Solution In order to construct the graph by use of these values, we select a scale such that 2 units on the X axis are equal to 10 units on the Y axis. The graph is shown in Fig. 13.2. The zeros appear to be -4.2, 0, and 4.2. By Descartes' rule of signs, there are only one negative and one positive root.

EXERCISE 13.4 Graphs and Real Roots

In Probs. 1 to 12, determine upper and lower bounds for the roots of the given equation.

1 $2x^3 - x^2 - 4x + 2 = 0$ 2 $3x^3 + x^2 - 9x - 3 = 0$
3 $2x^3 - 3x^2 - 26x + 39 = 0$ 4 $3x^3 + 3x^2 - 36x - 35 = 0$
5 $x^4 + 3x^3 - 15x^2 - 9x + 31 = 0$ 6 $3x^4 - 20x^3 + 28x^2 + 19x - 13 = 0$
7 $6x^4 + 11x^3 - 25x^2 - 33x + 21 = 0$ 8 $6x^4 + x^3 - 43x^2 - 7x + 7 = 0$
9 $2x^3 + x^2 + 8x + 4 = 0$ 10 $x^3 + 3x^2 + x - 4 = 0$
11 $6x^4 + 23x^3 + 25x^2 - 9x - 5 = 0$ 12 $x^4 + 5x^3 + 7x^2 - 3x - 9 = 0$

Determine the number of positive roots and number of negative roots of the equation in each of Probs. 13 to 24 by use of Descartes' rule of signs.

13 $x^3 - 4x^2 - 5x + 2 = 0$ 14 $-2x^3 - 3x^2 + 5x + 7 = 0$
15 $5x^3 + 3x^2 + 6x + 1 = 0$ 16 $2x^3 + 5x^2 - x + 2 = 0$
17 $2x^4 + 5x^3 - 3x^2 + x + 2 = 0$ 18 $x^4 - 4x^3 + 12x^2 + 24x + 24 = 0$
19 $-3x^4 - 5x^3 + 8x^2 - 2x + 6 = 0$ 20 $3x^4 + 8x^3 - 2x^2 + 5x - 1 = 0$
21 $4x^5 + 3x^4 - 2x^3 + x^2 - x + 4 = 0$ 22 $3x^5 - 3x^3 + 2x^2 + 5x - 1 = 0$
23 $3x^5 + 2x^2 + 5x - 1 = 0$ 24 $2x^6 - 3x^4 + x^3 - 3 = 0$

In Probs. 25 to 36, locate each real root between two consecutive integers.

25 $3x^3 + 10x^2 - 2x - 4 = 0$ 26 $3x^3 - 19x^2 + 21x - 4 = 0$
27 $2x^3 - 19x^2 + 50x - 28 = 0$ 28 $3x^3 - 28x^2 + 54x + 20 = 0$
29 $x^4 - 8x^3 + 12x^2 + 16x - 16 = 0$ 30 $x^4 + 4x^3 - 15x^2 - 66x - 54 = 0$
31 $3x^4 - 12x^3 - 9x^2 + 16x + 4 = 0$ 32 $3x^4 - 12x^3 - 6x^2 + 36x - 9 = 0$
33 $2x^3 - 11x^2 + 18x - 14 = 0$ 34 $3x^3 - 13x^2 + 19x - 5 = 0$

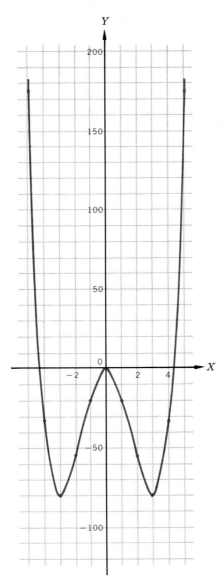

Figure 13.2

35 $x^4 - 4x^2 - 8x - 4 = 0$ **36** $x^4 - 8x^3 + 21x^2 - 20x - 6 = 0$

The equation in each of Probs. 37 to 40 has two roots between consecutive integers. Locate these roots by use of a value halfway between the consecutive integers. Also locate every other root between two consecutive integers.

37 $3x^3 - 4x^2 - 7x - 2 = 0$ **38** $3x^3 - 5x^2 - 6x + 10 = 0$

39 $9x^3 - 15x^2 - 12x + 20 = 0$ **40** $9x^3 - 24x^2 - 48x + 128 = 0$

41   Show that $2x^6 + 3x^4 + x^2 + 5 = 0$ has six imaginary roots.

42   Show that $x^5 + x^3 + 2x + 1 = 0$ has four imaginary roots.

43   Show that $x^7 + 5x + 3 = 0$ has six imaginary roots.

44   Show that $x^3 - x^2 - 1 = 0$ has two imaginary roots.

45   Show that a quadratic equation with real coefficients has either two real roots or none, counting multiplicity.

46   Show that a cubic equation with real coefficients has either three or one real roots, counting multiplicity.

47   Show that a fourth-degree equation with real coefficients has four, two, or no real roots, counting multiplicity.

48   Show that a fifth-degree equation with real coefficients has five, three, or one real roots, counting multiplicity.

Sketch the graphs of the following polynomial functions in the given intervals.

49   $y = x^3 + 2x^2 - x - 3;$ -3 to 3   50 $y = x^3 - 4x^2 + x + 6;$ -1 to 5
51   $y = x^3 - 3x^2 - 6x + 8;$ -2 to 4   52 $y = x^3 - 7x^2 + 2x - 8;$ 0 to 6
53   $y = x^3 - 5x^2 + 5x + 3;$ -3 to 4   54 $y = x^3 - 9x^2 + 21x - 5;$ -1 to 6
55   $y = x^3 - x^2 - 8x - 8;$ -3 to 4   56 $y = x^3 - 10x^2 + 25x - 6;$ 0 to 7
57   $y = -x^3 + 8x^2 - 3x - 46;$ -5 to 3   58 $y = -x^3 + 7x^2 - 5x - 13;$ -6 to 2
59   $y = -x^3 + 3x^2 + 14x - 12;$ -7 to 1
60   $y = -x^3 + 9x^2 - 11x - 21;$ -2 to 5
61   $y = x^4 - 10x^3 + 27x^2 - 10x - 26;$ -3 to 5
62   $y = x^4 - 6x^3 + 8x^2 + 4x - 4;$ -4 to 4
63   $y = x^4 - 22x^2 + 49;$ -5 to 3
64   $y = x^4 - 8x^3 + 14x^2 + 12x - 24;$ -2 to 6
65   $y = -x^4 - 4x^3 + 12x^2 + 32x + 17;$ -6 to 4
66   $y = -x^4 + 4x^3 + 12x^2 - 32x + 17;$ -4 to 6
67   $y = x^4 + x^3 - 18x^2;$ -5 to 4
68   $y = x^4 + 9x^3 + 12x^2 - 28x - 48;$ -7 to 2

13.12 / RATIONAL ROOTS OF A POLYNOMIAL EQUATION

If a polynomial equation has one or more rational roots, the work involved in finding the others is greatly reduced if the rational roots are found first. The process of identifying a rational root is a matter of trial. Hence, the following theorem on rational roots is very useful because it enables us to find a set of numbers which includes the rational roots.

THEOREM ON RATIONAL ROOTS

If the coefficients of

$$a_0 x^n + a_1 x^{n-1} + \cdots + a_{n-1} x + a_n = 0 \qquad (1)$$

are integers, then each of the rational roots, after being reduced to lowest terms, has a factor of a_n for its numerator and a factor of a_0 for its denominator.

Proof   We begin our proof of the theorem by assuming that q/p is a rational root of (1) and that q and p do not have a common factor greater than 1. Now, if we substitute q/p for x in (1) and multiply by p^n, we obtain

$$a_0 q^n + a_1 q^{n-1} p + \cdots + a_{n-1} q p^{n-1} + a_n p^n = 0 \qquad (2)$$

By adding $-a_n p^n$ to each member and then dividing by q, we have

$$a_0 q^{n-1} + a_1 q^{n-2} p + \cdots + a_{n-1} p^{n-1} = -\frac{a_n p^n}{q} \qquad (3)$$

The left member of (3) is made up of the sum, product, and integral powers of integers; hence, it is an integer. Therefore, the right member must be an integer. Consequently, q is a factor of a_n since by hypothesis q and p have no common factor greater than 1.

If we add $-a_0 q^n$ to each member of (2) and divide by p, we obtain

$$a_1 q^{n-1} + \cdots + a_{n-1} q p^{n-2} + a_n p^{n-1} = -\frac{a_0 q^n}{p} \qquad (4)$$

Hence, p is a factor of a_0, since q and p have no common factor greater than 1 and the left member of (4) is an integer.

Corollary of theorem on rational roots
   If $a_0 = 1$ in (1), we get the following corollary:

Each rational root of an equation

$$x^n + a_1 x^{n-1} + \cdots + a_{n-1} x + a_n = 0 \qquad (5)$$

with integral coefficients is an integer and a factor of a_n.

EXAMPLE   Find the set of possible rational roots of $2x^4 + x^3 - 9x^2 - 4x + 4 = 0$.

Solution   In the equation $2x^4 + x^3 - 9x^2 - 4x + 4 = 0$, the numerators of the rational roots must be factors of 4, and the denominators factors of 2. Hence the possibilities for the rational roots are $\pm 1, \pm 2, \pm 4, \pm \frac{1}{2}, \pm \frac{2}{2}, \pm \frac{4}{2}$. If we eliminate repetitions, this set of quotients becomes $\{-4, -2, -1, -\frac{1}{2}, \frac{1}{2}, 1, 2, 4\}$. We can use synthetic division and the remainder theorem in order to determine which of these possibilities are actually roots.

13.13 / THE DEPRESSED EQUATION

If r_1 is a root of $F(x) = 0$, then by the factor theorem,

$$F(x) = F_1(x)(x - r_1) = 0$$

Depressed equation and $F_1(x) = 0$ is called the *depressed equation* corresponding to r_1. Furthermore, the degree of $F_1(x)$ is one less than the degree of $F(x)$. Now if r_2 is a root of $F_1(x) = 0$, then r_2 is also a root of $F(x) = 0$ and we have

$$F(x) = (x - r_1)(x - r_2)F_2(x) = 0$$

We can use $F_2(x) = 0$ in seeking the remaining roots.

This operation is continued after each rational root is found; frequently we obtain a depressed equation that is a quadratic, and then the remaining roots can be found by one of the methods of Chap. 7.

We illustrate the use of the depressed equation by obtaining all roots of the equation in the example of Sec. 13.12. We first use synthetic division to determine whether 2 is a root and get

$$
\begin{array}{r}
2 + 1 - \ 9 - 4 + 4 \lfloor 2 \\
+ 4 + 10 + 2 - 4 \\
\hline
2 + 5 + \ 1 - 2 \quad 0
\end{array}
$$

Since the remainder is zero, 2 is a root. Furthermore, all roots of the given equation $2x^4 + x^3 - 9x^2 - 4x + 4 = 0$, except possibly $x = 2$, are roots of the depressed equation $2x^3 + 5x^2 + x - 2 = 0$. If 2 is a multiple root of the given equation, it is also a root of the depressed equation. We next try $x = -2$ in the depressed equation above and obtain

$$
\begin{array}{r}
2 + 5 + 1 - 2 \lfloor -2 \\
- 4 - 2 + 2 \\
\hline
2 + 1 - 1 \quad 0
\end{array}
$$

Since the remainder is zero, -2 is a root. The depressed equation, corresponding to $x = 2$ and $x = -2$, is the quadratic $2x^2 + x - 1 = 0$, which we may solve by factoring and get

$$(2x - 1)(x + 1) = 0$$
$$x = \tfrac{1}{2}, -1$$

Consequently, the solution set of the given equation is $\{2, -2, -1, \tfrac{1}{2}\}$. It should be noticed that the degree of the equation is 4, and there are four numbers in the solution set.

13.14 / PROCESS OF OBTAINING ALL RATIONAL ROOTS

We now outline the steps that should be followed in determining the rational roots of a polynomial equation.

Steps in process of obtaining all rational roots

1 Write the set of rational numbers that contains the rational roots (Sec. 13.12) in the order of the magnitude of their numerical values.

2 Test the smallest positive integer in the set, then the next larger, and so on, until each integral root or a bound of the roots is found. (*a*) If an upper bound is found, discard all larger numbers in the set. (*b*) If a root is found, use the depressed equation in further calculations.

3 Test the fractions that remain in the set after considering any bound that has been found.

4 Repeat steps 2 and 3 for negative roots.

NOTE: If a quadratic is obtained by use of the depressed equation, its roots can be found by use of any of the methods for solving quadratics.

EXAMPLE Find the solutions of $4x^4 - 4x^3 - 25\,x^2 + x + 6 = 0$

Solution The possible numerators of rational roots are $\pm 6, \pm 3, \pm 2, \pm 1$, and the possible denominators are $\pm 4, \pm 2, \pm 1$. The set of possible rational roots is $\{\pm\frac{1}{4}, \pm\frac{1}{2}, \pm\frac{3}{4}, \pm 1, \pm\frac{3}{2}, \pm 2, \pm 3, \pm 6\}$. We now test the positive integral possibilities.

$$
\begin{array}{r}
4 - 4 - 25 + 1 + 6\,\underline{|1} \\
+4 0 - 25 - 24 \\
\hline
4 0 - 25 - 24 - 18
\end{array}
$$

Since the remainder is not zero and the signs are neither all plus in the third line nor alternately plus and minus, 1 is neither a root nor a bound.

$$
\begin{array}{r}
4 - 4 - 25 + 1 + 6\,\underline{|2} \\
+8 + 8 - 34 - 66 \\
\hline
4 + 4 - 17 - 33 - 60
\end{array}
$$

Therefore, 2 is neither a root nor a bound.

$$
\begin{array}{r}
4 - 4 - 25 + 1 + 6\,\underline{|3} \\
+12 + 24 - 3 - 6 \\
\hline
4 + 8 - 1 - 2 0
\end{array}
$$

Consequently, 3 is a root and the corresponding depressed equation is $4x^3 + 8x^2 - x - 2 = 0$, which has all the roots of the original equation, with the possible exception of 3. Since the constant term in the depressed equation is -2 and the coefficient of x^3 is 4, the set of possible rational roots of the depressed equation is $\{\pm\frac{1}{4}, \pm\frac{1}{2}, \pm 1, \pm 2\}$. Not all of these need be considered, since we found that 1 and 2 are not roots of the original equation.

It is readily seen that $\frac{1}{4}$ is not a root, but $\frac{1}{2}$ is a root with $4x^2 + 10x + 4 = 0$ as the corresponding depressed equation. This is a quadratic that can be solved by factoring. Its roots are -2 and $-\frac{1}{2}$. Hence, the solution set of the original equation is $\{3, \frac{1}{2}, -\frac{1}{2}, -2\}$.

EXERCISE 13.5 Rational Roots

Find all roots of the equation in each of Probs. 1 to 28.

1	$x^3 - x^2 - 4x + 4 = 0$	2	$x^3 - 4x^2 + x + 6 = 0$
3	$x^3 - 7x + 6 = 0$	4	$x^3 + 2x^2 - 5x - 6 = 0$
5	$2x^3 + x^2 - 13x + 6 = 0$	6	$3x^3 + 19x^2 + 16x - 20 = 0$
7	$4x^3 - 7x - 3 = 0$	8	$6x^3 + 13x^2 + x - 2 = 0$
9	$12x^3 - 4x^2 - 3x + 1 = 0$	10	$8x^3 - 12x^2 - 18x + 27 = 0$
11	$4x^3 - 8x^2 - 15x + 9 = 0$	12	$6x^3 - x^2 - 31x - 10 = 0$
13	$x^3 - 4x^2 + 3x + 2 = 0$	14	$x^3 + x^2 - 8x - 6 = 0$
15	$2x^3 + 3x^2 - 4x - 5 = 0$	16	$2x^3 + 3x^2 - 6x - 8 = 0$
17	$2x^3 - x^2 + 2x - 1 = 0$	18	$3x^3 + 7x^2 + 8x + 2 = 0$
19	$2x^3 - 3x^2 + 2x - 3 = 0$	20	$2x^3 + x^2 + 8x + 4 = 0$
21	$6x^4 + x^3 - 22x^2 - 11x + 6 = 0$	22	$6x^4 + 17x^3 - 14x^2 - 27x + 18 = 0$
23	$x^4 + x^3 - 15x^2 + 23x - 10 = 0$	24	$x^4 - x^3 - 13x^2 + x + 12 = 0$
25	$x^4 - x^3 - x^2 - x - 2 = 0$	26	$x^4 - 2x^3 + x^2 + 2x - 2 = 0$
27	$2x^4 + 2x^3 + \frac{1}{2}x^2 + 2x - \frac{3}{2} = 0$		

HINT: Multiply by the lowest common denominator.

28   $x^4 - \frac{11}{6}x^3 + \frac{7}{3}x^2 - \frac{22}{3}x - \frac{20}{3} = 0$

29   Show that $\frac{1}{2}$ is a root of multiplicity 2 of $4x^3 + 8x^2 - 11x + 3 = 0$.

30   Show that $-\frac{2}{3}$ is a root of multiplicity 2 of $9x^3 + 3x^2 - 8x - 4 = 0$.

31   Show that $-\frac{1}{2}$ is a root of multiplicity 2 of $4x^4 + 4x^3 - 3x^2 - 4x - 1 = 0$.

32   Show that $-\frac{1}{3}$ is a root of multiplicity 3 of $27x^4 - 27x^3 - 45x^2 - 17x - 2 = 0$.

33   Show that there are no rational roots of $x^3 - 7x^2 + 2x - 1 = 0$.

34   Show that there are no rational roots of $x^3 + 13x^2 - 6x - 2 = 0$.

35   Show that there are no rational roots of $x^4 - 2x^3 + 10x^2 - x + 1 = 0$.

36   Show that there are no rational roots of $3x^4 - 9x^3 - 2x^2 - 15x - 5 = 0$.

Show that the number in each of Probs. 37 to 44 is irrational. HINT: To show $x = 1 + \sqrt{2}$ *is irrational, form the equivalent equations* $x - 1 = \sqrt{2}$, $(x - 1)^2 = 2$, *and* $x^2 - 2x - 1 = 0$, *and show that the last equation has no rational roots.*

37 $\sqrt{6}$ 38 $\sqrt[3]{5}$ 39 $2 - \sqrt{5}$ 40 $\frac{2}{3} + \sqrt[3]{18}$

41 $\sqrt{2} + \sqrt{7}$ 42 $\sqrt{5} - \sqrt{2}$ 43 $6^{2/3}$

44 $1 + \sqrt{2} + \sqrt{3}$

Note that if $x = 1 + \sqrt{2} + \sqrt{3}$, then also $x - 1 = \sqrt{2} + \sqrt{3}$.

13.15 / APPROXIMATION OF IRRATIONAL ROOTS

Since the general method for solving polynomial equations of degrees 3 and 4 is long and tedious and general methods for solving polynomial equations of degree greater than 4 do not exist, we must resort to some method of approximation to obtain the irrational roots of such equations. Several methods exist, and we shall use one **Basis for** that depends upon the rule for locating roots (Sec. 13.10) and the following **approximating** fact. If $f(x)$ is a polynomial and $y_1 = f(x_1)$ and $y_2 = f(x_2)$, then if x_1 and x_2 are sufficiently near each other, the portion of the graph of $y = f(x)$ between (x_1, y_1) and (x_2, y_2) will lie very near the straight line, or secant, that connects these two points. Consequently, in Fig. 13.3a, if the graph crosses the X axis at $(a, 0)$ and the straight line crosses it at $(b, 0)$, the root a of $f(x) = 0$ will be very near to b.

We can calculate the value of b by use of Fig. 13.3a, where the lines CD and EC are parallel to the X and Y axes, respectively, and the coordinates of C, D, E, F, and B are as indicated. The triangles EFB and ECD are similar.

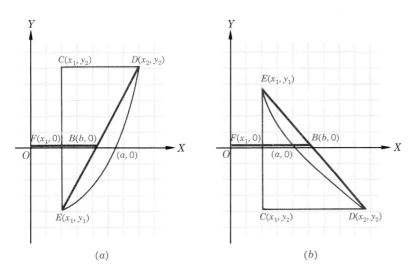

(a)   (b)   **Figure 13.3**

Hence

$$\frac{FB}{CD} = \frac{EF}{EC} \tag{1}$$

By the distance formula,

$$FB = b - x_1 \quad CD = x_2 - x_1 \quad EF = 0 - y_1 = -y_1 \quad EC = y_2 - y_1$$

Substituting these values in (1), we obtain

$$\frac{b - x_1}{x_2 - x_1} = \frac{-y_1}{y_2 - y_1} \tag{2}$$

Now we solve (2) for b and get

$$\boldsymbol{b = x_1 - \frac{y_1(x_2 - x_1)}{y_2 - y_1}} \tag{13.5}$$

We obtain the same result using Fig. 13.3b.

If the graph of $y = f(x)$ is to the right of the secant, as in Fig. 13.3a, then $b < a$, but if the graph is to the left of the secant, as in Fig. 13.3b, then $b > a$. Furthermore, if $f(x_1)$ and $f(b)$ have the same sign, then $b < a$, but if $f(x_1)$ and $f(b)$ have different signs, then $b > a$.

Methods that depend on calculus are usually employed if approximations correct to more than three decimal places are desired.

EXAMPLE 1 Approximate the largest positive root of

$$2x^3 - 3x^2 - 12x + 6 = 0 \tag{3}$$

Solution The synthetic divisions

$$
\begin{array}{l}
2 - 3 - 12 + 6 \underline{| 3} \\
 6 + 9 - 9 \\
\hline
2 + 3 - 3 - 3
\end{array}
\qquad
\begin{array}{l}
2 - 3 - 12 + 6 \underline{| 4} \\
 8 + 20 + 32 \\
\hline
2 + 5 + 8 + 38
\end{array}
$$

show that 4 is an upper bound of the roots and there is a root between 3 and 4, since $f(3) = -3 < 0$ and $f(4) = 38 > 0$. Equation (13.5) with $x_1 = 3$, $y_1 = -3$, $x_2 = 4$, $y_2 = 38$ gives, to one decimal place,

$$b = 3 - \frac{-3(4 - 3)}{38 - (-3)} = 3 + \frac{3}{41} = 3.1$$

We shall therefore repeat the process for values of x near 3.1. Synthetic division gives

$$
\begin{array}{l}
2 - 3 - 12 + 6 \quad \underline{| 3.1} \\
 6.2 + 9.92 - 6.448 \\
\hline
2 + 3.2 - 2.08 - 0.448
\end{array}
\qquad
\begin{array}{l}
2 - 3 - 12 + 6 \quad \underline{| 3.2} \\
 6.4 + 10.88 - 3.584 \\
\hline
2 + 3.4 - 1.12 + 2.416
\end{array}
$$

Hence there is a root between 3.1 and 3.2. Using Eq. (13.5) with $x_1 = 3.1$, $y_1 = -0.448$, $x_2 = 3.2$, $y_2 = 2.416$ gives, to three decimal places,

$$b = 3.1 - \frac{-0.448(3.2 - 3.1)}{2.416 - (-0.448)} = 3.116$$

At this stage we may use either two or three decimal places of 3.116. Often, results are accurate enough to skip the two-decimal step; in this case they are, since synthetic division shows that $f(3.116) = -0.0110$ and $f(3.117) = 0.0165$. Since $f(3.116)$ is closer to 0 than $f(3.117)$ is, we conclude that the desired root of (3) to three decimal places is 3.116.

It helps to know whether the function is increasing or decreasing near the desired root. This can usually be readily determined by comparing two or three nearby function values.

EXAMPLE 2   To enable the reader to check his understanding of the method, we give below the results in each step of the approximation to three decimal places of the root of (3) that is between 0 and 1.

1   $f(0) = 6$, $f(1) = -7$. If $(x_1, y_1) = (0, 6)$ and $(x_2, y_2) = (1, -7)$, then Eq. (13.5) yields $b_1 = 0.5$.

2   $f(0.5) = -0.5$ and $f(0.4) = 0.848$. Hence, for the second approximation, we use $(x_1, y_1) = (0.4, 0.848)$ and $(x_2, y_2) = (0.5, -0.5)$, substitute in Eq. (13.5), and get

$$b_2 = 0.463 \qquad \text{to three decimal places}$$

3   $f(0.463) = -0.0006$ to four decimal places. Since $f(x_1) = f(0.4) = 0.848$, 0.463 is greater than the root. Hence we try 0.462 and find that $f(0.462) = 0.0218$. Therefore, since $f(0.463)$ is nearer to zero than $f(0.462)$, we conclude that the root to three decimal places is 0.463.

The general method for solving third-degree equations is very involved, but the method for fourth-degree equations, surprisingly, is somewhat less involved, though still formidable. We illustrate the method in the following example.

EXAMPLE 3   Solve $$x^4 + 2x^3 - x^2 + x + \tfrac{1}{4} = 0 \tag{4}$$

Solution   By adding $x^2 - x - \tfrac{1}{4}$ to each member of (4), we have

$$x^4 + 2x^3 = x^2 - x - \tfrac{1}{4}$$

Completing the square of the left member by adding x^2, we get

$$(x^2 + x)^2 = 2x^2 - x - \tfrac{1}{4}$$

Now by adding $(x^2 + x)y + \tfrac{1}{4}y^2$ to each member, we have

$$(x^2 + x)^2 + (x^2 + x)y + \tfrac{1}{4}y^2 = 2x^2 - x - \tfrac{1}{4} + (x^2 + x)y + \tfrac{1}{4}y^2$$

or
$$(x^2 + x + \tfrac{1}{2}y)^2 = x^2(2 + y) + x(-1 + y) - \tfrac{1}{4} + \tfrac{1}{4}y^2 \qquad (5)$$

Since the left member is a perfect square, the right member must be also.
Hence its discriminant $b^2 - 4ac$ is zero. Thus,

$$(-1 + y)^2 - 4(2 + y)(-\tfrac{1}{4} + \tfrac{1}{4}y^2) = 0$$
$$1 - 2y + y^2 + 2 + y - 2y^2 - y^3 = -y^3 - y^2 - y + 3 = 0 \qquad (6)$$

We now find *any* root of (6) since (5) is a perfect square for any root of
(6). By Sec. 13.12 on rational roots, one root of (6) is 1. Substituting $y = 1$
in (5) gives

$$(x^2 + x + \tfrac{1}{2})^2 = 3x^2$$

Therefore, by taking the square root of each member, we obtain

$$x^2 + x + \tfrac{1}{2} = \sqrt{3}x \qquad \text{and} \qquad x^2 + x + \tfrac{1}{2} = -\sqrt{3}x$$

Consequently, the roots of (4) are those of

$$x^2 + (1 - \sqrt{3})x + \tfrac{1}{2} = 0 \qquad \text{and} \qquad x^2 + (1 + \sqrt{3})x + \tfrac{1}{2} = 0$$

Hence, by the quadratic formula,

$$x = \frac{-1 + \sqrt{3} \pm \sqrt{1 - 2\sqrt{3} + 3 - 2}}{2}$$

$$\text{and} \qquad \frac{-1 - \sqrt{3} \pm \sqrt{1 + 2\sqrt{3} + 3 - 2}}{2}$$

$$x = \frac{-1 + \sqrt{3} \pm \sqrt{2 - 2\sqrt{3}}}{2} \qquad \text{and} \qquad \frac{-1 - \sqrt{3} \pm \sqrt{2 + 2\sqrt{3}}}{2}$$

$$x = \frac{-1 + \sqrt{3} \pm \sqrt{2}\sqrt{1 - \sqrt{3}}}{2} \qquad \text{and} \qquad \frac{-1 - \sqrt{3} \pm \sqrt{2}\sqrt{1 + \sqrt{3}}}{2}$$

Approximations of square roots to three places give $x = -0.197$ and -2.535
as two of the four roots. The other two roots are $0.366 \pm 0.605i$.

EXERCISE 13.6 Irrational Roots

In Probs. 1 to 8, find the real root to two decimal places.

1 $x^3 - 6x^2 + 12x - 12 = 0$ 2 $x^3 - 3x^2 + 3x - 4 = 0$

3 $x^3 + 9x^2 + 27x + 7 = 0$ 4 $x^3 + 12x^2 + 48x + 55 = 0$

5 $x^3 + x - 4 = 0$

6 $x^3 + 3x - 18 = 0$

7 $x^3 + 5x + 16 = 0$

8 $x^3 + 9x + 13 = 0$

In Probs. 9 to 16, find the value of the least positive root to two decimal places.

9 $x^3 + 8x^2 + 15x - 2 = 0$

10 $x^3 - x^2 + 4x - 2 = 0$

11 $x^3 - 8x + 5 = 0$

12 $3x^3 + 9x^2 - 3x - 16 = 0$

13 $x^3 + 4x^2 - x - 6 = 0$

14 $2x^3 - 9x^2 + x + 11 = 0$

15 $x^3 + 3x^2 - 6x - 3 = 0$

16 $x^3 - 4x^2 + 3x + 1 = 0$

In Probs. 17 to 24, find the value of each irrational root to three decimal places.

17 $x^4 - 2x^3 - 4x - 4 = 0$

18 $x^4 - 8x^3 + 12x^2 - 8x + 11 = 0$

19 $x^4 - 6x^3 + 4x^2 - 6x + 3 = 0$

20 $x^4 + 4x^3 - 2x^2 + 4x - 3 = 0$

21 $x^3 + 3x^2 - 3x - 7 = 0$

22 $x^3 + 6x^2 - 28 = 0$

23 $x^3 - 6x^2 + 6x + 2 = 0$

24 $2x^3 - 11x^2 + 12x + 7 = 0$

In Probs. 25 to 32, use the method of Example 3 to find all four roots of the equation. In each problem, find the largest real root to three decimal places.

25 $x^4 - 4x^3 - 2x^3 + 12x + 8 = 0$

26 $x^4 - 2x^3 - 8x^2 + 12x - 4 = 0$

27 $x^4 - 2x^3 - 16x^2 + 22x + 7 = 0$

28 $x^4 + 2x^3 - 35x^2 - 22x + 36 = 0$

29 $x^4 - 8x^3 + 14x^2 - 8x + 13 = 0$

30 $x^4 - 4x^3 + x^2 - 12x - 6 = 0$

31 $x^4 + 6x^3 + 5x^2 + 24x + 4 = 0$

32 $x^4 + 4x^3 + 2x^2 + 20x - 15 = 0$

EXERCISE 13.7 **REVIEW**

1 Find the remainder if $x^3 + 3x^2 - x - 4$ is divided by $x - 2$.

2 Find the remainder if $-2x^3 + 3x^2 + 7x - 40$ is divided by $x + 3$.

3 What is the remainder when $x^4 - 2x^3 + 5x + 2$ is divided by $x - d$?

4 Show that $x + 4$ is a factor of $x^3 + 4x^2 - x - 4$.

5 Show that $x^3 - 3x^2 - 2x + 6$ is divisible by $x - 3$.

6 Show that $\frac{3}{2}$ is a root of $2x^3 - 3x^2 + 6x - 9 = 0$.

7 Show that $\sqrt{2}$ is a root of $x^3 + 4x^2 - 2x - 8 = 0$.

8 Show that $3i$ is a root of $x^3 - 4x^2 + 9x - 36 = 0$.

9 Find the quotient and remainder if $2x^3 + 5x - 6$ is divided by $x - 2$.

10 Find the quotient and remainder if $3x^4 + 5x^3 + 2x^2 + 6x - 3$ is divided by $x + 2$.

11 Find all roots of $x^3 - x^2 - 8x + 12$ if 2 is a root of multiplicity 2.

12 Find the other two roots of $x^4 + 6x^3 + 11x^2 + 12x + 18$ if -3 is a root of multiplicity 2.

13 Find a polynomial of least possible degree with real coefficients whose zeros include 2 and $3 + i$.

14 Find a polynomial of least possible degree with rational coefficients whose zeros include 3 and $1 + \sqrt{3}$.

15 Find a polynomial of degree 4 with real coefficients that has $x - 1 + 2i$ and $x + 2 - i$ as factors.

16 Find A and B so that $\dfrac{7x + 11}{(x + 3)(x - 2)} = \dfrac{A}{x + 3} + \dfrac{B}{x - 2}$.

17 Find A, B, and C so that $\dfrac{3x^2 + 8x - 3}{(x^2 + 3)(2x - 3)} = \dfrac{Ax + B}{x^2 + 3} + \dfrac{C}{2x - 3}$.

18 Find A, B, and C so that $\dfrac{-x^2 + 2x - 10}{(x^2 + x + 4)(x + 2)} = \dfrac{Ax + B}{x^2 + x + 4} + \dfrac{C}{x + 2}$.

19 What are the sum and the product of the roots of $x^3 - x^2 + x - 1 = 0$?

20 What are the sum and the product of the roots of $x^4 - 4x^3 + 8x^2 - 16x + 16 = 0$?

21 Show that all real roots of $2x^3 - x^2 + 4x + 3 = 0$ are between -1 and 1.

22 Show that there is exactly one positive root of $x^4 + 2x^3 + 4x - 5 = 0$.

23 Sketch the graph of $y = x^3 + x^2 - 4x + 3$ for $-3 \le x \le 3$.

24 Sketch the graph of $y = x^4 - 3x^3 - 5x^2 - 15x - 30$ for $-2 \le x \le 5$.

25 Sketch the graph of $y = x^5 - 2x^4 + x^3 - 6x^2 - 10x - 3$ for $-1 \le x \le 3$.

26 Find all roots of the equation $x^3 - 2x^2 + x - 2 = 0$.

27 Find all roots of the equation $4x^4 + 12x^3 + 25x^2 + 48x + 36 = 0$.

28 Find all roots of $6x^5 - 17x^4 - 7x^3 + 58x^2 - 78x + 20 = 0$.

29 Show that the arithmetic mean of the roots of $9x^4 - 24x^3 - 35x^2 - 4x + 4 = 0$ equals the arithmetic mean of the roots of $36x^3 - 72x^2 - 70x - 4 = 0$.

30 Show that $\sqrt[4]{6}$ is irrational.

31 Show that $\sqrt{3} - \sqrt{2}$ is irrational.

32 Is $\sqrt{3} + \sqrt{12} - \sqrt{27}$ irrational?

33 Find the real root of $x^3 + 6x^2 + 12x + 2 = 0$ to two decimal places.

34 Find both real roots of $x^4 + 2x^3 + x^2 + 6x - 6 = 0$ to two decimal places.

35 Find any real root of $x^4 - x^3 - 4x^2 + 3x + 3 = 0$ to three decimal places.

36 Find the exact value of each root of $x^4 - x^3 - 4x^2 + 3x + 3 = 0$.

CHAPTER FOURTEEN

Progressions

The inventor of chess, so it is said, asked that he be rewarded with one grain of wheat for the first square of the board, two grains for the second, four for the third, and so on for the 64 squares. Fortunately, this apparently modest request was examined before it was granted. By the twentieth square, the reward would have amounted to more than a million grains of wheat; by the sixty-fourth square the number called for would have been astronomical and the amount would have far exceeded all the grain in the kingdom.

This story deals with a sequence of numbers, and such relationships have a great many important applications. If a sequence of numbers is such that each term can be obtained from the preceding one by the operation of some law, the sequence is called a *progression*.

Progression

14.1 / ARITHMETIC PROGRESSIONS

An *arithmetic progression* is a sequence of numbers so related that each term after the first can be obtained from the preceding term by adding a fixed quantity called the *common difference*. The following examples illustrate the meaning and application of these terms.

Common difference

371

If the first term is 2 and the common difference is 5, then the first eight terms of the arithmetic progression are 2, 7, 12, 17, 22, 27, 32, 37.

In the sequence 16, $14\frac{1}{2}$, 13, $11\frac{1}{2}$, 10, $8\frac{1}{2}$, each term after the first is $1\frac{1}{2}$ less than the preceding. Hence, this is an arithmetic progression with the common difference equal to $-1\frac{1}{2}$.

Most problems in arithmetic progressions deal with three or more of the following five quantities: the first term, the last term, the number of terms, the common difference, and the sum of all the terms. We shall now derive formulas which enable us to determine any one of these five quantities if we know the values of three of the others.

We shall let†

$a =$ the first term in the progression
$l =$ the last term
$d =$ the common difference
$n =$ the number of terms
$s =$ the sum of all the terms

14.2 / LAST TERM OF AN ARITHMETIC PROGRESSION

In terms of the above notation, the first four terms of an arithmetic progression are

$$a \qquad a + d \qquad a + 2d \qquad a + 3d$$

We notice that d enters with the coefficient 1 in the second term and that this coefficient increases by 1 as we move from each term to the next. Hence the coefficient of d in any term is 1 less than the number of that term in the progression. Therefore, the sixth term is $a + 5d$, the ninth is $a + 8d$, and finally the last, or nth, term is $a + (n - 1)d$. Hence we have the formula

$$l = a + (n - 1)d \qquad\qquad (14.1)$$

EXAMPLE 1   If the first three terms of an arithmetic progression are 2, 6, and 10, find the eighth term.

Solution   Since the first and second terms, as well as the second and third, differ by 4, it follows that $d = 4$. Furthermore, $a = 2$ and $n = 8$. Hence if we substitute these values in Eq. (14.1), we have

$$l = 2 + (8 - 1)4 = 2 + 28 = 30$$

---

† Note that these symbols can be arranged so that they spell the word "lands."

EXAMPLE 2 If the first term of an arithmetic progression is -3 and the eighth and last term is 11, find d and write the eight terms of the progression.

Solution In this problem, $a = -3$, $n = 8$, and $l = 11$. If these values are substituted in Eq. (14.1), we have

$$11 = -3 + (8 - 1)d$$
$$11 = -3 + 7d \qquad \text{performing indicated operations}$$
$$-7d = -14 \qquad \text{adding } -11 - 7d \text{ to each member}$$
$$d = 2 \qquad \text{solving for } d$$

Therefore, since $a = -3$, the first eight terms of the desired progression are $-3, -1, 1, 3, 5, 7, 9, 11$.

14.3 / SUM OF AN ARITHMETIC PROGRESSION

The sum of the arithmetic progression 1, 5, 9, 13 may be written as $1 + (1 + 4) + (1 + 2 \times 4) + (1 + 3 \times 4)$ or, starting with the last term, $13 + (13 - 4) + (13 - 2 \times 4) + (13 - 3 \times 4)$. Similarly we may write the sum of any arithmetic progression as

$$s = a + (a + d) + (a + 2d) + \cdots + [a + (n - 2)d] + [a + (n - 1)d]$$

or as

$$s = l + (l - d) + (l - 2d) + \cdots + [l - (n - 2)d] + [l - (n - 1)d]$$

Adding corresponding terms of these two equations gives

$$2s = (a + l) + [(a + d) + (l - d)] + \cdots + [a + (n - 1)d + l - (n - 1)d]$$
$$= (a + l) + (a + l) + \cdots + (a + l) = n(a + l)$$

Hence, dividing by 2, we obtain the formula

$$s = \frac{n}{2}(a + l) \qquad (14.2)$$

By (14.1), $a + l = a + [a + (n - 1)d] = 2a + (n - 1)d$, and thus

$$s = \frac{n}{2}[2a + (n - 1)d] \qquad (14.3)$$

is a second formula for s.

If any three of the quantities l, a, n, d, and s are known, the other two can be found by use of Eqs. (14.1) through (14.3), either separately or by solving two of them simultaneously.

EXAMPLE 1   Find the sum of all the even integers from 2 to 1,000, inclusive.

Solution   Since the even integers 2, 4, 6, etc., taken in order, form an arithmetic progression with $d = 2$, we can use Eq. (14.2), with $a = 2$, $n = 500$, and $l = 1,000$, to obtain the desired sum. The substitution of these values in Eq. (14.2) yields

$$s = \tfrac{500}{2} \times (2 + 1,000)$$
$$= 250 \times 1,002 = 250,500$$

EXAMPLE 2   If $a = 4$, $n = 10$, and $l = 49$, find d and s.

Solution   Since each of Eqs. (14.1) and (14.2) contains a, n, and l, we can find d and s by using the formulas separately. If we substitute the given values for a, n, and l in Eq. (14.1), we get

$$49 = 4 + (10 - 1)d$$
$$49 = 4 + 9d \qquad \text{performing indicated operations}$$
$$-9d = -45 \qquad \text{adding } -49 - 9d \text{ to each member}$$
$$d = 5 \qquad \text{solving for } d$$

Similarly, by substituting in Eq. (14.2), we have

$$s = \tfrac{10}{2} \times (4 + 49)$$
$$= 5 \times 53$$
$$= 265$$

EXAMPLE 3   If $l = 23$, $d = 3$, and $s = 98$, find a and n.

Solution   If we substitute these values in Eqs. (14.1) and (14.2), we obtain

$$23 = a + (n - 1)3 \qquad (1)$$

from the former, and

$$98 = \frac{n}{2}(a + 23) \qquad (2)$$

from the latter. Each of these equations contains the two desired variables a and n. Hence we complete the solution by solving (1) and (2) simultaneously. If we solve (1) for a, we get

$$a = 23 - (n - 1)3$$
$$= 26 - 3n \qquad (3)$$

Upon substituting the above expression for a in (2), we obtain

$$98 = \frac{n}{2}(26 - 3n + 23)$$

$196 = n(49 - 3n)$	multiplying by 2 and combining
$196 = 49n - 3n^2$	performing indicated operations
$3n^2 - 49n + 196 = 0$	adding $3n^2 - 49n$ to each member
$n = 9\frac{1}{3}$ and 7	solving for n

Since n cannot be a fraction, we discard $9\frac{1}{3}$ and have

$$n = 7$$

If we substitute 7 for n in (3), we obtain

$$a = 26 - (3 \times 7)$$
$$= 5$$

Hence, the progression consists of the seven terms 5, 8, 11, 14, 17, 20, and 23.

_____

EXERCISE 14.1 Arithmetic Progressions

Write the n terms of the arithmetic progression described in each of Probs. 1 to 12.

1 $a = 2, d = 4, n = 5$ **2** $a = 4, d = -3, n = 4$

3 $a = 5, d = -2, l = -3$ **4** $a = 3, d = 2, l = 11$

5 $a = 4$, second term 7, $n = 4$ **6** $a = -3$, third term 3, $n = 5$

7 $a = 2$, third term -8, $l = -13$ **8** $a = 4$, fourth term -2, $l = -6$

9   Second term 6, third term 2, $n = 5$

10   Second term 5, fourth term -7, $l = -19$

11   Third term 4, fifth term 6, $n = 6$

12   Second term 9, fifth term 0, $l = -6$

In Probs. 13 to 24, use the three given quantities to find the one at the far right.

13 $a = 4, n = 7, d = -2; l$ **14** $l = 5, a = 2, n = 6; d$

15 $n = 7, d = -2, l = -9; a$ **16** $a = -1, d = \frac{1}{2}, l = 1; n$

17 $a = 2, l = 10, n = 8; s$ **18** $a = 3, l = -5, s = -10; n$

19 $s = 20, n = 10, a = 8; l$ **20** $n = 9, l = 1, s = 36; a$

21 $a = 4, d = 2, s = 88; n$ **22** $s = 12, n = 6, a = -3; d$

23 $n = 10, d = \frac{1}{2}, s = \frac{45}{2}; a$ **24** $a = 3, s = \frac{105}{2}, d = \frac{3}{2}; n$

Find the two quantities of l, a, n, d, and s that are missing in each of Probs. 25 to 36.

25 $a = 2, d = 3, n = 5$ **26** $a = 17, d = -3, n = 7$

27 $a = 6, n = 6, l = 1$ **28** $a = 16, l = -19, d = -7$

29 $a = -2, l = 13, s = 33$ **30** $n = 6, l = 21, s = 66$

31 $d = -4, a = 11, s = -7$ **32** $a = 11, d = -3, s = 14$

33 $d = -3, l = -7, s = 3$ 34 $d = 5, l = 7, s = -2$
35 $l = 16, d = 3, s = 51$ 36 $l = -10, d = -4, s = 0$

37   Find the sum of all even integers between 5 and 29.

38   Find the sum of all multiples of 3 between 2 and 43.

39   Find the sum of the first n positive multiples of 4.

40   Find the sum of the first n integral multiples of 5.

41   Find the value of x if $2x + 1$, $x - 2$, and $3x + 4$ are consecutive terms of an arithmetic progression.

42   Find the value of x if $3x - 1$, $1 - 2x$, and $2x - 5$ are consecutive terms of an arithmetic progression.

43   Find the values of x and y if $3x - y$, $2x + y$, $4x + 3$, and $3x + 3y$ are consecutive terms of an arithmetic progression.

44   Show that if a, b, c and x, y, z are two arithmetic progressions, then $a + x$, $b + y$, $c + z$ is an arithmetic progression.

45   How many times will a clock strike in 24 hours if it strikes only at the hours?

46   If a compact body falls vertically 16 ft during the first second, 48 ft during the next second, 80 ft during the third, and so on, how far will it fall during the seventh second?   During the first seven seconds?

47   A bomb was dropped from an altitude of 10,000 ft.   Neglecting air resistance, find the time required for it to reach the ground. (See Prob. 46.)

48   A student made a grade of 64 on the first test, and did 7 points better on each succeeding test than on the previous one.   What score was made on the fifth test, and what was the average grade of the five tests?

49   A machine that cost $5,800 depreciated 15 percent the first year, 13.5 percent the second, 12 percent the third, and so on.   What was its value at the end of nine years if all percentages apply to the original cost?

50   If a man buys a painting on June 12, 1975 for $7,000 and sells it on June 12, 1983 for $15,400, and the increase in value each year is $100 more than that of the previous year, find the value of the painting on June 12, 1981.

51   Find the approximate length of a motion picture film 7.01 cm thick if it is wound on a reel 6 cm in diameter that has a central core 2 cm in diameter.   Consider the film as being wound in concentric circles.

52   A display of cans has 18 cans on the bottom row, 17 on the row above, 16 on the next row, and so on.   If there are 12 rows of cans, how many cans are there in all?

14.4 GEOMETRIC PROGRESSIONS

A *geometric progression* is a sequence of numbers so related that each term after the first can be obtained from the preceding term by multiplying it by a fixed constant called the *common ratio*. The following are geometric progressions with the indicated common ratios:

Common ratio

$$2, \quad 6, \quad 18, \quad 54, \quad 162 \qquad \text{common ratio 3}$$
$$3, \quad -3, \quad 3, \quad -3, \quad 3 \qquad \text{common ratio } -1$$
$$96, \quad 24, \quad 6, \quad \tfrac{3}{2}, \quad \tfrac{3}{8} \qquad \text{common ratio } \tfrac{1}{4}$$

In order to obtain formulas for dealing with a geometric progression, we shall let†

$$a = \text{the first term}$$
$$l = \text{the last term}$$
$$r = \text{the common ratio}$$
$$n = \text{the number of terms}$$
$$s = \text{the sum of the terms}$$

Note that the symbols for geometric progressions are the same as those for arithmetic progressions except for the use of r, the common ratio, in place of d, the common difference.

14.5 LAST TERM OF A GEOMETRIC PROGRESSION

In terms of the above notation, the first six terms of a geometric progression in which the first term is a and the common ratio is r are

$$a \quad ar \quad ar^2 \quad ar^3 \quad ar^4 \quad ar^5$$

We notice here that the exponent of r in the second term is 1 and that this exponent increases by 1 as we proceed from each term to the next. Hence, the exponent of r in any term is 1 less than the number of that term in the progression. Therefore the nth term is ar^{n-1}. Thus, we have the formula

$$l = ar^{n-1} \tag{14.4}$$

EXAMPLE   Find the seventh term of the geometric progression $36, -12, 4, \ldots$.

† Note that the letters used here can be arranged to spell the word "snarl."

Solution   In this progression, each term after the first is obtained by multiplying the preceding term by $-\frac{1}{3}$.† Hence, $r = -\frac{1}{3}$. Also $a = 36$, $n = 7$, and the seventh term is l. If we substitute these values in Eq. (14.4), we have

$$l = 36 \times (-\tfrac{1}{3})^{7-1} = \frac{36}{(-3)^6} = \frac{36}{729} = \frac{4}{81}$$

14.6 / SUM OF A GEOMETRIC PROGRESSION

If we add the terms of the geometric progression $a, ar, ar^2, \ldots, ar^{n-2}, ar^{n-1}$, we have

$$s = a + ar + ar^2 + \cdots + ar^{n-2} + ar^{n-1} \tag{1}$$

However, by use of an algebraic device, we can obtain a more compact formula for s. First, we multiply each member of (1) by r and get

$$rs = ar + ar^2 + ar^3 + \cdots + ar^{n-1} + ar^n \tag{2}$$

Next, we notice that if we subtract the corresponding members of (1) and (2) and combine like terms, we have

$$s - rs = a + ar + \cdots + ar^{n-1} - (ar + ar^2 + \cdots + ar^{n-1} + ar^n) = a - ar^n$$

or $$s(1 - r) = a - ar^n$$

By solving this equation for s, we obtain

$$s = \frac{a - ar^n}{1 - r} \qquad r \neq 1 \tag{14.5}$$

If we multiply each member of Eq. (14.4) by r, we get $rl = ar^n$. Now, if we replace ar^n by rl in Eq. (14.5), we have

$$s = \frac{a - rl}{1 - r} \qquad r \neq 1 \tag{14.6}$$

If any three of the quantities s, n, a, r, and l are known, the other two can be found by using Eqs. (14.4) through (14.6).

EXAMPLE 1   Find the sum of the first six terms of the progression $2, -6, 18, \ldots$.

Solution   In this progression, $a = 2$, $r = -3$, and $n = 6$. Hence, if we substitute these values in Eq. (14.5), we have

$$s = \frac{2 - [2 \times (-3)^6]}{1 - (-3)} = \frac{2 - (2 \times 729)}{1 + 3} = -364$$

† Given any two consecutive terms of a geometric progression, we find r by dividing the second of the two by the first; in this case $-12 \div 36 = -\frac{1}{3}$.

EXAMPLE 2   The first term of a geometric progression is 3; the fourth term is 24.   Find the tenth term and the sum of the first 10 terms.

Solution   In order to find either the tenth term or the sum, we must have the value of r.   We can obtain this value by considering the progression as made up of the first four terms defined above.   Then we have $a = 3$, $n = 4$, and $l = 24$.   If we substitute these values in Eq. (14.4), we get

$$24 = 3r^{4-1}$$
$$3r^3 = 24 \qquad \text{performing indicated operations}$$
$$r^3 = 8 \qquad \text{dividing both members by 3}$$
$$r = 2 \qquad \text{solving for } r$$

Now, by using Eq. (14.4) again with $a = 3$, $r = 2$, and $n = 10$, we get

$$l = 3 \times 2^{10-1} = 3 \times 512 = 1{,}536$$

Hence, the tenth term is 1,536.
    In order to obtain s, we use Eq. (14.5), with $a = 3$, $r = 2$, and $n = 10$, and get

$$s = \frac{3 - (3 \times 2^{10})}{1 - 2} = \frac{3 - (3 \times 1{,}024)}{-1} = \frac{3 - 3{,}072}{-1} = 3{,}069$$

EXAMPLE 3   If $s = 61$, $l = 81$, and $n = 5$, find a and r.

Solution   If $s = 61$, $l = 81$, and $n = 5$, we can find a and r by use of any two of Eqs. (14.4) to (14.6).   However, the work is easier if we use Eqs. (14.4) and (14.6).   By substituting the given values in these, we get

$$81 = ar^4 \qquad (3)$$

$$61 = \frac{a - 81r}{1 - r} \qquad (4)$$

We solve (3) and (4) simultaneously by first solving (4) for a in terms of r and then substituting in (3).

$$61 - 61r = a - 81r \qquad \text{multiplying each member by } 1 - r$$
$$a = 61 + 20r \qquad \text{solving for } a \qquad (5)$$
$$81 = (61 + 20r)r^4 \qquad \text{substituting the value of } a \text{ in (3)}$$
$$20r^5 + 61r^4 - 81 = 0 \qquad \text{removing parentheses and adding}$$
$$\text{} \qquad -81 \text{ to each member}$$

Now, by use of the methods of Chap. 13, we find the rational roots of this equation to be -3 and 1.   The solution $r = 1$ must be discarded since, for this value, (4) is meaningless.   However, if $r = -3$, then by (5), $a = 1$. Hence, the solution is $a = 1$, $r = -3$.

EXERCISE 14.2 Geometric Progressions

Write the n terms of each geometric progression described in Probs. 1 to 8.

1 $a=2, r=3, n=5$ 2 $a=4, r=2, n=6$

3 $a=8, r=\frac{1}{4}, n=5$ 4 $a=32, r=\frac{1}{5}, n=5$

5 $a=1$, second term $=3$, $n=5$ 6 $a=\frac{1}{3}$, third term $=\frac{4}{3}$, $l=\frac{32}{3}$

7 Second term $=1$, fourth term $=\frac{9}{4}$, $l=\frac{81}{16}$

8 Second term $=\frac{2}{3}$, fifth term $=\frac{9}{32}$, $l=\frac{9}{32}$

In Probs. 9 to 20, use the three given quantities to find the one at the far right.

9 $a=4, r=3, n=7; l$ 10 $a=5, r=2, n=6; l$

11 $r=\frac{1}{2}, n=6, l=\frac{3}{32}; a$ 12 $l=162, a=\frac{1}{8}, r=6; n$

13 $a=2, r=3, n=4; s$ 14 $a=\frac{1}{5}, r=6, n=5; s$

15 $r=3, n=6, s=91; a$ 16 $a=\frac{1}{3}, s=455, r=4; n$

17 $a=2, r=3, l=162; s$ 18 $a=4, r=\frac{3}{2}, l=20.25; s$

19 $a=5, r=1, s=35; l$ 20 $a=6, l=96, s=186; r$

In Probs. 21 to 36, find the two of the quantities s, n, a, r, and l that are missing.

21 $a=3, r=2, n=5$ 22 $a=2, r=3, n=5$

23 $a=4, n=6, l=-128$ 24 $a=2, r=3, l=162$

25 $a=343, r=\frac{1}{7}, l=1$ 26 $n=4, r=3, l=3$

27 $n=9, a=256, l=1$ 28 $n=7, a=\frac{1}{8}, l=8$

29 $s=242, a=2, r=3$ 30 $s=781, n=5, r=\frac{1}{5}$

31 $s=400, r=\frac{1}{7}, l=1$ 32 $s=1, n=7, l=1$

33 $l=12, s=9, n=3$ 34 $r=-\frac{3}{2}, l=\frac{81}{2}, s=\frac{55}{2}$

35 $a=5, n=3, s=65$ 36 $a=8, l=\frac{1}{8}, s=\frac{127}{8}$

37 Find the sum of all of the integral powers of 2 between 5 and 500.

38 Find the sum of the first seven integral powers of 3 beginning with 3.

39 Find the sum $1+2+4+8+\cdots+2^n$.

40 Find the sum $\dfrac{1}{4}+\dfrac{1}{16}+\dfrac{1}{64}+\cdots+\dfrac{1}{4^n}$.

41 For which values of k are $2k$, $5k+2$, and $20k-4$ consecutive terms of a geometric progression?

42 If 1, 4, and 19 are added to the first, second, and third terms, respectively, of an arithmetic progression with $d=3$, a geometric progression is obtained. Find the arithmetic progression and the common ratio of the geometric progression.

43 Show that if $\dfrac{1}{y-x}$, $\dfrac{1}{2y}$, and $\dfrac{1}{y-z}$ form an arithmetic progression, then x, y, and z form a geometric progression.

44 Show that if *a, b, c* and *x, y, z* are two geometric progressions, then *ax, by, cz* is also a geometric progression.

45 Twelve men are fishing. If the first is worth \$1,000, the second \$2,000, the third \$4,000, and so on, how many of the men are millionaires?

46 Mrs. Timken willed one-third of her estate to one person, one-third of the remainder to a second, and so on, until the fifth received \$1,600. What was the value of the estate?

47 The number of bacteria in a culture doubles every 2 hours. If there were *n* bacteria present at noon one day, how many were there at noon the next day?

48 The first stroke of a pump removes one-fourth of the air from a bell jar, and each stroke thereafter removes one-fourth of the remaining air. What part of the original amount is left after six strokes?

49 Mr. Hughes bets \$1 on the first poker hand, \$2 on the second, \$4 on the third, and so on. If he loses nine hands in a row, how much does he lose? If he then wins the tenth hand, what is his net profit or loss?

50 If \$100 is put in a savings account at the end of each 6 months, how much money is in the account at the end of 6 years if the bank pays 6 percent interest per year compounded semiannually?

51 If there were no duplications, how many ancestors did a pair of twins have in the immediately preceding seven generations?

52 Each year, a \$30,000 machine depreciates by 20 percent of its value at the beginning of that year. Find its value at the end of the fifth year.

14.7 / INFINITE GEOMETRIC PROGRESSIONS

In this section we shall find the limit of the sum of a geometric progression with $|r| < 1$ as *n* increases indefinitely. By the statement

$$\lim_{n \to \infty} s(n) = s$$

we mean that, by taking *n* sufficiently large, the value of $s(n)$ differs from *s* by an amount that is less than any positive number chosen in advance. If we use $s(n)$ to represent the sum of the first *n* terms of a geometric progression, and if we factor the right member of Eq. (14.5), we have

$$s(n) = \frac{a}{1-r}(1 - r^n) \tag{1}$$

Since $|r| < 1$, it follows that $|r| > |r^2| > \cdots > |r^n|$, and it can be proved†
that r^n can be made arbitrarily small by taking n sufficiently large. Thus

$$\lim_{n \to \infty} s(n) = \frac{a}{1 - r}$$

and we follow the usual procedure by writing

$$s = \frac{a}{1 - r} \qquad |r| < 1 \tag{14.7}$$

EXAMPLE 1 Find the sum of the geometric progression $1, \frac{1}{2}, \frac{1}{4}, \ldots$, where the dots indicate that there is no end to the progression.

Solution In this progression, $a = 1$ and $r = \frac{1}{2}$. Hence, by Eq. (14.7),

$$s = \frac{1}{1 - \frac{1}{2}} = \frac{1}{\frac{1}{2}} = 2$$

A nonterminating, repeating decimal fraction is an illustration of an infinite geometric progression with $-1 < r < 1$. For example,

$$0.232323 \cdots = 0.23 + 0.0023 + 0.000023 + \cdots$$

The sequence of terms on the right is a geometric progression, with $a = 0.23$ and $r = \frac{1}{100}$.

By use of Eq. (14.7), we can express any repeating decimal fraction as a common fraction by the method illustrated in the following example. In fact, a decimal is repeating if and only if it is equal to a rational number.

EXAMPLE 2 Show that $0.333 \cdots = \frac{1}{3}$. $\hspace{3cm}$ (2)

Solution The decimal fraction $0.333 \cdots$ can be expressed as the progression

$$0.3 + 0.03 + 0.003 + \cdots$$

in which $a = 0.3$ and $r = 0.1$. Hence, by Eq. (14.7), the sum s is

$$s = \frac{0.3}{1 - 0.1} = \frac{0.3}{0.9} = \frac{3}{9} = \frac{1}{3}$$

---

† In order to see how large n must be so that $|r|^n$ is less than some positive number p, we write

$$|r|^n < p$$
$$\log |r|^n < \log p$$
$$n \log |r| < \log p$$
$$n > \frac{\log p}{\log |r|}$$

The sense of the inequality was changed in the last step since we were dividing by $\log |r|$, and $\log |r|$ is negative if $0 < |r| < 1$. The student should calculate $(\log p)/\log |r|$ for a few choices of p and $|r|$, say $p = 10^{-20}$ and $r = \frac{3}{10}$.

EXAMPLE 3   Express $3.2181818\cdots$ as a mixed fraction.

Solution   The given number can be expressed as 3.2 plus an infinite geometric progression with $a = 0.018$ and $r = 0.01$. Thus, from Eq. (14.7),

$$3.2181818\cdots = 3.2 + 0.018 + 0.00018 + 0.0000018 + \cdots$$

$$= 3.2 + \frac{0.018}{1 - 0.01} = 3.2 + \frac{0.018}{0.99}$$

$$= 3.2 + \tfrac{1}{55}$$

$$= 3\tfrac{12}{55}$$

EXERCISE 14.3 Infinite Geometric Progressions

Find the sum of the infinite geometric progression whose elements are given in each of Probs. 1 to 12.

1	$a = 3,\ r = \tfrac{1}{2}$	**2**	$a = 3,\ r = \tfrac{1}{4}$
3	$a = 3,\ r = -\tfrac{1}{4}$	**4**	$a = 2,\ r = \tfrac{2}{5}$
5	$a = 8,\ r = \tfrac{1}{3}$	**6**	$a = 18,\ r = -\tfrac{1}{2}$
7	$a = 22,\ r = -\tfrac{5}{6}$	**8**	$a = \tfrac{1}{10},\ r = \tfrac{1}{10}$
9	$a = 8$, second term $= 2$	**10**	$a = 3$, third term $= \tfrac{4}{3}$

11   Second term $= \tfrac{5}{2}$, fourth term $= \tfrac{1}{10}$

12   Second term $= \tfrac{1}{6}$, fifth term $= \tfrac{1}{48}$

13   Find r if $a = 2$ and $s = 6$.

14   Find r if $a = \tfrac{1}{3}$ and $s = \tfrac{1}{5}$.

15   Find a if $r = -\tfrac{3}{4}$ and $s = \tfrac{1}{7}$.

16   Find a if $r = \tfrac{1}{3}$ and $s = 1$.

If n is a positive integer, find the sum of all the numbers of the form given in each of Probs. 17 to 20.

17 $\left(\tfrac{1}{3}\right)^n$	**18** $\left(\tfrac{2}{3}\right)^n$	**19** $\left(-\tfrac{2}{5}\right)^n$	**20** $\left(-\tfrac{3}{4}\right)^n$

Express the repeating decimal in each of Probs. 21 to 28 as a rational number in lowest terms.

21	$0.444\cdots$	**22**	$0.2424\cdots$	**23**	$2.343434\cdots$
24	$1.414141\cdots$	**25**	$4.1222\cdots$	**26**	$2.2111\cdots$
27	$6.54848\cdots$	**28**	$2.0124124\cdots$		

Problems 29 to 32 give four different ways of showing that $0.999\cdots = 1$.

29   Treat $0.999\cdots$ as in Probs. 21 to 28.

30   Let $x = 0.999 \cdots$, so $10x = 9.999 \cdots$. Then subtract and solve for x.

31   Show that if $(a + b)/2 = a$, then $a = b$. Apply this with $a = 0.999 \cdots$ and $b = 1.000 \cdots$.

32   Multiply both members of (2) in Example 2 by 3.

In each of Probs. 33 to 36, find the sum of the series. Verify by calculation that the sum of an odd number of terms is larger than s, and the sum of an even number of terms is less than s.

33 $1 - \dfrac{1}{3} + \dfrac{1}{3^2} - \dfrac{1}{3^3} + \cdots$ **34** $1 - \dfrac{1}{4} + \dfrac{1}{4^2} - \dfrac{1}{4^3} + \cdots$

35 $\frac{2}{5} - (\frac{2}{5})^2 + (\frac{2}{5})^3 - (\frac{2}{5})^4 + \cdots$ **36** $2 - 1 + \frac{1}{2} - \frac{1}{4} + \cdots$

37   If the first arc made by the tip of a pendulum is 12 cm and each arc thereafter is 0.995 as long as the one just before it, how far does the tip move before coming to rest?

38   If a ball rebounds $\frac{3}{5}$ as far as it falls, how far will it travel before coming to rest if it is dropped from a height of 30 meters?

39   The motion of a particle through a certain medium is such that it moves $\frac{2}{3}$ as far each second as in the preceding second. If it moves 6 meters the first second, how far will it move before coming to rest?

40   An alumna gave an oil field to her university. If the university received \$230,000 from the field the first year and $\frac{2}{3}$ as much each year thereafter as during the immediately preceding year, how much did the college realize?

41   Assume that potatoes shrink one-half as much each week as during the previous week. If a dealer stores 1,000 kg when the price is n cents per kilogram, and if the weight decreases to 950 kg during the first week, for which values of n can he afford to hold the potatoes until the price rises to $n + 1$ cents per kilogram?

42   A hamster receives a dose of 3 mg of a compound and then $\frac{2}{3}$ as much as the previous dose at the end of every 3 hours. What is the maximum amount of the compound it will receive?

43   A series of squares is drawn by connecting the midpoints of the sides of a given square, then the midpoints of the square thus drawn, and so on. Find the sum of the areas of all the squares if the original square had sides of 10 in.

44   Find the sum of the perimeters of the squares of Prob. 43.

Find the replacement set for x in each of Probs. 45 to 48 in order for the sum to exist. Also find the sum.

45 $\dfrac{1}{2x - 1} + \dfrac{1}{(2x - 1)^2} + \dfrac{1}{(2x - 1)^3} + \cdots$

46 $\dfrac{2}{3x-2} + \dfrac{4}{(3x-2)^2} + \dfrac{8}{(3x-2)^3} + \cdots$

47 $\dfrac{1}{x^2+2} + \dfrac{1}{(x^2+2)^2} + \dfrac{1}{(x^2+2)^3} + \cdots$

48 $\dfrac{3}{x^2+1} + \dfrac{3}{(x^2+1)^2} + \dfrac{3}{(x^2+1)^3} + \cdots$

14.8 / ARITHMETIC MEANS

The terms between the first and last terms of an arithmetic progression are called *arithmetic means*. If the progression contains only three terms, the middle term is called *the arithmetic mean* of the first and last term. If the progression consists of three terms a, m, and l, then by the definition of an arithmetic progression, $m - a = l - m$, $2m = a + l$, and

$$m = \frac{a+l}{2} \tag{14.8}$$

Rule for finding arithmetic mean

Therefore, *the arithmetic mean of two numbers is equal to one-half their sum.*

EXAMPLE   Insert five arithmetic means between 6 and −10.

Solution   Since we are to find five means between 6 and −10, we shall have seven terms in all. Hence, $n = 7$, $a = 6$, and $l = -10$. Thus, by Eq. (14.1), we have

$$-10 = 6 + (7-1)d$$
$$-6d = 16 \qquad \text{adding } -6d + 10 \text{ to each member}$$
$$\qquad\qquad \text{and combining terms}$$
$$d = -\tfrac{16}{6} = -\tfrac{8}{3} \qquad \text{solving for } d$$

and the progression is 6, $\tfrac{10}{3}$, $\tfrac{2}{3}$, $-\tfrac{6}{3}$, $-\tfrac{14}{3}$, $-\tfrac{22}{3}$, $-\tfrac{30}{3}$.

14.9 / GEOMETRIC MEANS

The terms between the first and last terms of a geometric progression are called *geometric means*. If the progression contains only three terms, the middle term is called *the geometric mean* of the other two. If the three terms in the progression are a, m, and l, then by the definition of a geometric progression, $m/a = l/m$, so $m^2 = al$. Thus the second term, or the geometric mean between a and l, is

$$m = \pm\sqrt{al} \tag{14.9}$$

Rule for finding geometric mean   Hence, *the geometric mean between two quantities is either the square root of their product or its negative.*

EXAMPLE 1   Find the five geometric means between 3 and 192.

Solution   A geometric progression starting with 3, ending with 192, and containing five intermediate terms has seven terms. Hence, $n = 7$, $a = 3$, and $l = 192$. Therefore, by Eq. (14.4),

$$192 = 3(r^{7-1})$$
$$r^6 = \frac{192}{3}$$
$$= 64$$
$$r = \sqrt[6]{64} = \pm 2 \qquad \text{solving for } r$$

Consequently, the two sets of geometric means of five terms each between 3 and 192 are 6, 12, 24, 48, 96, and -6, 12, -24, 48, -96.

EXAMPLE 2   Find the geometric mean of $\frac{1}{2}$ and $\frac{1}{8}$.

Solution   By Eq. (14.9), the geometric mean of $\frac{1}{2}$ and $\frac{1}{8}$ is $\pm\sqrt{\frac{1}{2} \times \frac{1}{8}} = \pm\sqrt{\frac{1}{16}} = \pm\frac{1}{4}$.

14.10 / HARMONIC PROGRESSIONS

The series formed by the reciprocals of the terms of an arithmetic progression is called a *harmonic progression.*[†] For example, since two quantities are reciprocals if and only if their product is 1 and since $-5, -3, -1, 1, 3, 5, 7$ is an arithmetic series, it follows that $-\frac{1}{5}, -\frac{1}{3}, -1, 1, \frac{1}{3}, \frac{1}{5}, \frac{1}{7}$ is a harmonic progression.

From the above definition, we can derive the following rule:

Rule for finding *n*th term of harmonic progression   *In order to determine the nth term of a harmonic progression, we write the corresponding arithmetic progression, find the nth term of the arithmetic progression, and take its reciprocal.*

Harmonic means   The terms between any two terms of a harmonic progression are called *harmonic means.*

EXAMPLE   What is the tenth term of a harmonic progression if the first and third terms are $\frac{1}{2}$ and $\frac{1}{6}$? What is the harmonic mean of $\frac{1}{2}$ and $\frac{1}{6}$?

[†] It is interesting that strings of the same weight and subjected to the same tension will produce a harmonious sound if their lengths are in harmonic progression.

Solution The first and third terms of the corresponding arithmetic progression are 2 and 6. Hence, $l = a + (n-1)d$ becomes $6 = 2 + 2d$, and consequently, $d = 2$. Therefore, when $n = 10$, $l = 2 + (10-1)2 = 20$. Taking the reciprocal of 20, we find that the tenth term of the harmonic progression is $\frac{1}{20}$. Since $d = 2$, the first three terms of the arithmetic progression are 2, 4, 6, so the harmonic mean of $\frac{1}{2}$ and $\frac{1}{6}$ is $\frac{1}{4}$.

EXERCISE 14.4 Means and Harmonic Progressions

1 Find three arithmetic means between 3 and 15.

2 Find five arithmetic means between 3 and 15.

3 Insert four arithmetic means between 10 and -10.

4 Insert six arithmetic means between 18 and 7.5.

5 What are the two geometric means between 2 and 54?

6 What are the four geometric means between $\frac{2}{9}$ and $\frac{27}{16}$?

7 Find two sets of three geometric means between 2 and 32.

8 Find two sets of five geometric means between 2 and $\frac{1}{32}$.

9 Insert four harmonic means between $\frac{1}{3}$ and $\frac{1}{13}$.

10 Insert five harmonic means between $\frac{2}{3}$ and $\frac{1}{9}$.

11 What are the three harmonic means between 2 and 5?

12 What are the two harmonic means between $-\frac{1}{5}$ and 1?

13 Find the sixth term of the harmonic progression $\frac{1}{4}, \frac{1}{8}, \frac{1}{12}, \ldots$

14 Find the eighth term of the harmonic progression $\frac{3}{5}, \frac{3}{7}, \frac{1}{3}, \ldots$

15 Find the seventh term of the harmonic progression $\frac{2}{3}, \frac{4}{9}, \frac{1}{3}, \ldots$

16 Find the sixth term of the harmonic progression $3, 1, \frac{3}{5}, \ldots$

17 What is the first term of a harmonic progression whose third term is $\frac{1}{5}$ and ninth term is $\frac{1}{8}$?

18 What is the eighth term of a harmonic progression whose second term is 2 and fifth term is -2?

19 What is the sixth term of a harmonic progression whose third term is -1 and eighth term is $\frac{1}{9}$?

20 What is the thirteenth term of a harmonic progression whose third term is 12 and seventh term is 2?

21 Show that the harmonic mean between a and b is $2ab/(a+b)$.

22 Show that if $0 < a < b$ and the plus sign is used in Eq. (14.9), then the harmonic mean of a and b is less than their geometric mean, and their geometric mean is less than their arithmetic mean.

23 Verify Prob. 22 for $a = 2$ and $b = 8$.

24 Verify Prob. 22 for $a = 3.4$ and $b = 5.9$.

Classify the progression in each of Probs. 25 to 32, and give the next two terms.

25 $\frac{2}{3}, \frac{4}{9}, \frac{8}{27}, \frac{16}{81}$

26 $\frac{2}{3}, \frac{4}{9}, \frac{1}{3}, \frac{4}{15}$

27 $\frac{2}{3}, \frac{4}{9}, \frac{2}{9}, 0$

28 $\frac{1}{20}, \frac{3}{10}, \frac{11}{20}, \frac{4}{5}$

29 $\frac{1}{4}, 1, -\frac{1}{2}, -\frac{1}{5}$

30 $2, 3, 4\frac{1}{2}, 6\frac{3}{4}$

31 $\frac{1}{6}, \frac{2}{3}, \frac{7}{6}, \frac{5}{3}$

32 $\frac{1}{7}, \frac{2}{11}, \frac{1}{4}, \frac{2}{5}$

EXERCISE 14.5 REVIEW

1 Write the seven terms of an arithmetic progression with $a = 3$ and $d = \frac{1}{2}$.

2 Write the six terms of a geometric progression with $a = \frac{3}{4}$ and $r = \frac{2}{3}$.

3 Write the six terms of a harmonic progression with the first two terms being 12 and 3.

4 If a, b, c, d is an arithmetic progression, then $1/a, 1/b, 1/c, 1/d$ is by definition a harmonic progression. Show that if x, y, z, w is a geometric progression, then $1/x, 1/y, 1/z, 1/w$ is also a geometric progression.

5 Find the sixth term and the sum of the six terms of the geometric progression that begins $2, \frac{2}{3}, \ldots$.

6 Find the seventh term and the sum of the seven terms of the arithmetic progression that begins $\frac{1}{12}, \frac{5}{24}, \ldots$.

7 If the second and fourth terms of an arithmetic progression are $\frac{3}{4}$ and $\frac{7}{4}$ and $n = 5$, find $a, d, l,$ and s.

8 If the second and fifth terms of a geometric progression are $\frac{1}{2}$ and 32 and $s = \frac{341}{8}$, find $a, r, l,$ and n.

9 Find the sum of all integers between 8 and 800 that are multiples of 7.

10 Find all values of k so that $2k + 2$, $5k - 11$, and $7k - 13$ is a geometric progression.

11 If \$1,000 is put in a savings account and left for 6 years, how much is in the account at the end of that time if the bank pays 6 percent interest compounded semiannually?

12 Express $0.5151 \cdots$ as a rational number.

13 Find the sum of all numbers of the form $(-1/6)^n$, where n is a positive integer.

14 Show that if a, b, c, d is an arithmetic progression and x is a real number, then $a + x, b + x, c + x, d + x$ is also an arithmetic progression.

15 Repeat Prob. 14 for ax, bx, cx, dx.

16 Show that p^2, q^2, r^2 is a geometric progression if p, q, r is.

17 Show that $\frac{1}{4}$, $\frac{3}{4}$, $\frac{9}{4}$, $\frac{27}{4}$ is a geometric progression and $\log\frac{1}{4}$, $\log\frac{3}{4}$, $\log\frac{9}{4}$, $\log\frac{27}{4}$ is an arithmetic progression.

18 Show that if a^2, b^2, c^2 is an arithmetic progression, then $a+b$, $a+c$, $b+c$ is a harmonic progression.

19 Find three geometric means between 4 and 16.

20 Find three arithmetic means between 4 and 16.

21 Find three harmonic means between 4 and 16.

22 In Prob. 22 of Exercise 14.4, a proof was requested that the arithmetic mean exceeds the geometric mean for positive, unequal numbers a and b. Show that the difference is $\frac{1}{2}(\sqrt{a}-\sqrt{b})^2$.

23 Let $f(x) = x^3 - 2x^2 - x + 2$ and $g(x) = 3x^2 - 4x - 1$. Show that the arithmetic average of the roots of $f(x) = 0$ equals the arithmetic average of the roots of $g(x) = 0$.

24 Find the infinite geometric progression whose sum is $\frac{4}{3}$ and the sum of whose first three terms is $\frac{3}{2}$.

Classify the progression in each of Probs. 25 to 29, and give the next two terms.

25 $\frac{1}{12}$, $\frac{1}{4}$, $\frac{5}{12}$, $\frac{7}{12}$

26 $\frac{3}{5}$, $\frac{2}{5}$, $\frac{4}{15}$, $\frac{8}{45}$

27 $\frac{1}{9}$, $\frac{5}{18}$, $\frac{4}{9}$, $\frac{11}{18}$

28 $\frac{3}{2}$, $\frac{6}{5}$, 1, $\frac{6}{7}$

29 $\frac{4}{3}$, $\frac{8}{7}$, 1, $\frac{8}{9}$

CHAPTER FIFTEEN # Mathematical Induction

All sciences, including mathematics, proceed by establishing generalizations, or laws. The usefulness of a generalization is that it frees us from the particular case; we can handle the new case as it arises simply by identifying it as falling in the class covered by the generalization.

One way to arrive at a generalization is to work first with the general case itself, just as we did in arriving at the quadratic formula. Then in applying the formula we need only identify the equation as being in the general form. That is, we work from the general to the particular, a method of logic called *deduction*.

Another way to arrive at a generalization is to examine a number of specific cases to see in what way they are related; the relation, once found, is stated as a generalization, or law. That is, we work from the particular to the general, a method of logic known as *induction*.

The danger of induction is that the specific cases, however large in number, may be special ones, and the generalization based on them may therefore be erroneous.† Science must often be content to work with this

† Erroneous induction is commonplace in everyday life; for example, it is the method of the incompetent mechanic, and it is one of the means by which we get our phobias and prejudices.

limitation, called *incomplete induction*, but in mathematics we are able to avoid the pitfall by a method of proof called *mathematical induction*, or *complete induction*.

15.1 / METHOD OF MATHEMATICAL INDUCTION

If we let $n = 0, 1, 2, 3$ in $q(n) = n^2 - n + 41$, we find that $q(n)$ becomes 41, 41, 43, and 47, respectively. These numbers are primes; that is, no one of them is divisible by any integer other than itself and unity and each is greater than 1. If we calculate the value of $q(n)$ for each integral value of n up to and including 40, we see that $q(n)$ represents the same type of integer, and this surely suggests that $q(n)$ represents a prime number for every integral value assigned to n. However, if n is equal to 41, $q(n) = 41^2 - 41 + 41 = 41^2$, which is not a prime.

On the other hand, $1 + 3 = 4 = 2^2$, $1 + 3 + 5 = 9 = 3^2$, $1 + 3 + 5 + 7 = 16 = 4^2$. These results suggest that the sum of the first n odd integers is n^2; that is,

$$1 + 3 + 5 + \cdots + (2n - 1) = n^2 \qquad \textbf{(15.1)}$$

Since repeated verification of the truth of Eq. (15.1) for particular values of n does not constitute a proof, we must find some other means of demonstrating its general validity. The type of reasoning involved in mathematical induction is illustrated by the following hypothetical example: Suppose that a certain goal can be reached by a sequence of steps that are successive but of unknown number. Suppose, further, that a person in the process of achieving this goal can be assured that it will always be possible for him to take the next step. Then, regardless of all other circumstances, he knows that he can ultimately attain the goal.

In order to apply this method of reasoning to the proof of the statement in Eq. (15.1), we assume that the statement is true for some definite but unknown integral value of n, that is, for $n = k$, where k is a number for which the statement can be verified. Then we show that it *necessarily* follows that the statement is true for the *next* integer, $k + 1$. Then, if we can show that the statement is true for some number, say $k = 3$, we know that it is true for the next integer, $k = 4$, and, by proceeding in the same manner, for all following integers. We shall show how this is done in Example 1.

EXAMPLE 1 Prove Eq. (15.1) by mathematical induction.

Solution We begin with Eq. (15.1), which is

$$1 + 3 + 5 + \cdots + (2n - 1) = n^2 \qquad (1)$$

We then assume that (1) is true for $n = k$ and obtain

$$1 + 3 + 5 + \cdots + (2k - 1) = k^2 \qquad (2)$$

Next we write (1) with $n = k + 1$ and get

$$1 + 3 + 5 + \cdots + (2k + 1) = (k + 1)^2 \qquad (3)$$

Now we shall prove that the truth of (3) necessarily follows from (2). The last term in the left member of (3) is the $(k + 1)$st term of (1); hence the next to the last is the kth term and, therefore, is $(2k - 1)$. Consequently, we can write (3) in the form

$$1 + 3 + 5 + \cdots + (2k - 1) + (2k + 1) = (k + 1)^2 \qquad (4)$$

In order to prove that (4) is true, provided we assume the truth of (2), we notice that the left member of (4) is the corresponding member of (2) increased by $(2k + 1)$, that is, by the $(k + 1)$st term of (1). Hence, we add $2k + 1$ to each member of (2), thus obtaining

$$1 + 3 + 5 + \cdots + (2k + 1) = k^2 + 2k + 1 = (k + 1)^2$$

which is the same as (4). Therefore, if (2) is true, (4) is true. That is, (1) is true for $n = k + 1$ if it is true for $n = k$.

Evidently, (1) is true for $n = 1$.

Since (4) has been proved true if (2) is true, (1) holds for $n = 1, 2, 3, 4, 5$, and so on.

The formal process of a proof by mathematical induction consists of the following five steps:

Steps in proof by mathematical induction

1 Assume that the theorem or statement to be proved is true for $n = k$, a particular but unspecified integer, and express this assumption in symbolic form.

2 Obtain a symbolic statement of the theorem for $n = k + 1$.

3 Prove that if the statement in step 1 is true, then the statement in step 2 is true also.

4 Verify the theorem for the least integral value q of n for which it has a meaning.

5 Using the conclusion in step 3, observe that the theorem is true for $n = q + 1$ since it is true for the integer q of step 4; furthermore, that it is true for $n = q + 2$ since it is true for $n = q + 1$; . . . ; and finally, that it is true regardless of the positive integral value of n if $n \geq q$.

Notice that the basic steps are steps 3 and 4. No general directions can be given for carrying out the work of step 3. However, the following additional examples illustrate a procedure that can frequently be followed.

EXAMPLE 2   Prove that De Moivre's theorem,

$$[r(\cos \theta + i \sin \theta)]^n = r^n(\cos n\theta + i \sin n\theta) \tag{5}$$

is true if n is a positive integer.

Solution   According to step 1, we assume that (5) is true for $n = k$, thus obtaining

$$[r(\cos \theta + i \sin \theta)]^k = r^k(\cos k\theta + i \sin k\theta) \tag{6}$$

We write (5) with $n = k + 1$, and we get

$$[r(\cos \theta + i \sin \theta)]^{k+1} = r^{k+1}[\cos (k + 1)\theta + i \sin (k + 1)\theta] \tag{7}$$

In order to prove that the truth of (7) follows from (6), we multiply each member of the latter by $r(\cos \theta + i \sin \theta)$ since this will give us a new equation whose left member is the same as that of (7). We thus have

$$[r(\cos \theta + i \sin \theta)]^{k+1} = [r(\cos \theta + i \sin \theta)][r^k(\cos k\theta + i \sin k\theta)]$$
$$= r^{k+1}[\cos (k + 1)\theta + i \sin (k + 1)\theta]$$

Hence, (5) is true for $n = k + 1$ if it is true for $n = k$.
We next verify (6) for $k = 2$ and obtain

$$[r(\cos \theta + i \sin \theta)]^2 = r^2(\cos \theta + i \sin \theta)^2$$
$$= r^2[(\cos^2 \theta - \sin^2 \theta) + i(2 \sin \theta \cos \theta)]$$
$$= r^2(\cos 2\theta + i \sin 2\theta)$$

Hence, since we have proved that (7) is true if (6) is true, (5) is true for $n = 3, 4, 5$, and so on.

EXAMPLE 3   Prove by mathematical induction that $4^n - 1$ is divisible by 3.

Solution   Supposing that the statement is true for $n = k$, we have $4^k - 1 = 3q$ for some integer q. Now for $n = k + 1$ we may write

$$
\begin{aligned}
4^{k+1} - 1 &= 4^{k+1} - 4^k + (4^k - 1) && \text{adding and subtracting } 4^k \\
&= 4^k(4 - 1) + 3q && \text{using the induction hypothesis} \\
&= 3(4^k + q)
\end{aligned}
$$

which shows that the statement is true for $n = k + 1$ if it is true for $n = k$. For $n = 1$, the statement is that $4 - 1$ is divisible by 3, and since this is certainly true, the proof by induction is complete.

EXERCISE 15.1 **Mathematical Induction**

By use of mathematical induction, show that the equation in each of Probs. 1 to 20 is true for all positive integers [for step 3, add the $(k + 1)$st term of the formula under consideration to each member of the equation].

1 $1 + 2 + 3 + \cdots + n = n(n + 1)/2$

2 $3 + 5 + 7 + \cdots + (2n + 1) = n(n + 2)$

3 $1 + 4 + 7 + \cdots + (3n - 2) = n(3n - 1)/2$

4 $5 + 9 + 13 + \cdots + (4n + 1) = n(2n + 3)$

5 $2 + 6 + 18 + \cdots + 2(3^{n-1}) = 3^n - 1$

6 $\frac{1}{2} + \frac{1}{4} + \frac{1}{8} + \cdots + 1/2^n = 1 - 1/2^n$

7 $7 + 7^2 + 7^3 + \cdots + 7^n = 7(7^n - 1)/6$

8 $1 + 6 + 6^2 + \cdots + 6^{n-1} = (6^n - 1)/5$

9 $2 + 6 + 12 + \cdots + n(n + 1) = n(n + 1)(n + 2)/3$

10 $4 + 10 + 18 + \cdots + n(n + 3) = n(n + 1)(n + 5)/3$

11 $2 + 16 + 36 \cdots + n(5n - 3) = n(n + 1)(5n - 2)/3$

12 $-3 - 2 + 3 + \cdots + n(2n - 5) = n(n + 1)(4n - 13)/6$

13 $1^2 + 2^2 + 3^2 + \cdots + n^2 = n(n + 1)(2n + 1)/6$

14 $1^2 + 3^2 + 5^2 + \cdots + (2n - 1)^2 = n(2n - 1)(2n + 1)/3$

15 $1^3 + 2^3 + 3^3 + \cdots + n^3 = [n(n + 1)/2]^2$

16 $1^3 + 3^3 + 5^3 + \cdots + (2n - 1)^3 = 2n^4 - n^2$

17 $(1)(3) + (2)(4) + (3)(5) + \cdots + (n)(n + 2) = n(n + 1)(2n + 7)/6$

18 $(1)(2)(3) + (2)(3)(4) + (3)(4)(5) + \cdots + n(n + 1)(n + 2) =$
$$n(n + 1)(n + 2)(n + 3)/4$$

19 $\dfrac{1}{(1)(2)} + \dfrac{1}{(2)(3)} + \dfrac{1}{(3)(4)} + \cdots + \dfrac{1}{n(n + 1)} = \dfrac{n}{n + 1}$

20 $\dfrac{1}{(1)(2)(3)} + \dfrac{1}{(2)(3)(4)} + \dfrac{1}{(3)(4)(5)} + \cdots +$
$$\dfrac{1}{(n)(n + 1)(n + 2)} = \dfrac{n(n + 3)}{4(n + 1)(n + 2)}$$

21 Show that $5^n - 1$ is divisible by 4.

22 Show that $3^{2n+1} + 1$ is divisible by 4.

23 Show that $4^{2n} - 1$ is divisible by 3.

24 Show that $2^{3n} - 1$ is divisible by 7.

25 Show that $x^{2n+1} + y^{2n+1}$ is divisible by $x + y$.

26 Show that $x^{2n-1} - y^{2n-1}$ is divisible by $x - y$.

27 Show that $x^{2n} - y^{2n}$ is divisible by $x - y$.

28 Show that $x^{2n} - y^{2n}$ is divisible by $x + y$.

29 Show that $n < 2^n$.

30 Show that $n^2 < 2^n$ if $n \geq 5$.

31   Show that $2^{n-1} \le (n)(n-1)(n-2) \cdots (3)(2)(1)$.

32   Show that if $n \ge 2$, then

$$\frac{1}{n+1} + \frac{1}{n+2} + \frac{1}{n+3} + \cdots + \frac{1}{2n} > \frac{1}{2}$$

33   Show that if $2 + 5 + 8 + \cdots + (3n-1) = [n(3n+1)+3]/2$ is true for $n = k$, then it is true for $n = k+1$. Nevertheless, try any positive integer n and observe that the formula fails.

34   Repeat Prob. 33 for $2 + 7 + 12 + \cdots + (5n-3) = [n(5n-1)+5]/2$.

35   Show that $1 + 3 + 5 + \cdots + (2n-1) = 3n - 2$ holds for $n = 1$ and $n = 2$, but not for all n.

36   Let $p_1 = 2$, $p_2 = 3$, $p_3 = 5, \ldots$ be the primes. Show that $p_1 p_2 \cdots p_n + 1$ is prime for $n = 1, 2, 3$, but not for all n.

37   Show that

$$a + (a+d) + (a+2d) + \cdots + [a+(n-1)d] = (n/2)[2a+(n-1)d].$$

38   Show that $a + ar + ar^2 + \cdots + ar^{n-1} = (a - ar^n)/(l-r)$.

39   Show that the number of subsets (counting the empty set) of a set with n elements is 2^n.

40   Show that

$$\frac{1}{n} + \frac{1}{n+1} + \frac{1}{n+2} + \cdots + \frac{1}{2n-1} = 1 - \frac{1}{2} + \frac{1}{3} - \frac{1}{4} + \cdots + \frac{1}{2n-1}$$

CHAPTER SIXTEEN *Permutations and Combinations*

In the theory of probability, in statistics, in industry, in science, and in government, it is often desirable or necessary to determine the number of ways in which the elements of a set can be arranged or combined into subsets. For example, a telephone company must provide each subscriber with a unique number, and a state government must provide each car with a unique number or combination of numbers and letters. We shall study problems of this nature in this chapter.

16.1 / DEFINITIONS

Element
Combination
Permutation

In this chapter we shall deal with collections and arrangements of symbols, objects, or events. Each symbol, object, or event will be called an *element;* furthermore, each set of elements will be called a *combination,* and each unique arrangement of the elements in a combination will be called a *permutation.* Note that a permutation is distinguished by the *order* of the elements that form it; a combination is a set of elements without regard to order. In either case, all the elements may be of the same kind, but they need not be. All the elements may be used in any permutation or combination, but it is not necessary to use all of them.

396

16.2 / THE FUNDAMENTAL PRINCIPLE

Our work with combinations and permutations will be based on the following principle:

Fundamental principle

If a thing can be done in h_1 ways and if, after it has been done, a second thing can be done in h_2 ways, then the two things can be done in h_1h_2 ways in the indicated order.

This principle is illustrated in the following examples.

EXAMPLE 1   In how many ways can a boy and a girl be selected from a group of five boys and six girls?

Solution   The boy can be selected in five ways, and after that is done, the girl can be selected in six ways; hence a boy and a girl can be selected in $5 \times 6 = 30$ ways.

We can extend the fundamental principle by thinking of the first two events as a single one that can happen in (h_1h_2) ways; hence, if after the first two events have occurred, a third can happen in h_3 ways, then the three can happen in $(h_1h_2)h_3 = h_1h_2h_3$ ways. If then a fourth can occur in h_4 ways, the four can happen in $(h_1h_2h_3)h_4 = h_1h_2h_3h_4$ ways. This procedure can be continued, event by event, until we find that n events can happen in $h_1h_2h_3 \cdots h_n$ ways in the order indicated by the subscripts.

EXAMPLE 2   How many automobile license plates can be made if the inscription on each contains two different letters followed by three different digits?

Solution   There are 26 letters; therefore, the first of the two letters can be chosen in 26 ways. Since the two letters must be different, there are only 25 ways in which the second letter can be chosen. The first digit can be selected in 10 ways; and since the three digits must be different, the second can be chosen in 9 ways and the third in 8. Consequently license plates of two letters and three digits can be made up in $26 \times 25 \times 10 \times 9 \times 8 = 468,000$ ways.

Elements may be drawn from a set with or without replacement. If they are drawn without replacement, we have a combination or permutation. If each element is replaced before the next is drawn, we have neither a combination nor a permutation, and the elements withdrawn are called a *sample*.

Sample   If a set has n elements and we choose a sample of r elements, we begin by choosing any one of the n elements and then replacing it in the set. The second element is chosen similarly, so by the fundamental principle we may choose the first two elements in $n \cdot n = n^2$ ways. This may be continued for r drawings, showing that

The number of samples of r elements that can be drawn from a set of n elements is

$$n^r \qquad\qquad (16.1)$$

if each element is replaced before the next is drawn

EXAMPLE 3   How many three-letter "words" may be formed for a code which uses the letters A, E, I, O, and U and the numbers 2, 3, and 5 if any symbol may be repeated?

Solution   We have $n = 8$ elements and use $r = 3$ of them for each "word," so the answer is $8^3 = 512$.

EXERCISE 16.1 **The Fundamental Principle**

1   How many license plates can be made if each has three different letters followed by three different digits?

2   How many committees consisting of an Indian, an Alaskan, and a Mexican can be chosen from 12 Indians, 14 Alaskans, and 19 Mexicans?

3   In how many ways may a coach choose an outfield from four left fielders, six center fielders, and three right fielders?

4   How many integers between 1,000 and 9,000 may be formed from the digits 2, 3, 5, 7, and 8 if there are no repetitions?

5   How many license plates can be made if each has three letters followed by three digits?

6   How many integers between 1,000 and 9,000 may be formed from the digits 2, 4, 5, 6, and 7?

7   How many three-letter "words" may be formed from the alphabet, excluding A, B, X, Y, and Z?

8   If each person has three initials, how large a population may a city have if everybody has different initials?

9   How many five-letter "words" may be made using the 12 letters of the Hawaiian alphabet?

10   How many four-letter "words" may be made using the last 10 letters of the Hawaiian alphabet if no letter is repeated?

11   How many integers between 2,000 and 8,000 may be formed from the digits 2, 4, 6, 7, and 9 if the second digit is a 2?

12 If a man has 4 suits, 6 shirts, 13 ties, 8 pairs of socks, 2 hats, and 5 pairs of shoes, in how many ways can he dress?

13 In how many ways can a Japanese, two Chinese, and an Australian be chosen from eight of each nationality to be put into single hotel rooms numbered 1135, 1241, 1382, and 1364?

14 If an automobile dealership has 14 salesmen, in how many ways may it choose a salesman of the month for 6 months?

15 How many four-digit even numbers can be formed from 7, 6, 4, and 2 if no digit is repeated?

16 How many four-digit even numbers can be formed from 7, 6, 4, and 2?

17 Five people always buy their food from any of three stores. In how many ways may they shop on a given day if each person shops in just one store?

18 How many results are possible if two standard dice are thrown?

19 In how many ways may a fortune-teller draw and replace a card three times from a standard deck?

20 In how many ways may a true-false test of 12 questions be answered?

In Probs. 21 to 24, show that the first number is larger than the second.

21 3^4, 4^3 22 2^5, 5^2 23 3^8, 8^3 24 4^{16}, 16^4

16.3 / PERMUTATIONS OF n DIFFERENT ELEMENTS TAKEN r AT A TIME

The definition of a permutation was given in Sec. 16.1 but is repeated here for ready reference. Each arrangement of a set of n elements is called a *permutation* of the set.

The arrangements *abc, acb, bac, bca, cab,* and *cba* constitute the six permutations of the letters *a, b,* and *c* taken three at a time. Furthermore, *ab, ba, ac, ca, bc,* and *cb* are the six permutations of the same three letters taken two at a time.

We introduce here a new notation which will allow a more compact form for certain expressions. The product of any positive integer n and all positive integers less than n is called factorial n or n factorial, and is written $n!$ For example,

$$3! = 3 \cdot 2 \cdot 1 = 6$$
$$5! = 5 \cdot 4 \cdot 3 \cdot 2 \cdot 1 = 120$$

Notice that $4! = 4 \cdot 3 \cdot 2 \cdot 1 = 4 \cdot 3 \cdot 2$. It is convenient to define $0! = 1$.

We shall let $P(n, r)$ represent the number of permutations of n elements taken r at a time, and we shall develop a formula for evaluating this number.† We can fill the position 1 in the arrangement in n ways. After position 1 has been filled, we have $n - 1$ choices for position 2, then $n - 2$ choices for position 3, and finally, $n - (r - 1)$ choices for position r. Hence, since $n - (r - 1) = n - r + 1$, we have

$$P(n, r) = n(n - 1)(n - 2) \cdots (n - r + 1)$$

Note that this is the product of r consecutive integers. If we multiply the right member of this equation by $(n - r)!/(n - r)!$, where $(n - r)!$ is the product of the integer $n - r$ and all the positive integers less than $n - r$, we get

$$\frac{n(n - 1)(n - 2) \cdots (n - r + 1)(n - r)!}{(n - r)!} = \frac{n!}{(n - r)!}$$

Permutation of n elements taken r at a time

Therefore,

$$P(n, r) = \frac{n!}{(n - r)!} \qquad \textbf{(16.2)}$$

In order to obtain the number of permutations of n elements taken n at a time, we let $r = n$ in Eq. (16.2). Thus, we obtain

Permutation of n elements taken n at a time

$$P(n, n) = n! \qquad \textbf{(16.3)}$$

since, by definition, $0! = 1$.

EXAMPLE 1   Find the number of four-digit numbers that can be formed from the digits 1, 2, 3, and 4 if no digit is repeated.

Solution   Here we have four elements to be taken four at a time; by Eq. (16.3), the number is 4!, or $4 \times 3 \times 2 \times 1 = 24$.

EXAMPLE 2   Six people enter a room that contains 10 chairs. In how many ways can they be seated?

Solution   Since only six of the chairs are to be occupied, the number of different seating arrangements is equal to the number of permutations of 10 elements taken 6 at a time and, by Eq. (16.2), is

$$P(10, 6) = \frac{10!}{(10 - 6)!} = \frac{10 \times 9 \times 8 \times 7 \times 6 \times 5 \times (4!)}{(4!)}$$

$$= 10 \times 9 \times 8 \times 7 \times 6 \times 5 = 151{,}200$$

---

† The symbols $_nP_r$ and P_r^n are also used.

16.4 / PERMUTATIONS OF n ELEMENTS, NOT ALL DIFFERENT, TAKEN n AT A TIME

If the n elements of a set are not all different, the problem of determining the number of permutations of the set presents a new aspect. For example, there are two permutations of the letters a and b, namely, ab and ba, but there is only one permutation of the letters a and a, since neither letter can be distinguished from the other and the two can therefore be put in only one *unique* arrangement. We shall suppose that s members of the set are alike and then designate the $n - s$ different elements by $t_1, t_2, \ldots, t_{n-s}$. The permutations, that is, the distinguishable arrangements of the n elements, will then depend only on the arrangement of the elements $t_1, t_2, \ldots, t_{n-s}$. We can therefore obtain all the permutations of the n elements by distributing $t_1, t_2, \ldots, t_{n-s}$ in the n positions in all possible ways and then filling in the vacant places with the s identical elements.† This amounts to distributing $n - s$ elements in the positions $1, 2, 3, \ldots, n$ in as many ways as possible, that is, to finding the number of permutations of n elements taken $n - s$ at a time. By Eq. (16.2), we have

$$P(n, n - s) = \frac{n!}{[n - (n - s)]!} = \frac{n!}{s!}$$

Hence, if s members of a set of n elements are alike, the number of permutations of the n elements taken n at a time is equal to the number of permutations of n things taken n at a time divided by the number of permutations of s things taken s at a time. By a repeated application of this principle, we can derive the following theorem:

THEOREM

Permutations of like elements

If, in a set of n elements, there are g groups, the first containing n_1 members all of which are alike; the second containing n_2 which are alike; the third, n_3 which are alike; and so on to the gth group, which has n_g members alike; then, the number of permutations of the n elements taken n at a time is given by

$$\frac{n!}{n_1! n_2! \cdots n_g!} \tag{16.4}$$

EXAMPLE 1 In how many ways can the letters of the word "abracadabra" be arranged?

---

† The reasoning here is the same as that in finding the number of ways in which two people can be seated in a row of five chairs: since when one person has been seated in any one of the five chairs, the other may be seated in any one of four, the permutations are $5 \times 4 = 20$. It makes no difference which chairs are *empty;* that is, it makes no difference where the particular members of the s like elements are located.

Solution   The solution of this problem involves two factors: (1) the number of permutations of 11 letters taken 11 at a time and (2) the letters b and r two
times and the letter a five times. Hence, the number of unique arrangements is given by

$$\frac{11!}{5!2!2!1!1!} = \frac{11 \times 10 \times 9 \times 8 \times 7 \times 6 \times 5!}{5! \times 2 \times 2} = 83,160$$

EXERCISE 16.2 Permutations of n Elements

Find the value of each symbol in Probs. 1 to 8.

1 $P(6,2)$	**2** $P(6,4)$	**3** $P(8,3)$	**4** $P(8,5)$
5 $P(5,5)$	**6** $P(5,1)$	**7** $P(7,2)$	**8** $P(4,2)$

Show that the equation in each of Probs. 9 to 12 is an identity.

9 $P(n,n) = P(n, n-1)$ **10** $P(n,n) = 6P(n, n-3)$

11 $P(n,r) = (n-r+1)P(n, r-1)$

12 $P(n,n) - P(n-1, n-2) = (n-1)^2 P(n-2, n-3)$

Solve the equation in each of Probs. 13 to 16.

13 $(n+3)! = 30(n+1)!$ **14** $P(n,2) = 56$

15 $3P(n,4) = P(n-1, 5)$ **16** $P(n,3) = 336$

*How many permutations can be made from the letters in the words in Probs. 17 to
20 if all letters are used in each permutation?*

17 Analysis   **18** Referee

19 Company   **20** Bookkeeper

21 How many three-letter "words" can be made from the first eight con-
sonants if no letter is repeated?

22 In how many ways can five people be arranged in a lineup?

23 In how many ways may an inspector spot-check 3 of 10 tax returns?

24 In how many ways may a student visit four of eight fraternity houses?

25 A Scrabble player tests all possible seven-letter words that can be made
with his letters, A E E M S S S. If he tests a word every second, how
long will it take him?

26 Repeat Prob. 25 for six-letter words.

27 In how many ways can six Russians, four Germans, and four Libyans be seated in a row so that all the Russians are on the left and all the Germans are in the middle?

28 A plane has four double seats on one side and four single seats on the other. In how many ways may 12 people be seated?

29 How many nine-digit social-security numbers may be assigned to an area where the first and second digits are 4 and 3 respectively?

30 After he has selected his nine players, how many batting orders can a little-league coach have if his pitcher bats fourth?

31 How many four-digit numbers may be made using 1, 3, 5, 7, and 8 if no digit is repeated?

32 How many ways are there to choose four of the 12 face cards and give one to each of four poker players?

33 In how many ways may *n* people be arranged about a round table?

34 In how many ways may six people be seated around an elliptical table?

35 In how many ways may eight different stones be put in a row on the outside of a ring?

36 In how many ways may five coins be placed so that one is at each vertex of a pentagon?

16.5 / COMBINATIONS

The definition of a combination is given in Sec. 16.1, but it will be repeated here for ready reference: A set of elements (taken without regard to the order in which the elements are arranged) is called a *combination.* According to this definition, the six permutations *xyz, xzy, yzx, yxz, zxy,* and *zyx* are only one combination of the letters *x, y,* and *z.*

There is only one combination of *n* elements taken *n* at a time, but the problem of determining the number of combinations of *n* elements taken *r* at a time, where $r < n$, is both interesting and important. We shall designate the combination of *n* elements taken *r* at a time by $C(n, r)$.† Since a permutation involves choosing *r* elements (a combination) and then arranging these *r* elements in one of the *r*! possible ways, the fundamental principle gives $P(n, r) = C(n, r) \cdot r!$. Hence, by (16.2),

$$C(n, r) = \frac{P(n, r)}{r!} = \frac{1}{r!} \cdot \frac{n!}{(n-r)!}$$

Therefore, *the number of combinations of n elements taken r at a time is*

† The symbols $_nC_r$ and C_r^n are also used.

| Combinations of n elements taken r at a time | $$C(n, r) = \frac{n!}{(n-r)!r!}$$ | (16.5) |

EXAMPLE 1 How many committees of 7 men can be formed from a group of 25 men?

Solution The number of committees is equal to the number of combinations of 25 elements taken 7 at a time. Hence it is

$$C(25, 7) = \frac{25!}{18!7!} = \frac{25 \times 24 \times 23 \times 22 \times 21 \times 20 \times 19 \times (18!)}{(18!) \times 7 \times 6 \times 5 \times 4 \times 3 \times 2 \times 1} = 480,700$$

EXAMPLE 2 A business firm wishes to employ six men and three boys. In how many ways can the selection be made if nine men and five boys are available?

Solution The six men can be selected from the nine in $C(9, 6)$ ways, and the three boys from the five in $C(5, 3)$. Hence the number of ways in which the selection of the employees can be made is

$$C(9, 6)C(5, 3) = \frac{9!}{(9-6)!6!} \frac{5!}{(5-3)!3!}$$

$$= \frac{9 \times 8 \times 7 \times (6!)}{3 \times 2 \times (6!)} \frac{5 \times 4 \times (3!)}{2 \times (3!)}$$

$$= 840$$

16.6 / ORDERED PARTITIONS

In this section we shall give another application of Eq. (16.4). An example will illustrate the idea.

EXAMPLE In how many ways may an English class split up so that four people discuss Bacon, six discuss Johnson, five discuss Emerson, and four discuss Pope?

Solution There are 19 people in the class, so that 4 may be chosen to discuss Bacon in $C(19, 4)$ ways. Out of the 15 remaining, the 6 to discuss Johnson may be chosen in $C(15, 6)$ ways, and the 5 to discuss Emerson in $C(9, 5)$ ways, leaving 4 to discuss Pope. By the fundamental principle, the answer is

$$C(19, 4)C(15, 6)C(9, 5) = \frac{19!}{4!15!} \frac{15!}{6!9!} \frac{9!}{5!4!} = \frac{19!}{4!6!5!4!}$$

In the example we were dividing a set of n elements into g subsets with

Ordered
partition

n_1 in the first subset, n_2 in the second, and so on to n_g in the gth or last subset, where $n_1 + n_2 + \cdots + n_g = n$. The subsets were distinguished before the division of the set was made. This gives an *ordered partition*, and a simple extension of the argument in the example shows that the number of ordered partitions is given by

$$\frac{n!}{n_1! n_2! \cdots n_g!} \qquad (16.4)$$

EXERCISE 16.3 Combinations, Ordered Partitions

Calculate the numbers in Probs. 1 to 4.

1 $C(10, 2)$ **2** $C(11, 4)$ **3** $C(8, 5)$ **4** $C(12, 8)$

5 It is true that $C(n, r) = C(n, n - r)$. Verify this for $n = 7$, $r = 2$ and for $n = 9$, $r = 3$.

6 It is true that $C(n + 1, r) = C(n, r) + C(n, r - 1)$. Verify this for $n = 8$, $r = 4$ and for $n = 6$, $r = 4$.

7 It is true that $C(n, 0) + C(n, 1) + \cdots + C(n, n) = 2^n$. Verify this for $n = 4$ and for $n = 7$.

8 Verify that $C(7, 4) = C(6, 4) + C(5, 3) + C(4, 2) + C(3, 1) + C(2, 0)$.

9 How many groups of four letters may be chosen without replacement from nine letters?

10 How many subsets of 3 points may be chosen from 10 points?

11 How many different bridge hands are there?

12 In how many ways may 5 finalists be chosen from 50 contestants?

13 How many five-card poker hands are there?

14 In how many ways may a buyer choose 12 dresses from a group of 18?

15 In how many ways may a teacher give 6 A's to a class of 22 students?

16 In how many ways may a bear catch half of a school of 10 fish?

17 In how many ways may two men and three women be chosen from five men and seven women?

18 Ten people are to ride in two cars, with three in one car and seven in the other. How many ways are there to do this?

19 In how many ways may a committee of two Jews, two Catholics, and two Protestants be chosen from eight of each?

20 Out of four Jerseys, five Holsteins, and seven orioles, how many groups of three cows and two birds may be chosen?

21 In how many ways may a teacher give six A's, five B's, seven C's, three D's, and one F in a class of 22?

22 In how many ways may nine women split into three groups of two, three, and four women, to meet in the red, green, and blue rooms, respectively?

23 How many ways are there for a football team to win seven games, lose two, and tie two?

24 If 13 people go to a play in three different cars, how many ways are there to put six in the Ford, four in the Chevrolet, and three in the Chrysler? How many if each owner drives his own car?

25 Show that the number of combinations of n things taken r at a time is the same as the number of ordered partitions of n things into two groups with r in the first group.

26 In how many ways may 27 people get five different diseases if at least five people get each disease?

27 Show that $C(8, 4) = C^2(4, 0) + C^2(4, 1) + C^2(4, 2) + C^2(4, 3) + C^2(4, 4)$.

28 How many groups of at most 3 flowers may be made from 10 flowers?

EXERCISE 16.4 REVIEW

In Probs. 1 to 4, find the value indicated.

1 $P(12, 3)$ 2 $P(8, 3)$ 3 $C(12, 3)$ 4 $C(8, 5)$

Verify the equations in Probs. 5 to 8.

5 $3^5 - 5^3 = C(10, 3) - 2$

6 $P(9, 4) - P(15, 3) = C(13, 3) + C(8, 1)$

7 $C(4, 1) + 2C(4, 2) + 3C(4, 3) + 4C(4, 4) = (4)2^3$

8 $C(4, 1) + 4C(4, 2) + 9C(4, 3) + 16C(4, 4) = (4)(5)2^2$

9 How many ways are there to choose three numbers from the set $\{\frac{1}{2}, \frac{1}{3}, \frac{2}{3}, \frac{1}{4}, \frac{3}{4}, \frac{1}{5}\}$ if numbers may be repeated?

10 Repeat Prob. 9 for the case in which a number may not be repeated.

11 Repeat Prob. 10 for the case in which numbers are chosen and then placed, one each, on a red, blue, or green circle.

12 In how many ways may a coach keep two centers, four forwards, and four guards from a group of tryouts of five centers, seven forwards, and nine guards?

13 How many meals may a person order from a menu with five appeti-

zers, three salads, six entrees, six vegetables, and four desserts if two vegetables and one of everything else is allowed?

14 Two cards are drawn from the 12 face cards without replacement. In how many ways may this be done with three decks?

15 Solve the equation $P(n, 3) = 6C(n, 3)$.

16 How many permutations of the letters in "seventeenth" are there?

17 In how many ways may eight diamonds be placed on a round silver tray?

18 In how many ways may a board of directors of eight people reach a majority decision?

19 Verify that $\dfrac{9!}{2!3!4!} = \dfrac{8!}{1!3!4!} + \dfrac{8!}{2!2!4!} + \dfrac{8!}{2!3!3!}$.

20 How many triangles are formed by n lines in a plane if no two of the lines are parallel and no three pass through a common point?

CHAPTER SEVENTEEN

The Binomial Theorem

In Chap. 3 we examined the general expression for the square of a binomial and later made use of the expression to solve equations. Obviously, a bionomial can be raised not only to the second power but to any power, and there may well be occasion to deal with such an expansion. It is the purpose of this chapter to develop a general formula for obtaining any integral power of a binomial.

17.1 / THE BINOMIAL FORMULA

In this section we shall develop a formula that will enable us to express any positive integral power of a binomial as a polynomial. This polynomial is called the *expansion* of the power of the binomial.

Expansion

By actual multiplication, we obtain the following expansions of the first, second, third, fourth, and fifth powers of $x + y$:

$$(x + y)^1 = x + y$$
$$(x + y)^2 = x^2 + 2xy + y^2$$
$$(x + y)^3 = x^3 + 3x^2y + 3xy^2 + y^3$$
$$(x + y)^4 = x^4 + 4x^3y + 6x^2y^2 + 4xy^3 + y^4$$
$$(x + y)^5 = x^5 + 5x^4y + 10x^3y^2 + 10x^2y^3 + 5xy^4 + y^5$$

By referring to the above expansions, we can readily verify the fact that the following properties of $(x+y)^n$ exist when $n = 1, 2, 3, 4,$ and 5:

Properties of the expansion of $(x+y)^n$

1 The first term in the expansion is x^n.

2 The second term is $nx^{n-1}y$.

3 The exponent of x decreases by 1 and the exponent of y increases by 1 as we proceed from term to term.

4 There are $n + 1$ terms in the expansion.

5 The $(n + 1)$st, or last, term is y^n.

6 The nth, or next to the last, term of the expansion is nxy^{n-1}.

7 If we multiply the coefficient of any term by the exponent of x in that term and then divide the product by the number of the term in the expansion, we obtain the coefficient of the next term.

8 The sum of the exponents of x and y in any term is n.

If we assume that these eight properties hold for all integral positive n, we can write the first five terms in the expansion of $(x+y)^n$ as follows:

$$\text{First term} = x^n \qquad \text{by Property 1}$$
$$\text{Second term} = nx^{n-1}y \qquad \text{by Property 2}$$
$$\text{Third term} = \frac{n(n-1)}{2}x^{n-2}y^2 \qquad \text{by Properties 7 and 3}$$
$$\text{Fourth term} = \frac{n(n-1)(n-2)}{3 \times 2}x^{n-3}y^3 \qquad \text{by Properties 7 and 3}$$
$$\text{Fifth term} = \frac{n(n-1)(n-2)(n-3)}{4 \times 3 \times 2}x^{n-4}y^4 \qquad \text{by Properties 7 and 3}$$

We can continue this process until we have

$$n\text{th term} = nxy^{n-1} \qquad \text{by Property 6}$$
$$(n + 1)\text{st term} = y^n \qquad \text{by Property 5}$$

We are now in a position to form the sum of the above terms and obtain the binomial formula. Notice that $4 \times 3 \times 2 = 4 \times 3 \times 2 \times 1 = 4!$, $3 \times 2 = 3 \times 2 \times 1 = 3!$, and $2 = 2 \times 1 = 2!$, and write

Binomial formula

$$(x + y)^n = x^n + nx^{n-1}y + \frac{n(n-1)}{2!}x^{n-2}y^2$$
$$+ \frac{n(n-1)(n-2)}{3!}x^{n-3}y^3$$
$$+ \frac{n(n-1)(n-2)(n-3)}{4!}x^{n-4}y^4$$
$$+ \cdots + nxy^{n-1} + y^n \qquad (17.1)$$

Binomial theorem Equation (17.1) is called the *binomial formula,* and the statement that it is true is called the *binomial theorem.*

EXAMPLE 1   Use the binomial formula to obtain the expansion of $(2a + b)^6$.

Solution   We first apply Eq. (17.1) with $x = 2a$, $y = b$, and $n = 6$.   Thus,

$$(2a + b)^6 = (2a)^6 + 6(2a)^5 b + \frac{6 \times 5}{2!}(2a)^4 b^2 + \frac{6 \times 5 \times 4}{3!}(2a)^3 b^3$$
$$+ \frac{6 \times 5 \times 4 \times 3}{4!}(2a)^2 b^4 + \frac{6 \times 5 \times 4 \times 3 \times 2}{5!}(2a) b^5$$
$$+ \frac{6 \times 5 \times 4 \times 3 \times 2 \times 1}{6!}b^6$$

Now simplifying the coefficients, and raising $2a$ to the indicated powers, we obtain

$$(2a + b)^6 = 64a^6 + 6(32a^5)b + 15(16a^4)b^2 + 20(8a^3)b^3$$
$$+ 15(4a^2)b^4 + 6(2a)b^5 + b^6$$

Finally, we perform the indicated multiplication in each term and get

$$(2a + b)^6 = 64a^6 + 192a^5 b + 240a^4 b^2 + 160a^3 b^3 + 60a^2 b^4 + 12ab^5 + b^6$$

The computation of the coefficients can, in most cases, be performed mentally by use of Property 7, and thus we can avoid writing the first step in the expansion in the above example.

EXAMPLE 2   Expand $(a - 3b)^5$.

Solution   The first term in the expansion is a^5, and the second is $5a^4(-3b)$.   To get the coefficient of the third, we multiply 5 by 4 and divide the product by 2, thus obtaining 10.   Hence, the third term is $10a^3(-3b)^2$.   Similarly, the fourth term is

$$\tfrac{3 \cdot 0}{3} a^2(-3b)^3 = 10a^2(-3b)^3$$

By continuing this process, we obtain the following expansion:

$$(a - 3b)^5 = a^5 + 5a^4(-3b) + 10a^3(-3b)^2 + 10a^2(-3b)^3 + 5a(-3b)^4 + (-3b)^5$$
$$= a^5 - 15a^4 b + 90a^3 b^2 - 270a^2 b^3 + 405ab^4 - 243b^5$$

It should be noted that we carry the second term of the binomial, $-3b$, through the first step of the expansion as a single term.   Then we raise $-3b$ to the indicated power and simplify the result.

EXAMPLE 3   Expand $(2x - 5y)^4$.

Solution   We shall carry through the expansion with $2x$ as the first term and $-5y$ as the second and get

$$(2x - 5y)^4 = (2x)^4 + 4(2x)^3(-5y) + 6(2x)^2(-5y)^2 + 4(2x)(-5y)^3 + (-5y)^4$$
$$= 16x^4 + 4(8x^3)(-5y) + 6(4x^2)(25y^2) + 4(2x)(-125y^3) + 625y^4$$
$$= 16x^4 - 160x^3y + 600x^2y^2 - 1{,}000xy^3 + 625y^4$$

17.2 / THE rth TERM OF THE BINOMIAL FORMULA

In the preceding examples, we explained the method for obtaining any term of a binomial expansion from the term just before it. However, by use of this method, it is impossible to obtain any specific term of the expansion without first computing all the terms which precede it. We shall now develop a formula for finding the general rth term without reference to the other terms.

In order to find the rth term, we shall prove the binomial theorem. By the definition of product,

$$(x + y)^n = (x + y)(x + y) \cdots (x + y) \qquad n \text{ factors of } (x + y) \qquad (1)$$

To multiply out the right member of (1), we must take either x or y from each of the n factors $(x + y)$. If y is chosen r times, and x chosen $n - r$ times, the resulting term will be $x^{n-r}y^r$. But by the definition of a combination (see Chap. 16), there are $C(n, r)$ ways of choosing r of the y's. Thus the term involving $x^{n-r}y^r$ is $C(n, r)x^{n-r}y^r$. This shows that

$$(x + y)^n = C(n, 0)x^n + C(n, 1)x^{n-1}y + C(n, 2)x^{n-2}y^2 + \cdots$$
$$+ C(n, r)x^{n-r}y^r + \cdots + C(n, n - 1)xy^{n-1} + C(n, n)y^n \qquad (17.2)$$

The student should show that (17.2) is the same as (17.1).

If we call $C(n, 0)x^n$ the first term, $C(n, 1)x^{n-1}y$ the second term, and so on, then the rth term will involve $C(n, r - 1)$. Using (16.5), we find the rth term to be

$$C(n, r - 1)x^{n-r+1}y^{r-1} = \frac{n!}{(r - 1)!(n - r + 1)!}x^{n-r+1}y^{r-1}$$
$$= \frac{n(n - 1)(n - 2) \cdots (n - r + 2)}{(r - 1)!}x^{n-r+1}y^{r-1} \qquad (17.3)$$

Notice that the numerator and denominator of (17.3) are each the product of $r - 1$ consecutive integers.

EXAMPLE Find the fourth term in the expansion of $(2a - b)^9$.

Solution In this problem, $x = 2a$, $y = -b$, $n = 9$, and $r = 4$. Therefore, $r - 1 = 3$, $n - r + 1 = 9 - 4 + 1 = 6$, and $n - r + 2 = 7$. Now, if we substitute these values in Eq. (17.3), we see that the fourth term is

$$\text{Fourth term} = \frac{9 \times 8 \times 7}{3 \times 2 \times 1}(2a)^6(-b)^3$$

$$= 84(64a^6)(-b^3)$$

$$= -5,376a^6b^3$$

EXERCISE 17.1 The Binomial Formula and the nth Term

Find the expansion of the binomial in each of Probs. 1 to 12 by use of the binomial formula.

1	$(a+b)^8$	**2**	$(a-x)^7$	**3**	$(b-y)^5$
4	$(x+y)^6$	**5**	$(a-2y)^5$	**6**	$(x+3b)^6$
7	$(2b+x)^7$	**8**	$(3a+b)^4$	**9**	$(2a-3b^2)^3$
10	$(3x-2b^3)^5$	**11**	$(2b^2+3x)^6$	**12**	$(5x^2+2y)^4$

Find the first four terms of the expansion of the binomial in each of Probs. 13 to 16.

13 $(a+y)^{33}$ **14** $(x-y)^{51}$ **15** $(m-2y)^{101}$ **16** $(b+3c)^{42}$

Find the indicated power of the number in each of Probs. 17 to 20, and round off to four decimal places.

17 $(1+.04)^5$ **18** 1.05^4 **19** 1.03^6 **20** 1.06^3

Find the specified term of the expansion in each of Probs. 21 to 32.

21 Fifth term of $(x-2y)^7$ **22** Fourth term of $(2a-c)^6$
23 Sixth term of $(3x+y)^9$ **24** Third term of $(x+4y)^8$
25 Fourth term of $(a-a^{-1})^7$ **26** Sixth term of $(2x-x^{-2})^9$
27 Seventh term of $(x^2+2y)^{11}$ **28** Fifth term of $(3x+y^3)^8$
29 Middle term of $(x+2y^{1/2})^6$ **30** Middle term of $(x-3y^{1/4})^8$
31 The term in $(x+2y)^{10}$ that involves x^7.
32 The term in $(3x-y^{1/2})^{13}$ that involves y^4.

Problems 33 to 36 give the outline of an alternative proof of the binomial theorem which is based on mathematical induction.

33 Show that (17.2) holds for $n=1$.

34 Assuming (17.2) holds for $n=k$, multiply both sides of the equation by $x+y$. Why is the rth term of $(x+y)^{k+1}$ equal to $C(k+1, r-1)x^{k-r+2}y^{r-1}$?

35 Continuing Prob. 34, show that after the right member of (17.2) is multiplied by $x+y$, the terms involving y^{r-1} are $C(k, r-1)x^{k-r+2}y^{r-1} + C(k, r-2)x^{k-r+2}y^{r-1}$.

36 Show that $C(k+1, r-1) = C(k, r-1) + C(k, r-2)$. How does this allow you to complete the proof by induction?

37 Show that $C(n, 0) + C(n, 1) + C(n, 2) + \cdots + C(n, n) = 2^n$ for every positive integer n. HINT: Use (17.2) with $x = y = 1$.

38 Show that $C(n, 0) - C(n, 1) + C(n, 2) - C(n, 3) + \cdots + (-1)^n C(n, n) = 0$ for every positive integer n.

39 Show that $C(n, 0) + 2C(n, 1) + 2^2 C(n, 2) + 2^3 C(n, 3) + \cdots + 2^n C(n, n) = 3^n$ for every positive integer n.

40 Show that $C(2n, n) = C^2(n, 0) + C^2(n, 1) + C^2(n, 2) + \cdots + C^2(n, n)$ for every positive integer n. HINT: Look at the middle term in $(x + y)^{2n} = (x + y)^n (x + y)^n$.

17.3 / BINOMIAL THEOREM FOR FRACTIONAL AND NEGATIVE EXPONENTS

The proof of the binomial formula for fractional and negative exponents is beyond the scope of this book; however, we shall point out some elementary applications of it. It should be noted that the expansion of $(x + y)^n$ for n not a positive integer has no last term since the coefficient never becomes zero; hence it is impossible to complete the series, and we must be content with any desired or indicated number of terms. The following fact can be established, although the proof will not be given.

The binomial expansion of $x + y$ for fractional and negative exponents is valid if and only if the value of y is between the values of x and $-x$.

EXAMPLE 1 What are the first four terms in the expansion of $(2 + x)^{1/2}$? In what interval is the expansion valid?

Solution The expansion is

$$(2 + x)^{1/2} = 2^{1/2} + (\tfrac{1}{2} \times 2^{-1/2} x) + \frac{\tfrac{1}{2} \times (-\tfrac{1}{2}) \times 2^{-3/2} x^2}{2!}$$

$$+ \frac{\tfrac{1}{2} \times (-\tfrac{1}{2}) \times (-\tfrac{3}{2}) \times 2^{-5/2} x^3}{3!} + \cdots$$

$$= \sqrt{2} + \frac{x}{2\sqrt{2}} - \frac{x^2}{16\sqrt{2}} + \frac{x^3}{64\sqrt{2}} - \cdots$$

$$= \sqrt{2}\left(1 + \frac{x}{4} - \frac{x^2}{32} + \frac{x^3}{128} - \cdots\right)$$

It is valid if and only if $-2 < x < 2$.

EXAMPLE 2 Determine an approximation to the square root of 10.

Solution

$$\sqrt{10} = 10^{1/2} = (9+1)^{1/2} = (3^2+1)^{1/2}$$

$$= (3^2)^{1/2} + \tfrac{1}{2} \times (3^2)^{-1/2} \times 1 + \frac{\tfrac{1}{2} \times (-\tfrac{1}{2}) \times (3^2)^{-3/2} \times 1^2}{2}$$

$$+ \frac{\tfrac{1}{2} \times (-\tfrac{1}{2}) \times (-\tfrac{3}{2}) \times (3^2)^{-5/2} \times 1^3}{2 \times 3} + \cdots$$

$$= 3 + (\tfrac{1}{2} \times 3^{-1}) - (\tfrac{1}{8} \times 3^{-3}) + (\tfrac{1}{16} \times 3^{-5}) - \cdots$$

$$= 3 + \tfrac{1}{6} - \tfrac{1}{216} + \tfrac{1}{3,888}$$

$$= 3 + 0.16667 - 0.00463 + 0.00026$$

$$= 3.16230$$

By comparing the four terms in the above expansion, we see that their values decrease very rapidly. The rate of this decrease increases as the expansion is carried further. In fact, the fifth term is -0.0000178, and when this is combined with the other four terms, we obtain $\sqrt{10} = 3.1622822$, or, rounded to four decimal places, 3.1623. Hence we conclude that this is the correct value of $\sqrt{10}$ to five figures. Obviously, the expansion can be extended until we obtain any desired degree of accuracy.

EXERCISE 17.2 Binomial Theorem for Fractional and Negative Exponents

Find the first four terms of the expansion of the binomial in each of Probs. 1 to 20. Find the range of the variable for which the expansion is valid in Probs. 13 to 20.

1 $(x+y)^{-2}$	2 $(a+b)^{-4}$	3 $(a-x)^{-5}$
4 $(b-y)^{-3}$	5 $(2a+x)^{-1}$	6 $(x+2a)^{-1}$
7 $(a^2-y)^{-3}$	8 $(x+a^2)^{-2}$	9 $\left(x-\dfrac{1}{x}\right)^{-3}$
10 $(x+x^{-2})^{-3}$	11 $(y+1)^{1/3}$	12 $(1+y)^{1/3}$
13 $(8-x)^{1/3}$	14 $(25+x)^{1/2}$	15 $(4-x)^{-1/2}$
16 $(9+x)^{-1/4}$	17 $(1+x)^{1/2}$	18 $(1+x)^{-1/2}$
19 $(1-x)^{1/2}$	20 $(1-x)^{-3/4}$	

In Probs. 21 to 32, find the number to three decimal places by using a binomial expansion.

21 $\sqrt{25.5} = (25+0.5)^{1/2}$	22 $(7.9)^{1/3}$	23 $(3.8)^{-1/2}$	24 $(9.1)^{1/4}$
25 $\sqrt[5]{31}$	26 $\sqrt[3]{28}$	27 $\sqrt[4]{83}$	28 $\sqrt[3]{123}$
29 1.03^{-4}	30 1.02^{-5}	31 0.97^{-3}	32 1.05^{-6}

33 In Prob. 17 we found that the first four terms of the binomial expan-

sion of $\sqrt{1+x}$ are $1+x/2-x^2/8+x^3/16$. We may thus get a linear approximation to $\sqrt{1+x}$ by using $f(x)=1+x/2$, and a quadratic one by using $g(x)=1+x/2-x^2/8$. Graph $y=\sqrt{1+x}$, $y=f(x)$, and $y=g(x)$ on the same set of axes. Notice that the closer x is to 0, the better the approximations are.

34 Repeat Prob. 33 for $\sqrt{1-x}$ (see Prob. 19).

35 Find $\sqrt{26}$ to three decimal places by writing it as $(25+1)^{1/2}=$ $[25(1+\frac{1}{25})]^{1/2}=5(1+\frac{1}{25})^{1/2}$ and using Prob. 17.

36 Repeat Prob. 35 for $\sqrt{62}=(64-2)^{1/2}=[64(1-\frac{1}{32})]^{1/2}=8(1-\frac{1}{32})^{1/2}$.

CHAPTER EIGHTEEN *Probability*

The word "probability" is used loosely by the layman to indicate a vague likelihood that something will happen. It is often used synonymously with "chance." The likelihood that something will or will not happen under specified circumstances has application in a wide variety of human pursuits. The gambler, the statistician, the economist, and the engineer are among those who make use of probability. It is the purpose of this chapter to attempt to find some order in the chaos caused by the loose use of the word "probability."

18.1 / A SAMPLE SPACE AND PROBABILITY

If a card is selected from a standard deck, there are 52 possible results. If a marksman shoots at a target, there are two possible results, since he may hit or miss the target. If a football game is played, either team may win or they may make the same score. In general, if an experiment is undertaken, there is a set of possible results associated with it. The set of all possible results of an experiment is called the *sample space* for the experiment. Each element of a sample space is called an *outcome* or *sample point*. Any subset of a sample space is called an

Sample space
Outcome, or
sample point

416

Event   *event.* If a die is cast, it may stop with 1, 2, 3, 4, 5, or 6 on top; hence, $\{1, 2, 3, 4, 5, 6\}$ is the sample space, and any one of these elements is an outcome or sample point. Furthermore, any combination of one or more of them is an event.

    If a die is made accurately and rolled honestly, it is as likely to stop with one number up as with another. Thus, each of the outcomes is *equally* Equally likely *likely*, and we say the outcome is *random.* We shall assume that all outcomes Random   in experiments discussed in this chapter are random in the sense that they are equally likely. Under those conditions, we say that the experiment is Random experiment   a *random experiment.*

    We shall use the following notation:

Symbol	Meaning
S	The sample space of random outcomes
$n(S)$	The number of elements in S
E	A set of specified outcomes in S; hence $E \subseteq S$
$n(E)$	The number of elements in E
$p(E)$	The probability that E will happen, or more briefly, the probability of E

    If a probability is predicted in advance of an experiment on the basis of an analysis of all possibilities of occurrence, we say that we have a Mathematical   *mathematical* or *a priori probability.* probability   The probability $p(E)$ of an event $E \subseteq S$ is defined as follows:

$$p(E) = \frac{n(E)}{n(S)}$$

Since $n(E)$ and $n(S)$ are nonnegative integers and $E \subseteq S$, then $n(E) \le n(S)$, and it follows that $p(E)$ is a number between 0 and 1 inclusive.

EXAMPLE 1   Find the probability of tossing a 4 in one toss of a die.

Solution   The sample space is $\{1, 2, 3, 4, 5, 6\}$, and the specified outcome is that the die falls with 4 up. Hence $E = \{4\}$. Therefore,

$$p(4) = \frac{n(4)}{n(S)} = \frac{1}{6}$$

EXAMPLE 2   If one card is drawn from a standard deck of 52 cards, find the probability that the card will be a jack.

Solution   Here $S = \{x \mid x$ is a card in a deck of 52 cards$\}$.   Hence $n(S) = 52$.   Furthermore, $E = \{$club jack, diamond jack, heart jack, spade jack$\}$, so $n(E) = 4$.

Therefore,
$$p(E) = \frac{n(E)}{n(S)} = \frac{4}{52} = \frac{1}{13}$$

EXAMPLE 3   Each of the three-digit numbers that can be formed using the integers
from 1 to 9, with no digit repeated, is written on a card.   The cards are
then stacked and shuffled.   If one card is drawn from the stack, find the
probability that the sum of the digits in the number on it will be 10.

Solution   Here $S = \{x \mid x$ is a card bearing a three-digit number formed with the
digits from 1 to 9 with no digit repeated$\}$.   The number of cards in S is
$P(9, 3) = 9 \cdot 8 \cdot 7 = 504$.   Therefore, $n(S) = 504$.   Furthermore,

$$E = \{x \mid x \text{ is a card in } S \text{ whose digit sum is 10}\}$$

The sets of three different digits whose sum is 10 are $\{1, 2, 7\}$, $\{1, 3, 6\}$,
$\{1, 4, 5\}$, and $\{2, 3, 5\}$.   Since the digits in each of these sets can be arranged in $P(3, 3) = 3! = 6$ ways and there are four sets, there are $4 \cdot 6 = 24$
cards in E.   Hence $n(E) = 24$.   Therefore,

$$p(E) = \frac{n(E)}{n(S)} = \frac{24}{504} = \frac{1}{21}$$

18.2 / EMPIRICAL PROBABILITY

Relative frequency

In interpreting results of certain
experiments and in the analysis of statistical data, the following ratio is
often used: If out of n trials an event has occurred h times, the *relative
frequency* of its occurrence is h/n.

It is assumed, and the assumption is justified by experience, that the
larger n becomes, the more nearly the relative frequency approaches the
mathematical probability.   We make use of this assumption to define a

Empirical probability

ratio, known as the *empirical probability,* that is used extensively in the
interpretation of statistics and in insurance.   If out of n trials, where n is a
large number, an event has occurred h times, then the probability that it
will happen in any one trial is defined as h/n.

Mortality table

Life-insurance companies make use of the ratio h/n applied to information tabulated in a *mortality table* in order to determine their rates.   One
of these tables is the American Experience Mortality Table (Table A.4,
Appendix), which was compiled from data gathered by several large life-
insurance companies.   Starting with 100,000 people ten years of age, it
shows the number alive at the age of eleven years, twelve years, thirteen
years, and so on to ninety-five years.

EXAMPLE 1   According to the American Experience Mortality Table, what is the probability that a man twenty years of age will be alive at thirty?

Solution   Table A.4 shows that, out of 92,637 people alive at twenty years of age, 85,441 are living at the age of thirty. Hence $h = 85,441$, $n = 92,637$, and the desired probability is $85,441/92,637 = 0.9223$.

EXAMPLE 2   A traffic census shows that out of 1,000 vehicles passing a junction point on a highway, 600 turned to the right. What is the probability of an automobile's turning to the right at this junction?

Solution   In this case, $n = 1,000$, $h = 600$. Hence the desired probability is $600/1,000 = 3/5$.

According to the definition, the empirical probability depends upon n and h, and in any situation it will probably change as n changes. If n is sufficiently large, the change in the probability due to a change in n should be small. Hence, in any actual situation, the number of trials observed should be as large as the opportunity for observation permits.

18.3 / BASIC THEOREMS

In this section, we shall derive and explain the use of several theorems on probability. If E_1 and E_2 represent two events, and if $n(E_1)$ and $n(E_2)$ denote the number of elements in sets E_1 and E_2, respectively, then

$$n(E_1) \geq 0 \quad \text{and} \quad n(E_2) \geq 0 \tag{1}$$
$$n(\varnothing) = 0 \tag{2}$$
$$n(E_1 \cup E_2) = n(E_1) + n(E_2) - n(E_1 \cap E_2) \tag{3}$$

The situation for (3) is shown in Fig. 18.1 provided E_1 and E_2 have elements in common, and in Fig. 18.2 for the case in which $E_1 \cap E_2 = \varnothing$. Quite often, we say that two events are *mutually exclusive* or *disjoint* to indicate that $E_1 \cap E_2 = \varnothing$. If two events are mutually exclusive, then $n(E_1 \cap E_2) = 0$ and (3) becomes

Mutually exclusive
Disjoint

$$n(E_1 \cup E_2) = n(E_1) + n(E_2) \tag{4}$$

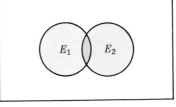

Figure 18.1

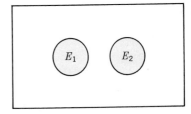

Figure 18.2

We shall now state and prove a theorem concerning the probability of two events.

THEOREM *If E_1 and E_2 are any two events, mutually exclusive or not, with probabilities $p(E_1)$ and $p(E_2)$, then the probability of E_1 or E_2 is*

$$p(E_1 \cup E_2) = p(E_1) + p(E_2) - p(E_1 \cap E_2) \qquad (18.1)$$

Proof   In the proof of this theorem, we begin by noting that

$$p(E_1 \cup E_2) = \frac{n(E_1 \cup E_2)}{n(S)} \qquad \text{definition of probability}$$

$$= \frac{n(E_1) + n(E_2) - n(E_1 \cap E_2)}{n(S)} \qquad \text{by (3)}$$

$$= \frac{n(E_1)}{n(S)} + \frac{n(E_2)}{n(S)} - \frac{n(E_1 \cap E_2)}{n(S)}$$

by dividing each term of the numerator separately by the denominator. Consequently, by use of the definition of probability, we have

$$p(E_1 \cup E_2) = p(E_1) + p(E_2) - p(E_1 \cap E_2)$$

as stated in the theorem.

EXAMPLE 1   If one card is drawn from a deck of 52 playing cards, find the probability that it will be red or a king.

Solution   If E_1 represents the set of red cards and E_2 represents the set of kings, then $n(E_1) = n(\text{reds}) = 26$, $n(E_2) = n(\text{kings}) = 4$, and $n(E_1 \cap E_2) = n(\text{red} \cap \text{king}) = 2$. Therefore, $p(E_1) = 26/52 = 1/2$, $p(E_2) = 4/52 = 1/13$, and $p(E_1 \cap E_2) = 2/52 = 1/26$. Consequently, the probability that the card that is drawn will be red or a king is

$$p(E_1 \cup E_2) = p(E_1) + p(E_2) - p(E_1 \cap E_2)$$
$$= 1/2 + 1/13 - 1/26$$
$$= 14/26 = 7/13$$

If the events considered in Eq. (18.1) are mutually exclusive, then $n(E_1 \cap E_2) = 0$, and by (4)

$$p(E_1 \cup E_2) = \frac{n(E_1 \cup E_2)}{n(S)}$$

$$= \frac{n(E_1) + n(E_2)}{n(S)}$$

$$= p(E_1) + p(E_2)$$

Consequently, we have shown:

If E_1 and E_2 are mutually exclusive events, then the probability that one of them will occur in a single trial is

$$p(E_1 \cup E_2) = p(E_1) + p(E_2) \tag{18.2}$$

EXAMPLE 2   If the probability of team A winning its conference championship in football is 1/8 and the probability of team B of the same conference winning the championship is 1/3, what is the probability that team A or team B will be the champion?

Solution   Since the events are mutually exclusive, the probability of A or B becoming the champion is the sum of the separate probabilities, as given by Eq. (18.2). Consequently,

$$p(A \cup B) = p(A) + p(B)$$
$$= 1/8 + 1/3 = 11/24$$

The theorem on mutually exclusive events can be extended to any finite number of events by considering $(E_1 \cup E_2)$ as an event and adding a third event E_3, then considering $(E_1 \cup E_2 \cup E_3)$ as an event and adding another. If this is continued, we reach the following conclusion:

The probability that some one of a set of mutually exclusive events will occur in a single trial is the sum of the probabilities of the separate events.

Symbolically,

$$p(E_1 \cup E_2 \cup \cdots \cup E_n) = p(E_1) + p(E_2) + \cdots + p(E_n) \tag{18.3}$$

EXAMPLE 3   In a race for mayor, four candidates A, B, C, and D have probabilities of 0.15, 0.18, 0.24, and 0.42 of winning. What is the probability that A, B, C, or D will win?

Solution   Only one of the candidates will win, so Eq. (18.3) is applicable and gives

$$P(A \cup B \cup C \cup D) = P(A) + P(B) + P(C) + P(D)$$
$$= 0.15 + 0.18 + 0.24 + 0.42 = 0.99$$

Notice that this probability is less than 1, so there is at least one other candidate in the race (a decided dark horse).

Both Eqs. (18.2) and (18.3) follow from (18.1), as does (18.4) below.

Complementary event   If E is an event and E' the *complementary event* (E' happens precisely when E does not happen), then E and E' are mutually exclusive, so (18.2) gives

$$P(E \cup E') = P(E) + P(E') \tag{5}$$

But if an experiment is performed, either E or E' must occur, so $E \cup E' = S$ and $P(E \cup E') = P(S) = 1$. Thus (5) becomes

$$1 = P(E) + P(E')$$

or
$$P(E') = 1 - P(E) \tag{18.4}$$

In Example 3, if we let $E = A \cup B \cup C$, then $E' = D \cup$ dark horse, and

$$P(E') = 1 - P(E) = 1 - (0.15 + 0.18 + 0.24) = 0.43$$

EXERCISE 18.1 Empirical Probability and Basic Theorems

1 If a card is drawn from a standard deck, what is the probability it will be a heart? A red card?

2 If a card is drawn from a standard deck, what is the probability it will be a king? An even number?

3 If a card is drawn from the face cards, what is the probability it is a jack? A black jack?

4 If a card is drawn from the red cards, what is the probability it is a king? The heart king?

5 Find the probability of throwing a 5 in one toss of a die.

6 Find the probability of throwing a number divisible by 4 in one toss of a die.

7 Find the probability of throwing a number with four letters in it in one toss of a die.

8 Find the probability of throwing a number x with $x^2 < 11$ in one toss of a die.

9 If two dice are tossed, what is the probability of getting a sum of 2? Of 5?

10 If two dice are tossed, what is the probability of getting a sum of 9? Of 12?

11 If two dice are tossed, what is the probability the sum will be 2, 3, or 4?

12 If two dice are tossed, how many ways are there for the sum to be 7?

13 A bag has 3 red, 4 green and 6 yellow balls. If one ball is drawn, what is the probability it is green? The probability it is not red?

14 In Prob. 13, what is the probability it is not green? The probability it is red or not yellow?

15 If three dice are tossed, what is the probability the sum will be 5?

16 A box contains seven $1 bills, six $5 bills, and eight $10 bills. If two bills are drawn simultaneously at random, find the probability that the sum drawn is $2. That the sum is $6.

17 A bridge hand of 13 cards is drawn from a deck. Find the probability of getting a hand with every card 8 or lower.

18 A poker hand of five cards is drawn from a deck. What is the probability that every card will be a 10, jack, queen, king, or ace?

19 Three people are to be chosen randomly from a group of 12 men and 8 women. What is the probability all three will be men?

20 Five dogs will be in the finals of a dog show containing 8 poodles, 10 terriers, and 14 bulldogs. What is the probability none of the finalists is a poodle?

21 Nine men and six women apply for jobs at a business concern. If eight people are hired, what is the probability it is three men and five women?

22 6 tenth-graders, 10 eleventh-graders, and 9 twelfth-graders are in a piano contest with five prizes. What is the probability that tenth-graders win two prizes and twelfth-graders win three?

23 A poker hand of five cards is drawn from a standard deck of 52 cards. What is the probability that all five cards are of the same suit (a flush)?

24 Find the probability of getting a full house in poker (three cards of one rank and two of another rank).

25 In a town of 8,600 there were 43 millionaires. What is the (empirical) probability of any one person in that town being a millionaire?

26 What is the probability in Prob. 25 if 100 people on welfare move into the city?

27 In a college of 16,000 students, each student takes five courses that meet three times per week, and there are usually about 15,000 absences per week. What is the probability of a student being absent a certain day from a certain class?

28 In Prob. 27, how many students could be expected to be absent on a given day in a class of 48 students?

29 In a survey of 35 people, 17 enjoyed classical music, 19 enjoyed popular music, and 9 enjoyed both. What is the probability a person in this survey enjoys at least one of the types of music?

30 In a roomful of people, 260 have attended a Methodist service, 370 have attended a Catholic service, 50 have attended both, and 240 have attended neither. What is the probability a person in that room has attended either type of service? Only a Methodist service?

31 At a college reunion it was discovered that 16 percent of the people

had climbed up a mountain, 51 percent had skied down a mountain, and 6 percent had done both. What is the probability a person has done neither?

32   In an election, 59 percent of the people voted for Proposition 1, 43 percent voted for Proposition 2, and 13 percent voted against both. What is the probability a person voted for both?

33   If the probabilities of getting an A, B, C, D, or F in a certain college are 0.26, 0.23, 0.29, 0.18, and 0.04, respectively, what is the probability of getting an A or a B? Of not failing?

34   A women's club is selecting officers. If there are 6 former presidents, 5 former vice-presidents, and 3 former treasurers among the 30 women, what is the probability the new secretary is a former president, vice-president, or treasurer?

35   The probabilities that teams L, T, and A will win a conference championship are 1/5, 1/6, and 1/10. What is the probability one of the three will win?

36   Suppose 20 percent of the population gets exactly one cold in January, 5/6 as many get exactly two colds as get one, and half as many get exactly three as get one. What is the probability of a person getting one, two, or three colds in January?

37   Albert's probability of vacationing this summer in Yellowstone is 0.22, in Cape Cod is 0.32, and in the Smoky Mountains is 0.42. What is his probability of doing none of the three?

38   What is the probability of getting a one-digit sum if two dice are tossed?

39   What is the probability that an integer between 1 and 75 inclusive has more than 3 letters in its English spelling?

40   What is the probability that the English spelling of a month will contain the letter R?

18.4 / DEPENDENT AND INDEPENDENT EVENTS

In this section we shall consider the probability of the occurrence of two events E and F. The event $E \cap F$ indicates that both E and F occur. The event $E \,|\, F$ (read "E given F") is the event E happening assuming that F has already happened.

Reduced sample space

In finding $P(E\,|\,F)$ we may use either of two viewpoints: we may consider the original sample space S or the *reduced sample space* F. From the reduced-sample-space point of view and the definition of probability we have

$$P(E|F) = \frac{n(E \cap F)}{n(F)} = \frac{n(E \cap F)/n(S)}{n(F)/n(S)} = \frac{P(E \cap F)}{P(F)} \qquad (18.5)$$

Thus,
$$P(E \cap F) = P(F)P(E|F) \qquad (18.5a)$$

We may interchange E and F, obtaining

$$P(E \cap F) = P(E)P(F|E) \qquad (18.5b)$$

This may also be extended to more than two events.

EXAMPLE 1   If two balls are drawn without replacement from a bag containing five white and three black balls, what is the probability that both will be white? That one will be black and one white?

Solution   A convenient way of listing all possibilities is a tree diagram (Fig. 18.1). On the first draw there are five white and three black balls, so the probability of a white ball on the first draw is $P(W_1) = 5/8$. Now if a white ball was drawn first, there are four white and three black left. The probability of white on the second draw, given that white was drawn first, is $P(W_2|W_1) = 4/7$. Thus the probability both are white is

$$P(W_1 \cap W_2) = P(W_1)P(W_2|W_1) = \frac{5}{8} \cdot \frac{4}{7} = \frac{5}{14}$$

The answer to the second question is $P(W_1 \cap B_2) + P(B_1 \cap W_2)$, since $W_1 \cap B_2$ and $B_1 \cap W_2$ are mutually exclusive events. Each can be calculated from the tree, giving

$$P(W_1 \cap B_2) + P(B_1 \cap W_2) = P(W_1)P(B_2|W_1) + P(B_1)P(W_2|B_1)$$
$$= \frac{5}{8} \cdot \frac{3}{7} + \frac{3}{8} \cdot \frac{5}{7} = \frac{15}{28}$$

EXAMPLE 2   If two cards are drawn without replacement from a deck, what is the probability the first is a spade and the second is a club?

Solution   The probability of a spade on the first draw is 13/52, and the probability then of a club on the second is 13/51, so the answer is

$$P(S_1 \cap C_2) = P(S_1)P(C_2|S_1) = \frac{13}{52} \cdot \frac{13}{51} = \frac{13}{204}$$

Independent   Two events E and F are *independent* if $P(E) = P(E|F)$. That is, the probability of E is unaffected by the occurrence or nonoccurrence of F. Using Eq. (18.5a) and this definition of independence gives

$$P(E \cap F) = P(F)P(E) \qquad (18.6)$$

for independent events E and F.

Dependent   If two events are not independent, they are called *dependent*.

EXAMPLE 3   Repeat Example 1 except that the first ball is replaced.

Solution   The answer to the first question is

$$P(W_1 \cap W_2) = P(W_1)P(W_2) = \frac{5}{8} \cdot \frac{5}{8} = \frac{25}{64}$$

and the answer to the second is

$$P(W_1 \cap B_2) + P(B_1 \cap W_2) = P(W_1)P(B_2) + P(B_1)P(W_2)$$

$$= \frac{5}{8} \cdot \frac{3}{8} + \frac{3}{8} \cdot \frac{5}{8} = \frac{15}{32}$$

EXAMPLE 4   Find the probability that all four women in a golf foursome will cook supper that night if their respective probabilities are 2/3, 3/5, 5/8, and 4/7 and the events are independent.

Solution   The answer is the product $\frac{2}{3} \cdot \frac{3}{5} \cdot \frac{5}{8} \cdot \frac{4}{7} = \frac{1}{7}$.

EXERCISE 18.2 **Dependent and Independent Events**

1   The probability Mr. Tomkins will plant tomatoes is 3/4, that they will grow if planted is 3/5, and that they will produce bountifully if they grow is 8/9. What is the probability of all three things happening?

2   What is the probability of Ross winning an election if the probabilities of winning the first and second primaries are 2/3 and 2/5 and of winning the general election is 3/8?

3   The probabilities that Sarah will graduate from college, then be accepted into medical school, and then become a doctor are 9/10, 2/5, and 3/5. Find the probability that she will be accepted into medical school. That she will become a doctor.

4   The probability that a committee will recommend Burnside for president of the university is 1/4, and that of being then selected is 1/3. Find the probability of his being recommended and selected. Of his being recommended and not selected.

5   If two balls are drawn without replacement from a bag with eight white and four black balls, what is the probability both are white?

6   If two balls are drawn without replacement from a bag with seven red

and six green balls, find the probability the first is red and the second is green.

7   If two balls are drawn without replacement from a bag with 10 black and 6 red balls, find the probability that one is red and one is black.

8   If two balls are drawn without replacement from a bag with eight red and six white balls, find the probability that both are the same color.

9   If three balls are drawn without replacement from a bag with 12 white and 6 black balls, what is the probability all three are white?

10   If three balls are drawn without replacement from a bag with eight white and four black balls, what is the probability that all three are white?   Is this larger than the answer to Prob. 9?

11   A bag has six red and four white balls, and a box has eight red and five white balls.   If one ball is drawn from each, find the probability that both balls are of the same color.

12   A bag has seven red and three white balls, and a box has six red and five black balls.   If exactly two balls are drawn from the bag without replacement, and one is drawn from the box, what is the probability that exactly two of the three balls are red?

13   If two cards are drawn without replacement from a standard deck, find the probability that both are sevens.

14   If two cards are drawn from a standard deck without replacement, find the probability that both are of the same rank (2 sevens, 2 kings, etc.).

15   If three cards are drawn without replacement from a standard deck, find the probability that all three are of the same suit.

16   If four cards are drawn without replacement from a standard deck, find the probability they are all of different suits.

17   The probabilities that Gene will marry and that Jean will win at Wimbleton are 7/8 and 2/9.   What is the probability that both will happen?

18   Find the probability of throwing a 5 with a pair of dice and choosing a heart from a deck of cards.

19   Three friends agree to meet in one year if all are alive.   What is the probability of their meeting if their probabilities of being alive are 0.95, 0.88, and 0.96?

20   What is the probability of throwing a 10 on one toss of a pair of dice and following it with a 4 on the next toss?

21   The probability that Webster, Hill, and Newhouse will eat at a certain restaurant one night is 3/8.   If the probability that Webster will is 3/4 and that Hill will is 2/3, what is the probability that Newhouse will?

22   If a coin is tossed three times, what is the probability of three heads?

23   If a coin is tossed three times, what is the probability of three heads if the first two tosses were heads?

24   Find the probability of throwing a 9 with a pair of dice and two heads in two tosses of a coin.

25   If two balls are drawn with replacement from a bag with eight white and four black balls, what is the probability both are white?

26   If two balls are drawn with replacement from a bag with seven red and six green balls, find the probability the first is red and the second is green.

27   A cup has four butterbeans and three red beans. If two beans are drawn with replacement, what is the probability of getting one of each kind?

28   In Prob. 27, what is the probability both are the same type of bean?

29   If a family is known to have two children, not both boys, what is the probability they have a boy?

30   If a family has three children not all boys, what is the probability they have exactly one boy?

31   A bag has six red and four white balls, and three balls are chosen without replacement. What is the probability the third ball is red if the first ball is white?

32   A pair of dice are thrown, and one of the dice shows a 4, 5, or 6. What is the probability the total is 8? Is 7?

33   Show that if E and F are independent events, then so are E and F'.

34   Show that E and S are always independent.

35   Show that $P(E)P(F|E) = P(F)P(E|F)$.

36   Show that if $P(E|F) = P(F|E)$, then $P(E) = P(F)$, assuming all four quantities are not zero.

18.5 / REPEATED TRIALS OF AN EVENT

If we know the probability of an event occurring in one trial, then the probability of its happening a given number of times in n trials is given by the following theorem:

THEOREM
If p is the probability that an event will occur in one trial, then the probability that it will occur exactly r times in n trials is equal to

$$C(n, r)p^r(1 - p)^{n-r} \tag{18.7}$$

Proof   We prove the above theorem as follows: The r trials can be selected from the n trials in $C(n, r)$ ways, by Sec. 16.6. The probability that the event will occur r times and fail the remaining $n - r$ times is $p^r(1-p)^{n-r}$, by Sec. 18.4, since the trials are independent and $1 - p$ is the probability of the event failing in any trial. By (18.3), the desired probability is therefore $C(n, r)p^r(1 - p)^{n-r}$. The theorem has the following corollary:

THEOREM   *If p is the probability that an event will occur in one trial, then the probability that it will occur at least r times in n trials is equal to*

$$p^n + C(n, n-1)p^{n-1}(1 - p) + C(n, n-2)p^{n-2}(1 - p)^2 + \cdots$$
$$+ C(n, r+1)p^{r+1}(1 - p)^{n-r-1} + C(n, r)p^r(1 - p)^{n-r} \qquad (18.8)$$

Proof   We prove the corollary as follows: The terms of the sum are the probabilities that the event will occur exactly n times, exactly $n - 1$ times, . . . , exactly $r + 1$ times, and exactly r times in n trials, and the events are mutually exclusive.

The reader should notice that the expression in this corollary is the first $n - r + 1$ terms of the expansion of binomial $(p + q)^n$, where $q = 1 - p$.

EXAMPLE 1   A bag contains three white and four red balls. The balls are drawn from the bag one at a time and are replaced after each drawing. What is the probability of drawing exactly three red balls in five trials?

Solution   The probability of drawing a red ball in one trial is 4/7. Therefore, the desired probability is

$$C(n, r)p^r(1 - p)^{n-r} = C(5, 3)\left(\frac{4}{7}\right)^3\left(1 - \frac{4}{7}\right)^{5-3}$$

$$= \left(\frac{5!}{3!2!}\right)\left(\frac{4^3}{7^3}\right)\left(\frac{3^2}{7^2}\right)$$

$$= \frac{5,760}{16,807}$$

EXAMPLE 2   Find the probability of throwing at least one 5 in three tosses of a die.

Solution   Since $P(5) = 1/6$, the answer, by Eq. (18.8), is

$$C(3, 1)\left(\frac{1}{6}\right)\left(\frac{5}{6}\right)^2 + C(3, 2)\left(\frac{1}{6}\right)^2\left(\frac{5}{6}\right) + C(3, 3)\left(\frac{1}{6}\right)^3 =$$

$$\frac{(3)(25)}{216} + \frac{(3)(5)}{216} + \frac{1}{216} = \frac{91}{216}$$

The answer could also have been found by Eq. (18.4). The probability of throwing no fives is $C(3, 0)(1/6)^0(5/6)^3 = (5/6)^3$, so the answer is

$$1 - (5/6)^3 = 1 - \frac{125}{216} = \frac{91}{216}$$

EXERCISE 18.3 Repeated Trials

1 Find the probability of throwing, in five tosses of a coin, (*a*) exactly three heads, (*b*) at least three heads.

2 Find the probability of throwing, in three tosses of a die, (*a*) exactly two 5s, (*b*) at least two 5s.

3 The probability that a boy will be on time for a meal is 0.2. Find the probability that he will be on time (*a*) exactly four times in two days, (*b*) at least four times.

4 Each of six boys tosses a die. Find the probability that at least four of them will throw a 2.

5 If each boy in Prob. 4 tosses the die twice, find the probability that exactly four of them will throw a 6 first and a 5 second.

6 A bag contains three white, four red, and five black balls. Five withdrawals of one ball each are made, and the ball is replaced after each. Find the probability that all five will be red.

7 In Prob. 6, find the probability that exactly three balls will be red.

8 In Prob. 6, find the probability that at least three balls will be red.

9 If the probability that a certain basketball team will win the conference championship is 2/3, find the probability that it will win exactly three championships in 5 years.

10 John and Tom are members of a Sunday-school class composed of five boys. The members sit on a bench. If they take seats at random and all five are present each of the four Sundays in June, find the probability that John and Tom will sit together (*a*) exactly three times, (*b*) at least three times.

11 Six history books, four mathematics books, and two civics books are on a table. If a book is removed and replaced, then another removed and replaced, and so on until six removals and replacements have been made, find the probability that a history book was removed and replaced (*a*) three times, (*b*) at least three times.

12 A student has made an A in 8 of the last 20 courses he has taken. Find the probability that he will make an A five times in the next seven courses he takes.

13   Each of three persons has a box containing six balls numbered from 1 to 6. Each person draws a ball from his box, replaces it, and then draws two balls simultaneously. Find the probability that as many as two persons will draw a 5 the first time, and balls whose sum is 5 the second time.

14   Six mathematics classes and four other classes meet at 7:30 A.M. in a certain building six times a week. There are six parking places near the entrance reserved for teachers. Find the probability that the mathematics teachers will park in the reserved places exactly four times in a week.

15   If the probability that a candidate will be elected to an office is 2/3, find the probability that he will be elected for four successive terms and then defeated for the fifth term.

16   If the probability that a certain person will tell the truth is 1/3, find the probability that he will tell 7 lies in answering 10 questions.

17   The probability of an event happening exactly twice in four trials is 18 times the probability of it happening exactly five times in six trials. Find the probability that it will occur in one trial.

18   If the probability of an event happening exactly four times in five trials is 10/243, find the probability that it will happen in one trial.

19   If the probability that an event will happen exactly three times in five trials is equal to the probability that it will happen exactly two times in six trials, find the probability that it will happen in one trial.

20   Find the probability that an event will occur in one trial if the probability of it happening exactly three times in six trials is equal to the probability that it will happen exactly twice in five trials.

EXERCISE 18.4 **REVIEW**

1   If a ball is drawn twice with replacement from a bag with 10 red and 4 white balls, what is the probability both balls are red?

2   Repeat Prob. 1 without replacement.

3   If the probability that a person will catch a certain disease is 5/7, what is the probability that both a man and his wife will catch the disease?

4   If the probability that a person will win something on a TV game show is 3/4, what is the probability that exactly three of four contestants will win something?

5   If a card is drawn from a standard deck, what is the probability it is either red or a three?

6   Repeat Prob. 5 with the condition that we know it is not a face card.

7 What is the probability of throwing a 6 with one die?

8 What is the probability of throwing a 6 with two dice?

9 What is the probability of throwing a 6 with three dice?

10 If an integer from 1 to 300 is chosen, what is the probability it is divisible by 2 or 3?

11 What is the probability that an integer between 1 and 100 inclusive is divisible by 3 or 4?

12 If four cards are drawn from a deck without replacement, what is the probability that they are all of the same suit?

13 What is the probability of hitting a target exactly three times out of four if the probability of missing it in one shot is 1/3?

14 What is the probability of at least two out of four people being on welfare in a city of 10,000 if 4,000 are on welfare?

15 What is the probability of at least three successes out of six trials if the probability of an individual success is 2/5? Is this larger than the answer to Prob. 14?

16 What is the probability of picking a committee of five people consisting of three men and two women, out of a group of 12 men and 8 women?

17 On a cruise ship 39 percent of the passengers enjoyed playing canasta, 52 percent enjoyed playing bridge, and 23 percent enjoyed neither. What is the probability a passenger enjoyed bridge but not canasta?

18 If the probability that a contract will be awarded to a firm in Kansas is 0.14, in Massachusetts is 0.18, in Tennessee is 0.21, and in Utah is 0.23, what is the probability it will not be awarded to a Tennessee or Kansas firm?

19 If a coin and two dice are thrown, what is the probability of a head and a product of 12?

20 If B is half as likely to happen as A and $P(A|B) = 2/5$, find $P(B|A)$.

APPENDIX *Reference Material from Trigonometry*

1/DEFINITIONS

ANGLE   If a ray is rotated about a fixed point on a stationary ray, an angle is formed.   The stationary ray is the *initial* side of the angle, the rotating ray is the *terminal* side, and the fixed point is the vertex.

The angle is positive or negative according as the direction of rotation of the terminal side is counterclockwise or clockwise.   (See Fig. A.1.)

THE SINE, COSINE, AND TANGENT OF AN ANGLE   If the angle θ is placed in the cartesian plane as indicated in Fig. A.2 and $P(x, y)$ is a point on the terminal

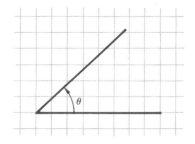 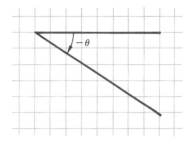

Figure A.1

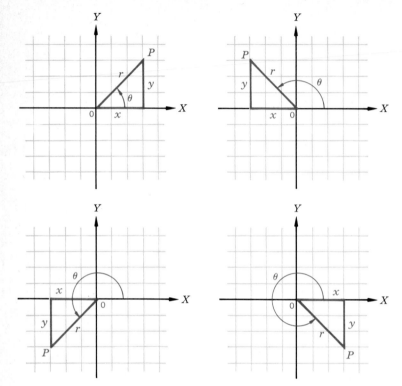

Figure A.2

side at a distance r from the vertex, then the sine, cosine, and tangent of the angle θ are defined and abbreviated as follows:

$$\sin \theta = \frac{y}{r} \qquad \cos \theta = \frac{x}{r} \qquad \tan \theta = \frac{y}{x}$$

ALGEBRAIC SIGNS OF SIN θ, COS θ, AND TAN θ If the terminal side of θ is in the quadrant specified below, then the algebraic signs of $\sin \theta$, $\cos \theta$, and $\tan \theta$ are as indicated.

> Quadrant 1: $x > 0$, $y > 0$, $r > 0$, $\sin \theta > 0$, $\cos \theta > 0$, $\tan \theta > 0$
> Quadrant 2: $x < 0$, $y > 0$, $r > 0$, $\sin \theta > 0$, $\cos \theta < 0$, $\tan \theta < 0$
> Quadrant 3: $x < 0$, $y < 0$, $r > 0$, $\sin \theta < 0$, $\cos \theta < 0$, $\tan \theta > 0$
> Quadrant 4: $x > 0$, $y < 0$, $r > 0$, $\sin \theta < 0$, $\cos \theta > 0$, $\tan \theta < 0$

ARC SINE v, ARC COSINE v, ARC TANGENT v The arc sine v, arc cosine v, and arc tangent v are defined and abbreviated as follows:

> arcsin v is an angle whose sine is v
> arccos v is an angle whose cosine is v
> arctan v is an angle whose tangent is v

2 / NUMERICAL VALUES OF THE SINE, COSINE, AND TANGENT OF 30°, 45°, 60°, 90°, 180°, 270°, AND 360°

$$\sin 0° = 0 = \cos 90° \qquad (a.1)$$

$$\tan 0° = 0 = \tan 180° \qquad (a.2)$$

$$\sin 30° = \tfrac{1}{2} = \cos 60° \qquad (a.3)$$

$$\tan 30° = \frac{\sqrt{3}}{3} \qquad (a.4)$$

$$\sin 45° = \frac{\sqrt{2}}{2} = \cos 45° \qquad (a.5)$$

$$\tan 45° = 1 \qquad (a.6)$$

$$\sin 60° = \frac{\sqrt{3}}{2} = \cos 30° \qquad (a.7)$$

$$\tan 60° = \sqrt{3} \qquad (a.8)$$

$$\sin 90° = 1 = \cos 0° \qquad (a.9)$$

$$\tan 90° = \infty \qquad (a.10)$$

$$\sin 180° = 0 = \cos 270° \qquad (a.11)$$

$$\tan 180° = 0 \qquad (a.12)$$

$$\sin 270° = -1 = \cos 180° \qquad (a.13)$$

$$\tan 270° = -\infty \qquad (a.14)$$

$$\sin 360° = 0 \qquad (a.15)$$

$$\cos 360° = 1 \qquad (a.16)$$

$$\tan 360° = 0 \qquad (a.17)$$

3 / RELATED ANGLE FORMULAS

$$\sin (180° \pm \theta) = \mp \sin \theta \qquad (a.18)$$

$$\cos (180° \pm \theta) = -\cos \theta \qquad (a.19)$$

$$\tan (180° \pm \theta) = \pm \tan \theta \qquad (a.20)$$

$$\sin (360° \pm \theta) = \pm \sin \theta \qquad (a.21)$$

$$\cos (360° \pm \theta) = \cos \theta \qquad (a.22)$$

$$\tan (360° \pm \theta) = \pm \tan \theta \qquad (a.23)$$

$$\sin (-\theta) = -\sin \theta \qquad (a.24)$$

$$\cos (-\theta) = \cos \theta \qquad (a.25)$$

$$\tan (-\theta) = -\tan \theta \qquad (a.26)$$

4 / IMPORTANT IDENTITIES

$$\sin^2 \theta + \cos^2 \theta = 1 \qquad (a.27)$$

$$\sin (\theta + \phi) = \sin \theta \cos \phi + \cos \theta \sin \phi \qquad (a.28)$$

$$\cos (\theta + \phi) = \cos \theta \cos \phi - \sin \theta \sin \phi \qquad (a.29)$$

$$\sin (\theta - \phi) = \sin \theta \cos \phi - \cos \theta \sin \phi \qquad (a.30)$$

$$\cos (\theta - \phi) = \cos \theta \cos \phi + \sin \theta \sin \phi \qquad (a.31)$$

TABLE A.1
COMMON LOGARITHMS

N	0	1	2	3	4	5	6	7	8	9
10	0000	0043	0086	0128	0170	0212	0253	0294	0334	0374
11	0414	0453	0492	0531	0569	0607	0645	0682	0719	0755
12	0792	0828	0864	0899	0934	0969	1004	1038	1072	1106
13	1139	1173	1206	1239	1271	1303	1335	1367	1399	1430
14	1461	1492	1523	1553	1584	1614	1644	1673	1703	1732
15	1761	1790	1818	1847	1875	1903	1931	1959	1987	2014
16	2041	2068	2095	2122	2148	2175	2201	2227	2253	2279
17	2304	2330	2355	2380	2405	2430	2455	2480	2504	2529
18	2553	2577	2601	2625	2648	2672	2695	2718	2742	2765
19	2788	2810	2833	2856	2878	2900	2923	2945	2967	2989
20	3010	3032	3054	3075	3096	3118	3139	3160	3181	3201
21	3222	3243	3263	3284	3304	3324	3345	3365	3385	3404
22	3424	3444	3464	3483	3502	3522	3541	3560	3579	3598
23	3617	3636	3655	3674	3692	3711	3729	3747	3766	3784
24	3802	3820	3838	3856	3874	3892	3909	3927	3945	3962
25	3979	3997	4014	4031	4048	4065	4082	4099	4116	4133
26	4150	4166	4183	4200	4216	4232	4249	4265	4281	4298
27	4314	4330	4346	4362	4378	4393	4409	4425	4440	4456
28	4472	4487	4502	4518	4533	4548	4564	4579	4594	4609
29	4624	4639	4654	4669	4683	4698	4713	4728	4742	4757
30	4771	4786	4800	4814	4829	4843	5857	4871	4886	4900
31	4914	4928	4942	4955	4969	4983	4997	5011	5024	5038
32	5051	5065	5079	5092	5105	5119	5132	5145	5159	5172
33	5185	5198	5211	5224	5237	5250	5263	5276	5289	5302
34	5315	5328	5340	5353	5366	5378	5391	5403	5416	5428
35	5441	5453	5465	5478	5490	5502	5514	5527	5539	5551
36	5563	5575	5587	5599	5611	5623	5635	5647	5658	5670
37	5682	5694	5705	5717	5729	5740	5752	5763	5775	5786
38	5798	5809	5821	5832	5843	5855	5866	5877	5888	5899
39	5911	5922	5933	5944	5955	5966	5977	5988	5999	6010
40	6021	6031	6042	6053	6064	6075	6085	6096	6107	6117
41	6128	6138	6149	6160	6170	6180	6191	6201	6212	6222
42	6232	6243	6253	6263	6274	6284	6294	6304	6314	6325
43	6335	6345	6355	6365	6375	6385	6395	6405	6415	6425
44	6435	6444	6454	6464	6474	6484	6493	6503	6513	6522
45	6532	6542	6551	6561	6571	6580	6590	6599	6609	6618
46	6628	6637	6646	6656	6665	6675	6684	6693	6702	6712
47	6721	6730	6739	6749	6758	6767	6776	6785	6794	6803
48	6812	6821	6830	6839	6849	6857	6866	6875	6884	6893
49	6902	6911	6920	6928	6937	6946	6955	6964	6972	6981
50	6990	6998	7007	7016	7024	7033	7042	7050	7059	7067
51	7076	7084	7093	7101	7110	7118	7126	7135	7143	7152
52	7160	7168	7177	7185	7193	7202	7210	7218	7226	7235
53	7243	7251	7259	7267	7275	7284	7292	7300	7308	7316
54	7324	7332	7340	7348	7356	7364	7372	7380	7388	7396
N	0	1	2	3	4	5	6	7	8	9

TABLE A.1
COMMON LOGARITHMS
(*Continued*)

N	0	1	2	3	4	5	6	7	8	9
55	7404	7412	7419	7427	7435	7443	7451	7459	7466	7474
56	7482	7490	7497	7505	7513	7520	7528	7536	7543	7551
57	7559	7566	7574	7582	7589	7597	7604	7612	7619	7627
58	7634	7642	7649	7657	7664	7672	7679	7686	7694	7701
59	7709	7716	7723	7731	7738	7745	7752	7760	7767	7774
60	7782	7789	7796	7803	7810	7818	7825	7832	7839	7846
61	7853	7860	7868	7875	7882	7889	7896	7903	7910	7917
62	7924	7931	7938	7945	7952	7959	7966	7973	7980	7987
63	7993	8000	8007	8014	8021	8028	8035	8041	8048	8055
64	8062	8069	8075	8082	8089	8096	8102	8109	8116	8122
65	8129	8136	8142	8149	8156	8162	8169	8176	8182	8189
66	8195	8202	8209	8215	8222	8228	8235	8241	8248	8254
67	8261	8267	8274	8280	8287	8293	8299	8306	8312	8319
68	8325	8331	8338	8344	8351	8357	8363	8370	8376	8382
69	8388	8395	8401	8407	8414	8420	8426	8432	8439	8445
70	8451	8457	8463	8470	8476	8482	8488	8494	8500	8506
71	8513	8519	8525	8531	8537	8543	8549	8555	8561	8567
72	8573	8579	8585	8591	8597	8603	8609	8615	8621	8627
73	8633	8639	8645	8651	8657	8663	8669	8675	8681	8686
74	8692	8698	8704	8710	8716	8722	8727	8733	8739	8745
75	8751	8756	8762	8768	8774	8779	8785	8791	8797	8802
76	8808	8814	8820	8825	8831	8837	8842	8848	8854	8859
77	8865	8871	8876	8882	8887	8893	8899	8904	8910	8915
78	8921	8927	8932	8938	8943	8949	8954	8960	8965	8971
79	8976	8982	8987	8993	8998	9004	9009	9015	9020	9025
80	9031	9036	9042	9047	9053	9058	9063	9069	9074	9079
81	9085	9090	9096	9101	9106	9112	9117	9122	9128	9133
82	9138	9143	9149	9154	9159	9165	9170	9175	9180	9186
83	9191	9196	9201	9206	9212	9217	9222	9227	9232	9238
84	9243	9248	9253	9258	9263	9269	9274	9279	9284	9289
85	9294	9299	9304	9309	9315	9320	9325	9330	9335	9340
86	9345	9350	9355	9360	9365	9370	9375	9380	9385	9390
87	9395	9400	9405	9410	9415	9420	9425	9430	9435	9440
88	9445	9450	9455	9460	9465	9469	9474	9479	9484	9489
89	9494	9499	9504	9509	9513	9518	9523	9528	9533	9538
90	9542	9547	9552	9557	9562	9566	9571	9576	9581	9586
91	9590	9595	9600	9605	9609	9614	9619	9624	9628	9633
92	9638	9643	9647	9652	9657	9661	9666	9671	9675	9680
93	9685	8689	9594	9699	9703	9708	9713	9717	9722	9227
94	9731	9736	9741	9745	9750	9754	9759	9763	9768	9773
95	9777	9782	9786	9791	9795	9800	9805	9809	9814	9818
96	9823	9827	9832	9836	9841	9845	9850	9854	9859	9863
97	9868	9872	9877	9881	9886	9890	9894	9899	9903	9908
98	9912	9917	9921	9926	9930	9934	9939	9943	9948	9952
99	9956	9961	9965	9969	9974	9978	9983	9987	9991	9996
N	0	1	2	3	4	5	6	7	8	9

TABLE A.2 TRIGONOMETRIC FUNCTIONS

Angles	Sines		Cosines		Tangents		Cotangents		Angles
	Nat.	Log.	Nat.	Log.	Nat.	Log.	Nat.	Log.	
0°00′	.0000	∞	1.0000	0.0000	.0000	∞	∞	∞ —	90°00′
10	.0029	7.4637	1.0000	0000	.0029	7.4637	343.77	2.5363	50
20	.0058	7648	1.0000	0000	.0058	7648	171.89	2352	40
30	.0087	9408	1.0000	0000	.0087	9409	114.59	0591	30
40	.0116	8.0658	.9999	0000	.0116	8.0658	85.940	1.9342	20
50	.0145	1627	.9999	0000	.0145	1627	68.750	8373	10
1°00′	.0175	8.2419	.9998	9.9999	.0175	8.2419	57.290	1.7581	89°00′
10	.0204	3088	.9998	9999	.0204	3089	49.104	6911	50
20	.0233	3668	.9997	9999	.0233	3669	42.964	6331	40
30	.0262	4179	.9997	9999	.0262	4181	38.188	5819	30
40	.0291	4637	.9996	9998	.0291	4638	34.368	5362	20
50	.0320	5050	.9995	9998	.0320	5053	31.242	4947	10
2°00′	.0349	8.5428	.9994	9.9997	.0349	8.5431	28.636	1.4569	88°00′
10	.0378	5776	.9993	9997	.0378	5779	26.432	4221	50
20	.0407	6097	.9992	9996	.0407	6101	24.542	3899	40
30	.0436	6397	.9990	9996	.0437	6401	22.904	3599	30
40	.0465	6677	.9989	9995	.0466	6682	21.470	3318	20
50	.0494	6940	.9988	9995	.0495	6945	20.206	3055	10
3°00′	.0523	8.7188	.9986	9.9994	.0524	8.7194	19.081	1.2806	87°00′
10	.0552	7423	.9985	9993	.0553	7429	18.075	2571	50
20	.0581	7645	.9983	9993	.0582	7652	17.169	2348	40
30	.0610	7857	.9981	9992	.0612	7865	16.350	2135	30
40	.0640	8059	.9980	9991	.0641	8067	15.605	1933	20
50	.0669	8251	.9978	9990	.0670	8261	14.924	1739	10
4°00′	.0698	8.8436	.9976	9.9989	.0669	8.8446	14.301	1.1554	86°00′
10	.0727	8613	.9974	9989	.0729	8624	13.727	1376	50
20	.0756	8783	.9971	9988	.0758	8795	13.197	1205	40
30	.0785	8946	.9969	9987	.0787	8960	12.706	1040	30
40	.0814	9104	.9967	9986	.0816	9118	12.251	0882	20
50	.0843	9256	.9964	9985	.0846	9272	11.826	0728	10
5°00′	.0872	8.9403	.9962	9.9983	.0875	8.9420	11.430	1.0580	85°00′
10	.0901	9545	.9959	9982	.0904	9563	11.059	0437	50
20	.0929	9682	.9957	9981	.0934	9701	10.712	0299	40
30	.0958	9816	.9954	9980	.0963	9836	10.385	0164	30
40	.0987	9945	.9951	9979	.0992	9966	10.078	0034	20
50	.1016	9.0070	.9948	9977	.1022	9.0093	9.7882	0.9907	10
6°00′	.1045	9.0192	.9945	9.9976	.1051	9.0216	9.5144	0.9784	84°00′
10	.1074	0311	.9942	9975	.1080	0336	9.2553	9664	50
20	.1103	0426	.9939	9973	.1110	0453	9.0098	9547	40
30	.1132	0539	.9936	9972	.1139	0567	8.7769	9433	30
40	.1161	0648	.9932	9971	.1169	0678	8.5555	9322	20
50	.1190	0755	.9929	9969	.1198	0786	8.3450	9214	10
7°00′	.1219	9.0859	.9925	9.9968	.1228	9.0891	8.1443	0.9109	83°00′
10	.1248	0961	.9922	9966	.1257	0995	7.9530	9005	50
20	.1276	1060	.9918	9964	.1287	1096	7.7704	8904	40
30	.1305	1157	.9914	9963	.1317	1194	7.5958	8806	30
40	.1334	1252	.9911	9961	.1346	1291	7.4287	8709	20
50	.1363	1345	.9907	9959	.1376	1385	7.2687	8615	10
8°00′	.1392	9.1436	.9903	9.9958	.1405	9.1478	7.1154	0.8522	82°00′
10	.1421	1525	.9899	9956	.1435	1569	6.9882	8431	50
20	.1449	1612	.9894	9954	.1465	1658	6.8269	8342	40
30	.1478	1697	.9890	9952	.1495	1745	6.6912	8255	30
40	.1507	1781	.9886	9950	.1524	1831	6.5606	8169	20
50	.1536	1863	.9881	9948	.1554	1915	6.4348	8085	10
9°00′	.1564	9.1943	.9877	9.9946	.1584	9.1997	6.3138	0.8003	81°00′
	Nat.	Log.	Nat.	Log.	Nat.	Log.	Nat.	Log.	

Angles	Cosines		Sines		Cotangents		Tangents		Angles

TABLE A.2 TRIGONOMETRIC FUNCTIONS (Continued)

Angles	Sines		Cosines		Tangents		Cotangents		Angles
	Nat.	Log.	Nat.	Log.	Nat.	Log.	Nat.	Log.	
9°00′	.1564	9.1943	.9877	9.9946	.1584	9.1997	6.3138	0.8003	81°00′
10	.1593	2022	.9872	9944	.1614	2078	6.1970	7922	50
20	.1622	2100	.9868	9942	.1644	2158	6.0844	7842	40
30	.1650	2176	.9863	9940	.1673	2236	5.9758	7764	30
40	.1679	2251	.9858	9938	.1703	2313	5.8708	7687	20
50	.1708	2324	.9853	9936	.1733	2389	5.7694	7611	10
10°00′	.1736	9.2397	.9848	9.9934	.1763	9.2463	5.6713	0.7537	80°00′
10	.1765	2468	.9843	9931	.1793	2536	5.5764	7464	50
20	.1794	2538	.9838	9929	.1823	2609	5.4845	7391	40
30	.1822	2606	.9833	9927	.1853	2280	5.3955	7320	30
40	.1851	2674	.9827	9924	.1883	2750	5.3093	7250	20
50	.1880	2740	.9822	9922	.1914	2819	5.2257	7181	10
11°00′	.1908	9.2806	.9816	9.9919	.1944	9.2887	5.1446	0.7113	79°00′
10	.1937	2870	.9811	9917	.1974	2953	5.0658	7047	50
20	.1965	2934	.9805	9914	.2004	3020	4.9894	6980	40
30	.1994	2997	.9799	9912	.2035	3085	4.9152	6915	30
40	.2022	3058	.9793	9909	.2065	3149	4.8430	6851	20
50	.2051	3119	.9787	9907	.2095	3212	4.7729	6788	10
12°00′	.2079	9.3179	.9781	9.9904	.2126	9.3275	4.7046	0.6725	78°00′
10	.2108	3238	.9775	9901	.2156	3336	4.6382	6664	50
20	.2136	3296	.9769	9899	.2186	3397	4.5736	6603	40
30	.2164	3353	.9763	9896	.2217	3458	4.5107	6542	30
40	.2193	3410	.9757	9893	.2247	3517	4.4494	6483	20
50	.2221	3466	.9750	9890	.2278	3576	4.3897	6424	10
13°00′	.2250	9.3521	.9744	9.9987	.2309	9.3634	4.3315	0.6366	77°00′
10	.2278	3575	.9737	9884	.2339	3691	4.2747	6309	50
20	.2306	3629	.9730	9881	.2370	3748	4.2193	6252	40
30	.2334	3682	.9724	9878	.2401	3804	4.1653	6196	30
40	.2363	3734	.9717	9875	.2432	3859	4.1126	6141	20
50	.2391	3786	.9710	9872	.2462	3914	4.0611	6086	10
14°00′	.2419	9.3837	.9703	9.9869	.2493	9.3968	4.0108	0.6032	76°00′
10	.2447	3887	.9696	9866	.2524	4021	3.9617	5979	50
20	.2476	3937	.9689	9863	.2555	4074	3.9136	5926	40
30	.2504	3986	.9681	9859	.2586	4127	3.8667	5873	30
40	.2532	4035	.9674	9856	.2617	4178	3.8208	5822	20
50	.2560	4083	.9667	9853	.2648	4230	3.7760	5770	10
15°00′	.2588	9.4130	.9659	9.9849	.2679	9.4281	3.7321	0.5719	75°00′
10	.2616	4177	.9652	9846	.2711	4331	3.6891	5669	50
20	.2344	4223	.9644	9843	.2742	4381	3.6470	5519	40
30	.2672	4269	.9636	9839	.2773	4430	3.6059	5570	30
40	.2700	4314	.9628	9836	.2805	4479	3.5656	5521	20
50	.2728	4359	.9621	9832	.2836	4527	3.5261	5473	10
16°00′	.2756	9.4403	.9613	9.9828	.2867	9.4575	3.4874	0.5425	74°00′
10	.2784	4447	.9605	9825	.2899	4622	3.4495	5378	50
20	.2812	4491	.9596	9821	.2931	4669	3.4124	5331	40
30	.2840	4533	.9588	9817	.2962	4716	3.3759	5284	30
40	.2868	4576	.9580	9814	.2994	4762	3.3402	5238	20
50	.2896	4618	.9572	9810	.3026	4808	3.3052	5192	10
17°00′	.2224	9.4659	.9563	9.9806	.3057	9.4853	3.2709	0.5147	73°00′
10	.2952	4700	.9555	9802	.3089	4898	3.2371	5102	50
20	.2979	4741	.9546	9798	.3121	4943	3.2041	5057	40
30	.3007	4781	.9537	9794	.3153	4987	3.1716	5013	30
40	.3035	4821	.9528	9790	.3185	5031	3.1397	4969	20
50	.3062	4861	.9520	9786	.3217	5075	3.1084	4925	10
18°00′	.3090	9.4900	.9511	9.9782	.3249	9.5118	3.0777	0.4882	72°00′
	Nat.	Log.	Nat.	Log.	Nat.	Log.	Nat.	Log.	

Angles	Cosines		Sines		Cotangents		Tangents		Angles

Angles	Sines		Cosines		Tangents		Cotangents		Angles
	Nat.	Log.	Nat.	Log.	Nat.	Log.	Nat.	Log.	
18°00′	.3090	9.4900	.9511	9.9782	.3249	9.5118	3.0777	0.4882	72°00′
10	.3118	4939	.9502	9778	.3281	5161	3.0475	4839	50
20	.3145	4977	.9492	9774	.3314	5203	3.0178	4797	40
30	.3173	5015	.9483	9770	.3346	5245	2.9887	4755	30
40	.3201	5052	.9474	9765	.3378	5287	2.9600	4713	20
50	.3228	5090	.9465	9761	.3411	5329	2.9319	4671	10
19°00′	.3256	9.5126	.9455	9.9757	.3443	9.5370	2.9042	0.4630	71°00′
10	.3283	5163	.9446	9752	.3476	5411	2.8770	4589	50
20	.3311	5199	.9436	9748	.3508	5451	2.8502	4549	40
30	.3338	5235	.9426	9743	.3541	5491	2.8239	4509	30
40	.3365	5270	.9417	9739	.3574	5531	2.7980	4469	20
50	.3393	5306	.9407	9734	.3607	5571	2.7725	4429	10
20°00′	.3420	9.5341	.9397	9.9730	.3640	9.5611	2.7475	0.4389	70°00′
10	.3448	5375	.9387	9725	.3673	5650	2.7228	4350	50
20	.3475	5409	.9377	9721	.3706	5689	2.6985	4311	40
30	.3502	5443	.9367	9716	.3739	5727	2.6746	4273	30
40	.3529	5477	.9356	9711	.3772	5766	2.6511	4234	20
50	.3557	5510	.9346	9706	.3805	5804	2.6279	4196	10
21°00′	.3584	9.5543	.9336	9.9702	.3839	9.5842	2.6051	0.4158	69°00′
10	.3611	5576	.9325	9597	.3872	5879	2.5826	4121	50
20	.3638	5609	.9315	9692	.3906	5917	2.5605	4083	40
30	.3665	5641	.9304	9687	.3939	5954	2.5386	4046	30
40	.3692	5673	.9293	9682	.3973	5991	2.5172	4009	20
50	.3719	5704	.9283	9677	.4006	6028	2.4960	3972	10
22°00′	.3746	9.5736	.9272	9.9672	.4040	9.6064	2.4751	0.3936	68°00′
10	.3773	5767	.9261	9667	.4074	6100	2.4545	3900	50
20	.3800	5798	.9250	9661	.4108	6136	2.4342	3864	40
30	.3827	5828	.9239	9656	.4142	6172	2.4142	3828	30
40	.3854	5859	.9228	9651	.4176	6208	2.3945	3792	20
50	.3881	5889	.9216	9646	.4210	6243	2.3750	3757	10
23°00′	.3907	9.5919	.9205	9.9640	.4245	9.6279	2.3559	0.3721	67°00′
10	.3934	5948	.9194	9635	.4279	6314	2.3369	3686	50
20	.3961	5978	.9182	9629	.4314	6348	2.3183	3652	40
30	.3987	6007	.9171	9624	.4348	6383	2.2998	3617	30
40	.4014	6036	.9159	9618	.4383	6417	2.2817	3583	20
50	.4041	6055	.9147	9613	.4417	6452	2.2637	3548	10
24°00′	.4067	9.6093	.9135	9.9607	.4452	9.6486	2.2460	0.3514	66°00′
10	.4094	6121	.9124	9602	.4487	6520	2.2286	3480	50
20	.4120	6149	.9112	9596	.4522	6553	2.2113	3447	40
30	.4147	6177	.9100	9590	.4557	6587	2.1943	3413	30
40	.4173	6205	.9088	9584	.4592	6620	2.1775	3380	20
50	.4200	6232	.9075	9579	.4628	6654	2.1609	3346	10
25°00′	.4226	9.6259	.9063	9.9573	.4663	9.6687	2.1445	0.3313	65°00′
10	.4253	6286	.9051	9567	.4699	6720	2.1283	3280	50
20	.4279	6313	.9038	9561	.4734	6752	2.1123	3248	40
30	.4305	6340	.9026	9555	.4770	6785	2.0965	3215	30
40	.4331	6366	.9013	9549	.4806	6817	2.0809	3182	20
50	.4358	6392	.9001	9543	.4841	6850	2.0655	3150	10
26°00′	.4384	9.6418	.8988	9.9537	.4877	9.6882	2.0503	0.3118	64°00′
10	.4410	6444	.8975	9530	.4913	6914	2.0353	3086	50
20	.4436	6470	.8962	9524	.4950	6946	2.0204	3054	40
30	.4462	6495	.8949	9518	.4986	6977	2.0057	3023	30
40	.4488	6521	.8936	9512	.5022	7009	1.9912	2991	20
50	.4514	6546	.8923	9505	.5059	7040	1.9768	2960	10
27°00′	.4540	9.6570	.8910	9.9499	.5095	9.7072	1.9626	0.2928	63°00′
	Nat.	Log.	Nat.	Log.	Nat.	Log.	Nat.	Log.	
Angles	Cosines		Sines		Cotangents		Tangents		Angles

TABLE A.2
TRIGONOMETRIC
FUNCTIONS
(Continued)

Angles	Sines		Cosines		Tangents		Cotangents		Angles
	Nat.	Log.	Nat.	Log.	Nat.	Log.	Nat.	Log.	
27°00′	.4540	9.6570	.8910	9.9499	.5095	9.7072	1.9626	0.2928	63°00′
10	.4566	6595	.8897	9492	.5132	7103	1.9486	2897	50
20	.4592	6620	.8884	9486	.5169	7134	1.9347	2866	40
30	.4617	6644	.8870	9479	.5206	7165	1.9210	2835	30
40	.4643	6668	.8857	9473	.5243	7196	1.9074	2804	20
50	.4669	6692	.8843	9466	.5280	7226	1.8940	2774	10
28°00′	.4695	9.6716	.8829	9.9459	.5317	9.7257	1.8807	0.2743	62°00′
10	.4720	7640	.8816	9453	.5354	7287	1.8676	2713	50
20	.4746	6763	.8802	9446	.5392	7317	1.8546	2683	40
30	.4772	6787	.8788	9439	.5430	7348	1.8418	2652	30
40	.4797	6810	.8774	9432	.5467	7378	1.8291	2622	20
50	.4823	6833	.8760	9425	.5505	7408	1.8165	2592	10
29°00′	.4848	9.6856	.8746	9.9418	.5543	9.7438	1.8040	0.2562	61°00′
10	.4874	6878	.8732	9411	.5581	7467	1.7917	2533	50
20	.4899	6901	.8718	9404	.5619	7497	1.7796	2503	40
30	.4924	6923	.8704	9397	.5658	7526	1.7675	2474	30
40	.4950	6946	.8689	9390	.5696	7556	1.7556	2444	20
50	.4975	6968	.8675	9383	.5735	7585	1.7437	2415	10
30°00′	.5000	9.6990	.8660	9.9375	.5774	9.7614	1.7321	0.2386	60°00′
10	.5025	7012	.8646	9368	.5812	7644	1.7205	2356	50
20	.5050	7033	.8631	9361	.5851	7673	1.7090	2327	40
30	.5075	7055	.8616	9353	.5890	7701	1.6977	2299	30
40	.5100	7076	.8601	9346	.5930	7730	1.6864	2270	20
50	.5125	7097	.8587	9338	.5969	7759	1.6753	2241	10
31°00′	.5150	9.7118	.8572	9.9331	.6009	9.7788	1.6643	0.2212	59°00′
10	.5175	7139	.8557	9323	.6048	7816	1.6534	2184	50
20	.5200	7160	.8542	9315	.6088	7845	1.6426	2155	40
30	.5225	7181	.8526	9308	.6128	7873	1.6319	2127	30
40	.5250	7201	.8511	9300	.6168	7902	1.6212	2098	20
50	.5275	7222	.8496	9292	.6208	7930	1.6107	2070	10
32°00′	.5299	9.7242	.8480	9.8284	.6249	9.7958	1.6003	0.2042	58°00′
10	.5324	7262	.8465	9276	.6289	7986	1.5900	2014	50
20	.5348	7282	.8450	9268	.6330	8014	1.5798	1986	40
30	.5373	7302	.8434	9260	.6371	8042	1.5697	1958	30
40	.5398	7322	.8418	9252	.6412	8070	1.5597	1930	20
50	.5422	7342	.8403	9244	.6453	8097	1.5497	1903	10
33°00′	.5446	9.7361	.8387	9.9236	.6494	9.8125	1.5399	0.1875	57°00′
10	.5471	7380	.8371	9228	.6536	8153	1.5301	1847	50
20	.5495	7400	.8355	9219	.6577	8180	1.5204	1820	40
30	.5519	7419	.8339	9211	.6619	8208	1.5108	1792	30
40	.5544	7438	.8323	9203	.6661	8235	1.5013	1765	20
50	.5568	7457	.8307	9194	.6703	8263	1.4919	1737	10
34°00′	.5592	9.7476	.8290	9.9186	.6745	9.8290	1.4826	0.1710	56°00′
10	.5616	7494	.8274	9177	.6787	8317	1.4733	1683	50
20	.5640	7513	.8258	9169	.6830	8344	1.4641	1656	40
30	.5664	7531	.8241	9160	.6873	8371	1.4550	1629	30
40	.5688	7550	.8225	9151	.6916	8398	1.4460	1602	20
50	.5712	7568	.8208	9142	.6959	8425	1.4370	1575	10
35°00′	.5736	9.7586	.8192	9.9134	.7002	9.8452	1.4281	0.1548	55°00′
10	.5760	7604	.8175	9125	.7046	8479	1.4193	1521	50
20	.5783	7622	.8158	9116	.7089	8506	1.4106	1494	40
30	.5807	7640	.8141	9107	.7133	8533	1.4019	1467	30
40	.5831	7657	.8124	9098	.7177	8559	1.3934	1441	20
50	.5854	7675	.8107	9089	.7221	8586	1.3848	1414	10
36°00′	.5878	9.7692	.8090	9.9080	.7265	9.8613	1.3764	0.1387	54°00′
	Nat.	Log.	Nat.	Log.	Nat.	Log.	Nat.	Log.	
Angles	Cosines		Sines		Cotangents		Tangents		Angles

TABLE A.2
TRIGONOMETRIC
FUNCTIONS
(Continued)

Angles	Sines		Cosines		Tangents		Cotangents		Angles
	Nat.	Log.	Nat.	Log.	Nat.	Log.	Nat.	Log.	
36°00′	.5878	9.7692	.8090	9.9080	.7265	9.8613	1.3764	0.1387	54°00′
10	.5901	7710	.8073	9070	.7310	9639	1.3680	1361	50
20	.5925	7727	.8056	9061	.7355	8666	1.3597	1334	40
30	.5948	7744	.8039	9052	.7400	8692	1.3514	1308	30
40	.5972	7761	.8021	9042	.7445	8718	1.3432	1282	20
50	.5995	7778	.8004	9033	.7490	8745	1.3351	1255	10
37°00′	.6018	9.7795	.7986	9.9023	.7536	9.8771	1.3270	0.1229	53°00′
10	.6041	7811	.7969	9014	.7581	8797	1.3190	1203	50
20	.6065	7828	.7951	9004	.7627	8824	1.3111	1176	40
30	.6088	7844	.7934	8995	.7673	8850	1.3032	1150	30
40	.6111	7861	.7916	8985	.7720	8876	1.2954	1124	20
50	.6134	7877	.7898	8975	.7766	8902	1.2876	1098	10
38°00′	.6157	9.7893	.7880	9.8965	.7813	9.8928	1.2790	0.1072	52°00′
10	.6180	7910	.7862	8955	.7860	8954	1.2723	1046	50
20	.6202	7926	.7844	8945	.7907	8980	1.2647	1020	40
30	.6225	7941	.7826	8935	.7954	9006	1.2572	0994	30
40	.6248	7957	.7808	8925	.8002	9032	1.2497	0968	20
50	.6271	7973	.7790	8915	.8050	9058	1.2423	0942	10
39°00′	.6293	9.7989	.7771	9.8905	.8098	9.9084	1.2349	0.0916	51°00′
10	.6316	8004	.7753	8895	.8146	9110	1.2276	0890	50
20	.6338	8020	.7735	8884	.8195	9135	1.2203	0865	40
30	.6361	8035	.7716	8874	.8243	9161	1.2131	0839	30
40	.6383	8050	.7698	8864	.8292	9187	1.2059	0813	20
50	.6406	8066	.7679	8853	.8342	9212	1.1988	0788	10
40°00′	.6428	9.8081	.7660	9.8843	.8391	9.9238	1.1918	0.0762	50°00′
10	.6450	8096	.7642	8832	.8441	9264	1.1847	0736	50
20	.6472	8111	.7623	8821	.8491	9289	1.1778	0711	40
30	.6494	8125	.7604	8810	.8541	9315	1.1708	0685	30
40	.6517	8140	.7585	8800	.8591	9341	1.1640	0659	20
50	.6539	8155	.7566	8789	.8642	9366	1.1571	0634	10
41°00′	.6561	9.8169	.7547	9.8778	.8693	9.9392	1.1504	0.0608	49°00′
10	.6583	8184	.7528	8767	.8744	9417	1.1436	0583	50
20	.6604	8198	.7509	8756	.8796	9443	1.1369	0557	40
30	.6626	8213	.7490	8745	.8847	9468	1.1303	0532	30
40	.6648	8227	.7470	8733	.8899	9494	1.1237	0506	20
50	.6670	8241	.7451	8722	.8952	9519	1.1171	0481	10
42°00′	.6691	9.8255	.7431	9.8711	.9004	9.9544	1.1106	0.0456	48°00′
10	.6713	8269	.7412	8699	.9057	9570	1.1041	0430	50
20	.6734	8283	.7392	8688	.9110	9595	1.0977	0405	40
30	.6756	8297	.7373	8676	.9163	9621	1.0913	0379	30
40	.6777	8311	.7353	8665	.9217	9646	1.0850	0354	20
50	.6799	8324	.7333	8653	.9271	9671	1.0786	0329	10
43°00′	.6820	9.8338	.7314	9.8641	.9325	9.9697	1.0724	0.0303	47°00′
10	.6841	8351	.7294	8629	.9380	9722	1.0661	0278	50
20	.6862	8365	.7274	8618	.9435	9747	1.0599	0253	40
30	.6884	8378	.7254	8606	.9490	9772	1.0538	0228	30
40	.6905	8391	.7234	8594	.9545	9798	1.0477	0202	20
50	.6926	8405	.7214	8582	.9601	9823	1.0416	0177	10
44°00′	.6947	9.8418	.7193	9.8569	.9657	9.9848	1.0355	0.0152	46°00′
10	.6967	8431	.7173	8557	.9713	9874	1.0295	0126	50
20	.6988	8444	.7153	8545	.9770	9899	1.0235	0101	40
30	.7009	8457	.7133	8532	.9827	9924	1.0176	0076	30
40	.7030	8469	.7112	8520	.9884	9949	1.0117	0051	20
50	.7050	8482	.7092	8507	.9942	9975	1.0058	0025	10
45°00′	.7071	9.8495	.7071	9.8495	1.0000	0.0000	1.0000	0.0000	45°00′
	Nat.	Log.	Nat.	Log.	Nat.	Log.	Nat.	Log.	

Angles	Cosines		Sines		Cotangents		Tangents		Angles

TABLE A.3 POWERS AND ROOTS	No.	Sq.	Sq. Root	Cube	Cube Root	No.	Sq.	Sq. Root	Cube	Cube Root
	1	1	1.000	1	1.000	51	2,601	7.141	132,651	3.708
	2	4	1.414	8	1.260	52	2,704	7.211	140,608	3.733
	3	9	1.732	27	1.442	53	2,809	7.280	148,877	3.756
	4	16	2.000	64	1.587	54	2,916	7.348	157,464	3.780
	5	25	2.236	125	1.710	55	3,025	7.416	166,375	3.803
	6	36	2.449	216	1.817	56	3,136	7.483	175,616	3.826
	7	49	2.646	343	1.913	57	3,249	7.550	185,193	3.849
	8	64	2.828	512	2.000	58	3,364	7.616	195,112	3.871
	9	81	3.000	729	2.080	59	3,481	7.681	205,379	3.893
	10	100	3.162	1,000	2.154	60	3,600	7.746	216,000	3.915
	11	121	3.317	1,331	2.224	61	3,721	7.810	226,981	3.936
	12	144	3.464	1,728	2.289	62	3.844	7.874	238,328	3.958
	13	169	3.606	2,197	2.351	63	3,969	7.937	250,047	3.979
	14	196	3.742	2,744	2.410	64	4,096	8.000	262,144	4.000
	15	225	3.873	3,375	2.466	65	4,225	8.062	274,625	4.021
	16	256	4.000	4,096	2.520	66	4,356	8.124	287,496	4.041
	17	289	4.123	4,913	2.571	67	4,489	8.185	300,763	4.062
	18	324	4.243	5,832	2.621	68	4,624	8.246	314,432	4.082
	19	361	4.359	6,859	2.668	69	4,761	8.307	328,509	4.102
	20	400	4.472	8,000	2.714	70	4,900	8.367	343,000	4.121
	21	441	4.583	9,261	2.759	71	5,041	8.426	357,911	4.141
	22	484	4.690	10,648	2.802	72	5,184	8.485	373,248	4.160
	23	529	4.796	12,167	2.844	73	5,329	8.544	389,017	4.179
	24	576	4.899	13,824	2.884	74	5,476	8.602	405,224	4.198
	25	625	5.000	15,625	2.924	75	5,625	8.660	421,875	4.217
	26	676	5.099	17,576	2.962	76	5,776	8.718	438,976	4.236
	27	729	5.196	19,683	3.000	77	5,929	8.775	456,533	4.254
	28	784	5.291	21,952	3.037	78	6,084	8.832	474,552	4.273
	29	841	5.385	24,389	3.072	79	6,241	8.888	493,039	4.291
	30	900	5.477	27,000	3.107	80	6,400	8.944	512,000	4.309
	31	961	5.568	29,791	3.141	81	6,561	9.000	531,441	4.327
	32	1,024	5.657	32,768	3.175	82	6,724	9.055	551,368	4.344
	33	1,089	5.745	35,937	3.208	83	6,889	9.110	571,787	4.362
	34	1,156	5.831	39,304	3.240	84	7,056	9.165	592,704	4.380
	35	1,225	5.916	42,875	3.271	85	7,225	9.220	614,125	4.397
	36	1,296	6.000	46,656	3.302	86	7,396	9.274	636,056	4.414
	37	1,369	6.083	50,653	3.332	87	7,569	9.327	658,503	4.431
	38	1,444	6.164	54,872	3.362	88	7,744	9.381	681,472	4.448
	39	1,521	6.245	59,319	3.391	89	7,921	9.434	704,969	4.465
	40	1,600	6.325	64,000	3.420	90	8,100	9.487	729,000	4.481
	41	1,681	6.403	68,921	3.448	91	8,281	9.539	753,571	4.498
	42	1,764	6.481	74,088	3.476	92	8,464	9.592	778,688	4.514
	43	1,849	6.557	79,507	3.503	93	8,649	9.644	804,357	4.531
	44	1,936	6.633	85,184	3.530	94	8,836	9.695	830,584	4.547
	45	2,025	6.708	91,125	3.557	95	9,025	9.747	857,375	4.563
	46	2,116	6.782	97,336	3.583	96	9,216	9.798	884,736	4.579
	47	2,209	6.856	103,823	3.609	97	9,409	9.849	912,673	4.595
	48	2,304	6.928	110,592	3.634	98	9,604	9.899	941,192	4.610
	49	2,401	7.000	117,649	3.659	99	9,801	9.950	970,299	4.626
	50	2,500	7.071	125,000	3.684	100	10,000	10.000	1,000,000	4.642

TABLE A.4
AMERICAN EXPERIENCE TABLE OF MORTALITY

Age	Number Living	Number Dying	Yearly Probability of Dying	Yearly Probability of Living	Age	Number Living	Number Dying	Yearly Probability of Dying	Yearly Probability of Living
10	100 000	749	0.007 490	0.992 510	53	66 797	1 091	0.016 333	0.983 667
11	99 251	746	0.007 516	0.992 484	54	65 706	1 143	0.017 396	0.982 604
12	98 505	743	0.007 543	0.992 457	55	64 563	1 199	0.018 571	0.981 429
13	97 762	740	0.007 569	0.992 431	56	63 364	1 260	0.019 885	0.980 115
14	97 022	737	0.007 596	0.992 404	57	62 104	1 325	0.021 335	0.978 665
15	96 285	735	0.007 634	0.992 366	58	60 779	1 394	0.022 936	0.977 064
16	95 550	732	0.007 661	0.992 339	59	59 385	1 468	0.024 720	0.975 280
17	94 818	729	0.007 688	0.992 312	60	57 917	1 546	0.026 693	0.973 307
18	94 089	727	0.007 727	0.992 273	61	56 371	1 628	0.028 880	0.971 120
19	93 362	725	0.007 765	0.992 235	62	54 743	1 713	0.031 292	0.968 708
20	92 637	723	0.007 805	0.992 195	63	53 030	1 800	0.033 943	0.966 057
21	91 914	722	0.007 855	0.992 145	64	51 230	1 889	0.036 873	0.963 127
22	91 192	721	0.007 906	0.992 094	65	49 341	1 980	0.040 129	0.959 871
23	90 471	720	0.007 958	0.992 042	66	47 361	2 070	0.043 707	0.956 293
24	89 751	719	0.008 011	0.991 989	67	45 291	2 158	0.047 647	0.952 353
25	89 032	718	0.008 065	0.991 935	68	43 133	2 243	0.052 002	0.947 998
26	88 314	718	0.008 130	0.991 870	69	40 890	2 321	0.056 762	0.943 238
27	87 596	718	0.008 197	0.991 803	70	38 569	2 391	0.061 993	0.938 007
28	86 878	718	0.008 264	0.991 736	71	36 178	2 448	0.067 665	0.932 335
29	86 160	719	0.008 345	0.991 655	72	33 730	2 487	0.073 733	0.926 267
30	85 441	720	0.008 427	0.991 573	73	31 243	2 505	0.080 178	0.919 822
31	84 721	721	0.008 510	0.991 490	74	28 738	2 501	0.087 028	0.912 972
32	84 000	723	0.008 607	0.991 393	75	26 237	2 476	0.094 371	0.905 629
33	83 277	726	0.008 718	0.991 282	76	23 761	2 431	0.102 311	0.897 689
34	82 551	729	0.008 831	0.991 169	77	21 330	2 369	0.111 064	0.888 936
35	81 822	732	0.008 946	0.991 054	78	18 961	2 291	0.120 827	0.879 173
36	81 090	737	0.009 089	0.990 911	79	16 670	2 196	0.131 734	0.868 266
37	80 353	742	0.009 234	0.990 766	80	14 474	2 091	0.144 466	0.855 534
38	79 611	749	0.009 408	0.990 592	81	12 383	1 964	0.158 605	0.841 395
39	78 862	756	0.009 586	0.990 414	82	10 419	1 816	0.174 297	0.825 703
40	78 106	765	0.009 794	0.990 206	83	8 603	1 648	0.191 561	0.808 439
41	77 341	774	0.010 008	0.989 992	84	6 955	1 470	0.211 359	0.788 641
42	76 567	785	0.010 252	0.989 748	85	5 485	1 292	0.235 552	0.764 448
43	75 782	797	0.010 517	0.989 483	86	4 193	1 114	0.265 681	0.734 319
44	74 985	812	0.010 829	0.989 171	87	3 079	933	0.303 020	0.696 980
45	74 173	828	0.011 163	0.988 837	88	2 146	744	0.346 692	0.653 308
46	73 345	848	0.011 562	0.988 438	89	1 042	555	0.395 863	0.604 137
47	72 497	870	0.012 000	0.988 000	90	847	385	0.454 545	0.545 455
48	71 627	896	0.012 509	0.987 491	91	462	246	0.532 468	0.467 532
49	70 731	927	0.013 106	0.986 894	92	216	137	0.634 259	0.365 741
50	69 804	962	0.013 781	0.986 219	93	79	58	0.734 177	0.265 823
51	68 842	1 001	0.014 541	0.985 459	94	21	18	0.857 143	0.142 857
52	67 841	1 044	0.015 389	0.984 611	95	3	3	1.000 000	0.000 000

Answers to
Selected Problems

1 $\{3, 6, 9, 12, 15\}$ **2** $\{v, w, l\}$
3 $\{$February, April, June, September, November$\}$ **5** $\{x \mid x = 2n, n = 1, 2, 3, 4\}$
6 $\{x \mid x = 2^n, n = 1, 2, 3, 4\}$ **7** $\{x \mid x$ is a president who died in office after 1900$\}$
9 $\{1, 2, 3, 5, 7, 10\}, \{5, 10\}, \{1, 3, 7\}$ **10** $B, \emptyset, \emptyset$ **11** $A, \emptyset, A$
13 $\{o, r, a, n, g, e, d\}, \{r, e\}, \{o, a, n, g\}$ **14** $A, B, \{1, 2, 4, 5, 7, 8\}$
15 $\{e, a, u\}, \{e, a\}, \{u\}$ **17** $A, C, B \cup C$ **18** $A - D, \{6\}, \emptyset$
19 $\{1, 3, 5, 7, 9, 11\}, B, A$ **21** $\emptyset, A, \{1, 3, 5, 7, 9, 11\}$ **22** $\{1, 3, 5, 7, 9, 11, 6\}$,
$\{2, 4, 8, 10\}, \emptyset$ **23** $\emptyset$ **25** $(4, 2), (4, 5), (4, 8), (7, 2), (7, 5), (7, 8)$
26 $(1, 1), (1, 4), (1, 8), (4, 1), (4, 4), (4, 8), (8, 1), (8, 4), (8, 8)$ **27** $(3, 6)$,
$(4, 6), (6, 6)$

1 Positive integers   **2** Negative integers   **3** Integers   **5** Fractions
6 Negative fractions   **7** Rational numbers   **9** Real numbers
10 Irrational numbers   **11** Real numbers   **13** Positive integers
14 Rational numbers   **15** Rational numbers   **17** Rational numbers
18 Positive integer   **19** $\emptyset$ **21** $>, <, <$ **22** $>, >, <$ **23** $>, <, >$
25 -1 **26** -5 **27** 5   **29** 13   **30** 5   **31** 11

EXERCISE 1.3

1 $\{x \mid x$ was a vice-president of the United States between 1944 and 1975$\}$
2 $\{x \mid x = 2^n, n = 1, 2, 3, 4, 5\}$ **3** $\{4, 8, 12, 16\}$ **5** $\{2, 3, 7, 9, 11, 13\}, \{2, 13\},$
$\{3, 7, 11\}$ **6** $\{c, a, m, e, l\}, \{m, e, a, l\}, \{c\}$ **7** $\{7\}, \{3, 4, 7, 11\}, \{6, 9\}$
10 $(2, 3), (2, 4), (2, 5), (7, 3), (7, 4), (7, 5)$ **11** $(5, 10), (5, 11), (6, 10),$
$(6, 11), (7, 10), (7, 11)$ **13** Rational   **14** $1, 5, 7$
15 $\sqrt{3} > 1.7, \frac{5}{9} < \frac{3}{5}, \sqrt{2} > 1.4$

EXERCISE 2.2

1 5 **2** 12 **3** 4 **5** 12 **6** 4 **7** 6 **9** $a - 4b + 4c$ **10** $6p - r$
11 $2x + 5z$ **13** $ab - ac$ **14** $4xy + 2xz - yz$ **15** $2pq$ **17** $4a - 7b + 9c$
18 $3x + y + 3z$ **19** $p + 2q$ **21** $9ab^2 + 6ab - 6a^2b$ **22** $-xy + 2xy^2 + 3xy^3$
23 $p^2q + 2pq + pq^2$ **25** 7 **26** 44 **27** 40 **29** $-a - b - 7c$
30 $a + b - c$ **31** $-2x + 4y - 9z$ **33** $a^2 + b^2 + c^2$ **34** $3x^2 + y^2 - z^2$
35 $7ax + ax^2 + a^2x$ **37** $3a - 6b$ **38** $x - 3y + 5z$ **39** $a - p + 2x$ **41** 4
42 2 **43** 4 **45** -6 **46** 2 **47** -1 **49** $6a - b$ **50** $5b - c$
51 $c + d$ **53** $-13e$ **54** $-e - 32f$ **55** $-12f - 5g$ **57** $-27h + 44i$
58 $-7i - 13j$ **59** $j + 14k + 3m$

EXERCISE 2.3

1 $12a^5$ **2** $8a^6$ **3** $-10a^8$ **5** $-6a^8$ **6** $-12a^7$ **7** $14a^{14}$ **9** $4x^4$
10 $9x^8$ **11** $125x^6y^3$ **13** $-8x^9y^6$ **14** $9x^8y^4$ **15** $-1024x^5$
17 $4x^4y^2 - 2x^3y^3$ **18** $9x^2y^3 - 6xy^4$ **19** $-15x^2y^2 + 5xy^3$
21 $-4x^4y^4 + 2x^3y^4 - 6x^3y^5$ **22** $-4x^4y^4 + 3x^5y^3 - 2x^6y^2$
23 $12x^2y^3 - 9x^3y^2 + 15x^3y^3$ **25** $-4xy^3 + 3x^3y$ **26** $-2x^2y + 9xy^2 + 2x^3y^2$
27 $-9x^7y^8z^6 - 2x^5y^7z^8$ **29** $6x^2 - xy - 12y^2$ **30** $12x^2 - xy - 6y^2$
31 $6x^2 + 7xy - 3y^2$ **33** $6x^3 - 13x^2y + 14xy^2 - 12y^3$ **34** $6x^3 - 5x^2y - 12xy^2 - 4y^3$
35 $10x^3 - x^2y - 17xy^2 - 6y^3$ **37** $2x^4 - 3x^3y - 8x^2y^2 + 15xy^3 - 6y^4$
38 $6x^4 - 5x^3y + 6x^2y^2 - xy^3 - 6y^4$ **39** $10x^4 - x^3y + 15x^2y^2 + 5xy^3 + 3y^4$
41 $6a + 2b$ **42** $5x + 4y$ **43** $2x + w + y$ **45** $4a - 2ab - 4bc - 2c$
46 $2x^3 + 10x^2$ **47** $4x - 7y - 3xy$

EXERCISE 2.4

1 a^5 **2** d^2 **3** b^2 **5** a^2b^3 **6** bc^2 **7** cd^5 **9** $3a^3b^3$ **10** $5a^4b^7$
11 $19a^4b^2$ **13** $7a^5 - 5a^2 - 2a$ **14** $8a^6 - 12a^4 + 5a$ **15** $-13x^5 - 11x^4 + 9x^2$
17 $7x^5y^3 - 5x^2y^4 + 4y^2$ **18** $3x^7y^7 + 2x^6y^6 - 4xy$ **19** $3x^5y^4 - 5xy - 7x^4y$
21 $x - 3$ **22** $x + 5$ **23** $2x - 1$ **25** $2x + 3$ **26** $3x - 1$ **27** $3x - 2$
29 $x^2 - x + 3$ **30** $x^2 + 2x - 1$ **31** $2x^2 - x - 3$ **33** $x^2 + 2x - 1, 3$
34 $2x^2 - 3x + 1, -5$ **35** $2x + 3, 3x - 1$ **37** $x^2 + 2x - 1, 2x - 3$
38 $x^2 - 3x - 2, 3x - 4$ **39** $x^3 + 5x^2 - 2x + 1, 0$

EXERCISE 2.5

1 A.4, A.5, M.5   **2** A.5, M.5   **3** A.1, M.1, M.5   **5** A.1, A.4, M.5
6 M.4, M.5   **7** A.1, A.4, A.5, M.1, M.4, M.5   **9** M.1, M.4, M.5
10 A.1, A.4, M.1, M.4, M.5   **11** M.1, M.5   **13** 2   **14** 3   **15** 5
17

$\oplus$	0	1	2
0	0	1	2
1	1	2	0
2	2	0	1

$\otimes$	0	1	2
0	0	0	0
1	0	1	2
2	0	2	1

EXERCISE 2.6

5 $3a + 9b$ **6** 0   **7** $a - 2p + 9r$ **9** $x^2 - 1$ **10** $4x^2 - 2x$ **11** 6
13 $-13y + 11z$ **14** $18x - 18$ **15** $6x^5$ **17** $16x^8$ **18** $-125x^9$
19 $6x^3y^2 - 3x^2y^3$ **21** $6x^2 - 7xy - 20y^2$ **22** $-10x^2 + 41xy - 21y^2$ **25** $3x^3y^3$
26 $4x^2 - 2x - 5x^3y^2$ **27** $-3x^3y^3 + 4xy - 6x^4y$ **29** $2x^2 - 3x + 7, -3x - 2$
30 $\{0, 1, 2, 3\}$

$\oplus$	0	1	2	3	4
0	0	1	2	3	4
1	1	2	3	4	0
2	2	3	4	0	1
3	3	4	0	1	2
4	4	0	1	2	3

$\otimes$	0	1	2	3	4
0	0	0	0	0	0
1	0	1	2	3	4
2	0	2	4	1	3
3	0	3	1	4	2
4	0	4	3	2	1

31 $b^2 - ba - a$; the commutative axiom does not hold.

EXERCISE 3.1

1 $x^2 + x - 6$ **2** $x^2 + 7x + 12$ **3** $x^2 - 9x + 20$ **5** $8x^2 + 22x + 15$
6 $6x^2 + x - 15$ **7** $10x^2 - 29x + 10$ **9** $8x^2 + 2xy - 21y^2$ **10** $21x^2 + 2xy - 8y^2$
11 $30x^2 + 17xy - 35y^2$ **13** $4a^2 + 4ab + b^2$ **14** $9a^2 + 12ab + 4b^2$
15 $36a^2 + 60ab + 25b^2$ **17** $a^4 + 18a^2b + 81b^2$ **18** $a^6 + 4a^3b + 4b^2$
19 $a^4 + 6a^2b^2 + 9b^4$ **21** $9a^2 - 12ab + 4b^2$ **22** $25a^2 - 40ab + 16b^2$

23 $25a^2 - 30ab + 9b^2$ **25** $a^4 - 12a^2b + 36b^2$ **26** $4a^4 - 12a^2b + 9b^2$
27 $4a^4 - 12a^2b^3 + 9b^6$ **29** 16   **30** 567   **31** 3,591   **33** $x^2 - 9$
34 $4x^2 - 1$ **35** $9x^2 - 4$ **37** $4x^2 - 25y^2$ **38** $9x^2 - 16y^2$ **39** $49x^2 - 36y^2$
41 $4a^4 - 9b^4$ **42** $16a^4 - 25b^4$ **43** $25a^4 - 49b^6$ **45** $x^2/4 - y^2/25$
46 $9x^2/16 - 16y^2/9$ **47** $4u^2/9v^2 - 25x^2/4w^2$ **49** $x^2 + y^2 + z^2 + 2xy + 2xz + 2yz$
50 $4x^2 + y^2 + z^2 - 4xy + 4xz - 2yz$ **51** $x^2 + y^2 + 9z^2 - 2xy + 6xz - 6yz$
53 $4x^2 + y^2 + z^2 + 9w^2 + 4xy + 4xz - 12xw + 2yz - 6yw - 6zw$
54 $x^2 + 4y^2 + 4z^2 + w^2 - 4xy + 4xz + 2xw - 8yz - 4yw + 4zw$
55 $x^6 + 4x^5 - 2x^3 + 16x^2 - 12x + 9$ **57** $6x^2 + 12xy + 6y^2 - 7x - 7y - 3$
58 $24a - 24ab + 6b^2 + 2a - b - 2$ **59** $60x^2 - 60xy + 15y^2 + 14x - 7y - 4$
61 $x^4 - 10x^2 + 9$ **62** $x^4 - 4x^3 + 4x^2 - 9$ **63** $4x^4 + y^4$ **65** $x^6 + x^4 + 3x^2 - 1$
66 $-x^6 + x^4 + x^2 - 1$ **67** $4x^8 - x^6 + 8x^5 - 4x^4 + x^2$

EXERCISE 3.2

1 $2(x - 5)$ **2** $3(x + 2)$ **3** $5(x + 3)$ **5** $x(x + 3)$ **6** $x(x - 7)$
7 $2x(x - 2)$ **9** $2(x^2 - 4x + 5)$ **10** $3(x^2 + 2x - 3)$ **11** $5(x^2 + 2x + 7)$
13 $(x + y)(x - y)$ **14** $(x + 2y)(x - 2y)$ **15** $(3x + y)(3x - y)$
17 $(2x + 3y)(2x - 3y)$ **18** $(3x + 5y)(3x - 5y)$ **19** $(5x + 7y)(5x - 7y)$
21 $(x + y^2)(x - y^2)$ **22** $(x^3 + y)(x^3 - y)$ **23** $(x^4 + y^3)(x^4 - y^3)$
25 $(x + 4y^4)(x - 4y^4)$ **26** $(3x^2 + y^5)(3x^2 - y^5)$ **27** $(2x^2 + 3y^3)(2x^2 - 3y^3)$
29 $(x - y)(x^2 + xy + y^2)$ **30** $(a + b)(a^2 - ab + b^2)$
31 $(m + n)(m^2 - mn + n^2)$ **33** $(a - 2b)(a^2 + 2ab + 4b^2)$
34 $(a - 3b)(a^2 + 3ab + 9b^2)$ **35** $(2a + 5b)(4a^2 - 10ab + 25b^2)$
37 $(3x^2 + 2y)(9x^4 - 6x^2y + 4y^2)$ **38** $(9x + 4y^3)(81x^2 - 36xy^3 + 16y^2)$
39 $(5x^2 - 3y^4)(25x^4 + 15x^2y^4 + 9y^{16})$ **41** $(x^2y + 4)(x^4y^2 - 4x^2y + 16)$
42 $(2x^3y^2 + 3)(4x^6y^4 - 6x^3y^2 + 9)$ **43** $(4x^4y^3 - 5)(16x^8y^6 + 20x^4y^3 + 25)$
45 $(x + y + 2z)(x + y - 2z)$ **46** $(x - 2y + 5z)(x - 2y - 5z)$
47 $(2x + y - 3z)(2x - y + 3z)$ **49** $(x + y - 1)(x^2 + 2xy + y^2 + x + y + 1)$
50 $(2x - y - 2)(4x^2 - 4xy + y^2 + 4 + 4x - 2y)$
51 $(3x + 2y + 3)(9x^2 + 12xy + 4y^2 - 9x - 6y + 9)$
53 $(4x^2 + y^2)(2x + y)(2x - y)$ **54** $(x^2 + 9y^2)(x + 3y)(x - 3y)$
55 $(9x^2 + 25y^2)(3x + 5y)(3x - 5y)$ **57** $(x^4 + y^2)(x^2 + y)(x^2 - y)$
58 $(x^4 + 9)(x^2 + 3)(x^2 - 3)$ **59** $(x + 1)(x^2 - x + 1)(x - 1)(x^2 + x + 1)$
61 $(x + y)(x^2 - xy + y^2)(x - y)(x^2 + xy + y^2)$ **62** $(x^2 + 3)(x^4 - 3x^2 + 9)$
63 $(7x^4 - 1)(49x^8 + 17x^4 + 1)$ **65** $(x + y)(x^4 - x^3y + x^2y^2 - xy^3 + y^4)$
66 $(x + y)(x^6 - x^5y + x^4y^2 - x^3y^3 + x^2y^4 - xy^5 + y^6)$
67 $(2x + y^2)(64x^6 - 32x^5y^2 + 16x^4y^4 - 8x^3y^6 + 4x^2y^8 - 2xy^{10} + y^{12})$
69 $(x - y)(x^6 + x^5y + x^4y^2 + x^3y^4 + xy^5 + y^6)$
70 $(x - 2y^2)(x^4 + 2x^3y^2 + 4x^2y^4 + 8xy^6 + 16y^8)$
71 $(3y^3 - 2y^2)(81y^{12} + 54x^9y^2 + 36x^6y^4 + 24x^3y^6 + 16y^8)$

EXERCISE 3.3

1 $(a - 1)(a - 1)$ **2** $(b - 3)(b - 3)$ **3** $(c + 2)(c + 2)$ **5** $(2a - 1)(2a - 1)$
6 $(3b + 1)(3b + 1)$ **7** $(5x + 1)(5x + 1)$ **9** $(2x - 3)(2x - 3)$

10 $(3x + 4)(3x + 4)$ **11** $(5x + 6)(5x + 6)$ **13** $(x - 2)(x + 3)$

14 $(x - 4)(x + 1)$ **15** $(x - 5)(x + 3)$ **17** $(x + 5)(x + 3)$

18 $(x + 7)(x + 2)$ **19** $(y + 5)(y + 7)$ **21** $(y - 3)(y - 2)$

22 $(y - 6)(y - 3)$ **23** $(y - 4)(y - 5)$ **25** $(3x + 1)(x + 2)$

26 $(2x + 1)(x + 2)$ **27** $(2x + 3)(3x - 1)$ **29** $(7x - 3)(x - 1)$

30 $(3x - 2)(x + 5)$ **31** $(2x - 1)(2x + 3)$ **33** $(7x - 2y)(4x + y)$

34 $(8x - 3y)(3x + y)$ **35** $(6x - 5y)(3x + y)$ **37** $(5x + 7y)(4x - 5y)$

38 $(12x - 5y)(2x + 3y)$ **39** $(7x - 5y)(6x + 5y)$ **41** $(6x - 7y)(5x + 7y)$

42 $(9x + 7y)(6x - 5y)$ **43** $(15x - 4y)(2x + 3y)$ **45** $(16x - 7y)(3x + 11y)$

46 $(7x - 12y)(6x + 7y)$ **47** $(5x - 11y)(2x + 7y)$ **49** $(x + y + 4)(x + y - 3)$

50 $(x + y - 2)(x + y + 4)$ **51** $(2x + y + 5)(2x + y - 3)$

53 $[5(3x - 4y) + 2][2(3x - 4y) + 3]$ **54** $[4x - 3(2y - 3z)][3x + 2(2y - 3z)]$

55 $[5(3x - 5y) - 3z][3(3x - 5y) + 8z]$ **57** $D^2 = 7^2,\ (2x - 1)(x + 3)$

58 $D^2 = 41$, not factorable **59** $D^2 = -104$, not factorable

61 $D^2 = -4$, not factorable **62** $D^2 = 11^2,\ (4x - 1)(x - 3)$

63 $D^2 = 19^2,\ (5x - 2)(2x + 3)$

EXERCISE 3.4

1 $(x + 2)(y + 1)$ **2** $(x - 3)(y + 2)$ **3** $(r - 4)(s - 3)$ **5** $(x + 3y)(2x - 1)$

6 $(2x - y)(x + 2)$ **7** $(3x - 2y)(2x + 3)$ **9** $(a - 3b)(c + d)$

10 $(a + 2b)(c - d)$ **11** $(a + b)(2c - d)$ **13** $(x - 1)(x + 1)(x + 3)$

14 $(x - 2)(x + 2)(2x - 1)$ **15** $(x - 3)(x + 3)(2x + 1)$ **17** $2r(r + 3)(2r - 1)$

18 $3s(s + 1)(s - 2)$ **19** $5t(t - 5)(2t + 1)$ **21** $3a(2a - 1)^2$ **22** $7b(2b + 1)^2$

23 $4c(3c + 4)^2$ **25** $(x + y)(x - y - z)$ **26** $(x + 2y)(x - y + z)$

27 $(x - 2y)(x + 3y + z)$ **29** $(x + 2y - z)(x + 2y + z)$

30 $(2x - y + 3z)(2x - y - 3z)$ **31** $(x + 3y - z)(x - 3y + z)$

33 $(x + y + 2)(x - y - 2)$ **34** $(x - 2y + 3z)(x + 2y - 3z)$

35 $(2x + 3y - 2w)(2x - 3y - 2w)$ **37** $(x^2 - 2x + 5)(x^2 - 2x - 5)$

38 $(x^2 - x - 1)(x^2 - x + 1)$ **39** $(y^2 + 4y + 3)(y^2 + 4y - 3)$

41 $(x^2 + 2xy + 3y^2)(x^2 - 2xy + 3y^2)$ **42** $(x^2 + xy - 5y^2)(x^2 - xy - 5y^2)$

43 $(2x^2 + 3xy + 4y^2)(2x^2 - 3xy + 4y^2)$ **45** $(x - y)(x^2 + xy + y^2 + x + y)$

46 $(x - y)(x^2 + xy + y^2 - x - y)$ **47** $(x + 2y)(x - 3y + x^2 - 2xy + 4y^2)$

EXERCISE 3.5

1 $2a^2 + 7a + 3$ **2** $3b^2 + 7b + 2$ **3** $4c^2 + 13c + 3$ **5** $3a^2 + 7ab + 2b^2$

6 $3b^2 - 7bc + 2c^2$ **7** $4r^2 + 11rs - 3s^2$ **9** $28r^2 - 5ru - 3u^2$

10 $-56t^2 + 3ts + 9s^2$ **11** $20c^2 + 13cd - 21d^2$ **13** $b^2 - 25$ **14** $9x^2 - 1$

15 $25x^2 - 1$ **17** $25x^2 - 9y^2$ **18** $9x^2 - 4y^2$ **19** $9x^2 - 16y^2$

21 $64x^2 - 25y^2$ **22** $9x^2 - 6x + 1$ **23** $49x^2 - 42x + 9$ **25** $m^2 + 6mn + 9n^2$

26 $36p^2 + 60pq + 25q^2$ **27** $x^2 + 12xy + 36y^2$ **29** $4x^6 - 12x^3y^2 + 9y^4$

30 $16p^6 + 40p^3q^2 + 25q^4$ **31** $9a^4 + 12a^2b^3 + 4b^6$ **33** $r^2 + 2rs + s^2 - 9t^2$

34 $u^2 + 6uv + 9v^2 - 4w^2$ **35** $a^2 + 6ab + 9b^2 - 4c^2$

37 $a^2 + 4b^2 + 9c^2 + d^2 + 4ab + 6ac - 2ad + 12bc - 4bd - 6cd$

38 $9a^2 + b^2 + c^2 + 4d^2 - 6ab - 6ac + 12ad + 2bc - 4bd - 4cd$

39 $x^6 + 6x^5 + 7x^4 - 10x^3 - 11x^2 + 4x + 4$ **41** $(x + 4)(x - 4)$

42 $(a + 3)(a - 3)$ **43** $(5a + 2b)(5a - 2b)$ **45** $3(3b^2 + 4c)(3b^2 - 4c)$

46 $3(2a^3 + 3b^2)(2a^3 - 3b^2)$ **47** $(5x + 2y + z)(5x + 2y - z)$

49 $5cd(3c - 4d)(3c + 4d)$ **50** $xy(9x^2 + 7y^3)(9x^2 - 7y^3)$

51 $xy^2(6x + 5y^2)(6x - 5y^2)$ **53** $(x + 4y)^2$ **54** $(x - 5y)^2$ **55** $(5m - 9n)^2$

57 $(7r + 9s)^2$ **58** $(8x + 5y)^2$ **59** $(6x^3 - 2y)^2$ **61** $(x - 3)(x - 5)$

62 $(y + 2)(y + 3)$ **63** $(3x + y)(2x - 3y)$ **65** $(4r + 5s)(3r - s)$

66 $(5r - s)(2r + 3s)$ **67** $2(3s - t)(s + 2t)$ **69** $(9r + 7s)(6r - 5s)$

70 $(6a - 5b)(a - 3b)$ **71** $(5a - 6b)(6a - 5b)$ **73** $(5x^2 + 2y^2)^2$

74 $(8r^2 - 7s^2)^2$ **75** $(9r^2 + 4q^2)^2$ **77** $(2x - y)(4x^2 + 2xy + y^2)$

78 $8(2a + b)(4a^2 - 2ab + b^2)$ **79** $(3a^2 + 1)(9a^4 - 3a^2 + 1)$

81 $(m^3 - 3n^4)(m^6 + 3m^3n^4 + 9m^8)$ **82** $(7c + 3d)(49c^2 - 21cd + 9d^2)$

83 $(5x - 6y)(25x^2 + 30xy + 36y^2)$

85 $(5c + 2d - 6)(25c^2 + 20cd + 4d^2 + 30c + 12d + 36)$

86 $(a - 2b)(7a^2 - ab + b^2)$ **87** $(x - y)(x^2 + xy + y^2 + x + y)$

89 $(2r^2 + r + 2s)(2r^2 - r - 2s)$ **90** $(a - 3b)(a^2 + 3ab + 9b^2 + a + 3b)$

EXERCISE 4.1

1 y^2/xy **2** bc/c^2 **3** $-u^2/-uv$ **5** $1/y$ **6** $4/s$ **7** $p/2$

9 $(3y - x)/(x - 4y)$ **10** $\dfrac{y - x}{5y - x}$ **11** $\dfrac{3x - 2y}{x - 6y}$ **13** $\dfrac{u^2 + 5u + 6}{u^2 - 4}$

14 $\dfrac{p^2 - 7p + 12}{p^2 - 16}$ **15** $\dfrac{q^2 + 10q + 25}{q^2 - 25}$ **17** $\dfrac{x^2 + 4x + 3}{(x - 2)(x + 1)}$ **18** $\dfrac{y^2 + 5y + 6}{(y - 1)(y + 3)}$

19 $\dfrac{2y^2 - 15y + 28}{(y + 3)(y - 4)}$ **21** $\dfrac{x - 1}{x + 5}$ **22** $\dfrac{x - 2}{x + 2}$ **23** $\dfrac{x + 3}{x + 1}$ **25** $4xy^2$ **26** $2y/x$

27 $3x^2/5y$ **29** $\dfrac{x - 2}{x - 1}$ **30** $\dfrac{x - 1}{x - 3}$ **31** $\dfrac{2x - 1}{2x + 1}$ **33** $\dfrac{(x - 2)^2}{(x - 1)(x + 3)}$

34 $\dfrac{x - 3}{x - 1}$ **35** $\dfrac{x - 2}{x + 1}$ **37** $\dfrac{c - d}{c + 2d}$ **38** $\dfrac{a + b}{a - b}$ **39** $\dfrac{c + d}{c - 2d}$ **41** $\dfrac{x + y}{x^2 + xy + y^2}$

42 $\dfrac{x - y}{x^2 - xy + y^2}$ **43** $\dfrac{x^4 + x^2y^2 + y^4}{x^2 + y^2}$ **45** $\dfrac{1}{x + 1}$ **46** $\dfrac{1}{x + 1}$ **47** $\dfrac{1}{2x + 1}$

49 $\dfrac{1}{2x + 1}$ **50** $\dfrac{1}{x - 1}$ **51** $\dfrac{1}{x + 1}$ **53** $\left\{ \dfrac{3xy}{x^2y}, \dfrac{2x}{x^2y}, \dfrac{4y}{x^2y} \right\}$ **54** $\left\{ \dfrac{y^2}{x^2y^2}, \dfrac{2xy}{x^2y^2}, \dfrac{3x^2}{x^2y^2} \right\}$

55 $\left\{ \dfrac{3y}{x^2y^2}, \dfrac{-2x}{x^2y^2}, \dfrac{xy}{x^2y^2} \right\}$ **57** $\left\{ \dfrac{2x^2 + xy - y^2}{x^2 - y^2}, \dfrac{x^2 - 3xy + 2y^2}{x^2 - y^2}, \dfrac{x + 2y}{x^2 - y^2} \right\}$

58 $\left\{ \dfrac{x^2 - y^2}{x^3 - y^3}, \dfrac{x^4 + x^2y^2 + y^4}{x^3 - y^3} \right\}$

59 $\left\{ \dfrac{x^2 - 2xy + y^2}{(x - y)(x + y)(x - 2y)}, \dfrac{x^2 + 2xy + y^2}{(x - y)(x + y)(x - 2y)}, \dfrac{x^2 - 4xy + 4y^2}{(x - y)(x + y)(x - 2y)} \right\}$

EXERCISE 4.2

1 $\frac{3}{8}$ **2** 1   **3** $\frac{3}{5}$ **5** $\dfrac{19x + 18}{18}$ **6** $-\dfrac{3x + 37}{30}$ **7** $\dfrac{6(-x + 2)}{7}$

9 $\dfrac{27b^2 - 6c^2 - 4a^2}{18abc}$ **10** $\dfrac{10c^2 - 9a^2 - 18b^2}{6abc}$ **11** $1/2b$ **13** $-a/7b$ **14** 0

15 $a/12b$ **17** $\dfrac{15x^2 + 9xy - 2y^2}{2y(3x + y)}$ **18** $\dfrac{3x^2 - 2y^2}{x(2x + y)}$ **19** $\dfrac{9x^2 + 4y^2}{2y(3x + 2y)}$

21 $\dfrac{6xy}{(x + y)(x - y)}$ **22** $\dfrac{5x^2 - 9xy + 3y^2}{(2x - y)(3x - y)}$ **23** $\dfrac{19x^2 - 62xy + 34y^2}{(3x - 8y)(2x - 5y)}$

25 $1/(r - s)$ **26** $(r + s)/s$ **27** $(r + s)/r$ **29** $\dfrac{2x + 3y}{y(x + y)}$ **30** $\dfrac{2x - 3y}{x(3x + 2y)}$

31 $\dfrac{2(x + 2y)}{y(x - 2y)}$ **33** $\dfrac{1}{(x - 2y)(x - y)}$ **34** $\dfrac{4(a + 2b)}{(a + b)(a - 2b)(a + 3b)}$

35 $\dfrac{7}{(a - 4b)(a + 5b)}$ **37** $\dfrac{-16x^3 - 29x^2y + 25xy^2 + 33y^3}{(3x + 2y)(4x - 3y)(2x - 3y)}$

38 $\dfrac{-5x^3 - 18x^2y - 18xy^2 + 13y^3}{(2x + 5y)(x + 3y)(x - 2y)}$ **39** $\dfrac{15x^3 + 20x^2y + 40xy^2 - 5y^3}{(x + 3y)(2x + y)(3x - y)}$

41 $\dfrac{x^2 + xy + 4y^2}{(x - y)(2x - y)(2x + 3y)}$ **42** $\dfrac{2(4x^2 + xy + y^2)}{(x + y)(x - 3y)(3x + y)}$

43 $\dfrac{x^2 - 9xy - 3y^2}{(x + 2y)(2x - y)(x - 2y)}$

EXERCISE 4.3

1 $\dfrac{y^2zw}{x^2}$ **2** $\dfrac{2xy^2z^3}{3w^5}$ **3** $\dfrac{25x^5y}{z^3w^2}$ **5** $\dfrac{w^4}{x^2y^5}$ **6** $\dfrac{4y^2z}{3w}$ **7** $\dfrac{2r}{9pq}$ **9** $\dfrac{37yx^7}{4z}$

10 $\dfrac{24a^2}{cd^2}$ **11** $\dfrac{5x^2z}{4y^4}$ **13** $\dfrac{z}{10}$ **14** $\dfrac{xz}{3y}$ **15** 1   **17** $\dfrac{y(2x - y)}{2x}$ **18** $\dfrac{y(x + 5)}{2}$

19 $\dfrac{x^2}{y(2x - 3y)}$ **21** $\dfrac{x^2}{2y}$ **22** $\dfrac{x^2}{y}$ **23** $xy^2(x + y)$ **25** $\dfrac{w^2}{z^2}$ **26** $\dfrac{q}{p^2}$

27 $\dfrac{3xz}{y^2(x - 7y)^2}$ **29** $\dfrac{y^3}{x}$ **30** $\dfrac{x(x + y)}{y - x}$ **31** $\dfrac{(x - y)(y + z)}{(x + z)(y - z)}$ **33** $2x$

34 $\dfrac{1}{x + 3y}$ **35** x **37** $\dfrac{x - 1}{x + 6}$ **38** $\dfrac{x}{2x + 3}$ **39** 1   **41** $x + 2$

42 $\dfrac{(x - 1)(3x - 1)}{(x + 1)(3x + 1)}$ **43** $\dfrac{x + 1}{x + 3}$

EXERCISE 4.4

1 $\frac{1}{4}$ **2** $\frac{27}{16}$ **3** $\frac{2}{23}$ **5** $\dfrac{x}{2x + 1}$ **6** $\dfrac{x - 3}{x}$ **7** $x + 4$ **9** $\dfrac{x^2 - 2}{x}$

10 $\dfrac{x^2 - 9}{5}$　　**11** $\dfrac{x}{5}$　　**13** $\dfrac{x - 5}{x - 4}$　　**14** $\dfrac{x - 3}{x + 2}$　　**15** $\dfrac{3x + 2}{x + 3}$　　**17** $\dfrac{y - x}{y + x}$

18 $\dfrac{y - x}{y + 2x}$　　**19** $\dfrac{3x + y}{2x + y}$　　**21** $\dfrac{a - 2}{a - 4}$　　**22** $\dfrac{2a + 3}{2a - 1}$　　**23** $\dfrac{2a + 3}{a - 2}$　　**25** $\dfrac{p}{p + 3}$

26 $\dfrac{1}{p + q}$　　**27** $\dfrac{2p}{3(1 - p)}$　　**29** $\dfrac{-1}{2x + 1}$　　**30** $\dfrac{x + 3}{-2(x + 1)}$　　**31** $\dfrac{-1}{x + 1}$

33 $\dfrac{y + 2x}{2x^2}$　　**34** $\dfrac{x - 1}{2y}$　　**35** $\dfrac{v}{w}$

EXERCISE 4.5

1 a^2/ab　　**2** ab/b^2　　**3** $b/3$　　**5** $\dfrac{2y - x}{y - 2x}$　　**6** $\dfrac{2y - x}{-x - 2y}$　　**7** $\dfrac{(a + 7)^2}{a^2 - 49}$

9 $\dfrac{3y^2 - 10y + 3}{(y + 2)(y - 3)}$　　**10** $\dfrac{x - 3}{x - 4}$　　**11** $2x/5y^2$　　**13** $\dfrac{x + 1}{x + 4}$　　**14** $x^2 - 4xy + 16y^2$

15 $\dfrac{2x + 1}{2x - 1}$　　**17** $\dfrac{x^2 + xy + y^2}{x + y}$　　**18** $\dfrac{x^3 + y^3}{x^6 + x^3y^3 + y^6}$　　**19** $\dfrac{1}{x + 1}$

21 $\left\{\dfrac{2x^2 - 7x + 3}{(x + 1)(x - 3)(2x - 1)}, \dfrac{6x^2 + 3x - 3}{(x + 1)(x - 3)(2x - 1)}, \dfrac{2x^2 - 4x - 6}{(x + 1)(x - 3)(2x - 1)}\right\}$

22 $\left\{\dfrac{3x^2 - 4x + 1}{(2x - 3)(x + 4)(3x - 1)}, \dfrac{4x^2 - 16x + 15}{(2x - 3)(x + 4)(3x - 1)}, \dfrac{3x^2 + 14x + 8}{(2x - 3)(x + 4)(3x - 1)}\right\}$

23 $-\frac{1}{10}$　　**25** $\dfrac{10a^2 - 6b^2 + 25c^2}{30abc}$　　**26** $\dfrac{2x^2 + 3xy - 3y^2}{3y(x - 3y)}$　　**27** $\dfrac{x^2 + 6xy - y^2}{(x - 2y)(x + y)}$

29 $\dfrac{x^2 + 3xy + 4y^2}{y(x^2 - 4y^2)}$　　**30** $\dfrac{x - 4}{(x - 2)(x + 1)(x + 2)}$　　**31** $\dfrac{-46x^3 + 47x^2y + 55xy^2 - 3y^3}{(4x + 3y)(2x + 3y)(3x - 2y)}$

33 $\dfrac{2xy}{w}$　　**34** $\dfrac{3w^2z^2}{2xy}$　　**35** $\dfrac{4zw^2}{y^2}$　　**37** $\dfrac{x(x + y)}{3(y + z)}$　　**38** $\dfrac{y}{x}$　　**39** $\dfrac{2x}{y}$　　**41** $\dfrac{x + y}{x}$

42 $\dfrac{x + 2y}{x + y}$　　**43** $\dfrac{x - 1}{x + 2}$　　**45** $\dfrac{5}{2}$　　**46** $\dfrac{2x + 5}{x}$　　**47** $\dfrac{2x - 5}{10}$　　**49** $\dfrac{x - 2}{x + 2}$

50 $\dfrac{(x + 1)(3x - 8)}{(x - 1)^2}$　　**51** $\dfrac{x + 3}{2x + 1}$　　**53** $\dfrac{4x - 5y}{(2x - 3y)(x + y)}$　　**54** $\dfrac{a}{2a - 1}$　　**55** $\dfrac{x + 2y}{x - 2y}$

EXERCISE 5.1

1 $\{2\}$　　**2** $\{3\}$　　**3** $\{7\}$　　**5** $\{-2\}$　　**6** $\{-1\}$　　**7** $\{3\}$　　**9** $\{1\}$　　**10** $\{-1\}$
11 $\{-3\}$　　**13** $\{10\}$　　**14** $\{8\}$　　**15** $\{12\}$　　**17** $\{6\}$　　**18** $\{12\}$　　**19** $\{18\}$
21 $\{\frac{1}{5}\}$　　**22** $\{\frac{1}{3}\}$　　**23** $\{36\}$　　**25** $\{6\}$　　**26** $\{4\}$　　**27** $\{5\}$　　**29** $\{4\}$
30 $\{6\}$　　**31** $\{12\}$　　**33** $\{1\}$　　**34** $\{-2\}$　　**35** $\{-3\}$

37 $\left\{\dfrac{a + b}{a}\right\}$　　**38** $\{-1\}$　　**39** $\left\{\dfrac{a + b}{a}\right\}$　　**41** $\left\{\dfrac{a + b}{a - b}\right\}$　　**42** $\{2a + b\}$　　**43** $\left\{\dfrac{1}{b - a}\right\}$

EXERCISE 5.2

1 $\{x \,|\, x > 4\}$ **2** $\{x \,|\, x < 5\}$ **3** $\{x \,|\, x > 5\}$ **5** $\{x \,|\, x > 2\}$ **6** $\{x \,|\, x > 3\}$
7 $\{x \,|\, x > 1\}$ **9** $\{x \,|\, x < 3\}$ **10** $\{x \,|\, x < 2\}$ **11** $\{x \,|\, x < 5\}$ **13** $\{x \,|\, x < 3\}$
14 $\{x \,|\, x < 1\}$ **15** $\{x \,|\, x > -2\}$ **17** $\{x \,|\, x < -6\}$ **18** $\{x \,|\, x > 1\}$
19 $\{x \,|\, x > 1\}$ **21** $\{x \,|\, x < -6\}$ **22** $\{x \,|\, x < -1\}$ **23** $\{x \,|\, x > -7\}$
25 $\{x \,|\, x < -1\} \cup \{x \,|\, x > 3\}$ **26** $\{x \,|\, x < 2\} \cup \{x \,|\, x > 4\}$
27 $\{x \,|\, x > -2\} \cup \{x \,|\, x < -3\}$ **29** $\{x \,|\, -\frac{3}{2} < x < \frac{2}{5}\}$ **30** $\{x \,|\, -\frac{2}{7} < x < \frac{7}{2}\}$
31 $\{x \,|\, \frac{3}{5} > x > -\frac{7}{2}\}$ **33** $\{x \,|\, x < -2\} \cup \{x \,|\, x > \frac{1}{2}\}$ **34** $\{x \,|\, x < -1\} \cup \{x \,|\, x > \frac{5}{3}\}$
35 $\{x \,|\, x < -3\} \cup \{x \,|\, x > \frac{5}{3}\}$ **37** $\{x \,|\, x < 1\} \cup \{x \,|\, x > \frac{5}{3}\}$
38 $\{x \,|\, x < -1\} \cup \{x \,|\, x > 1\}$ **39** $\{x \,|\, -\frac{1}{2} < x < \frac{1}{2}\}$
41 $\{x \,|\, x > \frac{1}{2}\} \cup \{x \,|\, -\frac{4}{3} < x < -1\}$ **42** $\{x \,|\, x > 3\} \cup \{x \,|\, -\frac{3}{2} < x < -\frac{1}{3}\}$
43 $\{x \,|\, x > \frac{7}{3}\} \cup \{x \,|\, -2 < x < -\frac{3}{7}\}$ **45** $\{x \,|\, x < -2\} \cup \{x \,|\, -\frac{1}{2} < x < \frac{5}{3}\}$
46 $\{x \,|\, x < -2\} \cup \{x \,|\, \frac{1}{5} < x < \frac{8}{3}\}$ **47** $\{x \,|\, x < -\frac{2}{7}\} \cup \{x \,|\, \frac{7}{2} < x < 4\}$

EXERCISE 5.3

1 $\{3\}$ **2** $\{2\}$ **3** $\{1\}$ **5** $\{7\}$ **6** $\{6\}$ **7** $\{5\}$ **9** $\{7\}$ **10** $\{12\}$
11 $\{8\}$ **13** $\{6\}$ **14** $\{8\}$ **15** $\{3\}$ **17** $\{11\}$ **18** $\{8\}$ **19** $\{8\}$

29 $F = 1.8C + 32$ **30** $d = \dfrac{Ak}{4\pi C}$ **31** $f = \dfrac{pq}{p + q}$ **33** $p = \dfrac{c - m(1 - d)}{c}$

34 $r = \dfrac{Ne - IR}{IN}$ **35** $a = \dfrac{S(1 - r)}{1 - r^n}$

EXERCISE 5.4

1 $\{-2, 2\}$ **2** $\{0\}$ **3** $\{-3, 3\}$ **5** $\{0, 2\}$ **6** $\{-10, 4\}$ **7** $\{-10, 2\}$
9 $\{-2, 3\}$ **10** $\{-\frac{14}{3}, 2\}$ **11** $\{-3, \frac{9}{5}\}$ **13** $\{2, 5\}$ **14** $\{-2, \frac{8}{3}\}$
15 $\{0, -\frac{5}{2}\}$ **17** $\{-\frac{1}{2}\}$ **18** $\{2\}$ **19** all x **21** $\{\frac{1}{3}, 9\}$ **22** $\{-10, -\frac{4}{5}\}$
23 $\{-1, 1\}$ **25** $\{x \,|\, -2 < x < 2\}$ **26** $\{x \,|\, -5 < x < 5\}$
27 $\{x \,|\, x < -3\} \cup \{x \,|\, x > 3\}$ **29** $\{x \,|\, -7 < x < 1\}$ **30** $\{x \,|\, -4 < x < 8\}$
31 $\{x \,|\, 2 < x < 8\}$ **33** $\{x \,|\, x > 3\} \cup \{x \,|\, x < -2\}$ **34** $\{x \,|\, x > 2\} \cup \{x \,|\, x < \frac{4}{3}\}$
35 $\{x \,|\, x > -\frac{2}{3}\} \cup \{x \,|\, x < -2\}$ **37** $\{x \,|\, -1 < x < 5\}$ **38** $\{x \,|\, -3 < x < 2\}$
39 $\{x < 1 \cup x > \frac{7}{3}\}$ **41** $\{x \,|\, x < 1\}$ **42** $\{x \,|\, x < 2\}$ **43** $\{x \,|\, \frac{2}{3} < x < 4\}$
45 $\{x \,|\, x < -\frac{1}{2}\} \cup \{x \,|\, x > 5\}$ **46** $\{\varnothing\}$ **47** $\{x \,|\, x < -2\} \cup \{x \,|\, x > \frac{1}{3}\}$

EXERCISE 5.5

1 24, 25, 26 **2** $48, $64 **3** 23,112 **5** $215, $410 **6** 5,376 **7** 48
9 7 hours at $1.25, 8 hours at $1.50 **10** Tim, 5 hours; Todd, 7 hours; Rick,
6 hours **11** 1,400 mi **13** 680 mi **14** $15,500 at 5 percent, $16,250 at
6 percent **15** $728 on living-room carpet; $760 on bedroom carpet
17 $200, $175 **18** 78 **19** 11, 14 **21** 10 mi **22** 46 mi/hour
23 48 mi/hour **25** 10 mi **26** $1\frac{1}{2}$ hours **27** 18 min **29** $3\frac{1}{4}$ hours

30 $\frac{3}{4}$ hour   **31** $17\frac{1}{2}$ min   **33** 10 min   **34** 10 hours   **35** $19\frac{1}{5}$ hours
37 4 at \$3.94; 3 at \$1.89   **38** 1,125 lb of 9.3 percent; 3,375 lb of 11.3 percent
39 24 cu yd   **41** 50 milliliters   **42** 112 mi   **43** Bus, 40 mi/hour; plane,
660 mi/hour   **45** 20 mi   **46** 1 hour and 24 min   **47** \$80

EXERCISE 5.6

1 $\{3\}$ **2** $\{-2\}$ **3** $\{2\}$ **5** $\{6\}$ **6** $\{8\}$ **7** $\{15\}$ **9** $\{-5\}$ **10** $\{\frac{60}{7}\}$
11 $\{x \mid x > 3\}$ **13** $\{x \mid x < 1\}$ **14** $\{x \mid x < 4\}$ **15** $\{x \mid -\frac{7}{3} < x < \frac{1}{2}\}$
17 $\{x \mid x < -\frac{9}{5}\} \cup \{x \mid x > \frac{5}{6}\}$ **18** $\{x \mid -\frac{3}{7} < x < \frac{7}{3}\}$
19 $\{x \mid x < -4.5\} \cup \{x \mid -2 < x < 4\}$ **21** 4   **22** 3   **23** 6   **25** $\{1, -7\}$
26 $\{-3, 4\}$ **27** $\{-1, 1\}$ **29** $\{x \mid -7 < x < 1\}$ **30** $\{x \mid x < -3\} \cup \{x \mid x > 4\}$
31 $\{x \mid -1 < x < 1\}$ **33** $5(F - 32)/9$ **34** $(2S - na)/n$

EXERCISE 6.1

1 $\{(2, 1), (3, 3), (4, 5), (5, 7)\}$ **2** $\{(3, 8), (5, 6), (7, 4), (9, 2)\}$
3 $\{(m, l), (a, m), (t, o), (h, l)\}$ **5** Function; no two pairs have the same first
element and different second elements.   **6** Relation; some points have the
same first element and different second elements.   **7** Relation; some points
have the same first element and different second elements.   **9** $\{(2, 4), (3, 7),$
$(4, 10), (5, 13), (6, 16)\}$; function   **10** $\{(-2, 3), (-1, 0), (0, -1), (1, 0), (2, 3)\}$;
function   **11** $\{(3, 0), (4, 4), (5, 10), (6, 18)\}$; function   **13** $\{(3, 0), (4, 1),$
$(5, \sqrt{2}), (6, \sqrt{3}), (4, -1), (5, -\sqrt{2}), (6, -\sqrt{3})\}$; not a function   **14** $\{(1, 1),$
$(7, 2), (15, 4\sqrt{2}), (31, 4), (1, -1), (7, -2), (15, -4\sqrt{2}), (31, -4)\}$; not a function
15 $\{0, 4), (\sqrt{7}, 3), (4, 0), (0, -4), (\sqrt{7}, -3)\}$; not a function   **17** $-7, 2$
18 $-3, 2$ **19** $-1, 9$ **21** $10, 2b^2 - 3b + 1$ **22** $a^2 + a - 3, 9$
23 $2, -b^2 + b + 2$ **25** $7, 3h + 1, 3h + 7$ **26** $3, 2h - 3, 2h + 3$ **27** $h^2 - h + 1, 1,$
$h^2 + h + 1$ **29** 3   **30** $2hx + h^2 - h$ **31** $4hx + 2h^2 + 3h$ **33** $t^2, 16$
34 $6t, 0$ **35** $t^2 + 1, 2$ **37** $\{(2, 4), (3, 10), (4, 16), (5, 22)\}$ **38** $\{(11, -8),$
$(13, 2), (15, 11), (17, 22)\}$ **39** $\{(-1, 6), (-\frac{1}{2}, 1), (0, -2), (\frac{1}{2}, -3)\}$
41 $\{1, 3, 5, 9\}$ **42** $\{-2, 4, 7, 16\}$ **43** $\{1, 4, 9, 16, 25\}$ **45** $\{(2, 1)\}$
46 $\varnothing$ **47** $\{(1, -3), (1.5, -2.5)\}$

EXERCISE 6.2

5 The X axis   **6** The bisector of the first quadrant   **7** The bisector of the
second and fourth quadrants   **9** The ray parallel to the Y axis and extending
up from $(1, 2)$ **10** The ray parallel to the X axis and extending to the left
from $(1, 3)$ **11** The ray parallel to the Y axis and extending down from $(-2, 0)$
37 $x = \frac{4}{3}, y = -4$ **38** $x = -\frac{3}{2}, y = -3$ **39** $x = \frac{5}{3}, y = 5$ **41** $x = -6, x = 6,$
$y = 6$ **42** $x = 3, y = 3$ **43** $x = -4, y = -4$

EXERCISE 6.3

1 The set of points is $\{(1,2), (1,4), (1,6), (2,2), (2,4), (2,6), (3,2),$ $(3,4), (3,6)\}$.   **2** The set of points is $\{(1,3), (1,5), (3,5)\}$.   **3** The set of points is $\{(1,1)\}$.   **17** $\{(x,y) \mid y = \pm\sqrt{4 - x^2}\}$, $R = \{x \mid -2 \le x \le 2\}$; a relation
18 $\{(x,y) \mid y = \pm\sqrt{18 - x^2}/\sqrt{2}$, $R = \{x \mid -3\sqrt{2} \le x \le 3\sqrt{2}\}$; a relation
19 $\{(x,y) \mid y = \pm\sqrt{16 + x^2}$, $R = \{x \mid -2\sqrt{5} \le x \le 2\sqrt{5}\}$; a relation
21 $\{(x,y) \mid y = x + 1\}$, $R = \{x \mid x \ge 1\}$; a function   **22** $\{(x,y) \mid y = (x - 3)/2\}$,
$R = \{x \mid x \ge 1\}$; a function   **23** $\{(x,y) \mid y = -x + 2\}$, $R = \{x \mid x \ge 1\}$; a function
25 $\{(x,y) \mid y = 3/(x - 1)\}$, $R = \{x \mid 1 < x < 4\}$; a function
26 $\{(x,y) \mid y = 2/(1 - x)\}$, $R = \{x \mid \frac{1}{2} < x < 1\}$; a function
27 $\{(x,y) \mid y = 2x/(x - 1)\}$, $R = \{x \mid 1 < x < 2\}$; a function   **29** 6   **30** $\frac{8}{1}$
31 $\frac{4}{25}$ **33** 145 lb   **34** 30 amperes   **35** The first produces $\frac{4}{9}$ as much
illumination as the second   **37** 54   **38** \$62.50   **39** 8 in.³

EXERCISE 6.4

1 Yes, only one value of y for each x **2** No, more than one value of y for some values of x **3** $\{(-2,-3), (-1,-1), (0,1), (1,3), (2,5), (3,7)\}$;
a function   **5** $\{0,\pm5), (3,\pm4), (4,\pm3), (5,0)\}$; not a function
6 $-7, -3, 3$ **7** 8, 4, 13   **9** $3h$ **10** $4t - 9, -1$ **11** $t^2 + 7t + 9, 53$
13 $\{0, 1, 4, 36, 81\}$ **14** $\{(8, 19)\}$ **15** $\{(1, 4), (-2, 19)\}$ **17** To the right
from $(-2, 2)$ and parallel to the X axis   **26** 128   **27** 1,875 lb   **29** $\frac{1}{8}$ in.
30 1.04 lb/sq in.   **31** 300 ergs

EXERCISE 7.1

1 128   **2** 243   **3** 15,625   **5** 5   **6** 27   **7** 1,296   **9** 64   **10** 4,096
11 6,561   **13** $\frac{81}{256}$ **14** $\frac{8}{125}$ **15** $\frac{8}{343}$ **17** 46,656   **18** 26,873,856
19 8,000,000   **21** $6x^5y^3$ **22** $-10x^2y^6$ **23** $12x^3y^5z^3$ **25** $2xy^4$ **26** $2x^2y$
27 $6xy$ **29** $8a^6b^3c^9$ **30** $9a^6b^4c^8$ **31** $625a^4b^8c^{16}$ **33** $a^6b^4/4c^6d^8$
34 $a^{12}b^9c^3/8d^6$ **35** $4x^8y^6/9w^4z^2$ **37** $324a^8b^6$ **38** $108a^{13}b^9$ **39** $200c^{10}d^{13}$
41 $uv/2w^2$ **42** $3b^6/4c$ **43** $4y^6/11x^6z^5$ **45** ac^2d^3 **46** c^4/b^{10} **47** $b^7c^6/108a^5$
49 $24b/cd^2$ **50** $16a^6b^4/c^2$ **51** $8z^{12}/3x^7y^2$ **53** b^7/c^2a^4 **54** $1/b^2c$ **55** c^2b^6/a^2d^7
57 $x^{2a-1}y^4$ **58** x^{a-5}/y^7 **59** xy^{2b-6} **61** a^9d^{3b} **62** a^4b^{2n-8} **63** $a^{5n}b^{7+2n}$
65 $1/a^{bc}$ **66** a^{2c} **67** $a^{2p-n}b^{2n+p}$

EXERCISE 7.2

1 $\frac{1}{9}$ **2** $\frac{1}{4}$ **3** $\frac{1}{16}$ **5** $\frac{1}{64}$ **6** $\frac{1}{64}$ **7** $\frac{1}{3}$ **9** 8   **10** 9   **11** $\frac{1}{8}$ **13** $\frac{1}{8}$
14 $\frac{1}{8}$ **15** 81   **17** $\frac{1}{5184}$ **18** $\frac{1}{144}$ **19** $\frac{81}{4}$ **21** $\frac{1}{5832}$ **22** $\frac{9}{2}$ **23** $\frac{1}{16}$
25 $\frac{16}{9}$ **26** $\frac{729}{8}$ **27** 5,184   **29** a^2b^2 **30** a^3b^{-1} **31** a^2b^{-3} **33** $a^{-1}bc^{-2}$
34 $a^{-3}b^{-6}c^{-6}$ **35** $3ab^5c^{-3}$ **37** b/a **38** $1/c^2d$ **39** $1/b^3$ **41** b^2/a

42 $1/a^3b^2$ **43** a^2b^3 **45** $\dfrac{18}{p^5q^2}$ **46** $\dfrac{q^2}{3p}$ **47** $\dfrac{6b^3}{a^5}$ **49** b^6/a^4 **50** $a^{12}b^3$

51 b^2/a^6 **53** $1/a^2b^4$ **54** a^4/b^2 **55** b^3/a^9 **57** $\dfrac{n}{a^4t^3}$ **58** $\dfrac{u^{10}}{b^6g^6}$ **59** $\dfrac{a^6}{s^8t^6}$

61 $2b^2$ **62** $1/a^2$ **63** $-c^3$ **65** $\dfrac{y^3-x^3}{xy}$ **66** $\dfrac{1}{y^2-x^2}$ **67** $\dfrac{1}{y+x}$

69 $y+x$ **70** $x+y$ **71** $1/(x+2y)$ **73** $\dfrac{-x-3}{(x-1)^3}$ **74** $\dfrac{-2x^2+3x-1}{(x-2)^4}$

75 $\dfrac{16x+2}{(4x-1)^2(2x+1)^2}$ **77** $\dfrac{9x+7}{(3x-2)^3(2x+3)^2}$ **78** $\dfrac{18x-8}{(3x-1)^3(1-2x)^2}$

79 $\dfrac{-51x+68}{(1-5x)^3}$

EXERCISE 7.3

1 8   **2** 4   **3** 2   **5** 0.5   **6** 0.6   **7** 0.2   **9** 4   **10** 8   **11** 27

13 0.008   **14** 0.09   **15** 0.027   **17** $\frac{1}{16}$ **18** $\frac{1}{4}$ **19** $\frac{1}{8}$ **21** $\frac{216}{125}$

22 $\frac{4}{9}$ **23** $\frac{27}{125}$ **25** $\frac{125}{216}$ **26** $\frac{9}{4}$ **27** $\frac{9}{16}$ **29** 7   **30** 5   **31** 2

33 16   **34** 512   **35** 16   **37** $4ab^2$ **38** $2a^2b^3$ **39** $2a^2b^3$ **41** $2a^2/b^3$

42 $2a^2/b^5$ **43** $3a^2/b^3$ **45** $\sqrt[3]{r}\sqrt[5]{s^2}$ **46** $\sqrt[4]{a^3}\sqrt[3]{b}$ **47** $\sqrt[7]{a^2b^3}$ **49** $\sqrt[5]{x^3/y^2}$

50 $\sqrt[4]{y^3/x}$ **51** $\sqrt[3]{x^2/y}$ **53** $a^{13/12}$ **54** $b^{7/6}$ **55** $c^{17/20}$ **57** $p^{2/21}$ **58** $q^{2/9}$

59 $r^{1/10}$ **61** $10x^{5/6}$ **62** $6x^{7/12}$ **63** $12x^{7/10}$ **65** $\dfrac{3y^{1/15}}{x^{1/4}}$ **66** $3x^{1/9}y$

67 $\dfrac{4x^{1/3}}{y^{3/2}}$ **69** $3x^2y^3$ **70** $4y^4$ **71** $2a^3$ **73** $\dfrac{5a^{1/7}}{b}$ **74** $\dfrac{a^2}{3b^{1/5}}$ **75** $\dfrac{a^3}{3}$

77 $\dfrac{4x^6}{9y^{8/9}}$ **78** $\dfrac{x^{1/6}}{3y^{1/12}}$ **79** $\dfrac{3y^{1/6}}{x^{3/4}}$ **81** $\frac{3}{2}$ **82** 20   **83** $\dfrac{2}{27a^8b^2}$ **85** $a^{3/2}-b$

86 $4a-b^{2/5}$ **87** $x-y$ **89** $\dfrac{3x+4}{(2x+3)^{1/2}}$ **90** $\dfrac{2-3x}{(5-2x)^{1/2}}$ **91** $\dfrac{21x-24}{(2x-5)^{1/4}}$

93 $\dfrac{14x+5}{(x+1)^{1/2}(2x-1)^{1/3}}$ **94** $\dfrac{13x+10}{(x+1)^{1/2}(3x+2)^{1/3}}$ **95** $\dfrac{9x+8}{(3x-1)^{1/2}(2x+3)^{3/4}}$

97 $x^{4/b}$ **98** x^a **99** x^{4ab} **101** x **102** x **103** x^2

EXERCISE 7.4

1 $3\sqrt{2}$ **2** $5\sqrt{2}$ **3** $5\sqrt{3}$ **5** $2\sqrt[3]{6}$ **6** $5\sqrt[3]{2}$ **7** $2\sqrt[4]{5}$ **9** $3ab^2\sqrt{b}$

10 $2ab^3\sqrt{2ab}$ **11** $4b\sqrt{6ab}$ **13** $2a^3b^4\sqrt{5a}$ **14** $4a^5b^4\sqrt{7ab}$ **15** $6b^4\sqrt{3b}$

17 $2a\sqrt[3]{2ab}$ **18** $3a^2b\sqrt{4b}$ **19** $5a^2b^2\sqrt[3]{3ab^2}$ **21** $2a^2b\sqrt[4]{3b}$ **22** $3ab\sqrt[4]{2b^3}$

23 $2ab\sqrt[5]{3ab^2}$ **25** $\dfrac{a^3}{2b^2}\sqrt{6a}$ **26** $\dfrac{2r^2}{3s^3}\sqrt{3r}$ **27** $\dfrac{4p}{5q^2}\sqrt{5p}$ **29** $\dfrac{b}{6a^2}\sqrt{21a}$

30 $\dfrac{2p^2}{5r^3}\sqrt{3r}$ **31** $\dfrac{3a^2}{2b^2}\sqrt[3]{ab}$ **33** $r^a s^a \sqrt[3]{s^a}$ **34** $r^a s^a \sqrt[4]{s^{3a}}$ **35** $p^{2a}q^{3a}$

37 $\sqrt{45}$ **38** $\sqrt{24}$ **39** $\sqrt[3]{81}$ **41** $\sqrt[3]{24x^4 y}$ **42** $\sqrt[4]{162x^6 y}$ **43** $\sqrt{12x^4 y^5}$

45 9 **46** 25 **47** 10 **49** 2 **50** 6 **51** $3\sqrt[5]{5}$ **53** $21x^2 y^4$

54 $5x^2 y^3 \sqrt{6}$ **55** $20x^2 y^3$ **57** $3xy\sqrt[3]{2y^2}$ **58** $4x^2 y^2 \sqrt[3]{5y^2}$ **59** $2x^2 \sqrt[4]{7y^3}$

61 $2x^2 \sqrt[5]{9y^4}$ **62** $6x^2 y \sqrt[5]{2y^2}$ **63** $2xy\sqrt[6]{27x^2}$ **65** $\sqrt{6}/2$ **66** $\sqrt[3]{10}/2$

67 $\sqrt[4]{6}/3$ **69** $x\dfrac{\sqrt{6xy}}{2y}$ **70** $\dfrac{\sqrt{14xy}}{7y^2}$ **71** $\dfrac{2x^2\sqrt{xy}}{5y^4}$ **73** $\dfrac{3x^3 y^2}{2}$ **74** $\dfrac{9x^2 y\sqrt{xy}}{7}$

75 $2r^4 y^3 \sqrt{y}$ **77** $\dfrac{3a^3 b^2 \sqrt[3]{b}}{4}$ **78** $\dfrac{ab\sqrt[3]{6ab}}{5}$ **79** $\dfrac{ab^2 \sqrt[4]{40}}{3}$ **81** $\sqrt{2}-1$

82 $\dfrac{3(\sqrt{5}-1)}{4}$ **83** $\dfrac{-5-3\sqrt{3}}{2}$ **85** $-7-3\sqrt{6}$ **86** $\dfrac{10-\sqrt{3}+\sqrt{5}-2\sqrt{15}}{2}$

87 $\sqrt{2}$

EXERCISE 7.5

1 $\sqrt[4]{3}$ **2** $\sqrt[6]{7}$ **3** $\sqrt[12]{5}$ **5** $\sqrt[6]{2a}$ **6** $\sqrt[9]{6b}$ **7** $\sqrt[15]{3c}$ **9** $\sqrt[4]{4x^2}$

10 $\sqrt[6]{25y^2}$ **11** $\sqrt[9]{125y^3}$ **13** $\sqrt[12]{81x^8 y^4}$ **14** $\sqrt[8]{16x^{12}y^8}$ **15** $\sqrt[8]{4x^2 y^6}$

17 $\sqrt[6]{a^3}, \sqrt[6]{a^2}, \sqrt[6]{a}$ **18** $\sqrt[12]{a^6}, \sqrt[12]{a^4}, \sqrt[12]{a^3}$ **19** $\sqrt[12]{16a^4 c^4}, \sqrt[12]{8a^3 c^6}, \sqrt[12]{4a^4 c^2}$

21 $\sqrt{4xy^2}$ **22** $\sqrt{2x^2 y}$ **23** $\sqrt[3]{6x^3 y^2}$ **25** $\underline{\sqrt{3x^2 y}}$ **26** $\sqrt[4]{4x^2 y^3}$ **27** $\sqrt[3]{2x^2 y^4}$

29 $4\sqrt{3}$ **30** $5\sqrt{2}$ **31** $\sqrt{3}$ **33** $4\sqrt[3]{3}$ **34** $\sqrt[3]{6}$ **35** $8\sqrt[4]{2}$

37 $8\sqrt[3]{2}-2\sqrt{2}$ **38** $13\sqrt{2}+\sqrt[3]{3}$ **39** $\sqrt{3}+13\sqrt{2}$ **41** $(2a-ab)\sqrt{ab}+b\sqrt[3]{a}$

42 $(2t+3s+5st^2)\sqrt{s}$ **43** $(2x-5xy+3y^2 z)\sqrt{3x}$ **45** $(5+2x)\sqrt{3x}/x$

46 $4\sqrt{x}/x$ **47** $2c\sqrt{2c}/d$ **49** $a\sqrt{b}+3b\sqrt{a}$

50 $(2b^2+b)\sqrt{5ab}+(b^2+ab)\sqrt{a}$ **51** $ab(3-2ab)\sqrt[3]{ab^2}+b(4ab-3)\sqrt[3]{a^2 b}$

53 $3\sqrt{2ab}$ **54** $3a\sqrt[3]{ab}$ **55** $-2ab\sqrt{b}$

EXERCISE 7.6

1 729 **2** 36 **3** $\frac{16}{81}$ **5** 1,259,712 **6** $15x^2 y^7$ **7** $4x^2 y^3$ **9** $8a^9 b^6$

10 a^2/b^4 **11** $a^{12}bc^6$ **13** $a^{3n}b^{n+5}$ **14** $\frac{1}{9}$ **15** 9 **17** $\frac{1}{2}$ **18** $a^3 b^2$

19 $1/a^3 b^4 c^3$ **21** $a^6 b^9$ **22** 5 **23** 0.8 **25** 3 **26** $5ab^3$ **27** $2a/b^2$

29 $6x^{7/12}$ **30** $15x^{8/15}$ **31** $5xy^2$ **33** $2/3y$ **34** x

35 $\dfrac{30x-19}{(2x-3)^{2/3}(3x+2)^{1/2}}$ **37** $4\sqrt{5}$ **38** $2\sqrt[3]{5}$ **39** $2\sqrt[5]{3}$ **41** 6

42 $a\sqrt[5]{a}$ **43** $10xy^2$ **45** $3x^4/4y^2$ **46** $4\sqrt{2}-5$ **47** $\sqrt[3]{3}$ **49** $\sqrt[5]{x^2}$

50 2 **51** 0 **53** $(3b+2a+3ab)\sqrt{ab}$ **54** $\sqrt{3ab}$ **55** -4

57 $u^{3/2}+v^{3/2}$ **58** $a-b+c+2\sqrt{ac}$

EXERCISE 8.1

1 $\{1, -1\}$ **2** $\{3, -3\}$ **3** $\{4, -4\}$ **5** $\{\frac{1}{2}, -\frac{1}{2}\}$ **6** $\{\frac{1}{3}, -\frac{1}{3}\}$ **7** $\{\frac{1}{5}, -\frac{1}{5}\}$
9 $\{\frac{3}{2}, -\frac{3}{2}\}$ **10** $\{\frac{2}{5}, -\frac{2}{5}\}$ **11** $\{\frac{9}{4}, -\frac{9}{4}\}$ **13** $\{2, 1\}$ **14** $\{3, 2\}$
15 $\{1, 3\}$ **17** $\{2, -3\}$ **18** $\{1, -2\}$ **19** $\{-2, -5\}$ **21** $\{3, \frac{1}{2}\}$
22 $\{2, \frac{1}{3}\}$ **23** $\{4, \frac{1}{5}\}$ **25** $\{2, -\frac{1}{2}\}$ **26** $\{3, -\frac{1}{3}\}$ **27** $\{5, -\frac{1}{5}\}$
29 $\{4, \frac{2}{3}\}$ **30** $\{3, \frac{2}{5}\}$ **31** $\{6, \frac{2}{5}\}$ **33** $\{5, -\frac{2}{7}\}$ **34** $\{3, -\frac{3}{4}\}$
35 $\{2, -\frac{2}{3}\}$ **37** $\{-3, \frac{2}{5}\}$ **38** $\{-2, \frac{3}{5}\}$ **39** $\{-1, \frac{2}{3}\}$ **41** $\{\frac{2}{3}, \frac{3}{2}\}$
42 $\{\frac{2}{3}, -\frac{3}{2}\}$ **43** $\{\frac{3}{4}, \frac{4}{3}\}$ **45** $\{\frac{2}{3}, \frac{3}{5}\}$ **46** $\{\frac{2}{7}, \frac{2}{5}\}$ **47** $\{\frac{5}{8}, \frac{3}{2}\}$
49 $\{\frac{4}{5}, -\frac{2}{3}\}$ **50** $\{\frac{2}{5}, \frac{3}{2}\}$ **51** $\{\frac{3}{8}, -\frac{5}{9}\}$ **52** $\{-\frac{3}{7}, \frac{4}{9}\}$ **53** $\{2, a\}$
54 $\{3, -b\}$ **55** $\{\frac{1}{3}, -2a\}$ **57** $\{2a/5, 3a/2\}$ **58** $\{5b/2, 2c/3\}$
59 $\{2p/q, 2q/p\}$

EXERCISE 8.2

1 $\{2, 1\}$ **2** $\{3, 2\}$ **3** $\{5, 3\}$ **5** $\{3, -2\}$ **6** $\{-2, 1\}$ **7** $\{-4, -2\}$
9 $\{2, \frac{1}{2}\}$ **10** $\{3, \frac{2}{3}\}$ **11** $\{5, \frac{3}{4}\}$ **13** $\{-1, \frac{2}{3}\}$ **14** $\{2, -\frac{1}{3}\}$
15 $\{5, -\frac{2}{7}\}$ **17** $\{\frac{2}{5}, \frac{5}{3}\}$ **18** $\{\frac{3}{5}, \frac{5}{4}\}$ **19** $\{\frac{2}{7}, \frac{7}{2}\}$ **21** $\{\frac{7}{3}, -\frac{3}{2}\}$
22 $\{-\frac{3}{5}, -\frac{2}{3}\}$ **23** $\{-\frac{6}{5}, \frac{5}{6}\}$ **25** $\{1 + \sqrt{2}, 1 - \sqrt{2}\}$ **26** $\{2 + \sqrt{3}, 2 - \sqrt{3}\}$
27 $\{3 + \sqrt{5}, 3 - \sqrt{5}\}$ **29** $\{4 + \sqrt{2}, 4 - \sqrt{2}\}$ **30** $\{2 + \sqrt{5}, 2 - \sqrt{5}\}$
31 $\{-3 + \sqrt{3}, -3 - \sqrt{3}\}$ **33** $\{4 + 2i, 4 - 2i\}$ **34** $\{3 + i. 3 - i\}$
35 $\{5 + 3i, 5 - 3i\}$ **37** $\{-2 + 3i, -2 - 3i\}$ **38** $\{-1 + 4i, -1 - 4i\}$
39 $\{-7 + 2i, -7 - 2i\}$ **41** $\{a, b\}$ **42** $\{2a, 3b\}$ **43** $\{a, 2a\}$

45 $\{b/a, b/2a\}$ **46** $\{a/c, -2a/c\}$ **47** $\left\{\dfrac{a+bi}{2}, \dfrac{a-bi}{2}\right\}$

EXERCISE 8.3

1 $\{3, 1\}$ **2** $\{2, 5\}$ **3** $\{4, 2\}$ **5** $\{-6, 1\}$ **6** $\{7, -2\}$ **7** $\{-3, -4\}$
9 $\{3, \frac{2}{5}\}$ **10** $\{2, \frac{1}{3}\}$ **11** $\{2, \frac{3}{7}\}$ **13** $\{5, -\frac{1}{2}\}$ **14** $\{6, -\frac{2}{3}\}$
15 $\{3, -\frac{4}{3}\}$ **17** $\{\frac{2}{3}, \frac{3}{4}\}$ **18** $\{\frac{3}{4}, \frac{4}{5}\}$ **19** $\{\frac{4}{5}, \frac{3}{7}\}$ **21** $\{-\frac{2}{5}, \frac{3}{4}\}$
22 $\{-\frac{3}{4}, \frac{2}{3}\}$ **23** $\{-\frac{2}{3}, -\frac{1}{2}\}$ **25** $\{1 + \sqrt{3}, 1 - \sqrt{3}\}$ **26** $\{2 + \sqrt{5}, 2 - \sqrt{5}\}$

27 $\{3 + \sqrt{2}, 3 - \sqrt{2}\}$ **29** $\left\{\dfrac{2+\sqrt{3}}{2}, \dfrac{2-\sqrt{3}}{2}\right\}$ **30** $\left\{\dfrac{3+\sqrt{2}}{3}, \dfrac{3-\sqrt{2}}{3}\right\}$

31 $\left\{\dfrac{2+\sqrt{5}}{3}, \dfrac{2-\sqrt{5}}{3}\right\}$ **33** $\{2 + 3i, 2 - 3i\}$ **34** $\{3 + 2i, 3 - 2i\}$

35 $\{4 + 2i, 4 - 2i\}$ **37** $\left\{\dfrac{3+i}{2}, \dfrac{3-i}{2}\right\}$ **38** $\left\{\dfrac{2+i}{3}, \dfrac{2-i}{3}\right\}$

39 $\left\{\dfrac{3+2i}{5}, \dfrac{3-2i}{5}\right\}$ **41** $\left\{\dfrac{1+i\sqrt{3}}{2}, \dfrac{1-i\sqrt{3}}{2}\right\}$ **42** $\left\{\dfrac{2+i\sqrt{5}}{3}, \dfrac{2-i\sqrt{5}}{3}\right\}$

43 $\left\{\dfrac{3+i\sqrt{2}}{3}, \dfrac{3-i\sqrt{2}}{3}\right\}$ **45** $\{a, 2a\}$ **46** $\{a, -3a\}$ **47** $\{a, b\}$

49 $\left\{\dfrac{a+ib}{2}, \dfrac{a-ib}{2}\right\}$ **50** $\left\{\dfrac{2a+ib}{3}, \dfrac{2a-ib}{3}\right\}$ **51** $\left\{\dfrac{3a+2ib}{4}, \dfrac{3a-2ib}{4}\right\}$

53 $\{x-1, x-2\}$ **54** $\{x+1, x-3\}$ **55** $\{x+2, x-3\}$ **57** $\{2x-1, x+2\}$
58 $\{3x+2, x-1\}$ **59** $\{3x-2, x+1\}$ **61** $\{3.732, 0.268\}$
62 $\{4.414, 1.586\}$ **63** $\{7.236, 2.764\}$ **65** $\{2.207, 0.793\}$
66 $\{1.244, 0.089\}$ **67** $\{1.411, 0.089\}$

EXERCISE 8.4

1 $\{-2, -1, 1, 2\}$ **2** $\{-3, -2, 2, 3\}$ **3** $\{-4, -1, 1, 4\}$ **5** $\{-2, 2, -4i, 4i\}$
6 $\{-1, 1, -2i, 2i\}$ **7** $\{-3, 3, -5i, 5i\}$ **9** $\{-\frac{1}{2}, -\frac{1}{3}, \frac{1}{3}, \frac{1}{2}\}$
10 $\{-\frac{1}{2}, -\frac{1}{5}, \frac{1}{5}, \frac{1}{2}\}$ **11** $\{-\frac{2}{3}, \frac{2}{3}, -i/2, i/2\}$ **13** $\{1, 2\}$ **14** $\{2, 3\}$
15 $\{-3, 1\}$ **17** $\{-2, -1, 1, 2, -2i, -i, i, 2i\}$

18 $\{-3, -1, 1, 3, -3i, -i, i, 3i\}$ **19** $\left\{-\frac{i}{2}, -1, 1, \frac{1}{2}, -\frac{i}{2}, -i, i, \frac{i}{2}\right\}$

21 $\{2, -2, \frac{3}{2}, -\frac{3}{2}\}$ **22** $\{\frac{4}{3}, -\frac{4}{3}, \frac{5}{2}, -\frac{5}{2}\}$ **23** $\{2i/5, -2i/5, -2, 2\}$ **25** $\{\pm 8, \pm 1\}$
26 $\{\pm 27, \pm 8\}$ **27** $\{\pm 64, \pm i\}$ **29** $\{\pm 32, \pm 243i\}$ **30** $\{\pm i, \pm 1,024\}$
31 $\{\pm 32i, \pm 3, 125\}$ **33** $\{\pm 8, \pm 27\}$ **34** $\{\pm \frac{27}{8}, \pm 64\}$ **35** $\{\pm \frac{8}{125}, \pm \frac{27}{8}\}$
37 $\{2, -2, 3, -3\}$ **38** $\{i\sqrt{2}, -i\sqrt{2}, 2, -2\}$ **39** $\{3, -3, 2, -2\}$
41 $\{1, 2, 4, -1\}$ **42** $\{1, 3, 2\}$ **43** $\{-3, -2, -4, -1\}$ **45** $\{-10, 4\}$
46 $\{-1, 0\}$ **47** $\{-9, -\frac{19}{7}\}$ **49** $\{-5, 1\}$ **50** $\{-6, 12\}$ **51** $\{-9, -\frac{1}{4}\}$

EXERCISE 8.5

1 2   **2** 1   **3** $\varnothing$ **5** 3   **6** $1, \frac{1}{5}$ **7** 4   **9** -2 **10** 0   **11** 5   **13** 4
14 3   **15** $\varnothing$ **17** 3   **18** -2 **19** 1   **21** 2   **22** 4   **23** -2
25 -3 **26** -2 **27** 1   **29** $4, 0$ **30** $7, -1$ **31** $5, 0$ **33** 2
34 $\frac{5}{16}$ **35** -2 **37** 1   **38** 2   **39** $4, \frac{5}{2}$ **41** 2   **42** 5   **43** 3
45 1   **46** 3   **47** 0   **49** 1   **50** 4   **51** 2   **53** 2   **54** 0   **55** 1
57 2   **58** 3   **59** -1 **61** 4   **62** $3, -6$ **63** -1 **65** $b, 2b$
66 a **67** $2t, 6t$

EXERCISE 8.6

1 Rational, unequal   **2** Rational, unequal   **3** Rational, unequal
5 Irrational, unequal   **6** Irrational, unequal   **7** Irrational, unequal
9 Irrational, unequal   **10** Irrational, unequal   **11** Irrational, unequal
13 Conjugate complex   **14** Conjugate complex   **15** Conjugate complex
17 Rational, equal   **18** Rational, equal   **19** Rational, equal
21 $-\frac{5}{3}, \frac{1}{3}$ **22** $-\frac{1}{5}, \frac{2}{5}$ **23** $\frac{3}{7}, \frac{4}{7}$ **25** $2, \frac{3}{2}$ **26** $-\frac{5}{4}, -\frac{1}{4}$
27 $\frac{1}{3}, \frac{5}{6}$ **29** $-1, 1/\sqrt{3}$ **30** $-3/\sqrt{2}, 2$ **31** $\sqrt{2}(1-\sqrt{2}), -3(1-\sqrt{2})$
33 $0, 8$ **34** $1, 5$ **35** $3, -4$ **37** $\frac{1}{4}$ **38** 1   **39** All k
41 $2, -2$ **42** $0, -2$ **43** $\frac{1}{2}$ **45** $x^2 - 7x + 12 = 0$ **46** $3x^2 - 5x - 2 = 0$
47 $4x^2 + 17x + 15 = 0$ **49** $x^2 - 4x + 13 = 0$ **50** $x^2 - 6x + 34 = 0$
51 $x^2 - 8x + 41 = 0$ **53** $x^2 - 8x + 11 = 0$ **54** $x^2 - 4x + 1 = 0$
55 $x^2 - 2x - 6 = 0$ **57** $3x - 1, 2x + 7$ **58** $5x + 2, 3x - 4$

59 $2x - 5, 7x + 3$ **61** $2x + 1 - \sqrt{6}, 2x + 1 + \sqrt{6}$ **62** $x - 2 + \sqrt{5}, x - 2 - \sqrt{5}$
63 $3x + 2 + \sqrt{5}, 3x + 2 - \sqrt{5}$ **65** $4x + 1 + 6i, 4x + 1 - 6i$
66 $7x + 2 + i, 7x + 2 - i$ **67** $3x + 4 + 2i, 3x + 4 - 2i$

EXERCISE 8.8

1 $\frac{3}{2}, -\frac{5}{4}$ **2** $-\frac{4}{3}, -6$ **3** $\frac{2}{5}, \frac{3}{2}$ **5** $\frac{2}{5}, -\frac{5}{3}$ **6** $-\frac{5}{4}, -\frac{2}{5}$ **7** $\frac{3}{2}, -\frac{5}{4}$
9 $(1 + 4i)/2, (1 - 4i)/2$ **10** $(4 + 3i)/3, (4 - 3i)/3$ **11** $2x - 3, -x - 2$
13 $-2, 2, -1, 1$ **14** $1, -3$ **15** $\sqrt{14}/14, -\sqrt{14}/14, \sqrt{10}/5, -\sqrt{10}/5$
17 $\frac{1}{8}, 27$ **18** $2, 1, 3$ **19** 2 **21** 5 **22** 2 **23** 3 **25** -2
26 $-\frac{7}{25}, 1$ **27** -3 **29** $\frac{3}{2}, -\frac{9}{2}$ **30** $-\frac{8}{5}, \frac{3}{5}$ **31** Rational, unequal
33 Irrational, unequal   **34** Conjugate complex   **35** $\frac{1}{4}$ **37** $x^2 - 6x + 4 = 0$

EXERCISE 9.2

1 $\{(5.2, 3.2), (0, 8, -1.2)\}$ **2** $\{(0, 0), (-2, -4)\}$ **3** $\{(6.9, 7.9), (0.1, 1.1)\}$
5 $\{(0.9, -0.9), (-0.9, 0.9)\}$ **6** $\{(0.5, -1.0), (-0.5, 1.0)\}$
7 $\{(0.9, 2.8), (-1.5, -2.0)\}$ **9** $\{(0.5, 1.0), (-0.5, -1.0)\}$ **10** $\{\varnothing\}$
11 $\{\varnothing\}$ **13** $\{(2.7, 1.3), (2.7, -1.3), (-2.7, 1.3), (-2.7, -1.3)\}$
14 $\{(1.8, 0.8), (1.8, -0.8), (-1.8, 0.8), (-1.8, -0.8)\}$ **15** $\{\varnothing\}$
17 $\{(0.2, 1.1), (0.2, -1.1)\}$ **18** $\{(0.9, 1.3), (0.9, -1.3)\}$
19 $\{(2.2, -4.8), (-2.2, -4.8)\}$ **21** $\{(2.2, 0.2)\}$ **22** $\{(0.7, -1.4)\}$
23 $\{(12.0, -7.0)\}$ **25** Dependent   **26** Dependent   **27** Dependent
29 $\{(1.5, -0.2)\}$ **30** $\{(3.2, 0.4)\}$ **31** $\{(1.8, -0.6)\}$ **33** Inconsistent
34 Inconsistent   **35** Inconsistent

EXERCISE 9.3

1 $(2, 1)$ **2** $(3, 1)$ **3** $(1, 2)$ **5** $(1, -1)$ **6** $(3, -2)$ **7** $(5, -3)$
9 $(\frac{1}{2}, \frac{1}{3})$ **10** $(\frac{1}{3}, -\frac{1}{2})$ **11** $(-1, -3)$ **13** $(\frac{1}{2}, \frac{1}{5})$ **14** $(\frac{1}{3}, \frac{2}{5})$
15 $(\frac{1}{4}, -\frac{1}{3})$ **17** $(1, 0), (-1, 0)$ **18** $(1, 1), (1, -1), (-1, 1), (-1, -1)$
19 $(1, 2), (1, -2), (-1, 2), (-1, -2)$ **21** $(1, 1), (1, -1), (-1, 1), (-1, -1)$
22 $(2, 3), (2, -3), (-2, 3), (-2, -3)$ **23** $(3, 0), (-3, 0)$
25 $(2, 1), (2, -1), (-\frac{11}{3}, \sqrt{26}/3), (-\frac{11}{3}, -\sqrt{26}/3)$ **26** $(3, 1), (3, -1)$
27 $(1, 1), (1, -1)$ **29** $(2, 5), (-2, 5)$ **30** $(1, 2), (-1, 2)$
31 $(2, 3), (-2, 3), (0, -1)$ **33** $(-1, 1), (-\frac{4}{5}, \frac{9}{5})$ **34** $(2, 5), (-7, -\frac{10}{7})$
35 $(-1, 1), (-\frac{1}{13}, \frac{109}{13})$ **37** $(1, 2), (-\frac{1}{6}, -\frac{3}{2})$ **38** $(-3, 2), (1, -2)$
39 $(1, 2), (-\frac{1587}{215}, -\frac{43}{5})$

EXERCISE 9.4

1 $\{(2, 1)\}$ **2** $\{(1, 2)\}$ **3** $\{(1, 3)\}$ **5** $\{(1, -2)\}$ **6** $\{(2, -1)\}$
7 $\{(2, 2)\}$ **9** $\{(2, 0)\}$ **10** $\{(3, -2)\}$ **11** $\{(4, -3)\}$

13 $\{(2,1),(\frac{5}{4},\frac{5}{2})\}$ **14** $\{(1,3),(\frac{6}{25},-\frac{4}{5})\}$ **15** $\{(1,1),(\frac{17}{16},\frac{5}{4})\}$

17 $\{(1,-1),(-\frac{17}{13},-\frac{3}{13})\}$ **18** $\{(2,-2),(-\frac{2}{3},-\frac{14}{3})\}$

19 $\{(1,-3),(\frac{31}{11},-\frac{23}{11})\}$ **21** $\{(3,2),(\frac{23}{3},-\frac{22}{3})\}$

22 $\{(-3,1),(-\frac{111}{43},-\frac{11}{43})\}$ **23** $\{(1,0),(-\frac{5}{3},-\frac{8}{9})\}$

25 $\{(10,2),(-2,-2),(5,\frac{3}{2}),(-4,-\frac{3}{2})\}$

26 $\{(2,3),(14,-3),(-6-2i\sqrt{5},i\sqrt{5}),(-6+2i\sqrt{5},-i\sqrt{5})\}$

27 $\{(20,2),(-4,-2),(9+9\sqrt{2},3\sqrt{2}/2),(9-9\sqrt{2},-3\sqrt{2}/2)\}$

29 $\{(4,-8),(-4,-24),(3i\sqrt{5}/2,\frac{45}{4}+3i\sqrt{5}),(-3i\sqrt{5}/2,\frac{45}{4}-3i\sqrt{5})\}$

30 $\{(0,0),(\sqrt{6},6+\sqrt{6}),(-\sqrt{6},6-\sqrt{6})\}$

31 $\{(4,1),(0,-1),(0,\frac{1}{3}),(-\frac{4}{3},-\frac{1}{3})\}$

33 $\{(-2,1),(-4,-1),(3+\sqrt{7},\sqrt{7}),(3-\sqrt{7},-\sqrt{7})\}$

34 $\left\{(10,2),(6,-2),\left(\dfrac{-17+i\sqrt{17}}{2},\dfrac{i\sqrt{17}}{2}\right),\left(\dfrac{-17-i\sqrt{17}}{2},\dfrac{-i\sqrt{17}}{2}\right)\right\}$

35 $\{(5,1),(-3,-1)(14+4\sqrt{14},\sqrt{14}),(14-4\sqrt{14},-\sqrt{14})\}$

37 $\{(1,2),(-1,-2),(2\sqrt{2},\sqrt{2}/2),(-2\sqrt{2},-\sqrt{2}/2)\}$

38 $\{(3,1),(-3,-1),(\sqrt{6}/2,\sqrt{6}),(-\sqrt{6}/2,-\sqrt{6})\}.$

39 $\{(1,2),(-1,-2),(-2i,i),(2i,-i)\}$

41 $\{(1,3),(-1,-3),(3\sqrt{6}/2,\sqrt{6}/3),(-3\sqrt{6}/2,-\sqrt{6}/3)\}$

42 $\{(2,-4),(-2,4),(2i\sqrt{3},4i\sqrt{3}/3),(-2i\sqrt{3},-4i\sqrt{3}/3)\}$

43 $\{(1,-3),(-1,3),(3i\sqrt{2}/2,i\sqrt{2}),(-3i\sqrt{2}/2,-i\sqrt{2})\}$

EXERCISE 9.5

1 $\{(3,3),(-3,-3),(2,-1),(-2,1)\}$

2 $\{(1,-1),(-1,1),(\sqrt{2}/2,-\sqrt{2}),(-\sqrt{2}/2,\sqrt{2})\}$

3 $\{(3,-1),(-3,1),(2i/3,5i/3),(-2i/3,-5i/3)\}$

5 $\{(3,-1),(-3,1),(4,2),(-4,-2)\}$ **6** $\{(1,1),(1,-2),(-1,-1),(-1,2)\}$

7 $\{(0,i),(0,-i),(2i,i),(-2i,-i)\}$ **9** $\{(3,2),(-3,-2),(1,-1),(-1,1)\}$

10 $\{(2,-1),(-2,1),(25i\sqrt{69}/69,14i\sqrt{69}/69),(-25i\sqrt{69}/69,-14i\sqrt{69}/69)\}$

11 $\{(2,-2),(-2,2),(8/\sqrt{41},2/\sqrt{41}),(-8/\sqrt{41},-2/\sqrt{41})\}$

13 $\{(4,-1),(-4,1),(\sqrt{2},\sqrt{2}),(-\sqrt{2},-\sqrt{2})\}$

14 $\{(2,-\frac{1}{3}),(-2,\frac{1}{3}),(1,-1),(-1,1)\}$

15 $\{(3,-\frac{1}{2}),(-3,\frac{1}{2}),(2,1),(-2,-1)\}$ **17** $\{(2,1),(-\frac{2}{5},\frac{1}{5})\}$

18 $\{(-1,2),(\frac{5}{2},-\frac{3}{2})\}$ **19** $\{(-2,-1),(\frac{1}{2},\frac{3}{2})\}$ **21** $\{(\frac{2}{3},-\frac{1}{3}),(\frac{1}{2},-\frac{1}{2})\}$

22 $\{(-1,-2),(-\frac{9}{13},-\frac{6}{13})\}$ **23** $\{(3,-1),(\frac{1}{2},\frac{3}{2})\}$ **25** $\{(2,1),(\frac{17}{5},-\frac{9}{5})\}$

26 $\{(3,2),(-1,-1)\}$ **27** $\{(4,1),(3,2)\}$ **29** $\{(4,1),(1,4),(2,5),(5,2)\}$

30 $\{(4,5),(5,4),(-3,6),(6,-3)\}$ **31** $\{(4,-2),(-2,4),(3,1),(1,3)\}$

33 $\{(3,2),(2,3),(3,1),(1,3)\}$ **34** $\{(-2,-6),(-6,-2),(4,0),(0,4)\}$

35 $\{(3,1),(1,3),(4,-2),(-2,4)\}$ **37** $\{(3,3)\}$ **38** $\{(2,-1),(-1,2)\}$

39 $\{(2,3),(3,2)\}$

EXERCISE 9.6

25 $(2, 2), (4, 0), (0, -3)$ **26** $(1, 3), (2, -1), (-3, 1)$
27 $(-1, 4), (3, 0), (1, -1)$ **29** $(-1, 4), (3, 0), (2, -3), (-5, -3)$
30 $(-4, 3), (1, 2), (2, -3), (-3, -4)$ **31** $(4, 1), (7, 0), (6, -4), (-3, -3)$

EXERCISE 9.7

1 $m = 1$ at $(0, 0)$, $M = 15$ at $(7, 0)$ **2** $m = -4$ at $(0, 2)$, $M = 9$ at $(1, 0)$
3 $m = 9$ at $(0, 1)$, $M = 19$ at $(-2, 1)$ **5** $m = -3$ at $(0, 0)$, $M = 5$ at $(0, 2)$
6 $m = 0$ at $(3, 1)$, $M = 12$ at $(-1, 1)$ **7** $m = -5$ at $(0, 2)$, $M = 5$ at $(1, -1)$
9 $m = 8$ at $(0, 0)$, $M = 19$ at $(1, -2)$ **10** $m = -13$ at $(-2, 2)$, $M = -1$ at $(0, 0)$
11 $m = 5$ at $(0, 1)$, $M = 17$ at $(2, -2)$ **13** $m = -\frac{40}{7}$ at $(\frac{23}{7}, -\frac{13}{7})$, $M = 16$ at $(2, 3)$
14 $m = -4$ at $(-1, 1)$, $M = 2$ at $(3, -1)$ **15** $m = -13$ at $(0, -3)$, $M = 4$ at $(2, 2)$
17 4 houses of type A, 14 houses of type B; profit \$9,600
18 5 houses of type A, 7 houses of type B; profit \$7,800
19 11 houses of type A, 22 houses of type B; profit \$19,800   **21** No rice,
100 acres of soybeans; profit \$4,700   **22** No rice, 100 acres of soybeans;
profit \$4,700   **23** No rice, 600/7 acres of soybeans; profit \$4,028.57   **25** No
pokers, 5 tongs   **26** peppers, 10 acres; rhubarb, 3 acres; tomatoes, 7 acres
27 A, 3 lb; B, 2 lb

EXERCISE 9.8

1 Sucker, \$0.10; apple, \$0.15   **2** 2 qt whipping cream, $\frac{1}{2}$ qt half-and-half
3 4, 2   **5** 54, 36   **6** 21 mi   **7** 10 one-bedroom, 7 two-bedroom   **9** 9, 5
10 1,880, 500   **11** 200 mi, 220 mi   **13** \$1, \$2   **14** 18, shorter hike; 17,
longer hike   **15** 6, 3   **16** \$65, \$80   **17** 333-6483   **18** Tom, 49 days;
Joe, 38 days   **19** Second, $2\frac{1}{2}$ hours; third, $1\frac{1}{2}$ hours   **21** Square, 15 by 15
ft; rectangle, 15 by 10 ft   **22** 40 rods, 40 rods, 70 rods   **23** Base, 6 by 6 ft;
depth, 5 ft   **25** 160 by 40 rods   **26** 800 by 300 rods, or 600 by 400 rods
27 80, \$12   **29** 14 by 20 ft, or $\frac{126}{47}$ by $\frac{2308}{47}$ ft   **30** \$2, \$3   **31** 16 in.,
14 in.; or 26.56 in., 7.28 in.

EXERCISE 9.9

9 $\{(0, 0), (1, 2)\}$ **10** $\{(2, 4), (-2, -4)\}$ **11** $\{(2, 1.7), (2, -1.7), (-2, 1.7),$
$(-2, -1.7)\}$ **13** Dependent   **14** Inconsistent   **15** $(1, 3)$ **17** $(2, 1)$
18 $(3, -2)$ **19** $(\frac{1}{2}, \frac{1}{3})$ **21** $(1, -1)$, $(\frac{109}{13}, -\frac{1}{13})$ **22** $(1, 4)$, $(-1, 0)$, $(\frac{1}{3}, 0)$,
$(-\frac{1}{3}, -\frac{4}{3})$ **23** $(\sqrt{3}, -\sqrt{3})$, $(-\sqrt{3}, \sqrt{3})$, $(i\sqrt{6}, i\sqrt{6}/2)$, $(-i\sqrt{6}, -i\sqrt{6}/2)$
25 $(2, 3), (3, 2)$ **27** $m = -4$ at $(0, 0)$; $M = 10$ at $(4, 2)$ **29** Smaller, 400;
larger, 300; profit \$9,900   **30** Smaller, 500; larger, 200; profit \$10,300

EXERCISE 10.1

1 $a = 4, b = 3, c = 2, d = -3$ **2** $a = 4, b = 5, c = -2, d = -3$ **3** $a = 4,$
$b = -2, c = 1, d = 1$ **5** $\begin{bmatrix} 9 & -3 & 6 \\ 12 & 0 & 15 \end{bmatrix}$ **6** $\begin{bmatrix} 0 & 0 & 0 \\ 0 & 0 & 0 \end{bmatrix}$ **7** $\begin{bmatrix} 2 & -3 & 0 \\ -1 & -5 & 4 \end{bmatrix}$

9 $\begin{bmatrix} 1 & 2 & 2 \\ 5 & 5 & 1 \end{bmatrix}$ **10** $\begin{bmatrix} 1 & 2 & 2 \\ 5 & 5 & 1 \end{bmatrix}$ **11** $\begin{bmatrix} 0 & 7 & 4 \\ 11 & 15 & -2 \end{bmatrix}$ **13** $\begin{bmatrix} 3 & 4 \\ -1 & 0 \\ 2 & 5 \end{bmatrix}$

14 $\begin{bmatrix} -2 & 1 \\ 3 & 5 \\ 0 & -4 \end{bmatrix}$ **15** Not defined   **25** $\begin{bmatrix} 7 & 2 \\ 0 & 1 \end{bmatrix}$ **26** $\begin{bmatrix} 7 & 4 \\ 0 & 1 \end{bmatrix}$ **27** $\begin{bmatrix} 0 & 5 \\ -1 & 0 \end{bmatrix}$

29 $\begin{bmatrix} -3 & 13 & -7 \\ -5 & 14 & -3 \\ 11 & -9 & 31 \end{bmatrix}$ **30** $\begin{bmatrix} -3 & -5 & 11 \\ 13 & 14 & -9 \\ -7 & -3 & 31 \end{bmatrix}$ **31** $\begin{bmatrix} 11 & -5 & -2 \\ -13 & 17 & 5 \\ 0 & 20 & 14 \end{bmatrix}$ **37** $\begin{bmatrix} 1 \\ 7 \\ 1 \end{bmatrix}$

38 $\begin{bmatrix} -1 \\ 18 \\ 19 \end{bmatrix}$ **39** $\begin{bmatrix} 21 & 1 \\ 1 & 10 \end{bmatrix}$

EXERCISE 10.2

1 $x = 2, y = 1$ **2** $x = 3, y = -2$ **3** $x = -2, y = -5$ **5** $x = 1, y = 2, z = 1$
6 $x = 1, y = 0, z = 3$ **7** $x = 2, y = -1, z = -2$ **9** $x = 0, y = 1, z = -1$
10 $x = 2, y = 2, z = -2$ **11** $x = \frac{1}{2}, y = \frac{1}{3}, z = \frac{5}{6}$ **13** $x = 1, y = 2, z = 1, w = 2$
14 $x = 3, y = 0, z = 1, w = -1$ **15** $x = 0, y = 2, z = 3, w = -4$ **17** $\frac{1}{5}\begin{bmatrix} 1 & 2 \\ -2 & 1 \end{bmatrix}$

18 $\frac{1}{11}\begin{bmatrix} 1 & 3 \\ 3 & -2 \end{bmatrix}$ **19** $\frac{1}{-7}\begin{bmatrix} -3 & -1 \\ -1 & 2 \end{bmatrix}$ **21** $\frac{-1}{18}\begin{bmatrix} -7 & -1 & 5 \\ -1 & 5 & -7 \\ 5 & -7 & -1 \end{bmatrix}$

22 $\frac{1}{64}\begin{bmatrix} 18 & 13 & 11 \\ 10 & -7 & -1 \\ 16 & 8 & -8 \end{bmatrix}$ **23** $\frac{-1}{19}\begin{bmatrix} 3 & 3 & -10 \\ -4 & -4 & 7 \\ -13 & 6 & -1 \end{bmatrix}$ **25** $\frac{1}{26}\begin{bmatrix} -1 & 7 & 3 \\ 3 & 5 & -9 \\ 11 & 1 & -7 \end{bmatrix}$

26 $\frac{1}{62}\begin{bmatrix} 6 & 10 & 2 \\ -20 & 8 & 14 \\ -18 & 1 & 25 \end{bmatrix}$ **27** $\frac{-1}{60}\begin{bmatrix} -6 & -9 & -3 \\ -8 & 8 & -4 \\ 4 & 1 & -3 \end{bmatrix}$ **29** $\begin{bmatrix} 1 & 1 & -1 & 0 \\ 0 & 1 & -1 & 1 \\ 1 & -1 & 0 & -1 \\ 1 & 0 & -1 & -1 \end{bmatrix}$

30 $\frac{1}{2}\begin{bmatrix} 2 & -5 & -4 & 5 \\ 0 & -1 & -2 & 3 \\ 0 & 2 & 2 & -2 \\ 0 & -1 & 0 & 1 \end{bmatrix}$ **31** $\frac{1}{2}\begin{bmatrix} 1 & 0 & 1 & 0 \\ -1 & 1 & -1 & 1 \\ 0 & -1 & 1 & 0 \\ 0 & 0 & 1 & -1 \end{bmatrix}$

EXERCISE 10.3

1 7   **2** −5   **3** 10   **5** 1   **6** 10   **7** 1   **9** −3   **10** −12   **11** 12
13 −10 + 7a   **14** 7a − 6b   **15** 3a + 2b   **17** ac − ab   **18** ad − bc
19 $a^2 + bc$ **21** 42   **22** −74   **23** −87   **25** 16   **26** −24   **27** 0
29 −101   **30** −12   **31** 40   **33** 4   **34** 14   **35** 49   **37** 0   **38** 0
39 0

EXERCISE 10.4

21 42   **22** −7   **23** −28   **25** 4   **26** 18   **27** −1, 2   **29** 1, 2   **30** 2
31 −3, 3   **33** 1, 5   **34** −3, 5   **35** −2, 2

EXERCISE 10.5

1 $\{(2, 1)\}$ **2** $\{(3, 2)\}$ **3** $\{(1, 3)\}$ **5** $\{(3, −1)\}$ **6** $\{(1, 2)\}$
7 $\{(2, 0)\}$ **9** $\{(−1, −3)\}$ **10** $\{(−2, −3)\}$ **11** $\{(−1, −4)\}$

13 $\left\{\left(n, \dfrac{2n − 4}{3}\right)\middle| n \text{ is any number}\right\}$ **14** $\{(n, 2 − 3n) \,|\, n \text{ is any number}\}$

15 $\varnothing$ **17** $\{(a, b)\}$ **18** $\{(a, a − b)\}$ **19** $\{(a − b, a + b)\}$ **21** $\{(1, 1, 2)\}$
22 $\{(2, 1, 3)\}$ **23** $\{(2, −1, 1)\}$ **25** $\{(1, −2, 3)\}$ **26** $\{(−2, −1, 5)\}$
27 $\{(2, 0, −1)\}$ **29** No unique solution   **30** $\{(1, −1, 2)\}$ **31** No solution
33 $\{(2, 3, −2)\}$ **34** $\{(4, 3, −5)\}$ **35** $\{(2, −1, −3)\}$ **37** $\{(a, a + b, a)\}$
38 $\{(−a, a − b, b)\}$ **39** $\{(a + b, a − b, a)\}$

EXERCISE 10.6

1 $x = 2, y = 3$ **2** $x = 1, y = 2, z = −1$ **3** $\begin{bmatrix} 0 & 2x & x \\ 3 & 1 & x \end{bmatrix}$ **5** $\begin{bmatrix} 1 & 0 & 1 \\ 2 & 0 & a \\ 3 & −1 & −1 \end{bmatrix}$

6 $\begin{bmatrix} −4 & 11 & 1 & 5 \\ −7 & 25 & −1 & 18 \end{bmatrix}$ **7** $\begin{bmatrix} 40 & −27 & 25 \\ −23 & 16 & −14 \\ −9 & 5 & −4 \end{bmatrix}$ **9** $\begin{bmatrix} 1 & 2 & 5 \\ 2 & 3 & 8 \\ −1 & 1 & 2 \end{bmatrix}$

13 15   **14** −24   **15** 0   **17** −1   **25** 1   **26** −4, 3   **27** $x = 2, y = −3$
29 $x = 1, y = −2, z = 5$ **30** $x = 3, y = 0, z = −4$ **31** $x = 1, y = −2, z = 5$

EXERCISE 11.1

1 $2.856(10^3)$ **2** $5.9437(10^4)$ **3** $6.48(10^2)$ **5** $2.13(10)$ **6** $4.021(10^2)$
7 $3.030(10^3)$ **9** $4.75(10^{-1})$ **10** $9.160(10^{-2})$ **11** $7.04(10^{-3})$
13 $8.430(10^3)$ **14** $6.700(10^3)$ **15** $3.060(10^3)$ **17** $8.43(10^3)$

18 6.70(10^3)   **19** 3.06(10^3)   **21** 596.3   **22** 67.13   **23** 7.844
25 380.3   **26** 29.92   **27** 0.01235   **29** 3.763(10^4)   **30** 4,974
31 2.346(10^4)   **33** 5,556   **34** 5.556(10^4)   **35** 4.836(10^4)   **37** 19
38 30   **39** 11   **41** 298   **42** 9.52   **43** 4.65(10^3)   **45** 62   **46** 1.8(10^2)
47 225   **49** 1.8(10)   **50** 1.9(10^3)   **51** 5.0(10^2)   **53** 2.6   **54** 1.3(10)
55 1.8(10)   **57** 1.5   **58** 2.3(10)   **59** 3.2(10)   **61** 1.73(10)
62 8.34(10)   **63** 7.24(10)   **65** 2.2   **66** 9.30   **67** 8.4

EXERCISE 11.2

1 $\log_2 8 = 3$ **2** $\log_3 81 = 4$ **3** $\log_4 1,024 = 5$ **5** $\log_5 \frac{1}{5} = -1$
6 $\log_6 \frac{1}{36} = -2$ **7** $\log_3 \frac{1}{243} = -5$ **9** $\log_{1/3} 27 = -3$ **10** $\log_{1/2} 32 = -5$
11 $\log_{3/5} \frac{125}{27} = -3$ **13** $\log_{16} 8 = \frac{3}{4}$ **14** $\log_{27} 9 = \frac{2}{3}$ **15** $\log_{64} 32 = \frac{5}{6}$
17 $3^2 = 9$ **18** $5^3 = 125$ **19** $2^6 = 64$ **21** $5^{-1} = \frac{1}{5}$ **22** $3^{-3} = \frac{1}{27}$
23 $5^{-2} = \frac{1}{25}$ **25** $27^{2/3} = 9$ **26** $36^{3/2} = 216$ **27** $8^{5/3} = 32$ **29** 2   **30** 3
31 2   **33** $\frac{3}{2}$ **34** $\frac{4}{3}$ **35** $\frac{3}{2}$ **37** 125   **38** 81   **39** 49   **41** 4
42 8   **43** 343   **45** 5   **46** 2   **47** 3   **49** $\frac{2}{5}$ **50** $\frac{2}{3}$ **51** $\frac{6}{7}$ **53** 10
54 11   **55** 17   **57** 21   **58** 22   **59** 19   **61** 10   **62** 6   **63** 6
65 9   **66** 16   **67** 26   **69** 4   **70** $\frac{13}{3}$ **71** $\frac{6}{5}$ **73** -1 **74** 1   **75** 1

EXERCISE 11.3

1 .3746, 1   **2** .4137, 8   **3** .5709, 2   **5** .3814, -3 **6** .6186, -4
7 .2333, -3 **9** 1   **10** 0   **11** 3   **13** 0   **14** 1   **15** 4   **17** -1
18 -2 **19** -4 **21** -1 **22** -1 **23** -3 **25** 2.8831   **26** 0.9047
27 1.7574   **29** 1.6739   **30** 1.4886   **31** 2.0294   **33** 1.4624   **34** 1.9823
35 0.6532   **37** 3.6704   **38** 2.2022   **39** 1.8947   **41** 2.4043   **42** 3.9359
43 1.5883   **45** 0.0927   **46** 2.9907   **47** 3.7281   **49** 8.4771 $-$ 10
50 8.8808 $-$ 10   **51** 7.9031 $-$ 10   **53** 9.9420 $-$ 10   **54** 8.9841 $-$ 10
55 9.6839 $-$ 10   **57** 9.0927 $-$ 10   **58** 8.6809 $-$ 10   **59** 8.9076 $-$ 10
61 9.9428 $-$ 10   **62** 9.9428 $-$ 10   **63** 8.6705 $-$ 10   **65** 8.3703 $-$ 10
66 8.3703 $-$ 10   **67** 9.7633 $-$ 10

EXERCISE 11.4

1 3.84   **2** 27.6   **3** 403   **5** 57.2   **6** 698   **7** 945   **9** 3.25(10^3)
10 8.42(10^4)   **11** 7.66(10^3)   **13** 8.02(10^{-1})   **14** 3.56(10^{-3})
15 6.55(10^{-2})   **17** 1.86(10^{-2})   **18** 6.43(10^{-4})   **19** 3.98(10^{-3})   **21** 20.1
22 377   **23** 5.90   **25** 4.21(10^4)   **26** 3.26(10^3)   **27** 1.89(10^3)
29 5.25(10^{-2})   **30** 3.28(10^{-3})   **31** 6.34(10^{-1})   **33** 272.8   **34** 39.76
35 1.890   **37** 6.513(10^4)   **38** 1.536(10^3)   **39** 1.707(10^5)   **41** 2.131(10^{-2})
42 2.149(10^{-3})   **43** 2.584(10^{-1})   **45** 4.171(10^{-2})   **46** 1.061(10^{-3})
47 2.752(10^{-2})

EXERCISE 11.5

1 $\log a - \log (b + x)$ **2** $\log (a - x) - \log b$ **3** $\log a + \log b + \log (c - x)$
5 $\log a + 4 \log x$ **6** $3 \log a + \frac{1}{2} \log (b + x)$
7 $\log b + \frac{1}{2} \log (x - a) - \log c - \frac{1}{2} \log (x + a)$ **9** $\log ab$ **10** $\log (b/c)$

11 $\log (ab/c)$ **13** $\log (ab^2/c)$ **14** $\log (ac^2/b^3)$ **15** $\log (a^2b^3/\sqrt{2a + 3b})$
17 81.9   **18** 809   **19** $6.12(10^3)$ **21** 2.85   **22** 0.358   **23** 0.695
25 206   **26** 62.4   **27** 1.86   **29** 0.104   **30** 2.99   **31** 0.0957   **33** 205
34 $1.70(10^3)$ **35** 800   **37** 214   **38** $2.22(10^3)$ **39** 0.102
41 $6.06(10^4)$ **42** $7.36(10^3)$ **43** $1.91(10^4)$ **45** 8.73   **46** 7.92
47 1.19   **49** 0.903   **50** 0.335   **51** 0.976   **53** 15.3   **54** 2.12
55 3.71   **57.** 20.9   **58.** 13.8   **59.** 0.266   **61** 175.0   **62** 931.3
63 0.006385   **65** 1.781   **66** 1.248   **67** 0.3993   **69** 52
70 $2.1(10^3)$ **71** $8.1(10^2)$ **73** 0.130   **74** 25.9   **75** 4.95   **77** 0.266
78 1.01   **79** 0.121   **81** 0.5018   **82** 1.528   **83** 0.9192

EXERCISE 11.6

1 0.164   **2** 6.52   **3** 2.44   **5** 3.15   **6** 2.20   **7** 2.10   **9** 2.34
10 1.78   **11** 1.96   **13** $9.05 - 10$ **14** 4.33   **15** 1.81   **17** 1.99
18 3.15   **19** 0.819   **21** 0.760   **22** 2.20   **23** 2.36

EXERCISE 11.7

1 $\ln y$ **2** $-\ln y$ **3** $\ln (y/2)]/3$ **5** $\dfrac{\log ay}{\log r}$ **6** $\dfrac{\log y - \log ar}{\log r}$ **7** $\dfrac{\log (y/2)}{\log (c + r)}$

9 $\dfrac{e^{2y} + 1}{2e^y}$ **10** $\dfrac{1 + \sqrt{9 + 4e^{2y}}}{2}$ **11** $\dfrac{e^{2y} - 1}{2e^y}$ **13** $\ln (y \pm \sqrt{y^2 - 1}), y \geq 1$

14 $\ln (y + \sqrt{y^2 + 1})$ **15** $\ln (e^y + \sqrt{e^{2y} + 1})$ **17** 4   **18** 3   **19** 6
21 1   **22** 2.5   **23** 2   **25** $-2, 2$ **26** $1, -4$ **27** $-1, 2$ **29** 0.246
30 4.061   **31** -2.300 **33** 2   **34** 0   **35** -1 **37** -2 **38** 5
39 10   **41** $(4, \frac{1}{2})$ **42** $(1, 1)$ **43** $(0.31, 0.88)$ **45** $(0.23, 0.035)$
46 $(0.528, 0.0176)$ **47** $(500, 20)$

EXERCISE 11.8

25 $(0.54, 1.45)$ **26** $(0.30, 1.40)$ **27** $(0.70, -0.16)$ **29** $(0.2, -0.7)$
30 $(1, 0), (0.14, -0.86)$ **31** $(0.54, -0.27)$

EXERCISE 11.9

1 61.02   **2** 85.31   **3** 17.04   **5** 16.36   **6** 54.76   **7** -17.51 **9** 181.481
10 292.338   **11** 973.53   **13** 3.21   **14** 0.350   **15** 0.732   **17** 2.85

18 314   **19** 7.98   **21** 2.7760   **22** 1.8639   **23** 9.7993 − 10   **25** 1.0332
26 1.9961   **27** 3.0837   **29** 9.97   **30** 111.05   **31** 2.05   **33** 6.11
34 20.49   **35** 15.80   **37** 562.36   **38** 89.13   **39** 5.79

EXERCISE 11.10

1 $7.863(10^3)$ **2** $5.91(10^{-2})$ **3** $4.0864(10^3)$ **5** $5.85(10^3)$ **6** $5.72(10^3)$
7 200   **9** 7.1   **10** 0.1690   **11** 12.7   **13** $\log_7 343 = 3$ **14** $\log_5 \frac{1}{25} = -2$
15 $3^4 = 81$ **17** 3   **18** $\frac{2}{3}$ **19** 81   **21** 6   **22** 9   **23** 1.9280
25 3.3724   **26** 22.01   **27** 393.29   **29** $1.03(10^3)$ **30** 645   **31** 209
33 29.1   **34** 0.845   **35** 3.68   **37** $\frac{1}{2}\log(y/5)$ **38** $-2, 3$ **39** 1.83
41 $\frac{22}{7}$ **42** $(5.3, 0.7)$ **43** $(0.4, 2.6)$

EXERCISE 12.1

1 $x = 2, y = 5$ **2** $x = 3, y = -2$ **3** $x = 4, y = 2$ **5** $x = 3, y = -5$
6 $x = -2, y = -4$ **7** $x = -5, y = 0$ **9** $(1, 7)$ **10** $(5, 4)$ **11** $(-3, 2)$
13 $-1 + 5i$ **14** $2 + 7i$ **15** 7   **17** $(1, -5)$ **18** $(2, 2)$ **19** $(0, 1)$
21 $3 - 2i$ **22** $-2 + 2i$ **23** $-2i$ **25** $(-13, 16)$ **26** $(1, 12)$
27 $(-8, 1)$ **29** $5 + 5i$ **30** 13   **31** $4 + 24i$ **33** $(\frac{13}{5}, \frac{6}{5})$

34 $(\frac{7}{26}, -\frac{17}{26})$ **35** $(\frac{15}{17}, \frac{8}{17})$ **37** $\frac{11}{29} - \frac{13i}{29}$ **38** $\frac{4}{29} + \frac{19i}{29}$ **39** $-\frac{4}{5} + \frac{3i}{5}$

41 $-1 - 8i$ **42** $-3 + 7i$ **43** $\frac{21 + i}{34}$ **45** $2 - 11i$ **46** $18 + 26i$ **47** -4

53 $-i, 1$

EXERCISE 12.2

21 $5, \arctan \frac{4}{3}$ **22** $13, \arctan \frac{12}{-5}$ **23** $10, \arctan \frac{-8}{-6}$ **25** $2\sqrt{2}, 45°$

26 $2, 300°$ **27** $2\sqrt{3}, 150°$ **29** $\sqrt{2}, 315°$ **30** $2\sqrt{2}, 225°$

31 $\sqrt{82}, \arctan \frac{-9}{1}$ **33** $3\sqrt{2} \text{ cis } 45°$ **34** $2\sqrt{2} \text{ cis } 135°$ **35** $5\sqrt{2} \text{ cis } 225°$

37 $2 \text{ cis } 30°$ **38** $2 \text{ cis } 330°$ **39** $2 \text{ cis } 120°$ **41** $5 \text{ cis } 90°$ **42** $6 \text{ cis } 0°$

43 $2 \text{ cis } 270°$ **45** $17 \text{ cis } 298°$ **46** $5 \text{ cis } 143°$ **47** $25 \text{ cis } 74°$ **49** $\sqrt{29} \text{ cis } 22°$

50 $\sqrt{29} \text{ cis } 248°$ **51** $\sqrt{17} \text{ cis } 166°$

EXERCISE 12.3

1 $2 \text{ cis } 217°$ **2** $2\sqrt{6} \text{ cis } 327°$ **3** $9 \text{ cis } 248°$ **5** $12 \text{ cis } 154°$ **6** $16 \text{ cis } 257°$
7 $9.36 \text{ cis } 270°$ **9** $2 \text{ cis } 90°$ **10** $4 \text{ cis } 135°$ **11** $12 \text{ cis } 150°$ **13** $4 \text{ cis } 29°$
14 $6 \text{ cis } 255°$ **15** $3 \text{ cis } 298°$ **17** $\frac{4}{3} \text{ cis } 11°$ **18** $4\sqrt{2} \text{ cis } 82°$

19 $(1/\sqrt{6})$ cis $90°$ **33** 4 cis $90° = 4i$ **34** 4 cis $180° = -4$
35 4 cis $165° = -3.86 + 1.04i$ **37** 19.42 cis $78° = 4.08 + 19.03i$
38 68.68 cis $132° = -45.96 + 51.04i$ **39** 8.29 cis $73° = 2.40 + 7.96i$
41 8 cis $240° = -4 - 4i\sqrt{3}$ **42** $16\sqrt{2}$ cis $45° = 16 + 16i$
43 $1,105$ cis $88° = 38.56 + 1,104i$ **45** 1 cis $270° = -i$ **46** 1 cis $630° = -i$
47 1 cis $540° = -1$ **49** 8 cis $360° = 8$ **50** 27 cis $180° = -27$
51 100 cis $180° = -100$ **53** 0.38 cis $30° = 0.33 + 0.19i$
54 0.68 cis $46° = 0.47 + 0.49i$ **55** 0.69 cis $10° = .68 + .12i$
57 $\sqrt{2}$ cis $195° = -1.37 - 0.37i$ **58** $2\sqrt{2}$ cis $45° = 2 + 2i$

59 $(\sqrt{2}/4)$ cis $(-135°) = -\dfrac{1}{2} - \dfrac{i}{2}$

EXERCISE 12.4

1 32 cis $90°$ **2** 81 cis $0°$ **3** 125 cis $180°$ **5** 128 cis $60°$ **6** 32 cis $150°$
7 4 cis $180°$ **9** 32 cis $120°$ **10** 8 cis $90°$ **11** $8\sqrt{2}$ cis $315°$
13 $28,561$ cis $92°$ **14** $4,913$ cis $174°$ **15** $3,125$ cis $85°$ **17** $34\sqrt{34}$ cis $183°$
18 $4,225$ cis $120°$ **19** $405,224$ cis $324°$ **21** cis $k120°$, $k = 0, 1, 2$
22 cis $(45° + k90°)$, $k = 0, 1, 2, 3$ **23** cis $(30° + k120°)$, $k = 0, 1, 2$
25 $2^{1/4}$ cis $(22\frac{1}{2}° + k180°)$, $k = 0, 1$ **26** $2^{1/10}$ cis $(45° + k72°)$, $k = 0, 1, 2, 3, 4$
27 $2^{1/18}$ cis $(15° + k40°)$, $k = 0, 1, 2, 3, 4, 5, 6, 7, 8$
29 $2^{1/3}$ cis $(10° + k120°)$, $k = 0, 1, 2$ **30** cis $(75° + k90°)$, $k = 0, 1, 2, 3$
31 2 cis $(40° + k60°)$, $k = 0, 1, 2, 3, 4, 5$ **33** $2^{3/10}$ cis $(9° + k72°)$, $k = 0, 1, 2, 3, 4$
34 $2^{2/5}$ cis $(24° + k36°)$, $k = 0, 1, 2, 3, 4, 5, 6, 7, 8, 9$
35 $2^{1/4}$ cis $(75° + k90°)$, $k = 0, 1, 2, 3$ **37** $\sqrt{5}$ cis $(27° + k180°)$, $k = 0, 1$
38 $\sqrt[16]{13}$ cis $(38° + k45°)$, $k = 0, 1, 2, 3, 4, 5, 6, 7$
39 $\sqrt[12]{10}$ cis $(27° + k60°)$, $k = 0, 1, 2, 3, 4, 5$
45 2 cis $(135° + k180°)$, $k = 0, 1$ **46** 4 cis $(30° + k120°)$, $k = 0, 1, 2$
47 cis $(22\frac{1}{2}° + k90°)$, $k = 0, 1, 2, 3$ **49** $1, 1 - i$ **50** $-2, -2 - i$
51 $-3 + i, 3 + i$

EXERCISE 12.5

1 $x = 3$, $y = 5$ **2** $x = \pm 3$, $y = \pm 1$ **3** $(3, 9)$ **5** $17 + i$ **6** $(13 + 11i)/10$
7 $(\frac{39}{61}, \frac{2}{61})$ **9** 15 cis $260°$ **10** $\frac{9}{2}$ cis $67°$ **11** $-16\sqrt{3} - 16i$ **23** 1 cis $30°$

EXERCISE 13.1

1 -3 **2** -26 **3** -49 **5** -2 **6** -108 **7** 13 **9** 22 **10** -2
11 7 **25** -1 **26** -1 **27** 23

EXERCISE 13.2

1 $x^2 + x + 6$, 13   **2** $x^2 - x$, 1   **3** $2x^2 + 3x + 4$, -1 **5** $2x^2 - 2x + 3$, -4
6 $-3x^2 - 2x - 3$, -4 **7** $-2x^2 - x + 4$, 4   **9** $x^3 + 4x^2 - x + 6$, -3
10 $-2x^3 + 9x^2 - 12x + 13$, -24 **11** $x^4 + 2x^3 - 5x^2 - 3x - 2$, -3
13 $4x^3 + 2x^2 - 4x - 2$, 0   **14** $6x^2 - 6x + 3$, -3 **15** $2x^2 + ax + 4a^2$, 0
17 2   **18** 31   **19** -20 **21** 0   **22** -3 **23** 9   **25** 1   **26** -2
27 a^3 **41** -10 **42** -6 **43** -2 **45** 1   **46** -1 **47** 1

EXERCISE 13.3

1 5; 2, 2; -3, 2; 4, 1   **2** 9; -3, 4; 2, 4; 7, 1   **3** 6; -1, 1; 2, 2; -4, 2; 5, 1
5 5; $-\frac{5}{2}$, 2; $\frac{1}{3}$, 3   **6** 7; $\frac{3}{4}$, 2; $-\frac{9}{2}$, 2; 4, 3   **7** 7; $-\frac{1}{3}$, 3; i, 1; $-i$, 1; -1, 2
9 1, 2, -3 **10** -1, 2, 4   **11** -2, -3, 4   **13** 1, 1, -1, -2
14 -1, -1, i, $-i$ **15** 1, 2, $2i$, $-2i$ **17** -1, i, $-i$ **18** $\frac{3}{2}$, $2i$, $-2i$
19 $-\frac{5}{3}$, $2i$, $-2i$ **21** $-\frac{3}{2}$, $1 + i$, $1 - i$ **22** $-\frac{1}{4}$, $-1 + 2i$, $-1 - 2i$
23 $\frac{3}{4}$, $2 - i$, $2 + i$ **25** $2i$, $-2i$, $-1 + 2i$, $-1 - 2i$
26 3, $-2 + i$, $-2 + i$, $-2 - i$, $-2 - i$ **27** $-1 + 2i$, $-1 + 2i$, $-1 - 2i$, $-1 - 2i$, i, $-i$
29 $3x^3 + x^2 + 12x + 4$ **30** $2x^3 - 7x^2 + 2x - 7$ **31** $x^3 - 5x^2 + 9x - 5$
33 $x^4 + 8x^3 + 26x^2 + 40x + 25$ **34** $x^4 + 4x^3 + 14x^2 + 20x + 25$
35 $x^4 - 8x^3 + 32x^2 - 64x + 64$ **37** $A = 2$, $B = -2$ **38** $A = 2$, $B = -1$
39 $A = -\frac{7}{5}$, $B = \frac{1}{5}$, $C = 2$ **41** $A = -2$, $B = 3$ **42** $A = 2$, $B = -1$, $C = 5$
43 $A = 1$, $B = -1$, $C = -3$ **46** $x^3 + x^2 - 7x + 65 = 0$
47 $x^4 - 10x^3 + 42x^2 - 82x + 65 = 0$

EXERCISE 13.4

1 2, -2 **2** 2, -2 **3** 5, -4 **5** 3, -6 **6** 7, -1 **7** 2, -4 **9** 1, -1
10 1, -3 **11** 1, -4 **13** 2 or 0 positive, 1 negative   **14** 1 positive, 2 or 0
negative   **15** 0 positive, 3 or 1 negative   **17** 2 or 0 positive, 2 or 0 negative
18 2 or 0 positive, 2 or 0 negative   **19** 3 or 1 positive, 1 negative   **21** 4, 2,
or 0 positive, 1 negative   **22** 3 or 1 positive, 2 or 0 negative   **23** 1 positive,
2 or 0 negative   **25** -4 and -3, -1 and 0, 0 and 1   **26** 0 and 1, 1 and 2,
4 and 5   **27** 0 and 1, 3 and 4, 5 and 6   **29** -2 and -1, 0 and 1, 3 and 4,
5 and 6   **30** -5 and -4, -3 and -2, -2 and -1, 4 and 5   **31** -2 and -1,
-1 and 0, 1 and 2, 4 and 5   **33** 3 and 4   **34** 0 and 1   **35** -1 and 0,
2 and 3   **37** -1 and $-\frac{1}{2}$, $-\frac{1}{2}$ and 0, 2 and 3   **38** -2 and -1, 1 and $\frac{3}{2}$. $\frac{3}{2}$ and 2
39 -2 and -1, 1 and $\frac{3}{2}$, $\frac{3}{2}$ and 2

EXERCISE 13.5

1 -2, 1, 2   **2** -1, 2, 3   **3** 1, 2, -3 **5** -3, 2, $\frac{1}{2}$ **6** -5, -2, $\frac{2}{3}$
7 -1, $-\frac{1}{2}$, $\frac{3}{2}$ **9** $-\frac{1}{2}$, $\frac{1}{3}$, $\frac{1}{2}$ **10** $-\frac{3}{2}$, $\frac{3}{2}$, $\frac{3}{2}$ **11** $-\frac{3}{2}$, $\frac{1}{2}$, 3   **13** 2, $1 - \sqrt{2}$,
$1 + \sqrt{2}$ **14** -3, $1 - \sqrt{3}$, $1 + \sqrt{3}$ **15** -1, $\frac{1}{4}(-1 + \sqrt{41})$, $\frac{1}{4}(-1 - \sqrt{41})$

17 $\frac{1}{2}, i, -i$ **18** $-\frac{1}{3}, -1+i, -1-i$ **19** $\frac{3}{2}, i, -i$ **21** $-1, 2, -\frac{3}{2}, \frac{1}{3}$
22 $-3, 1, \frac{2}{3}, -\frac{3}{2}$ **23** $-5, 1, 1, 2$ **25** $-1, 2, i, -i$ **26** $-1, 1, 1+i, 1-i$
27 $-\frac{3}{2}, \frac{1}{2}, i, -i$

EXERCISE 13.6

1 3.59   **2** 2.44   **3** -0.29 **5** 1.38   **6** 2.24   **7** -1.88 **9** 0.12
10 0.53   **11** 0.66   **13** 1.18   **14** 1.42   **15** 1.67   **17** $2.732, -0.732$
18 6.236, 1.764   **19** 5.449, 0.551   **21** $-3.260, -1.340, 1.602$ **22** $-4.769,$
$-3.116, 1.884$ **23** $-0.262, 1.660, 4.600$ **25** $1+\sqrt{3}, 1-\sqrt{3}, 1+\sqrt{5},$
$1-\sqrt{5}; 3.236$ **26** $-1+\sqrt{3}, -1-\sqrt{3}, 2+\sqrt{2}, 2-\sqrt{2}; 3.414$ **27** $3+\sqrt{2},$
$3-\sqrt{2}, -2+\sqrt{3}, -2-\sqrt{3}; 4.414$ **29** $i, -i, 4-\sqrt{3}, 4+\sqrt{3}; 5.732$
30 $i\sqrt{3}, -i\sqrt{3}, 2-\sqrt{6}, 2+\sqrt{6}; 4.449$ **31** $2i, -2i, -3-\sqrt{8}, -3+\sqrt{8};$
-0.172

EXERCISE 13.7

1 14   **2** 20   **3** $d^4 - 2d^3 + 5d + 2$ **9** $2x^2 + 4x + 13; 20$
10 $3x^3 - x^2 + 4x - 2; 1$ **11** $2, 2, -3$ **13** $x^3 - 8x^2 + 22x - 20$
14 $x^3 - 5x^2 + 4x + 6$ **15** $x^4 + 2x^3 + 2x^2 + 10x + 25$ **17** $A = 0, B = 4, C = 3$
18 $A = 2, B = 1, C = -3$ **19** Sum $= 1$, product $= 1$ **26** $2, i, -i$
27 $-\frac{3}{2}, -\frac{3}{2}, 2i, -2i$ **33** -0.18 **34** $0.73, -2.73$
35 $-0.618, 1.618, 1.732, -1.732$

EXERCISE 14.1

1 $2, 6, 10, 14, 18$ **2** $4, 1, -2, -5$ **3** $5, 3, 1, -1, -3$ **5** $4, 7, 10, 13$
6 $-3, 0, 3, 6, 9$ **7** $2, -3, -8, -13$ **9** $10, 6, 2, -2, -6$ **10** $11, 5, -1,$
$-7, -13, -19$ **11** $2, 3, 4, 5, 6, 7$ **13** $l = -8$ **14** $d = \frac{3}{5}$ **15** $a = 3$
17 $s = 48$ **18** $n = 10$ **19** $l = -4$ **21** $n = 8$ **22** $d = 2$ **23** $a = 0$
25 $l = 14, s = 40$ **26** $l = -1, s = 56$ **27** $s = 21, d = -1$ **29** $d = 3, n = 6$
30 $a = 1, d = 4$ **31** $l = -13, n = 7$ **33** $a = 8, n = 6$ **34** $a = -8, n = 4$
35 $a = 1, n = 6$ **37** 204   **38** 315   **39** $2n(n+1)$ **41** -3 **42** $\frac{8}{9}$
43 $x = 2, y = 3$ **45** 156   **46** 208 ft, 784 ft   **47** 25 sec   **49** \$1,102
50 \$11,500   **51** 802π cm

EXERCISE 14.2

1 $2, 6, 18, 54, 162$ **2** $4, 8, 16, 32, 64, 128$ **3** $8, 2, \frac{1}{2}, \frac{1}{8}, \frac{1}{32}$ **5** $1, 3, 9, 27,$
81   **6** $\frac{1}{3}, \frac{2}{3}, \frac{4}{3}, \frac{8}{3}, \frac{16}{3}, \frac{32}{3}$ **7** $\frac{2}{3}, 1, \frac{3}{2}, \frac{9}{4}, \frac{27}{8}, \frac{81}{16}$ **9** 2.916   **10** 160
11 3   **13** 80   **14** 311   **15** $\frac{1}{4}$ **17** 242   **18** 52.75   **19** 5
21 $l = 48, s = 93$ **22** $l = 162, s = 242$ **23** $r = -2, s = -84$ **25** $s = 400,$
$n = 4$ **26** $a = \frac{1}{9}, s = \frac{40}{9}$ **27** $r = \frac{1}{2}, s = 511$ **29** $l = 162, n = 5$
30 $a = 625, l = 1$ **31** $a = 343, n = 4$ **33** $r = -2, a = 3$ **34** $a = 8, n = 5$

35 $r = 3$, $l = 45$; $r = -4$, $l = 80$ **37** 504   **38** 3,279   **39** $2^{n+1} - 1$
41 $2, -\frac{2}{15}$ **42** 2, 5, 8; $r = 3$ **45** 2   **46** \$24,300   **47** 4,096n **49** \$511,
\$1 profit   **50** \$1,419.20   **51** 254

EXERCISE 14.3

1 6   **2** 4   **3** $\frac{12}{5}$ **5** 12   **6** 12   **7** 12   **9** $\frac{32}{3}$ **10** 9   **11** $\frac{125}{8}$
13 $\frac{2}{3}$ **14** $-\frac{2}{3}$ **15** $\frac{1}{4}$ **17** $\frac{1}{2}$ **18** 2   **19** $-\frac{2}{7}$ **21** $\frac{4}{9}$ **22** $\frac{8}{33}$
23 $\frac{232}{99}$ **25** $\frac{371}{90}$ **26** $\frac{199}{90}$ **27** $\frac{2161}{330}$ **33** $\frac{3}{4}$ **34** $\frac{4}{5}$ **35** $\frac{2}{7}$
37 2,400 cm   **38** 120 meters   **39** 18 meters   **41** $n < 9$ **42** 9 mg

43 200 sq in.   **45** $\{x \mid x < 0\} \cup \{x \mid x > 1\}, \dfrac{1}{2x - 2}$

46 $\{x \mid x < 0\} \cup \{x \mid x > \frac{4}{3}\}, \dfrac{2}{3x - 4}$ **47** All x, $\dfrac{1}{x^2 + 1}$

EXERCISE 14.4

1 6, 9, 12   **2** 5, 7, 9, 11, 13   **3** 6, 2, -2, -6 **5** 6, 18   **6** $\frac{1}{3}, \frac{1}{2}, \frac{3}{4}, \frac{9}{8}$
7 4, 8, 16 and -4, 8, -16 **9** $\frac{1}{5}, \frac{1}{7}, \frac{1}{9}, \frac{1}{11}$ **10** $\frac{1}{3}, \frac{2}{9}, \frac{1}{6}, \frac{2}{15}$ **11** $\frac{40}{17}, \frac{20}{7}, \frac{40}{11}$
13 $\frac{1}{24}$ **14** $\frac{3}{19}$ **15** $\frac{1}{6}$ **17** $\frac{1}{4}$ **18** $-\frac{2}{3}$ **19** $\frac{1}{5}$ **25** $\frac{32}{243}, \frac{64}{729}$, geometric
26 $\frac{2}{9}, \frac{4}{21}$, harmonic   **27** $-\frac{2}{9}, -\frac{4}{9}$, arithmetic   **29** $-\frac{1}{8}, -\frac{1}{11}$, harmonic
30 $\frac{81}{8}, \frac{243}{16}$, geometric   **31** $\frac{13}{6}, \frac{8}{3}$, arithmetic

EXERCISE 14.5

1 $3, \frac{7}{2}, 4, \frac{9}{2}, 5, \frac{11}{2}, 6$ **2** $\frac{3}{4}, \frac{1}{2}, \frac{1}{3}, \frac{2}{9}, \frac{4}{27}, \frac{8}{81}$ **3** $12, 3, \frac{12}{7}, \frac{6}{5}, \frac{12}{13}, \frac{3}{4}$
5 $\frac{2}{243}$, sum $= \frac{728}{243}$ **6** $\frac{5}{6}$, sum $= \frac{77}{24}$ **7** $a = \frac{1}{4}, d = \frac{1}{2}, l = \frac{9}{4}, s = \frac{25}{4}$ **9** 45, 878
10 $\frac{21}{11}$, 7   **11** \$1,425.76   **13** $-\frac{1}{7}$ **19** $4\sqrt{2}, 8, 8\sqrt{2}$ **21** $\frac{64}{13}, \frac{32}{5}, \frac{64}{7}$
25 $\frac{3}{4}, \frac{11}{12}$, arithmetic   **26** $\frac{16}{135}, \frac{32}{405}$, geometric   **27** $\frac{7}{9}, \frac{17}{18}$, arithmetic
29 $\frac{4}{5}, \frac{8}{11}$, harmonic

EXERCISE 16.1

1 11,232,000   **2** 3,192   **3** 72   **5** 17,576,000   **6** 625   **7** 9,261
9 248,832   **10** 5,040   **11** 100   **13** 3,584   **14** 7,529,536   **15** 18
17 243   **18** 36   **19** 140,608

EXERCISE 16.2

1 30   **2** 360   **3** 336   **5** 120   **6** 5   **7** 42   **13** 3   **14** 8   **15** 10
17 10,080   **18** 105   **19** 5,040   **21** 336   **22** 120   **23** 720   **25** 7 min
26 15 min   **27** 210,210   **29** 10^7 **30** 40,320   **31** 120   **33** $(n - 1)!$
34 120   **35** 5,040

EXERCISE 16.3

1 45 **2** 330 **3** 56 **9** $C(9, 4) = 126$ **10** $C(10, 3) = 120$
11 $C(52, 13) = 635{,}013{,}559{,}600$ **13** $C(52, 5) = 2{,}598{,}960$
14 $C(18, 12) = 18{,}564$ **15** $C(22, 6) = 74{,}613$ **17** $C(5, 2)C(7, 3) = 350$

18 $C(10, 3) = 120$ **19** $[C(8, 2)]^3 = 21{,}952$ **21** $\dfrac{22!}{6!5!7!3!1!}$

22 $\dfrac{9!}{2!3!4!} = 1{,}260$ **23** $\dfrac{11!}{7!2!2!} = 1{,}980$

26 $C(5, 1)\,\dfrac{27!}{5!5!5!5!7!} + C(5, 2)\,\dfrac{27!}{5!5!5!6!6!}$

EXERCISE 16.4

1 1,320 **2** 336 **3** 220 **9** 216 **10** 20 **11** 120 **13** 5,400
14 287,496 **15** All integers $n \geq 3$ **17** 5,040
18 $C(8, 5) + C(8, 6) + C(8, 7) + C(8, 8) = 93$

EXERCISE 17.1

1 $a^8 + 8a^7b + 28a^6b^2 + 56a^5b^3 + 70a^4b^4 + 56a^3b^5 + 28a^2b^6 + 8ab^7 + b^8$
2 $a^7 - 7a^6x + 21a^5x^2 - 35a^4x^3 + 35a^3x^4 - 21a^2x^5 + 7ax^6 - a^7$
3 $b^5 - 5b^4y + 10b^3y^2 - 10b^2y^3 + 5by^4 - y^5$
5 $a^5 - 10a^4y + 40a^3y^2 - 80a^2y^3 + 80ay^4 - 32y^5$
6 $x^6 + 18x^5b + 135x^4b^2 + 540x^3b^3 + 1{,}215x^2b^4 + 1{,}458xb^5 + 729b^6$
7 $128b^7 + 448b^6x + 672b^5x^2 + 560b^4x^3 + 280b^3x^4 + 84b^2x^5 + 14bx^6 + x^7$
9 $8a^3 - 36a^2b^2 + 54ab^4 - b^6$
10 $243x^5 - 810x^4b^3 + 1{,}080x^3b^6 - 720x^2b^9 + 240xb^{12} - 32b^{15}$
11 $64b^{12} + 576b^{10}x + 2{,}160b^8x^2 + 4{,}320b^6x^3 + 4{,}860b^4x^4 + 2{,}916b^2x^5 + 729x^6$
13 $a^{33} + 33a^{32}y + 528a^{31}y^2 + 5{,}456a^{30}y^3$ **14** $x^{51} - 51x^{50}y + 1{,}275x^{49}y^2 - 20{,}825x^{48}y^3$
15 $m^{101} - 202m^{100}y + 20{,}200m^{99}y^2 - 1{,}333{,}200m^{98}y^3$ **17** 1.2167 **18** 1.2155
19 1.1941 **21** $560x^3y^4$ **22** $-160a^3c^3$ **23** $10{,}206x^4y^5$ **25** $-35a$
26 $-2{,}016x^{-6}$ **27** $29{,}568x^{10}y^6$ **29** $160x^3y^{3/2}$ **30** $5{,}670x^4y$ **31** $960x^7y^3$

EXERCISE 17.2

1 $x^{-2} - 2x^{-3}y + 3x^{-4}y^2 - 4x^{-5}y^3$ **2** $a^{-4} - 4a^{-5}b + 10a^{-6}b^2 - 20a^{-7}b^3$

3 $a^{-5} + 5a^{-6}x + 15a^{-7}x^2 + 35a^{-8}x^3$ **5** $\dfrac{1}{2a} - \dfrac{x}{4a^2} + \dfrac{x^2}{8a^3} - \dfrac{x^3}{16a^4}$

6 $\dfrac{1}{x} - \dfrac{2a}{x^2} + \dfrac{4a^2}{x^3} - \dfrac{8a^3}{x^4}$ **7** $\dfrac{1}{a^6} + \dfrac{3y}{a^8} + \dfrac{6y^2}{a^{10}} + \dfrac{10y^3}{a^{12}}$ **9** $\dfrac{1}{x^3} + \dfrac{3}{x^5} + \dfrac{6}{x^7} + \dfrac{10}{x^9}$

10 $\dfrac{1}{x^3} - \dfrac{3}{x^6} + \dfrac{6}{x^9} - \dfrac{10}{x^{12}}$ **11** $y^{1/3} + \dfrac{y^{-2/3}}{3} - \dfrac{y^{-5/3}}{9} + \dfrac{5y^{-8/3}}{81}$

13 $2 - \dfrac{x}{12} - \dfrac{x^2}{288} - \dfrac{5x^3}{20,736}$, $-8 < x < 8$ **14** $5 + \dfrac{x}{10} - \dfrac{x^2}{1,000} + \dfrac{x^3}{50,000}$, $-25 < x < 25$

15 $\dfrac{1}{2} + \dfrac{x}{16} + \dfrac{3x^2}{256} + \dfrac{5x^3}{2,048}$, $-4 < x < 4$ **17** $1 + \dfrac{x}{2} - \dfrac{x^2}{8} + \dfrac{x^3}{16}$, $-1 < x < 1$

18 $1 - \dfrac{x}{2} + \dfrac{3x^2}{8} - \dfrac{5x^3}{16}$, $-1 < x < 1$ **19** $1 - \dfrac{x}{2} - \dfrac{x^2}{8} - \dfrac{x^3}{16}$, $-1 < x < 1$ **21** 5.050

22 1.992   **23** 0.513   **25** 1.987   **26** 3.037   **27** 3.018   **29** 0.888
30 0.906   **31** 1.096   **35** 5.099

EXERCISE 18.1

1 1/4, 1/2   **2** 1/13, 5/13   **3** 1/3, 1/6   **5** 1/6   **6** 1/6   **7** 1/3   **9** 1/36,
1/9   **10** 1/9, 1/36   **11** 1/6   **13** 4/13, 10/13   **14** 9/13, 7/13   **15** 1/36
17 $C(28, 13)/C(52, 13)$ **18** $C(20, 5)/C(52, 5)$ **19** $C(12, 3)/C(20, 3)$
21 $C(9, 3)C(6, 5)/C(15, 8)$ **22** $C(6, 2)C(9, 3)/C(25, 5)$
23 $C(4, 1)C(13, 5)/C(52, 5)$ **25** 1/200   **26** 43/8,700   **27** 1/16   **29** 27/35
30 29/41, 21/82   **31** 39/100   **33** 0.49, 0.96   **34** 7/15   **35** 7/15
37 0.04   **38** 5/6   **39** 71/75

EXERCISE 18.2

1 2/5   **2** 1/10   **3** 9/25, 27/125   **5** 14/33   **6** 7/26   **7** 1/2   **9** 55/204
10 14/55, no   **11** 34/65   **13** 1/221   **14** 1/17   **15** 22/425   **17** 7/36
18 1/36   **19** 0.80256   **21** 3/4   **22** 1/8   **23** 1/2   **25** 4/9   **26** 42/169
27 24/49   **29** 2/3   **30** 3/7   **31** 2/3

EXERCISE 18.3

1 $\dfrac{5}{16}, \dfrac{1}{2}$ **2** $\dfrac{5}{72}, \dfrac{2}{27}$ **3** 0.01536, 0.01696   **5** $\dfrac{6,125}{725,594,112}$ **6** $\dfrac{1}{243}$ **7** $\dfrac{40}{243}$

9 $\dfrac{80}{243}$ **10** $\dfrac{96}{625}, \dfrac{112}{625}$ **11** $\dfrac{5}{16}, \dfrac{21}{32}$ **13** $\dfrac{133}{91,125}$ **14** $\dfrac{43,681}{5,717,741,400,000}$

15 $\dfrac{16}{243}$ **17** $\dfrac{1}{3}$ **18** $\dfrac{1}{3}$ **19** 0.451

EXERCISE 18.4

1 25/49   **2** 45/91   **3** 25/49   **5** 7/13   **6** 11/20   **7** 1/6   **9** 5/108
10 2/3   **11** 1/2   **13** 8/27   **14** 328/625   **15** 1,424/3,125, no   **17** 19/50
18 0.65   **19** 1/18

Index

Index

Paul K. Rees | Fred W. Sparks | Charles Sparks Rees

Instructor's Manual to accompany

COLLEGE ALGEBRA
Seventh Edition

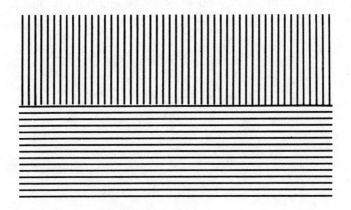

PAUL K. REES
Louisiana State University

FRED W. SPARKS
Texas Tech University

CHARLES SPARKS REES
The University of New Orleans

McGRAW-HILL BOOK COMPANY
New York   St. Louis   San Francisco   Auckland   Düsseldorf
Johannesburg   Kuala Lumpur   London   Mexico   Montreal
New Delhi   Panama   Paris   São Paulo
Singapore   Sydney   Tokyo   Toronto

INTRODUCTION

The authors fully realize that the amount of material that can be covered satisfactorily in a course with a specified number of non-quiz meetings depends to a considerable extent on the industry and ability of the members of the class and on the standard that is maintained. Even so, it may be worthwhile for us to point out what we consider to be the normal number of lessons required by each chapter in the book. With this information at hand, the instructor should be able to decide readily what chapters can be covered in his course. There is enough material in the book for a six-semester or ten term-hour course. We realize that in many cases the instructor will want to make selections from the material and use the book for only a term or a semester. There are ordinarily about thirty-eight non-examination class meetings in a three-semester-hour course and nearly forty-two in a five term-hour course. Our estimate of the number of lessons required for each chapter is given below, but we realize that circumstances may be such that more or less than that number may be desirable.

Exercise 1.1, page 7

4 [Johnson, Nixon, Ford]   **8** [x|x is a color of the rainbow]
12 [2, 3, 5, 7, 11, 14], [2, 5, 11], [3, 7]
16 A, B [x|x is a male student under five feet tall at K College]
20 D, ϕ, [1, 5, 7, 11, 3, 6, 9]   **24** ϕ **28** (3, 4), (3, 6), (6, 4), (6, 6)

Exercise 1.2, page 11

4 Integers   **8** Rational numbers   **12** Real numbers
16 Irrational numbers   **20** Real numbers   **24** $<, <, >$ **28** 2
32 2

Exercise 1.3, page 17

4 [1, 5, 7, 11]   **8** [4, 7, 11], [6, 9], [3, 6, 7, 9]   **12** Positive integer

Exercise 2.2, page 34

4 -3 **8** 0   **12** $a + 6b + 2x$ **16** 0   **20** $3r + 2s + t$
24 $r^2s + 2s^2$ **28** 11   **32** $-a + k + 6p$ **36** $2pq^2 + pq$
40 $-y - a$ **44** 3   **48** 18   **52** $2c - 2d$ **56** $-58g$
60 $11m + 6p - 13q$

Exercise 2.3, page 43

4 $-20a^9$ **8** $30a^9$ **12** $256x^{12}y^8$ **16** $625x^{12}$
20 $-9x^3y^3 + 12x^2y^4$ **24** $15x^5y^5 - 10x^4y^6 - 25x^3y^7$
28 $2x^{12}y^{10}z^{11} + x^{13}y^8z^9 + 6x^6y^6z^{11}$ **32** $5x^2 + 8xy - 21y^2$
36 $6x^3 - 11x^2y + 7xy^2 - 4y^3$ **40** $x^4 - 8x^2y^2 + 9xy^3 - 2y^4$
44 $a + 2b + c$ **48** $2x^3 + 15x^2 - 165x$

Exercise 2.4, page 50

4 c^6 **8** d^6e^3 **12** $5ab^3$ **16** $3x^3 - 5x^2 + 2x$
20 $4x^5y - 7xy^2 - 3x^4y^2$ **24** $3x + 1$ **28** $2x + 3$
32 $3x^2 + 2x + 4$ **36** $3x - 2, 4x - 3$ **40** $2x^3 + 6x^2 - x - 5, -3$

Exercise 2.5, page 56

4 A.5, M.4, M.5   **8** A.1. A.4, A.5, M.1, M.4, M.5   **12** A.1, A.5   **16** 0
20 (a) $b + a + \sqrt{ab}$, commutative axiom holds
    (b) $b + \sqrt{a} + ba$, commutative axiom does not hold
    (c) $ba (b + a)$, commutative axiom holds
    (d) $ba^2 (b^2 + a)$, commutative axiom does not hold

Exercise 2.6, page 56

8 $4y$ 12 1 16 $-10x^7$ 20 $6x^3y^4 - 8x^2y^5$ 24 a^2b^3
28 $2x - 1, 2$

Exercise 3.1, page 60

4 $x^2 + 2x - 8$ 8 $12x^2 - 17x + 6$ 12 $42x^2 - 19xy - 35y^2$
16 $49a^2 + 56ab + 16b^2$ 20 $4a^6 + 20a^3b^2 + 25b^4$ 24 $49a^2 - 70ab + 25b^2$
28 $9a^4 - 42a^2b^2 + 49b^4$ 32 2695 36 $25x^2 - 36$
40 $64x^2 - 25y^2$ 44 $36a^6 - 49b^4$ 48 $25x^2/16y^2 - 4y^2/9x^2$
52 $4x^2 + y^2 + z^2 + 4xy - 4xz - 2yz$ 56 $4x^6 + 12x^5 + 21x^4 + x^3 - 3x^2 - 12x + 4$
60 $72x^2 - 96xy + 32y^2 + 78x - 52y + 15$ 64 $x^4 - 3x^2y^2 + 9y^4$
68 $x^{10} - 15x^6 + 6x^3 + 9x^2 - 1$

Exercise 3.2, page 65

4 $7(x - 3)$ 8 $3x^3(x - 4)$ 12 $7(3x^2 - x + 1)$ 16 $(4x + y)(4x - y)$
20 $(6x + 11y)(6x - 11y)$ 24 $(x^3 + y^7)(x^3 - y^7)$
28 $(4x^4 + 9y^2)(2x^2 + 3y)(2x^2 - 3y)$ 32 $(p - q)(p^2 + pq + q^2)$
36 $(3a + 7b)(9a^2 - 21ab + 49)$ 40 $(6x^3 - 5y^2)(36x^6 + 30x^3y^2 + 25y^4)$
44 $(8x^3y^5 - 7)(64x^6y^{10} + 56x^3y^5 + 49)$ 48 $(9x + 4y - 5z)(9x - 4y + 5z)$
52 $(x - 3y - 4)(x^2 - 6xy + 9y^2 + 4x - 12y + 16)$
56 $(4x^2 + 9y^2)(2x + 3y)(2x - 3y)$
60 $(x^2 + 1)(x^4 - x^2 + 1)(x + 1)(x - 1)(x^4 + x^2 + 1)$
64 $(x + y)(x^2 - xy + y^2)(x^6 - x^3y^3 + y^6)(x - y)(x^2 + xy + y^2)(x^6 + x^3y^3 + y^6)$
68 $(x^3 2y^3)(x^{12} + 2x^9y^3 + 4x^6y^6 + 8x^3y^9 + 16y^{12})$
72 $(5x^4 - 3y^2)(25x^8 \ 15x^4y^2 + 9y^4)$

Exercise 3.3, page 67

4 $(d + 3)(d + 3)$ 8 $(4x - 1)(4x - 1)$ 12 $(8x - 5)(8x - 5)$
16 $(x + 6)(x - 2)$ 20 $(y + 2)(y + 8)$ 24 $(y - 7)(y - 3)$
28 $(4x - 3)(x - 4)$ 32 $(5x - 4)(x + 2)$ 36 $(4x - 3y)(3x + 5y)$
40 $(9x - 5y)(2x + 3y)$ 44 $(6x - 11y)(5x + 9y)$ 48 $(9x - 10y)(7x - 6y)$
52 $[(m - 2n) - 2][(m - 2n) + 3]$ 56 $[4(5x - 2y) + 5z][3(5x - 2y) - 4z]$
60 $D^2 = 19^2, (3x - 2)(2x + 5)$ 64 $D^2 = 217$, not factorable

Exercise 3.4, page 69

4 $(a + 3)(b - 5)$ 8 $(5x - 2y)(x - 3)$ 12 $(a - b)(3c + 2d)$
16 $(x - 1)(x + 1)(3x + 2)$ 20 $2u(u + 2)(2u + 3)$ 24 $5d(2d - 3)^2$
28 $(x - 3y)(2x - y + 2z)$ 32 $(3x + 2y + z)(3x - 2y - z)$
36 $(3x - 4y - 2w)(3x + 4y - 2w)$ 40 $(y^2 - y + 2)(y^2 - y - 2)$
44 $(5x^2 + 2xy + 2y^2)(5x^2 - 2xy + 2y^2)$ 48 $(2x - y)(x + 2y + 4x^2 + 2xy + y^2)$

2

Exercise 3.5, page 69

4 $6d^2 + 11d + 3$ **8** $6t^2 - 13tu - 5u^2$ **12** $28s^2 - 27st - 36t^2$

16 $9x^2 - y^2$ **20** $49x^2 - 25y^2$ **24** $4x^2 - 20x + 25$

28 $25q^2 - 40qr + 16r^2$ **32** $u^2 + 2uv + v^2 - w^2$

36 $4a^2 - 9b^2 + 24bc - 16c^2$ **40** $r^6 - 4r^5 - 2r^4 + 20r^3 - 7r^2 - 24r + 16$

44 $(7x + 6y)(7x - 6y)$ **48** $(3a - 5b + 2c)(3a - 5b - 2c)$

52 $2xy^2(6x^2y + 7)(6x^2y - 7)$ **56** $(3p - 8q)^2$ **60** $(5a^2 - 3b^4)^2$

64 $(5x - y)(2x - y)$ **68** $(5a + 7b)(7a - 3b)$ **72** $(4a^2 + 3b^2)^2$

76 $(3x + y)(9x^2 - 3xy + y^2)$ **80** $(2 + y^2)(4 - 2y^2 + y^4)$

84 $(3 - 3a - b)(9 + 9a + 3b + 9a^2 + 6ab + b^2)$ **88** $(m - n - 1)(m + n)$

Exercise 4.1, page 77

4 $\dfrac{st}{-t^2}$ **8** $\dfrac{1}{2x}$ **12** $\dfrac{4y - 5x}{3y - 2x}$ **16** $\dfrac{r^2 - 5r + 6}{r^2 - 9}$

20 $\dfrac{3w^2 + 5w - 2}{(w + 1)(w + 2)}$ **24** $\dfrac{2x - 1}{x + 4}$ **28** $\dfrac{4x^2}{9yz^2}$ **32** $\dfrac{3x + 2}{2x + 3}$ **36** $\dfrac{x + 1}{x - 1}$

40 $\dfrac{a - 3b}{3a - b}$ **44** $\dfrac{x^6 - x^3y^3 + y^6}{(x - y)(x^2 + xy + y^2)}$ **48** $\dfrac{1}{3x + 1}$ **52** $\dfrac{1}{x - 1}$

56 $[\ \dfrac{5y^2}{x^3y^3}, \dfrac{-4xy}{x^3y^3}, \dfrac{-3x^2}{x^3y^3}\ ]$

60 $[\ \dfrac{3x^2 - 10xy + 3y^2}{(x \cdot 2y(x + y)(3x - y)}, \dfrac{2x^2 - 5xy + 2y^2}{(x - 2y)(x + y)(3x - y)}, \dfrac{x^2 - y^2}{(x - 2y)(x + y)(3x - y)}\ ]$

Exercise 4.2, page 79

4 $-\dfrac{1}{4}$ **8** $\dfrac{22x + 9}{12}$ **12** $\dfrac{7}{9a}$ **16** $\dfrac{-b}{36a}$ **20** $\dfrac{2(x - 5y)(3x + y)}{5y(2x - y)}$

24 $\dfrac{29(x^2 - 2y^2)}{(5x - 3y)(2x - 7y)}$ **28** $\dfrac{(r + 2s)}{r}$ **32** $\dfrac{2x + 1}{x + 3}$ **36** $\dfrac{1}{(a + b)(a + 2b)}$

40 $\dfrac{-11x^3 - 26x^2y - 31xy^2 - 21y^3}{(x + 2y)(4x + 3y)(x + 3y)}$ **44** $\dfrac{4(x^2 + 2xy + 3y^2)}{(2x + 5y)(x + y)(x - y)}$

Exercise 4.3, page 83

4 $2xy^3zw^3$ **8** $\dfrac{16y^7z^9}{15x^3w^5}$ **12** $\dfrac{12x^2}{y^8z^3}$ **16** $\dfrac{3s}{qr}$ **20** $\dfrac{x + y}{x}$

24 $\dfrac{y^5}{6x^3}$ **28** $\dfrac{xy^2(3x + y)}{2(x - 2y)}$ **32** $\dfrac{y(x + z)}{z^2}$ **36** $\dfrac{x + 3y}{x - 3y}$ **40** $x + 1$

44 $\dfrac{x + 2}{x + 3}$

Exercise 4.4, page 87

4 $\dfrac{11}{5}$ **8** $\dfrac{x}{2x - 3}$ **12** $\dfrac{8 - x}{5}$ **16** $\dfrac{3x - 2}{4x - 1}$ **20** $\dfrac{y + 2x}{y - x}$

3

$24\ \dfrac{a}{a+1}$ $28\ \dfrac{2}{p-2}$ $32\ \dfrac{-1}{x+1}$ $36\ \dfrac{(x-2)(x+1)}{2x}$

Exercise 4.5, page 88

$4\ \dfrac{-xy}{(-y^2)}$ $8\ \dfrac{b^2+5b+6}{b^2+b-2}$ $12\ \dfrac{3x^2y^2}{5z^2}$ $16\ \dfrac{a+b}{a-2b}$ $20\ \dfrac{1}{x+2}$

$24\ \dfrac{11x-4}{12}$ $28\ \dfrac{a}{b(a-b)}$ $32\ \dfrac{5x^2-3xy+5y^2}{(2x-y)(x-2y)(x+3y)}$

$36\ \dfrac{9zy^2}{2w^2x}$ $40\ \dfrac{(x-2y)(x+2y)}{(x-y)(x-3y)}$ $44\ \dfrac{1}{2}$ $48\ \dfrac{2x-1}{2x+5}$ $52\ \dfrac{1}{(x+1)}$

Exercise 5.1, page 100

$4\ \{5\}$ $8\ \{9\}$ $12\ \{2\}$ $16\ \{-12\}$ $20\ \left\{\dfrac{3}{2}\right\}$ $24\ \left\{\dfrac{-3}{2}\right\}$

$28\ \{8\}$ $32\ \{3\}$ $36\ \{5\}$ $40\ \left\{\dfrac{(3a-5b)}{(a+b)}\right\}$ $44\ \left\{\dfrac{ab}{b-3a}\right\}$

Exercise 5.2, page 106

$4\ \{x|x<3\}$ $8\ \{x|x>4\}$ $12\ \{x|x<3\}$ $16\ \{x|x<-3\}$

$20\ \left\{x|x<\dfrac{-17}{7}\right\}$ $24\ \{x|x>-6\}$ $28\ \{x|x<2\}\cup\{x|x>5\}$

$32\ \left\{x|\dfrac{-5}{3}<x<\dfrac{3}{2}\right\}$ $36\ \left\{x|x<\dfrac{-7}{4}\right\}\cup\left\{x|x>\dfrac{3}{2}\right\}$

$40\ \left\{x|\dfrac{5}{9}<x<\dfrac{4}{3}\right\}$ $44\ \{x|x>3\}\cup\left\{x|\dfrac{-5}{2}<x<\dfrac{3}{2}\right\}$

$48\ \{x|x<-1\}\cup\left\{x|\dfrac{7}{4}<x<\dfrac{7}{2}\right\}$

Exercise 5.3, page 108

$4\ \{0\}$ $8\ \{3\}$ $12\ \{8\}$ $16\ \{5\}$ $20\ \left\{\dfrac{17}{2}\right\}$ $24\ \phi$

$28\ \{6\}$ $32\ q=\dfrac{p\mathcal{R}}{2p-\mathcal{R}}$ $36\ f=\dfrac{25L}{FM-L}$

Exercise 5.4, page 112

$4\ \{-5,5\}$ $8\ \{-1,5\}$ $12\ \{0,3\}$ $16\ \left\{-1,\dfrac{10}{3}\right\}$ $20\ \{1\}$

$24\ \left\{-\dfrac{1}{4},3\right\}$ $28\ \{x|x<-7\}\cup\{x|x>7\}$ $32\ \{x|-12<x<-2\}$

$36\ \text{all } x$ $40\ \{x|x>1\}\cup\left\{x|x<\dfrac{1}{5}\right\}$ $44\ \{x|-1<x<1\}$

$48\ \left\{x|x<-\dfrac{1}{4}\right\}\cup\{x|x>3\}$

Exercise 5.5, page 119

4 1760 yd.	**8** $700	**12** 557	**16** 30 days	**20** 25
24 625 mph	**28** 33 min., 20 sec.	**32** 5:05 P.M.	**36** 18	
40 100	**44** 12 mph	**48** 12 yd., $6.50 per yd.		

Exercise 5.6, page 124

4 $\{5\}$ **8** $\left\{-\dfrac{154}{167}\right\}$ **12** $\{x|x<2\}$ **16** $\left\{x|x<-\dfrac{1}{3}\right\}\cup\left\{x|x>\dfrac{8}{5}\right\}$

20 $\left\{x|x>\dfrac{5}{2}\right\}\cup\left\{x|-\dfrac{8}{3}<x<-\dfrac{3}{5}\right\}$ **24** 12   **28** $\left\{-\dfrac{4}{7},4\right\}$

32 $\{x|-1<x<5\}$

Exercise 6.1, page 131

4 $\{(1, 7), (3, 9), (5, 11), (4, 10), (8, 2) (12, 6)\}$
8 Function, no two points have the same first element and different second elements.
12 $\{(-1, 9), (0, 1), (1, 1), (2, 9), (3, 25)\}$, function
16 $\{(0, 16), (\sqrt{7}, 9), (4, 0)\}$, function   **20** $-3, 13$ **24** $-1, 3x^2 + 2x - 1$
28 $40, 2h^2 + 3h - 4, 2h^2 + 19h + 40$ **32** 4h   **36** $2t^2, 2$
40 $\{(-1, 1), (2, 19), (4, 89), (14, 271)\}$ **44** $\{2, 3, 6, 11\}$
48 $\{(1, -6), (2, -5)\}$

Exercise 6.2, page 140

8 The line parallel to and 3 units to the right of the y axis
12 The ray parallel to the x axis and extending to the right from $(2, -4)$

40 $x = -\dfrac{7}{4}, y = 7$ **44** $x = -7, x = 7, y = -7$

Exercise 6.3, page 149

4 The set of points is $\{(2,1)\}$
20 $\{(x, y)|y = \pm\sqrt{4 - x^2}/2, R = \{x|-2 \leqslant x \leqslant 2\}$, a relation
24 $\{(x, y)|y = (x + 1)/3\}, R = \{x|x \geqslant 1\}$, a function
28 $\{(4, y)|y = (1 + 4x)/(1 - x)\}, R = \{x|0 < x < 1\}$, a function
32 161 ft per sec.   **36** 6000 lb   **40** (10,000/9) lb.

Exercise 6.4, page 151

4 $\{(-1 \pm \sqrt{2}), (0, \pm\sqrt{3}), (1, \pm 2), (2, \pm\sqrt{5}), (3, \pm\sqrt{6})\}$, not a function
8 $2hx + h^2 - h$ **12** $\{-13, -7, -1, 5, 11\}$
16 Down from $(-1, 3)$ and parallel to the y axis.   **28** .054 in.

5

Exercise 7.1, page 155

4 64   **8** 64   **12** 1,953,125   **16** 32/243   **20** 36,905,625
24 $6x^5y^4z^5$ **28** $4x^2y^2$ **32** $343\,a^9b^6c^{12}$ **36** $81x^{12}y^{20}/16p^8q^{12}$
40 $16,875x^{14}y^{13}$ **44** $2b^7/3cd^5$ **48** $27/2b^2c^7$ **52** $4d^8/9d^{10}c^8$
56 c^9/a^2b^3 **60** $x^{2a}y^5$ **64** $a^{6n}b^{6n\,+\,1}$ **68** $a^{2p\,-\,2n}b^{3p\,-\,3n}$

Exercise 7.2, page 161

4 1   **8** 128   **12** 25   **16** 1   **20** 4608   **24** $\dfrac{9}{256}$

28 $\dfrac{64}{729}$ **32** a^{-2} **36** $2^{-1}xy^{-1}$ **40** $\dfrac{1}{a^3b^2}$ **44** $\dfrac{b^4}{a^3}$

48 $\dfrac{2a^2}{3b}$ **52** $\dfrac{b^8}{a^{12}}$ **56** $\dfrac{b^9}{a^{12}}$ **60** $\dfrac{\ell^2}{a^6p^2}$ **64** 0

68 $\dfrac{1}{y^2 - xy + x^2}$ **72** $\dfrac{1}{(y - x)}$ **76** $\dfrac{18x + 5}{(2x - 1)^3\,(3x + 1)^2}$

80 $\dfrac{18x - 1}{(2 - 3x)^3\,(4x + 1)^2}$

Exercise 7.3, page 164

4 3   **8** 0.3   **12** 49   **16** 0.04   **20** $\dfrac{1}{125}$ **24** $\dfrac{8}{27}$

28 $\dfrac{32}{243}$ **32** 7   **36** 27   **40** $3a^2b^4$ **44** $\dfrac{27a^4}{b^3}$ **48** $\sqrt[5]{rs^2}$

52 $\sqrt[7]{\dfrac{x^3}{y^5}}$ **56** $d^{11/15}$ **60** $s^{7/12}$ **64** $20x^{8/15}$ **68** $\dfrac{x^{5/6}}{y}$

72 $\dfrac{2b}{a^2}$ **76** $\dfrac{x^2}{5y^{2/7}}$ **80** $\dfrac{5a^2}{2b^{1/3}}$ **84** $\dfrac{1}{2a^2b^3}$ **88** $a^2 + b^2$

92 $\dfrac{30x + 17}{(3x + 4)^{3/4}}$ **96** $\dfrac{17x + 9}{(3x + 5)^{1/3}\,(4x - 3)^{1/4}}$ **100** $x^{b/(a + b)}$

104 $x^{a\,+\,b}$

Exercise 7.4, page 168

1 $7\sqrt{6}$ **8** $2\sqrt[5]{5}$ **12** $5a^4b\sqrt{5ab}$ **16** $5a^3b^3\sqrt{7b}$ **20** $7a^3b^3\sqrt[3]{2b}$
24 $3a^2b\sqrt[5]{2ab}$ **28** $\dfrac{3r^2}{4s^4}\sqrt{2r}$ **32** $\dfrac{d^2}{3c^3}\sqrt[3]{10c^2d^2}$ **36** $t^au^{2d}\sqrt[5]{t^a}$

40 $\sqrt[4]{1250}$ **44** $\sqrt[5]{729x^7y^{11}}$ **48** 12   **52** $14\sqrt[4]{3}$ **56** $6x^2y\sqrt{3y}$
60 $6x^2y^2\sqrt[4]{5y}$ **64** $2xy\sqrt[7]{5y}$ **68** $\dfrac{\sqrt{21}}{3}$ **72** $\dfrac{3x^3\sqrt{6xy}}{4y^3}$ **76** $\dfrac{4x^2y\sqrt{2}}{7}$

80 $\dfrac{ab\sqrt[5]{48a^2b}}{3}$ **84** $-2 - \sqrt{7}$ **88** $-6\sqrt{6} - 7\sqrt{7}$

Exercise 7.5, page 174

4 $\sqrt[20]{6}$ **8** $\sqrt[8]{13b}$ **12** $\sqrt[8]{81x^2}$ **16** $\sqrt[20]{81x^8y^{16}}$

20 $\sqrt[30]{a^{15}b^{15}}$, $\sqrt[30]{a^{20}b^{10}}$, $\sqrt[30]{a^6b^{12}}$ **24** $\sqrt{2xy^2}$ **28** $\sqrt[5]{6x^3y^2}$

32 $4\sqrt{5}$ **36** 0 **40** $15\sqrt{2} - 5\sqrt{3}$ **44** $(2uv - 3u^2 + 2v^2)\sqrt{2u}$

48 $(x^2 + 3x^{-1})\sqrt{3x/y}$ **52** $2xy(2x - 1)\sqrt[3]{x^2y} + x(2y - 3x^2\sqrt[3]{xy^2}$

56 $\sqrt[5]{ab}$

Exercise 7.6, page 175

4 64 **8** a^4b^8 **12** $\dfrac{32a}{c^2d}$ **16** $\dfrac{1}{81}$ **20** $\dfrac{6b^3}{a^4}$ **24** 2

28 $\dfrac{26a^2}{b}$ **32** $\dfrac{4a^{1/5}}{b}$ **36** $3\sqrt{3}$ **40** 4 **44** $3x^3y^2\sqrt{10}$

48 $\sqrt[3]{4}$ **52** $3\sqrt{a} + 2\sqrt{b} + 3\sqrt[3]{a}$ **56** $7\sqrt{10} - 3$

Exercise 8.1, page 179

4 $\left\{6, -6\right\}$ **8** $\left\{\dfrac{1}{7}, -\dfrac{1}{7}\right\}$ **12** $\left\{\dfrac{7}{6}, -\dfrac{7}{6}\right\}$ **16** $2, 5$

20 $\left\{1, -3\right\}$ **24** $\left\{3, \dfrac{1}{4}\right\}$ **28** $\left\{6, -\dfrac{1}{6}\right\}$ **32** $\left\{4, \dfrac{3}{4}\right\}$

36 $\left\{2, -\dfrac{3}{5}\right\}$ **40** $\left\{-1, \dfrac{4}{5}\right\}$ **44** $\left\{\dfrac{3}{4}, -\dfrac{4}{3}\right\}$ **48** $\left\{\dfrac{3}{7}, -\dfrac{7}{2}\right\}$

60 $\left\{\dfrac{3a}{2}, \dfrac{2c}{3}\right\}$

Exercise 8.2, page 184

4 $\left\{4, 1\right\}$ **8** $\left\{-3, -5\right\}$ **12** $\left\{1, \dfrac{3}{5}\right\}$ **16** $\left\{-3, -\dfrac{2}{5}\right\}$

20 $\left\{\dfrac{3}{2}, -\dfrac{4}{5}\right\}$ **24** $\left\{-\dfrac{3}{2}, -\dfrac{4}{5}\right\}$ **28** $\left\{5 + \sqrt{3}, 5 - \sqrt{3}\right\}$

32 $\left\{-1 + \sqrt{7}, -1 - \sqrt{7}\right\}$ **36** $\left\{3 + 5i, 3 - 5i\right\}$ **40** $\left\{-2 + 7i, -2 - 7i\right\}$

44 $\left\{2b, -b\right\}$ **48** $\left\{\dfrac{a + 2bi}{3}, \dfrac{a - 2bi}{3}\right\}$

Exercise 8.3, page 187

4 $\left\{-5, 3\right\}$ **8** $\left\{-2, -5\right\}$ **12** $\left\{4, \dfrac{1}{2}\right\}$ **16** $\left\{2, -\dfrac{3}{7}\right\}$

20 $\left\{\dfrac{3}{7}, \dfrac{2}{3}\right\}$ **24** $\left\{-\dfrac{2}{5}, -\dfrac{3}{5}\right\}$ **28** $\left\{5 + \sqrt{7}, 5 - \sqrt{7}\right\}$

32 $\left\{ \dfrac{5+\sqrt{3}}{2}, \dfrac{5-\sqrt{3}}{2} \right\}$ **36** $\left\{ 5+3i, 5-3i \right\}$ **40** $\left\{ \dfrac{5+4i}{2}, \dfrac{5-4i}{2} \right\}$

44 $\left\{ \dfrac{4+i\sqrt{7}}{2}, \dfrac{4-i\sqrt{7}}{2} \right\}$ **48** $\left\{ -a, 2b \right\}$ **52** $\left\{ \dfrac{4a+2ib}{3}, \dfrac{4a-2ib}{3} \right\}$

56 $\left\{ x-2, x+1 \right\}$ **60** $\left\{ 2x+5, x-7 \right\}$ **64** $\left\{ 4.646, -0.646 \right\}$

68 $\left\{ 2.609, 0.724 \right\}$

Exercise 8.4, page 190

4 $\left\{ -5, -2, 2, 5 \right\}$ **8** $\left\{ -1, 1, -3i, 3i \right\}$ **12** $\left\{ -\dfrac{5}{2}, \dfrac{5}{2}, -\dfrac{3i}{4}, \dfrac{3i}{4} \right\}$

16 $\left\{ -2, -1 \right\}$ **20** $\left\{ -1, -\dfrac{2}{3}, \dfrac{2}{3}, 1, -i, -\dfrac{2i}{3}, \dfrac{2i}{3}, i \right\}$

24 $\left\{ -\dfrac{2}{3}, \dfrac{2}{3}, \dfrac{3i}{4}, -\dfrac{3i}{4} \right\}$ **28** $\left\{ \pm 8i, \pm 1 \right\}$ **32** $\left\{ \pm 1024i, \pm 1 \right\}$

36 $\left\{ \pm \dfrac{27}{64}, \pm 8 \right\}$ **40** $\left\{ 0, 2, -2 \right\}$ **44** $\left\{ 1, -3, -1+2\sqrt{6}, -1, -2\sqrt{6} \right\}$

48 $\left\{ \dfrac{1}{12}, \dfrac{2}{7} \right\}$ **52** $\left\{ -\dfrac{5}{2}, \dfrac{1}{2} \right\}$

Exercise 8.5, page 194

4 3   **8** -1 **12** 6   **16** 2   **20** 5   **24** -1 **28** 0
32 3   **36** 3   **40** $-3, 8$ **44** -5 **48** 4   **52** 3
56 -1 **60** 0   **64** 0   **68** a

Exercise 8.6, page 201

4 Rational, unequal   **8** Irrational, unequal   **12** Irrational, unequal

16 Conjugate complex   **20** Rational, equal   **24** $-2, \dfrac{-5}{3}$

28 $\dfrac{3}{2}, -3$ **32** $-2\sqrt{3} (4-\sqrt{3} / 13, -4(4-\sqrt{3}) / 13$ **36** $4, -4$

40 4   **44** $-2, -1$ **48** $x^2 - 14x + 49 = 0$ **52** $x^2 - 4x + 7 = 0$
56 $x^2 - 6x + 7 = 0$ **60** $3x+4, 6x-1$ **64** $5x+4+\sqrt{5}, 5x+4-\sqrt{5}$
68 $5x-3+5i, 5x-3-5i$

Exercise 8.8, page 207

4 $\dfrac{4}{3}, -\dfrac{3}{4}$ **8** $2-\sqrt{11}, 2+\sqrt{11}$ **12** $3x+2, -4x+5$

16 $1, 243, -1, -243$ **20** 3   **24** 7   **28** 9   **32** Rational, equal

36 $\dfrac{1}{3}$

Exercise 9.2, page 220

4 $\{(-0.2, -1.4), (-1.3, -3,6)\}$ **8** $\{(0, 2.0), (1.3)\}$
12 $\{(1.6, 4.9), (-1.6, -4.9)\}$ **16** $\{\phi\}$ **20** $\{(1.9, -1.2), (-1.9, -1.2)\}$
24 $\{(9, 7)\}$ **28** Dependent
32 $(1, -2), (-1, -2), (\sqrt{761}/145, 8/145), (-\sqrt{761}/145, 8/145)$
36 Inconsistent

Exercise 9.3, page 228

4 $(1, 1)$ **8** $(-2, -1)$ **12** $(-1, 2)$ **16** $(-\frac{1}{2}, \frac{2}{3})$
20 $(2, 3), (2, -3), (-2, 3), (-2, -3)$ **24** $(0, 2), (0, -2)$
28 $(2, 2), (2, -2), (\frac{16}{5}), 2\sqrt{46}/5), (\frac{16}{5}), -2\sqrt{46}/5)$

32 $(1, -2), (-1, -2)$ **36** $(1, 4), (2, 2.5)$ **40** $(3, 1), (-\frac{47}{21}, -\frac{7}{3})$

Exercise 9.4, page 232

4 $\{(1, -1)\}$ **8** $\{(-1, -2)\}$ **12** $\{(2, 5)\}$ **16** $\{(1, -2), (-\frac{5}{9}, \frac{8}{3})\}$
20 $\{(-1, 2), (\frac{49}{29}, \frac{32}{29})\}$ **24** $\{(1, 3), (-\frac{7}{17}, \frac{39}{17})\}$
28 $\{(0, 10), (200, -10), (-1 - 10i, i), (-1 + 10i, -i)\}$
32 $\{(2, 14), (-2, 10), (2i - 12 + 2i), (-2i, -12 - 2i)\}$
36 $\{(0, 1), (-2, -1), (1 + \sqrt{3}, \sqrt{3}), (1 - \sqrt{3}, -\sqrt{3})\}$
40 $\{(1, 2), (-1, -2), (i\sqrt{3}, -2i\sqrt{3}/3), (-i\sqrt{3}, 2i\sqrt{3}/3)\}$
44 $\{(5, -1), (-5, 1), (\sqrt{2}, -5\sqrt{2}/2), (-\sqrt{2}, 5\sqrt{2}/2)\}$

Exercise 9.5, page 238

4 $\{(1, -\frac{2}{3}), (-1, \frac{2}{3}), (0, i\sqrt{3/3}), (0, -i\sqrt{3/3})\}$

8 $\{(3, \frac{1}{2}), (-3, -\frac{1}{2}), (2i, 2.5i), (-2i, -2.5i)\}$

12 $\{(2\sqrt{3}, 5\sqrt{3}), (-2\sqrt{3}, -5\sqrt{3}), (\sqrt{3}, -2\sqrt{3}), (-\sqrt{3}, 2\sqrt{3})\}$

16 $\{(2, 1), (-2, -1), (\sqrt{3}, \sqrt{3}), (-\sqrt{3}, -\sqrt{3})\}$ **20** $\{(3, 1), (\frac{21}{10}, -\frac{17}{10})\}$

24 $\{(3, 2), (-1, 2)\}$ **28** $\{(1, 1)\}$ **32** $\{(1, 2), (2, 1), (3, 2), (2, 3)\}$
36 $\{(6, 2), (2, 6), (7, -1), (-1, 7)\}$ **40** $\{(2, 2)\}$

9

Exercise 9.6, page 241

28 $(-3, 1), (0, 3), (-2, -4)$ **32** $(9, -1), (5, 2), (3, -2), (6, -4)$

Exercise 9.7, page 244

4 $m = -14$ at $(2, -2)$, $M = -2$ at $(2, 2)$
8 $m = -14$ at $(-1, 1)$, $M = 20$ at $(3, 2)$
12 $m = 4$ at $(-1, -1)$, $M = 19$ at $(2, 2)$
16 $m = -12$ at $(0, 1)$, $M = 20$ at $(2, -3)$
20 4 houses of type A, 14 houses of type B, profit \$18,000
24 30 acres of rice, 75 acres of soybeans, profit \$4020
28 A, 25 units; B, 30 units; C, 110 units

Exercise 9.8, page 249

4 3 organ, 5 piano   **8** 165 mi at 55 mph, 110 mi at 40 mph
12 \$2.20, \$.55   **16** \$65 per mo, \$80 per mo
20 19, 10, 3 of ages 18, 19, 20, respectively   **24** 6 ft, 8 ft
28 200 shares at \$55 per share   **32** 8 yd, 16 yd

Exercise 9.9, page 253

12 $[(1.0, 0.2) (-1.0, 0.2)]$ **16** $(1, -1)$
20 $(1, 2), (1, -2), (-1, 2), (-1, -2)$ **24** $(3, 2), (2, 3), (3, 1), (1, 3)$
28 $m = 7$ at $(0, 2)$, $M = 41$ at $(5, 3)$

Exercise 10.1, page 258

4 $a = 1, b = 3, c = -2, d = 0$ **8** $\begin{bmatrix} 6 & -2 & 4 \\ 8 & 0 & 10 \end{bmatrix}$ **12** $\begin{bmatrix} 17 & -15 & 6 \\ 8 & -20 & 31 \end{bmatrix}$
16 Not defined   **28** $\dfrac{1}{3}\begin{bmatrix} 0 & -1 \\ 1 & -6 \end{bmatrix}$ **32** $\begin{bmatrix} 3 & 6 & -3 \\ -3 & 5 & 1 \\ 11 & 1 & 27 \end{bmatrix}$ **40** $\begin{bmatrix} 2 & -4 \\ 11 & -1 \\ 15 & 0 \end{bmatrix}$

Exercise 10.2, page 263

4 $x = -1, y = -3$ **8** $x = 3, y = -2, z = -3$ **12** $x = \dfrac{1}{3}, y = \dfrac{1}{4}, z = -\dfrac{1}{8}$
16 $x = 1, y = 1, z = -1, w = 0$ **20** $\begin{bmatrix} -2 & 4 \\ -3 & 5 \end{bmatrix}$
24 $\dfrac{1}{5}\begin{bmatrix} -28 & 21 & -8 \\ 8 & -6 & 3 \\ 35 & -25 & 10 \end{bmatrix}$ **28** $\dfrac{1}{84}\begin{bmatrix} 38 & 7 & 5 \\ 14 & 7 & -7 \\ 23 & -3.5 & -2.5 \end{bmatrix}$

32
$$\begin{bmatrix} -5 & 3 & 3 & 2 \\ 3 & -1 & -2 & -1 \\ 3 & -2 & -1 & -1 \\ -2 & 1 & 1 & 1 \end{bmatrix}$$

Exercise 10.3, page 268

4 -2 **8** -2 **12** 11   **16** $ab + 6$ **20** $ab - c^2$ **24** -12
28 0   **32** 130   **36** -6 **40** $m(a+1)(m-a)$

Exercise 10.4, page 273

24 6   **28** $\frac{1}{2}, -4$ **32** $1, \frac{3}{2}$ **36** 2, 6

Exercise 10.5, page 280

4 $\{(2, 5)\}$ **8** $\{(-1, 2)\}$ **12** $\{(3, -5)\}$ **16** ϕ
20 $\{(a + 2b, b - 2a)\}$ **24** $\{(3, 2, -2)\}$ **28** $\{(-1, -2, 3)\}$
32 $\{(-3z + 15, 4z - 18, z)\}$ **36** $\{(5, -2, -3)\}$
40 $\{(2a - b, a + 2b, -a)\}$

Exercise 10.6, page 281

4 $\begin{bmatrix} -12 & 10 & 36 & 24 \\ -9 & 6 & 21 & 15 \\ 6 & -3 & -10 & -8 \end{bmatrix}$ **8** $\begin{bmatrix} -12 & 10 & 36 & 24 \\ -9 & 6 & 21 & 15 \\ 6 & -3 & -10 & -8 \end{bmatrix}$ **12** 10
16 48   **24** 7   **28** $x = 3, y = 5$ **32** $x = 3, y = 0, z = -4$

Exercise 11.1, page 289

4 $1.7 (10)$ **8** $8.263 (10^2)$ **12** $7.429(10^{-1})$ **16** $5.000 (10^3)$
20 $5.00 (10^3)$ **24** 0.5845 **28** 0.001235 **32** $9.759 (10^5)$
36 483.6 **40** 32   **44** $4.44 (10^3)$ **48** $4.89 (10^2)$
52 $4.3(10^2)$ **56** $1.1 (10)$ **60** $9.1(10)$ **64** $6.49(10)$ **68** $9.88 (10)$
68 $9.88 (10)$

Exercise 11.2, page 293

4 $\log_7 49 = 2$ **8** $\log_2 \frac{1}{16} = -4$ **12** $\log_{\frac{4}{7}} \frac{49}{16} = -2$

16 $\log_{25} 125 = \frac{3}{2}$ **20** $7^3 = 343$ **24** $10^{-3} = 0.001$ **28** $32^{6/5} = 64$

32 6   **36** $\frac{3}{4}$ **40** 6   **44** 32   **48** 7   **52** $\frac{3}{2}$ **56** 23

60 12   **64** 5   **68** 36   **72** $\frac{13}{4}$ **76** $\frac{1}{2}$

Exercise 11.3, page 299

4 .5028, 0   **8** .7667, − 2   **12** 2   **16** 1
20 − 2   **24** − 3   **28** 1.8055   **32** 0.4031
36 0.7782   **40** 3.7763   **44** 2.6479   **48** 3.8160
52 7.8583 − 10   **56** 7.7760 − 10   **60** 9.9427 − 10
64 7.1329 − 10   **68** 7.7342 − 10

Exercise 11.4, page 301

4 271   **8** 1.37   **12** $4.31\,(10^3)$ **16** $2.77\,(10^{-4})$
20 $4.73\,(10^{-1})$ **24** 153   **28** $2.12\,(10^4)$ **32** $2.51\,(10^{-4})$
36 15.05   **40** $5.022\,(10^5)$ **44** $8.308\,(10^{-4})$ **48** $1.716\,(10^{-4})$

Exercise 11.5, page 307

4 $\log(b − x) + \log(c − y)$ **8** $2\log a + 5\log x − \log b − (\frac{1}{2})\log(ax − b)$

12 $\log(\frac{a}{bc})$ **16** $\log(a^3\sqrt[3]{a − b})/\sqrt{b}$ **20** 820   **24** 2.26

28 11.6   **32** 1.01   **36** $2.69(10^5)$ **40** $9.62(10^{-1})$ **44** 35.8
48 2.68   **52** 0.945   **56** 2.19   **60** 0.418   **64** 4294
68 0.2899   **72** 53   **76** 0.0919   **80** 3.07   **84** 3.546

Exercise 11.6, page 310

4 3.78   **8** 7.21   **12** 2.98   **16** 3.07   **20** 2.34   **24** 1.59

Exercise 11.7, page 313

4 $\ln(\frac{3}{y})$ **8** $\dfrac{\log(3/y)}{\log(c+r)}$ **12** $\dfrac{e^{2y} + 2}{2e^{2y} + 1}$ **16** $\partial n(e^y \pm \sqrt{e^{2y} − 1})$
20 3   **24** 2   **28** 1, 3   **32** 0.595   **36** 2   **40** 1, 3
44 (0.42, 1.58)   **48** $(10\sqrt{5}, 2\sqrt{5})$

Exercise 11.8, page 317

28 (0, 1), (−.87, .24)   **32** (1, 0), (.21, − .67)

Exercise 11.9, page 318

4 116.019   **8** − 11.6   **12** 277.623   **16** 54.8   **20** 18.5
24 3.9932   **28** 8.8416 − 10   **32** 6.44   **36** 1.21   **40** 1.38

Exericse 11.10, page 319

4 585	8 69	12 -0.66	16 $64^{1/3} = 4$	20 343
24 $9.8535 - 10$	28 0.0140	32 3.509	36 2.80	40 1

Exercise 12.1, page 325

4 $x = 1, y = -3$ 8 $x = 4, y = -3$ 12 $(-1, 0)$ 16 i

20 $(-6, -1)$ 24 2i   28 $(-15, 9)$ 32 $-26 - 7i$

36 $(\dfrac{-2}{5}, \dfrac{-1}{5})$ 40 $\dfrac{31}{53} + \dfrac{29i}{53}$ 44 $\dfrac{1 - i}{2}$ 48 $-119 - 120i$

Exercise 12.2, page 329

24 17, $\arctan \dfrac{-8}{15}$ 28 $4, 210°$ 32 $\sqrt{34}$, $\arctan \dfrac{5}{3}$ 36 $3\sqrt{2}$ cis $315°$

40 2 cis $240°$ 44 5 cis $180°$ 48 61 cis $260°$ 52 $\sqrt{58}$ cis $293°$

Exercise 12.3, page 333

4 5.7608 cis $270°$ 8 120 cis $37°$ 12 1.5 cis $135°$ 16 .875 cis $54°$

20 100 cis $100°$ 36 4 cis $360° = 4$ 40 16.12 cis $82° = 2.27 + 15.96i$

44 16.25 cis $136° = -11.69 + 11.29i$ 48 1 cis $900° = -1$

52 4 cis $180° = -4$ 56 1.29 cis $37° = 1.03 + .77i$

60 2 cis $150° = -\sqrt{3} + 1$

Exercise 12.4, page 337

4 343 cis $0°$ 8 16 cis $240°$ 12 256 cis $240°$ 16 625 cis $32°$

20 $73\sqrt{73}$ cis $27°$ 24 2 cis $(18 + k\,72°)$, $k = 0, 1, 2, 3, 4$

28 $2^{1/6}$ cis $(15° + k\,120°)$, $k = 0, 1, 2$

32 $2^{1/8}$ cis $(15° + k\,45°)$, $k = 0, 1, 2, 3, 4, 5, 6, 7$

36 $2^{1/10}$ cis $(60° + k\,72°)$, $k = 0, 1, 2, 3, 4$

40 $\sqrt[4]{13}$ cis $(17° + k\,90°)$, $k = 0, 1, 2, 3$

48 2 cis $(54° + k\,72°)$, $k = 0, 1, 2, 3, 4$ 52 $1 - i, -1 - i$

Exercise 12.5, page 338

4 $(5, 6)$ 8 $(23, -2)$ 12 $2^{1/6}$ cis $(10° + k\,60°)$, $k = 0, 1, 2, 3, 4, 5$

24 $\dfrac{\sqrt{3}}{2}$ cis $(\arctan \sqrt{2})$

Exercise 13.1, page 342

4 4   8 1   12 -1 28 14

13

Exercise 13.2, page 346

4 $3x^2 + 4, -1$ **8** $-2x^2 + x, -2$ **12** $2x^4 - 3x^2 - x + 5, -12$
16 $2x^3 - 2a^2x + a^3, 4a^4$ **20** -8 **24** 0 **28** a^4 **44** -1
48 2

Exercise 13.3, page 351

4 $8; 4, 3; -2, 2; 1, 2; -10, 1$ **8** $9; \dfrac{7}{4}, 2; 3i, 2; -3i, 2; \dfrac{1}{2}, 3$

12 $-1, 2, 3$ **16** $-1, 3, 3i, -3i$ **20** $\dfrac{3}{4}, i\sqrt{6}, -i\sqrt{6}$

24 $\dfrac{5}{2}, -3 - 2i, -3 + 2i$ **28** $-1, 2 + 2i, 2 + 2i, 2 - 2i, 2 - 2i$

32 $4x^3 - 21x^2 + 22x + 30$ **36** $x^6 + 12x^4 + 48x^2 + 64$
40 $A = 1, B = 2, C = 21$ **44** $A = 1, B = 2, C = -5$
48 $2x^5 - 23x^4 + 106x^3 - 262x^2 + 392x - 255 = 0$

Exercise 13.4, page 358

4 $4, -4$ **8** $3, -3$ **12** $1, -5$ **16** 2 or 0 positive, 1 negative
20 3 or 1 positive, 1 negative   **24** 3 or 1 positive, 1 negative
28 -1 and 0, 3 and 4, 6 and 7   **32** -2 and -1, 0 and 1, 1 and 2, 3 and 4

36 -1 and 0, 4 and 5   **40** -3 and -2, 2 and $\dfrac{5}{2}$, $\dfrac{5}{2}$ and 3

Exercise 13.5, page 364

4 $-1, 2, -3$ **8** $-2, -\dfrac{1}{2}, \dfrac{1}{3}$ **12** $-2, -\dfrac{1}{3}, \dfrac{5}{2}$

16 $-2, \left(\dfrac{1}{4}\right)(1 + \sqrt{33}), \left(\dfrac{1}{4}\right)(1 - \sqrt{33})$

20 $-\dfrac{1}{2}, 2i, -2i$ **24** $-3, -1, 1, 4$ **28** $-\dfrac{2}{3}, \dfrac{5}{2}, 2i, -2i$

Exercise 13.6, page 368

4 -1.92 **8** -1.23 **12** 1.24 **16** 1.45 **20** $-4.646, 0.646$
24 $-0.414, 2.414, 3.500$ **28** $-4 + \sqrt{7}, -4 - \sqrt{7}, 3 - \sqrt{5}, 3 + \sqrt{5}, 5.236$
32 $i\sqrt{5}, -i\sqrt{5}, -2 - \sqrt{7}, -2 + \sqrt{7}; 0.646$

Exercise 13.7, page 369

12 $i\sqrt{2}, -i\sqrt{2}$ **16** $A = 2, B = 5$ **20** sum = 4, product = 16

14

28 $\dfrac{5}{2}$, -2, $\dfrac{1}{3}$, $1 + i$, $1 - i$ **32** No, it is 0.

36 $\dfrac{1 + \sqrt{5}}{2}$, $\dfrac{1 - \sqrt{5}}{2}$, $\sqrt{3}$, $-\sqrt{3}$

Exercise 14.1, page 375

4 3, 5, 7, 9, 11   **8** 4, 2, 0, -2, -4, -6 **12** 12, 9, 6, 3, 0, -3, -6
16 n = 5   **20** a = 7   **24** n = 7   **28** n = 6, s = -9
32 $\ell = -7$, n = 7   **36** a = 10, n = 6   **40** $\dfrac{5n(n+1)}{2}$ **48** 92,78
52 150

Exercise 14.2, page 380

4 32, 6.4, 1.28, .256, .0512   **8** $\dfrac{8}{9}$, $\dfrac{2}{3}$, $\dfrac{1}{2}$, $\dfrac{3}{8}$, $\dfrac{9}{32}$ **12** 5   **16** 3

20 2   **24** s = 242, n = 5   **28** r = 2, s = $\dfrac{127}{8}$ **32** a = 1, r = -1

36 r = $\dfrac{1}{2}$, n = 7   **40** $\dfrac{1}{3}(1 - \dfrac{1}{4^n})$ **48** $\dfrac{729}{4096}$ **52** \$9,830.40

Exercise 14.3, page 383

4 $\dfrac{10}{3}$ **8** $\dfrac{1}{9}$ **12** $\dfrac{2}{3}$ **16** $\dfrac{2}{3}$ **20** $-\dfrac{3}{7}$ **24** $\dfrac{140}{99}$

28 $\dfrac{10,052}{4,995}$ **36** $\dfrac{4}{3}$ **40** \$690,000   **44** $40(2 + \sqrt{2})$ in.

48 $[x|x \neq 0]$, $\dfrac{3}{x^2}$

Exercise 14.4, page 387

4 16.5, 15, 13.5, 12, 10.5, 9
8 1, $\dfrac{1}{2}$, $\dfrac{1}{4}$, $\dfrac{1}{8}$, $\dfrac{1}{16}$ and -1, $\dfrac{1}{2}$, $-\dfrac{1}{4}$, $\dfrac{1}{8}$, $-\dfrac{1}{16}$ **12** $-\dfrac{1}{3}$, -1

16 $\dfrac{3}{11}$ **20** $\dfrac{8}{9}$ **28** $\dfrac{21}{20}$, $\dfrac{13}{10}$, arithmetic   **32** 1, -2, harmonic

Exercise 14.5, page 388

8 a = $\dfrac{1}{8}$, r = 4, ℓ = 32, n = 5   **12** $\dfrac{17}{33}$ **20** 7, 10, 13

24 2, -1, $\dfrac{1}{2}$, $-\dfrac{1}{4}$ **28** $\dfrac{3}{4}$, $\dfrac{2}{3}$, harmonic

Exercise 16.1, page 398

4 120   **8** 17,576   **12** 24,960   **16** 192   **20** 4096

Exercise 16.2, page 402

4 6720   **8** 12   **16** 8   **20** 151,200   **24** 1680   **28** 12!
32 11,880   **36** 24

Exercise 16.3, page 405

4 495   **12** C(50, 5) = 2,118,760   **16** C(10, 5) = 252

20 C(9, 3)C(7, 2) = 1764   **24** $\dfrac{13!}{6!4!3!}$ = 60,060, $\dfrac{10!}{5!3!2!}$ = 2,520

28 176

Exercise 16.4, page 406

4 56   **12** C(5, 2)C(7, 4)C(9, 4) = 44,100   **16** 415,800   **20** C(n, 3)

Exercise 17.1, page 412

4 $x^6 + 6x^5y + 15x^4y^2 + 20x^3y^3 + 15x^2y^4 + 6xy^5 + y^6$
8 $81a^4 + 108a^3b + 54a^2b^2 + 108ab^3 + b^4$
12 $625x^8 + 1000x^6y + 600x^4y^2\ 160x^2y^3 + 16y^4$
16 $b^{42} + 126b^{41}c + 7749b^{40}c^2 + 309960b^{39}c^3$ **20** 1.1910
24 $448x^6y^2$ **28** $5670x^4y^{12}$ **32** $312741x^5y^4$

Exercise 17.2, page 414

4 $b^{-3} + 3b^{-4}y + 6b^{-5}y^2 + 10b^{-6}y^3$ **8** $\dfrac{1}{x^2} - \dfrac{2a^2}{x^3} + \dfrac{3a^4}{x^4} - \dfrac{4a^6}{x^5}$

12 $1 + \dfrac{y}{3} - \dfrac{y^2}{9} + \dfrac{5y^3}{81}$

16 $\dfrac{1}{\sqrt{3}} - \dfrac{x}{36\sqrt{3}} + \dfrac{5x^2}{2592\sqrt{3}} - \dfrac{5x^3}{31,104\sqrt{3}}$, $-9 < x < 9$

20 $1 + \dfrac{3x}{4} + \dfrac{21x^2}{32} + \dfrac{77x^3}{128}$, $-1 < x < 1$

24 1.737   **28** 4.973   **32** 0.746   **36** 7.874

Exercise 18.1, page 422

4 $\dfrac{1}{13}$, $\dfrac{1}{26}$ **8** $\dfrac{1}{2}$ **12** 6   **16** $\dfrac{1}{10}$, $\dfrac{1}{5}$

20 $\dfrac{C(24, 5)}{C(32, 5)}$ **24** $\dfrac{C(13, 1)C(4,3)C(12, 1)C(4, 2)}{C(52, 5)}$ **28** 3 **32** $\dfrac{3}{20}$

36 $\dfrac{7}{15}$ **40** $\dfrac{2}{3}$

Exercise 18.2, page 426

4 $\dfrac{1}{12}$, $\dfrac{1}{6}$ **8** $\dfrac{43}{91}$ **12** $\dfrac{7}{15}$ **16** $\dfrac{2197}{20825}$ **20** $\dfrac{1}{144}$

24 $\dfrac{1}{36}$ **28** $\dfrac{25}{49}$ **32** $\dfrac{1}{6}$, $\dfrac{1}{6}$

Exercise 18.3, page 430

4 $\dfrac{203}{23328}$ **8** $\dfrac{17}{81}$ **12** $\dfrac{6048}{78125}$ **16** $\dfrac{5120}{19683}$ **20** $\dfrac{1}{2}$

Exercise 18.4, page 431

4 $\dfrac{27}{64}$ **8** $\dfrac{5}{36}$ **12** $\dfrac{44}{4165}$ **16** $\dfrac{C(12, 3)C(8, 2)}{C(20, 5)}$ **20** $\dfrac{1}{5}$

(10.4) $\quad D_n = (-1)^{i+1} a_{i1} m(a_{i1}) + (-1)^{i+2} a_{i2} m(a_{i2})$
$\qquad + (-1)^{i+3} a_{i3} m(a_{i3}) + \cdots$
$\qquad + (-1)^{i+j} a_{ij} m(a_{ij}) + \cdots$
$\qquad + (-1)^{i+n} a_{in} m(a_{in})$

(10.5) $\quad D_n = a_{i1} A_{i1} + a_{i2} A_{i2} + a_{i3} A_{i3} + \cdots$
$\qquad + a_{ij} A_{ij} + \cdots + a_{in} A_{in}$

(10.6) $\quad D_n = (-1)^{1+j} a_{1j} m(a_{1j}) + (-1)^{2+j} a_{2j} m(a_{2j})$
$\qquad + (-1)^{3+j} a_{3j} m(a_{3j}) + \cdots$
$\qquad + (-1)^{i+j} a_{ij} m(a_{ij}) + \cdots$
$\qquad + (-1)^{n+j} a_{nj} m(a_{nj})$

(10.7) $\quad D_n = a_{1j} A_{1j} + a_{2j} A_{2j} + a_{3j} A_{3j} + \cdots$
$\qquad + a_{ij} A_{ij} + \cdots + a_{nj} A_{nj}$

(11.2) $\quad \log_b N = L$ if and only if $b^L = N$

(11.5) $\quad \log_b MN = \log_b M + \log_b N$

(11.6) $\quad \log_b M/N = \log_b M - \log_b N$

(11.7) $\quad \log_b M^k = k \log_b M$

(11.8) $\quad \log_a N = (\log_b N)/\log_b a$

(11.9) $\quad \log_e N = (\log_{10} N)/\log_{10} e$
$\qquad = 2.3026 \log_{10} N$

(12.1) $\quad a + ib = c + id$ if and only if $a = c$
$\qquad$ and $b = d$

(12.2) $\quad (a, b) + (c, d) = (a+c, b+d)$

(12.3) $\quad (a, b)(c, d) = (ac - bd, ad + bc)$

(12.4) $\quad (a, 0) = a$

(12.5) $\quad (a, b) = a + ib$

(12.6) $\quad (a, b) - (c, d) = (a - c, b - d)$

(12.7) $\quad \dfrac{(a, b)}{(c, d)} = \left(\dfrac{ac + bd}{c^2 + d^2}, \dfrac{bc - ad}{c^2 + d^2} \right)$

(12.8) $\quad |V| = r = \sqrt{x^2 + y^2}$

(12.9) $\quad \theta = \arctan(y/x)$

(12.10) $\quad z = r(\cos\theta + i \sin\theta) = r \operatorname{cis} \theta$

(12.11) $\quad (r \operatorname{cis} \theta)(R \operatorname{cis} \phi) = rR \operatorname{cis}(\theta - \phi)$

(12.12) $\quad \dfrac{r \operatorname{cis} \theta}{R \operatorname{cis} \phi} = \dfrac{r}{R} \operatorname{cis}(\theta + \phi)$

(12.13) $\quad (r \operatorname{cis} \theta)^n = r^n \operatorname{cis} n\theta$

(13.2) $\quad f(x) = Q(x)(x - r) + R$

(13.3) $\quad f(x) = Q(x)(x - r)$ if $R = 0$

(13.4) $\quad f(x) = a_0(x - r_1)(x - r_2) \cdots (x - r_n)$

(13.5) $\quad b = x_1 - \dfrac{y_1(x_2 - x_1)}{y_2 - y_1}$

(14.1) $\quad l = a + (n-1)d$

(14.2) $\quad s = \dfrac{n}{2}(a + l)$

(14.3) $\quad s = \dfrac{n}{2}[2a + (n-1)d]$

(14.4) $\quad l = ar^{n-1}$

(14.5) $\quad s = \dfrac{a - ar^n}{1 - r}$

(14.6) $\quad s = \dfrac{a - rl}{1 - r}$

(14.7) $\quad s = \dfrac{a}{1 - r}; \ |r| < 1$

(14.8) $\quad m = (a + l)/2$

(14.9) $\quad m = \sqrt{al}$

(16.2) $\quad P(n, r) = n!/(n - r)!$

(16.3) $\quad P(n, n) = n!$

(16.5) $\quad C(n, r) = n!/(n - r)!r!$

(17.1) $\quad (x + y)^n = x^n + nx^{n-1}y + \dfrac{n(n-1)}{2!}x^{n-2}y^2$
$\qquad + \dfrac{n(n-1)(n-2)}{3!}x^{n-3}y^3$
$\qquad + \dfrac{n(n-1)(n-2)(n-3)}{4!}x^{n-4}y^4$
$\qquad + \cdots + nxy^{n-1} + y^n$

(17.2) $\quad (x + y)^n = x^n + C(n, 1)x^{n-1}y$
$\qquad + C(n, 2)x^{n-2}y^2 + \cdots$
$\qquad + C(n, n-1)xy^{n-1} + C(n, n)y^n$

(18.1) $\quad p(E_1 \cup E_2) = p(E_1) + p(E_2)$
$\qquad - p(E_1 \cap E_2)$

(18.3) $\quad p(E_1 \cup E_2 \cup \cdots \cup E_{n-1} \cup E_n) =$
$\qquad p(E_1) + p(E_2) + \cdots + p(E_n)$

(18.4) $\quad p(E^1) = 1 - p(E)$

(18.5) $\quad P(E|F) = \dfrac{n(E \cap F)}{n(F)} = \dfrac{n(E \cap F)/n(S)}{n(F)/n(S)}$
$\qquad = \dfrac{P(E \cap F)}{P(F)}$

(18.5a) $\quad p(E \cap F) = p(E)p(F|E)$

(18.6) $\quad p(E \cap F) = p(E)p(F); E, F$
$\qquad$ independent